Collins

BIOLOGY

AQA A-level Year 1 and AS

Student Book

Mary Jones

Lesley Higginbottom

William Collins' dream of knowledge for all began with the publication of his first book in 1819. A self-educated mill worker, he not only enriched millions of lives, but also founded a flourishing publishing house. Today, staying true to this spirit, Collins books are packed with inspiration, innovation and practical expertise. They place you at the centre of a world of possibility and give you exactly what you need to explore it.

Collins. Freedom to teach

HarperCollins Publishers
1 London Bridge Street
London SE1 9GF

Browse the complete Collins catalogue at
www.collins.co.uk

First edition 2015

10 9 8 7 6 5 4 3 2

ISBN 978-0-00-759016-2

A catalogue record for this book is available from the British Library

Authored by Mary Jones and Lesley Higginbottom
Commissioned by Emily Pither
Project managed by 4science
Edited by Mark Gadd and Gina Walker
Proofread by Alison Walters, Laura Booth and Amanda Harman
Typeset by Jouve
Cover design by We are Laura
Illustrations by Geoff Jones

Printed by Grafica Veneta S.p.A.

The publisher would like to thank Keith Hirst, Mike Bailey, David Wright, Sue Fletcher and Alastair Michael Laing.

Approval message from AQA

This textbook has been approved by AQA for use with our qualification. This means that we have checked that it broadly covers the specification and we are satisfied with the overall quality. Full details for our approval process can be found on our website.

We approve textbooks because we know how important it is for teachers and students to have the right resources to support their teaching and learning. However, the publisher is ultimately responsible for the editorial control and quality of this book.

Please note that when teaching the AS and A-level Biology course, you must refer to AQA's specification as your definitive source of information. While this book has been written to match the specification, it cannot provide complete coverage of every aspect of the course.

A wide range of other useful resources can be found on the relevant subject pages of our website: www.aqa.org.uk

CONTENTS

To the student v

Practical work in biology 1

1 – Water and carbohydrates 4
- 1.1 Biological molecules 5
- 1.2 Water 6
- 1.3 Carbohydrates 10

2 – Lipids and proteins 20
- 2.1 Lipids 21
- 2.2 Proteins 26

3 – Enzymes 33
- 3.1 Biological catalysts 34
- 3.2 Enzymes and chemical reactions 34
- 3.3 How enzymes work 35
- 3.4 Factors affecting enzyme activity 37

4 – Nucleotides 53
- 4.1 Structure of DNA and RNA 54
- 4.2 DNA replication 59
- 4.3 ATP 61

5 – Cells 67
- 5.1 Cells and living organisms 68
- 5.2 Structure of eukaryotic cells 68
- 5.3 Prokaryotic cells and viruses 77
- 5.4 Methods of studying cells 79
- 5.5 Making new cells 85

6 – Cell membranes 98
- 6.1 Structure of cell membranes 99
- 6.2 Diffusion and facilitated diffusion 101
- 6.3 Osmosis 105
- 6.4 Active transport 113

7 – The immune system 120
- 7.1 Cell-surface antigens 121
- 7.2 Phagocytosis 122
- 7.3 The immune response 123
- 7.4 HIV/AIDS 133
- 7.5 Monoclonal antibodies 136

8 – Exchange with the environment 142
- 8.1 Surface area : volume ratio 143
- 8.2 Gas exchange 145
- 8.3 The human gas exchange system 150
- 8.4 Digestion and absorption 157

9 – Mass transport 165
- 9.1 Mass flow 166
- 9.2 Oxygen transport in mammals 166
- 9.3 The heart and circulatory system 170
- 9.4 Blood vessels 176
- 9.5 Water transport in plants 185
- 9.6 Transport of organic substances in plants 191

10 – DNA and protein synthesis 199
- 10.1 Genes and chromosomes 200
- 10.2 The genetic code 203
- 10.3 Protein synthesis 205

11 – Genetic diversity 215
- 11.1 Mutation 216
- 11.2 Meiosis 219
- 11.3 Natural selection 223

12 – Taxonomy and biodiversity 239

12.1 The species concept 240

12.2 Biodiversity 244

12.3 Investigating diversity 248

13 – Maths techniques in biology 260

13.1 Handling numbers 260

13.2 Recording and displaying data 262

13.3 Analysing and interpreting data 264

13.4 Statistics 266

Answers 270

Glossary 282

Index 288

Acknowledgements 302

TO THE STUDENT

The aim of this book is to help make your study of advanced biology interesting and successful. It includes examples of modern issues, developments and applications that reflect the continual evolution of scientific knowledge and understanding. We hope it will encourage you to study science further when you complete your course.

USING THIS BOOK

Biology is a fascinating, but complex subject – underpinned by some demanding ideas and concepts, and by a great deal of experimental data ('facts'). This mass of information can sometimes make its study daunting. So don't try to achieve too much in one reading session and always try to keep the bigger picture in sight.

There are a number of features in the book to help with this:

- Each chapter starts with a brief example of how the biology you will learn has been applied somewhere in the world, followed by a short outline of what you should have learned previously and what you will learn through the chapter.
- Important words and phrases are given in bold when used for the first time, with their meaning explained. There is also a glossary at the back of the book. If you are still uncertain, ask your teacher or tutor because it is important that you understand these words before proceeding.
- Throughout each chapter there are many questions, with the answers at the back of the book. These questions enable you to make a quick check on your progress through the chapter.
- Similarly, throughout each chapter there are checklists of key ideas that summarise the main points you need to learn from what you have just read.
- Where appropriate, worked examples are included to show how calculations are done.
- There are many assignments throughout the book. These are tasks relating to pieces of text and data that show how biological ideas have been developed or applied. They are not required knowledge for an A-level examination. Rather, they provide opportunities to apply the science you have learned to new contexts, practise your maths skills and practise answering questions about scientific methods and data analysis.
- Some chapters have information about the 'required practical' activities that you need to carry out during your course. These sections provide the necessary background information about the apparatus, equipment and techniques that you need to be prepared to carry out the required practical work. There are questions that give you practice in answering questions about equipment, techniques, attaining accuracy, and data analysis.
- At the end of each chapter are practice questions. These are examination-style questions which cover all aspects of the chapter.

This book covers the requirements of AS Biology and the first year of A-level Biology. There are a number of sections, questions, assignments and practice questions that have been labelled 'Stretch and challenge', which you should try to tackle if you are studying for A-level. In places these go beyond what is required for the specification but they will help you build upon the skills and knowledge you acquire and better prepare you for further study beyond advanced level.

Good luck and enjoy your studies. We hope this book will encourage you to study biology further when you complete your course.

PRACTICAL WORK IN BIOLOGY

Practical work is a vital part of biology. Biologists apply their practical skills in a wide variety of contexts – from conservation to food production; from tracking invasive species to controlling disease. In your AS or A-level Biology course you need to learn, practise and demonstrate that you have acquired these skills.

WRITTEN EXAMINATIONS

Your practical skills will be assessed in the written examinations at the end of the course. Questions on practical skills will account for about 15% of your marks at AS and 15% at A-level. The practical skills assessed in the written examinations are:

Independent thinking

- solve problems set in practical contexts
- apply scientific knowledge to practical contexts

Use and application of scientific methods and practices

- comment on experimental design and evaluate scientific methods
- present data in appropriate ways
- evaluate results and draw conclusions with reference to measurement uncertainties and errors
- identify variables including those that must be controlled

Numeracy and the application of mathematical concepts in a practical context

- plot and interpret graphs
- process and analyse data using appropriate mathematical skills
- consider margins of error, accuracy and precision of data

Instruments and equipment

- know and understand how to use a wide range of experimental and practical instruments, equipment and techniques appropriate to the knowledge and understanding included in the specification

Throughout this book there are questions and longer assignments that will give you the opportunity to develop and practise these skills. The contexts of some of the exam questions will be based on the 'required practical activities'.

Figure 1 Biologists often use techniques and apparatus in the field as well as in the laboratory.

Figure 2 It is important to be able to interpret and analyse data – this doctor is analysing an electrocardiogram; understanding anomalies is crucial when making a diagnosis.

Figure 3 You will need to use a variety of equipment correctly and safely.

ASSESSMENT OF PRACTICAL SKILLS

Some practical skills can only be practised when you are doing experiments. For A-level, these **practical competencies** will be assessed by your teacher:

- follow written procedures
- apply investigative approaches and methods when using instruments and equipment
- safely use a range of practical equipment and materials
- make and record observations
- research, reference and report findings

You must show your teacher that you consistently and routinely demonstrate the competencies listed above during your course. The assessment will not contribute to your A-level grade, but will appear as a 'pass' alongside your grade on the A-level certificate.

These practical competencies must be demonstrated by using a specific range of **apparatus and techniques**:

- use appropriate apparatus to record a range of quantitative measurements (to include mass, time, volume, temperature, length and pH)
- use appropriate instrumentation to record quantitative measurements, such as a colorimeter or potometer
- use laboratory glassware apparatus for a variety of experimental techniques to include serial dilutions
- use a light microscope at high power and low power, including use of a graticule
- produce scientific drawing from observation with annotations
- use qualitative reagents to identify biological molecules
- separate biological compounds using thin layer/paper chromatography or electrophoresis
- safely and ethically use organisms to measure plant or animal responses, and physiological functions
- use microbiological aseptic techniques, including the use of agar plates and broth
- safely use instruments for dissection of an animal organ, or plant organ
- use sampling techniques in fieldwork
- use ICT such as computer modelling or data logger to collect data, or use software to process data

For AS, the above will not be assessed but you will be expected to use these skills and these types of apparatus to develop your manipulative skills and your understanding of the processes of scientific investigation.

REQUIRED PRACTICAL ACTIVITIES

During the A-level course you will need to carry out 12 **required practical** activities. These are the main sources of evidence that your teacher will use to award you a pass for your competency skills. If you are doing the AS, you will need to carry out the first six in this list.

Figure 4 *Aseptic techniques must be used when handling agar plates and broth.*

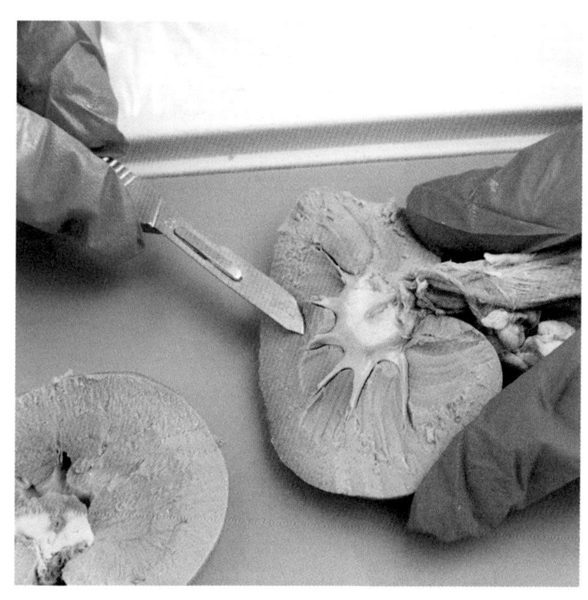

Figure 5 *Dissection tools – such as this scalpel being used to cut a sheep kidney – must be handled with care; they are often surgically sharp, but not always surgically clean!*

1. Investigation into the effect of a named variable on the rate of an enzyme-controlled reaction
2. Preparation of stained squashes of cells from plant-root tips; set-up and use of an optical microscope to identify the stages of mitosis in these stained squashes and calculation of a mitotic index
3. Production of a dilution series of a solute to produce a calibration curve with which to identify the water potential of plant tissue
4. Investigation into the effect of a named variable on the permeability of cell-surface membranes
5. Dissection of animal or plant gas exchange or mass transport system or of organ within such a system
6. Use of aseptic techniques to investigate the effect of antimicrobial substances on microbial growth
7. Use of chromatography to investigate the pigments isolated from leaves of different plants, for example leaves from shade-intolerant and shade-intolerant plants, or leaves of different colours
8. Investigation into the effect of a named factor on the rate of dehydrogenase activity in extracts of chloroplasts
9. Investigation into the effect of a named variable on the rate of respiration of cultures of single-celled organisms
10. Investigation into the effect of an environmental variable on the movement of an animal using either a choice chamber or a maze
11. Production of a dilution series of a glucose solution and use of colorimetric techniques to produce a calibration curve with which to identify the concentration of glucose in an unknown 'urine' sample
12. Investigation into the effect of a named environmental factor on the distribution of a given species

Information about the apparatus, techniques and analysis of required practicals 1 to 6 are found in the relevant chapters of this book, and 7 to 12 in Book 2.

You will be asked some questions in your written examinations about skills developed as result of carrying out these required practicals.

Practical skills are really important. Take time and care to learn, practise and use them.

1 WATER AND CARBOHYDRATES

PRIOR KNOWLEDGE

You have probably learned that many substances are made up of molecules, which in turn are formed by groups of atoms bonded together. You may also know that each atom consists of a central nucleus carrying a positive charge, surrounded by a cloud of electrons each with a negative charge. In a neutral atom, the number of negative and positive charges is equal, but an atom that has gained extra electrons, or lost some, has a negative or positive charge, and is called an ion. You may also remember the kinetic theory of matter, which helps us to understand the properties of different states of matter (solids, liquids and gases) by describing how their particles move around and interact with one another.

LEARNING OBJECTIVES

In this chapter, we will use these concepts to explain the structure and roles of some of the most important types of molecules that make up the bodies of all living organisms. You will also learn why water is essential to all life.

(Specification 3.1.1, 3.1.7, 3.1.2)

Many scientists believe there is a strong possibility that there is life elsewhere in the universe, not just here on Earth. Perhaps there are other planets similar to ours out there somewhere, circling one of the innumerable stars in our galaxy or in other galaxies far away. Nearer to home, could our neighbouring planet, Mars, harbour life?

Biologists searching for life on distant planets tend to begin by looking for water. Although we have no reason to think that life on other planets would look anything at all like the living organisms on Earth, it is very difficult to predict how any life form could exist without water. Water is a unique substance that seems to be essential for the existence of life.

There is now no doubt that there is at least some water on Mars, and that in the past there was much more. NASA's Martian rovers, Curiosity and Opportunity, have found that Mars had a warmer and wetter climate long ago. Lakes used to exist on Mars, and there may even be liquid water present now. There used to be hot springs and flowing streams.

The next question is to find out if there has ever been life there, and the rovers will continue to search for traces of past or present living organisms. They will look for the presence of carbon-containing compounds that are normally produced only by living things. Even if there is no life on the surface now – where conditions seem to be too inhospitable for even microorganisms to live – there is an outside possibility that something may have survived deep underground. Perhaps by the time you read this, a new discovery will have been made.

1.1 BIOLOGICAL MOLECULES

What are living organisms made of? All living organisms, including yourself, plants, bacteria and every other kind of living thing, are made of **molecules**. The study of these molecules is called molecular biology, and the way in which the molecules behave is biochemistry. All life on Earth has similar biochemistry.

Molecules are unimaginably small particles, which themselves are made up of even smaller particles, called **atoms**. In a molecule, atoms are joined by bonds between them, which hold them tightly together (see Figure 1).

Joining together different atoms in different quantities, or even the same atoms in different arrangements, produces molecules that behave in very different ways.

There are more than 100 different kinds of atoms, and an infinite number of different kinds of molecule that can exist. In living things, however, most of the atoms are of just six different kinds – hydrogen, oxygen, carbon, nitrogen, sulfur and phosphorus. Other kinds of atoms are less common, but they are nevertheless vitally important. Atoms of calcium, iron, zinc, magnesium, sodium, chlorine and potassium are all important ingredients in the body of every living organism.

Some of these atoms are important not only when they are part of molecules, but also on their own, particularly when they exist in their charged form as an ion. Whereas uncharged atoms have equal numbers of positively charged protons and negatively charged electrons, ions have lost or gained electrons, leaving them with a positive or negative charge. Iron ions are an important component of the oxygen-transporting substance haemoglobin, and phosphate ions are found in DNA and other molecules. You will come across these and other examples as you work through your biology course.

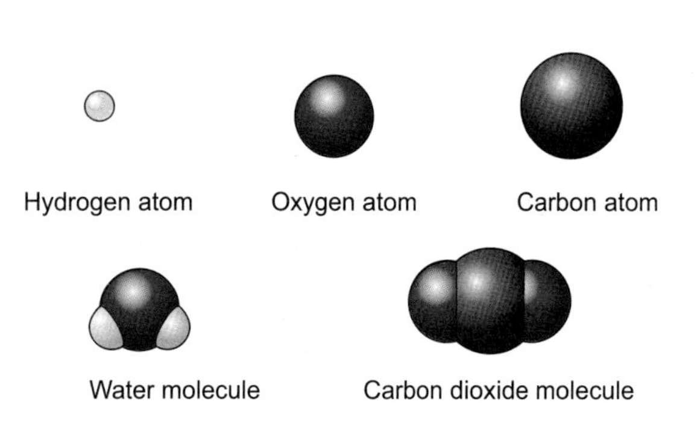

Figure 1 *Water and carbon dioxide molecules: molecules are made of atoms held firmly together.*

In this chapter and the next, we will look at some of the most important molecules that make up living organisms.

Overview of biological molecules

Water (see Figure 1) is a major component of the body of every living organism; in humans, for example, about 70% of our body weight is made up of water.

Water is so important to life that, when searching for life on planets other than Earth, astrobiologists tend to look first for water – it is difficult to imagine any life form that could exist without it. You will find out why this is so when you read the next section in this chapter.

Water has very small molecules, but most molecules found in organisms are much larger. One important class of molecules in living organisms is carbohydrates. The smallest carbohydrate molecules are called sugars, and they are made of several carbon atoms, several hydrogen atoms and several oxygen atoms all joined together. There are many different kinds of sugars, each with molecules in which different numbers of these atoms are linked in slightly different ways. Sugar molecules can join together in long chains, producing huge molecules such as starch, glycogen and cellulose. Sugars, starch, glycogen and cellulose are all **carbohydrates**.

Protein molecules are also made up of long chains of smaller molecules. These smaller molecules are called **amino acids**. Like the molecules that make up carbohydrates, amino acid molecules contain atoms of carbon, hydrogen and oxygen, but they also contain atoms of nitrogen. You will find out more about proteins in Chapter 2.

A third type of molecule found in every living organism is **nucleotides**. These include DNA and RNA. The structures and functions of nucleotides are described in Chapter 4.

Lipids (also known as fats) often get a bad press, but they are an essential component of all living things. Like carbohydrates, their molecules contain carbon, hydrogen and oxygen atoms, but they are arranged in a very different way. You will learn more about lipids in Chapter 2.

Monomers and polymers

The largest molecules in living organisms are made by joining together long chains of smaller molecules. For example, starch is made by joining together very long chains of glucose molecules (see Figure 2). The glucose molecules are **monomers**, and the long chain of them joined together forms a **polymer**.

Protein molecules are also polymers. Here, the monomers are amino acids.

Nucleotides also form polymers, joining together into long chains to produce polynucleotides.

QUESTIONS

1. Arrange these in order of size, smallest first: protein molecule; carbon atom; amino acid molecule; animal cell.

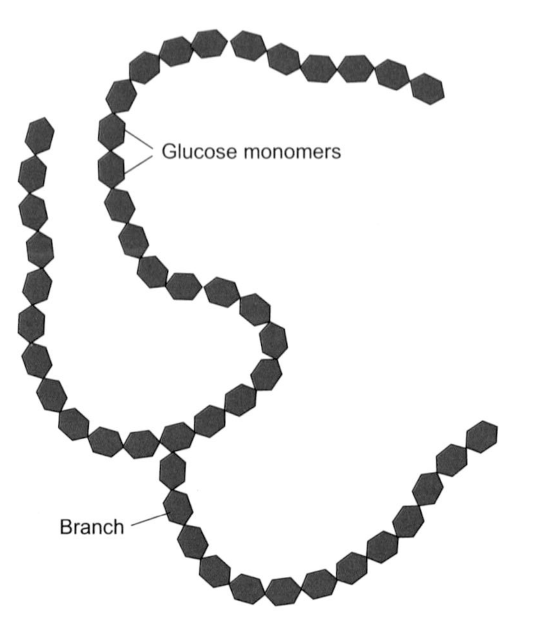

Figure 2 *Part of a starch molecule. Starch is a polymer of glucose. Unlike most polymers, starch molecules have branches along their chains of monomers.*

KEY IDEAS

- Living organisms are made up of many different kinds of molecules, most of which contain atoms of carbon, hydrogen, oxygen and nitrogen.
- Many biological molecules are polymers, made up of long chains of smaller molecules – called monomers – joined together.

1.2 WATER

Water is an amazing substance. It has a set of properties that are not found in any other substance. Without water, there would be no life on Earth.

The structure of water molecules

To understand why water behaves in the way that it does, we need to understand the structure of its molecules (Figure 3). Here, the symbols of the atoms are used instead of drawing them as balls, as in Figure 1. Lines between the symbols represent bonds that hold them together. A water molecule is made up of two hydrogen atoms joined by **covalent bonds** to a single oxygen atom, so its formula is H_2O. This is a very small molecule compared with most molecules found in living organisms.

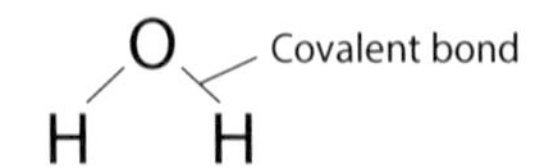

Water molecule H_2O

Figure 3 *The atoms in water molecules are held tightly together by strong covalent bonds.*

Covalent bonds are very strong bonds, formed when two atoms share electrons with each other. It is very difficult to split the hydrogen and oxygen atoms apart – a lot of energy is required to make this happen. As you will see if you continue your studies to A-level, plants have evolved a way to split water molecules using energy from light, which they do in photosynthesis.

Hydrogen bonds

Atoms contain electrons, and an electron has a small negative charge. In a water molecule, the electrons in the chemical bond are not shared equally between the oxygen and the hydrogen atoms. This results in the oxygen atom having a tiny negative charge, and each hydrogen atom having a tiny positive charge. The water molecule is said to have a **dipole**. Although its overall charge is zero, it has a small positive charge in some places, and a small negative charge in another. These small charges are written $\delta-$ and $\delta+$. The symbol δ is the Greek letter delta, so you can say 'delta minus' and 'delta plus'.

Negative charges and positive charges are attracted to one another. This means that each water molecule is attracted to its neighbours. The negative charge on one water molecule is attracted to the positive charge on another water molecule. These attractions are called **hydrogen bonds** (see Figure 4). Though they are important, hydrogen bonds are much weaker than the strong covalent bonds that hold the oxygen and hydrogen atoms firmly together.

a

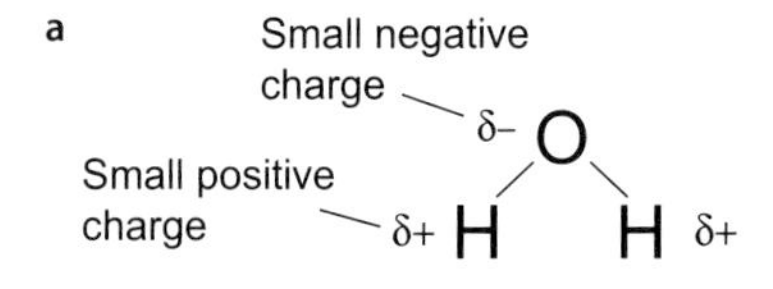

b

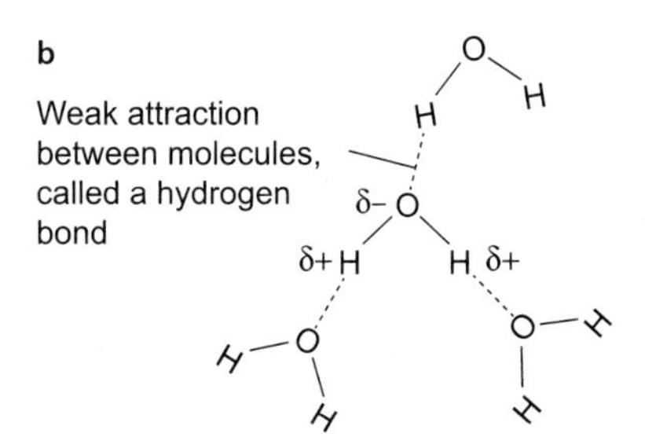

***Figure 4** Water molecules are weakly attracted to each other because of their dipoles.*

Water helps keep temperatures stable

Molecules are in constant motion. In a solid, they stay in the same place and just vibrate on the spot. In a liquid, they move around more freely, but remain in contact with one another. In a gas, they move completely freely, only rarely making contact.

Temperature is a measure of the kinetic energy of molecules. Kinetic energy is movement energy. A substance in which the molecules are moving very fast has a higher temperature than one where the molecules are moving very slowly.

We can make molecules move around faster by adding heat (thermal) energy to them. For example, you could put some water into a beaker and heat it over a Bunsen flame. As the heat energy from the flame is transferred to the water, the water molecules move faster and faster. The temperature of the water rises.

However, we find that the temperature of water does not rise as much as it would if a different liquid was heated, such as hexane (C_6H_{14}). This is because of the hydrogen bonds between the water molecules. Hydrogen bonds are weak bonds, and they are easily broken, but it still takes energy to break them. And, although each hydrogen bond between the water molecules in a beaker is very weak, there are billions of them. A lot of the heat energy from the Bunsen flame is transferred in breaking these hydrogen bonds, separating the water molecules from each other. Only the remaining heat energy is transferred to kinetic energy, making the water molecules move around faster and increasing the temperature.

So, to increase the temperature of a beaker of water by 1°C, we have to put in much more heat energy than to increase the temperature of a beaker of hexane by 1°C. We say that water has a relatively high **heat capacity**. A lot of heat energy has to be transferred to the water to raise its temperature significantly.

This is very useful to living organisms. Because our bodies are mostly made of water, we have a relatively high heat capacity. The water in and around our cells absorbs a lot of heat energy without its temperature rising very much. The water 'buffers' (tones down) heat changes.

This is also very helpful to aquatic organisms. Large bodies of water, such as the sea or a lake, or even a pond, do not change temperature as quickly or as greatly as the air (Figure 5). As air temperature rises during the daytime, water temperature rises only a little. Similarly, as air temperature drops at night, water

***Figure 5** The water in the sea is much cooler than the air on a hot, sunny day.*

temperature drops by much less. So, aquatic organisms live in an environment where temperature changes are much smaller, and happen much more gradually, than the environment of terrestrial (land-living) organisms.

Water helps keep organisms cool

In a body of water, not all of the water molecules are moving at the same speed. Some have more kinetic energy than others. Some will have enough kinetic energy to escape from the water and shoot off into the air. They have become a gas. This is called **evaporation**. (Do not confuse evaporation with boiling, which happens only when the temperature of a liquid rises to its boiling point. Boiling also happens throughout the liquid, while evaporation happens only at the surface.)

As these fast-moving molecules are lost from the water, their energy goes with them. As more and more of them leave, the average kinetic energy of the molecules that are left behind in the water gradually decreases. The water cools down.

The energy that is lost from the liquid water as it evaporates is called **latent heat of vaporisation**. Water has a relatively high latent heat of vaporisation, so the evaporation of quite small quantities of water has a large cooling effect. We make use of this when we sweat. Sweat, which is mostly water, lies on the skin surface and evaporates. As it evaporates, the skin surface is cooled. Plants, too, are cooled down when water evaporates from their leaves in transpiration.

QUESTIONS

2. Fish are not able to regulate their body temperature in the way that mammals do. Explain how hydrogen bonding in water helps fish to have a relatively constant body temperature.

Stretch and challenge

3. When climbing big mountains, mountaineers often set up a base camp before attempting the final climb to the summit.
   a. Why do mountaineers find that, when cooking at altitude, their water boils more quickly than when they are at base camp, but it takes longer to cook their food?
   b. When sterilising water, why do the mountaineers need to boil it for longer at high altitudes?

Water as a solvent

Water is an excellent solvent. A **solvent** is a liquid in which other substances – called **solutes** – can dissolve. Many different substances are able to dissolve in water, and this is all down to its hydrogen bonds.

Figure 6 shows how salt (sodium chloride) dissolves in water. Sodium chloride is made up of positively charged sodium ions, Na^+, and negatively charged

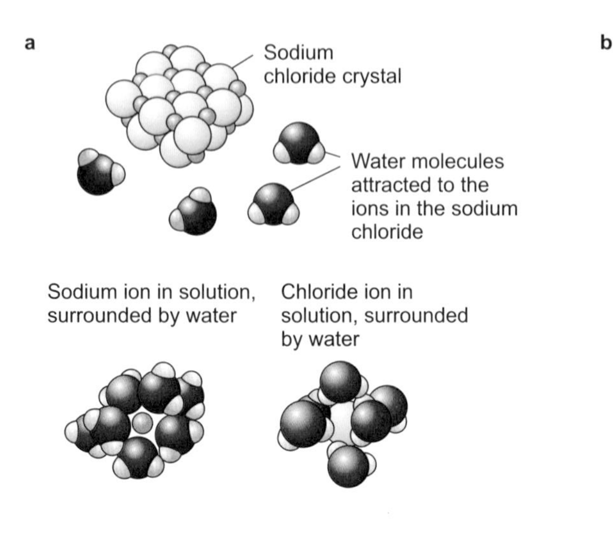

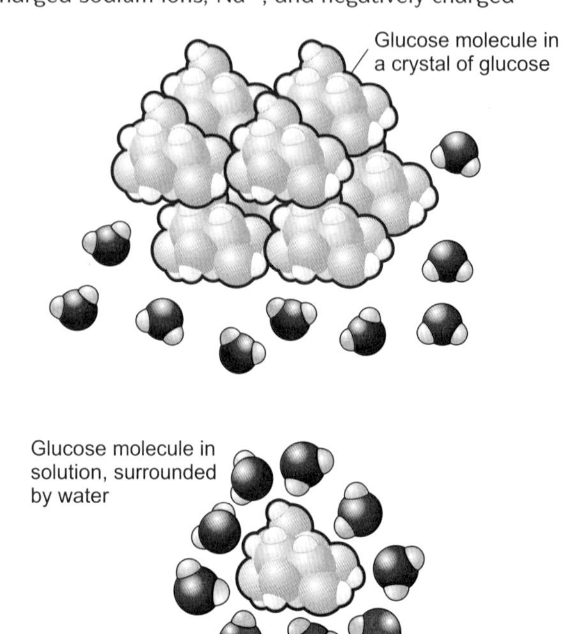

Figure 6 *(a) How sodium chloride dissolves in water; (b) how glucose dissolves in water*

chloride ions, Cl^-. These are attracted to the $\delta-$ and $\delta+$ charges on the water molecules. The sodium ions and chloride ions are therefore separated from each other, and spread out in between the water molecules. This is why salt seems to 'disappear' when you stir it into water.

Water can even dissolve quite large molecules, such as sugars. As you will see later in this chapter, sugar molecules have tiny positive and negative charges on parts of their molecules, and these are attracted to the negative and positive charges on the water molecules. When you stir sugar into a cup of tea, the sugar molecules spread out in between the water molecules.

When substances are dissolved in water, their molecules or ions are free to move around and to react with other molecules or ions that are also in solution. The metabolic reactions that take place in your body, and in the bodies of every other living organism, only take place because the reactants are dissolved in water.

Water's solvent properties are also important in transporting substances around the body. Blood plasma is mostly water, and it carries a huge range of different substances – such as glucose, vitamins, urea and carbon dioxide – in solution. Plants transport mineral ions and sugars dissolved in water in their xylem and phloem vessels. We also use water to dissolve substances to be excreted from the body, especially urea. Urine is a solution of urea and other solutes in water.

Water molecules stay together

We have seen that water molecules are attracted to each other. These attractions help to hold water molecules together, so that they can flow as a continuous stream. You see this happening when you watch water flowing from a tap, or along a river. This movement of a whole mass of water together is an example of **mass flow**.

In plants, water moves up xylem vessels as a continuous stream, by mass flow. Water can move like this all the way to the top of the tallest trees. The tall, continuous columns of water inside the tree's trunk and branches can only hold together because of the attraction between the water molecules. The way in which the water molecules tend to 'stick' together is called **cohesion**. You can read more about this in Chapter 9.

Normally, a water molecule is attracted to other water molecules on all of its sides. But at the surface of a body of water, the top layer of water molecules do not have any water molecules above them, so the net attraction is downwards. This produces **surface tension**, making the water behave as though there is a thin 'skin' where it meets air. Surface tension allows small animals to walk on the water surface (Figure 7).

Water as a metabolite

So far, all of the properties of water described in this chapter are physical properties. This means that they involve whole water molecules, which do not break up. The water may change state, but it is still water.

However, water molecules can – and do – break up when they react with other substances. In these chemical reactions, the covalent bonds between the hydrogen and oxygen molecules are broken, and the atoms that made up the water molecule form new bonds with other atoms.

There are many metabolic reactions in living organisms that include water as one of the reactants. For example, digestion – reactions in which large molecules such as starch are broken down into smaller ones such as glucose – requires water as a reactant. These reactions are called **hydrolysis** reactions. Water is also produced when small molecules are joined together to produce larger ones, such as glucose molecules joining together to form starch. These are **condensation** reactions. There are several examples of hydrolysis and condensation reactions discussed later in this chapter and also in Chapter 2.

Figure 7 *The water's surface tension makes it possible for this pond skater to walk on the pond.*

KEY IDEAS

- Water molecules have dipoles – that is, they have a small negative charge on some parts of the molecule and a small positive charge on others.
- Water molecules are attracted to each other because of these charges. The weak attractions are called hydrogen bonds.
- Hydrogen bonding causes water to have a high heat capacity, buffering temperature changes.
- Water has a high latent heat of vaporisation, which provides a cooling effect when it evaporates.
- Water is an excellent solvent, in which metabolic reactions can take place and substances can be transported.
- Cohesion between water molecules enables tall columns of water to be supported in the xylem vessels of plants.
- Water is a metabolite in many metabolic reactions.

1.3 CARBOHYDRATES

All carbohydrates contain the elements carbon, hydrogen and oxygen. The hydrogen and oxygen atoms are generally in the ratio 2:1, giving carbohydrates the general formula CH_2O.

There are three basic types of carbohydrate molecule: **monosaccharides**, **disaccharides** and **polysaccharides**.

Monosaccharides

Monosaccharides are simple sugars – small molecules that dissolve in water and taste sweet. These are the monomers from which all larger carbohydrates are made.

Glucose is a monosaccharide. Its structural formula is shown in Figure 8. Five of the carbon atoms (numbered 1–5 in the diagram) form a ring. Glucose can exist in two slightly different forms, depending on the orientation of the hydrogen and hydroxyl (–OH) groups on carbon atom 1. These are known as **α-glucose** (alpha glucose) and **β-glucose** (beta glucose). They are isomers of one another – that is, their molecules are made of the same numbers of carbon, hydrogen and oxygen atoms, but these atoms are arranged differently.

Fructose, another monosaccharide, is shown in Figure 9a. A third commonly occurring monosaccharide is **galactose**, which is shown in Figure 9b. All of these monosaccharides are isomers. They all have the formula $C_6H_{12}O_6$ but have slightly different properties from one another.

All monosaccharides are soluble in water.

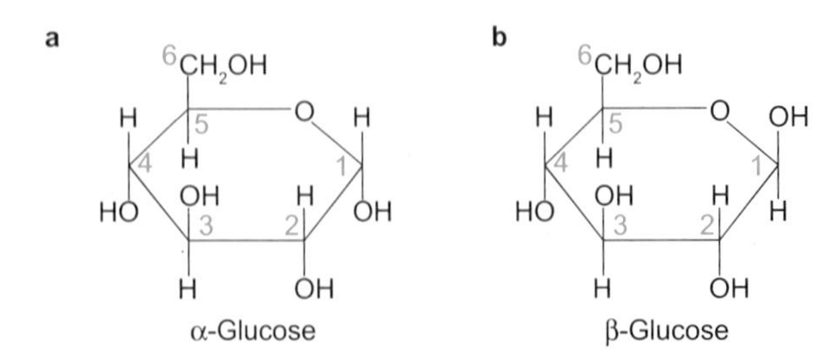

Figure 8 The molecular structures of α-glucose and β-glucose

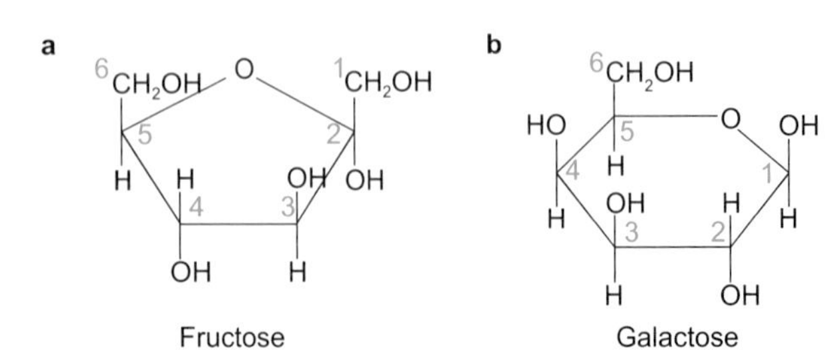

Figure 9 The molecular structures of fructose and galactose

ASSIGNMENT 1: USING CALIBRATION CURVES

(MS 1.3, MS 1.11, MS 3.1, MS 3.2, PS 3.1, PS 3.2, PS 3.3)

A calibration curve can be used to determine the concentration of an unknown solution. Calibration curves can be produced for any substance. In this assignment, we will look at the use of a calibration curve to determine the concentration of glucose in an unknown solution.

If a patient is suspected of having diabetes, a sample of blood is taken after the patient has fasted and the concentration of glucose in the blood is measured. In a person without diabetes, this would be in the range of 80–120 mg of glucose per 100 cm^3 of blood. The blood glucose concentration in a diabetic person would be much higher.

Many people like to know how sweet the wine they are buying is, and a glucose calibration curve can be used to measure the amount of glucose in wine. Medium sweet wines have a sugar content between 18 g dm^{-3} and 45 g dm^{-3}, while those with a concentration greater than 45 g dm^{-3} are classified as sweet.

Manufacturers of sugar-free and low sugar foods and drinks will regularly test their products to ensure they fall within a narrow range of acceptable values.

The glucose concentration of a solution can be measured using potassium permanganate solution, which is pink. Glucose easily donates electrons to permanganate ions in a redox reaction. The rate of this reaction depends on the concentration of glucose in solution. When potassium permanganate is reduced by accepting electrons, it becomes colourless, and by measuring the time taken for different known concentrations of glucose to turn the solution colourless you can generate the data to produce a graph of *glucose concentration* against *time*.

To produce a glucose calibration curve, you first need to make up a range of glucose solutions of known concentration. For example, to prepare a 1% solution, you dissolve 1 g of glucose in 100 cm^3 of water. If you make up a range of different concentrations, and time how long it takes for a standard volume of standard concentration potassium permanganate solution to be decolourised, you can then plot a calibration curve.

Glucose concentration/ mg 100 cm^{-3}	Time taken for potassium permanganate to go colourless/seconds
0	5
25	12
50	18
75	24
100	30
125	36
150	42
175	48
200	54

Table A1 *A set of data for a calibration curve*

After generating a calibration curve using a standard glucose solution and potassium permanganate, you can then determine the concentration of glucose in an unknown solution.

Questions

A1. Plot the data in Table A1 to produce a calibration curve.

A2. Use your curve to determine the glucose concentration of a sample which took:

a. 32 seconds to go colourless

b. 10 seconds to go colourless.

A3. How long would it take a glucose solution with a concentration of 160 mg 100 cm^{-3} to go colourless?

A4. Suggest the main source of error in determining the data shown in Table A1.

A5. Suggest why using a glucose calibration curve to estimate the concentration of sugar in wine might give an underestimate.

Disaccharides

Disaccharides have molecules made of two monosaccharides joined together. Like monosaccharides, all disaccharides taste sweet and are soluble in water.

Examples include **maltose** (made from two glucose molecules), **sucrose** (made from a glucose molecule and a fructose molecule) and **lactose** (made from a glucose molecule and a galactose molecule).

Maltose, or malt sugar, is found in germinating seeds. Sucrose is the sugar we use to add to drinks, or in cooking. Lactose is found in milk.

Figure 10 shows how two α-glucose molecules can join together to form maltose. Figure 11 shows how an α-glucose molecule and a fructose molecule join together to form sucrose.

Both of these reactions are **condensation reactions**, and you can see that a water molecule is formed. The bond that is formed between the two sugar molecules is called a **glycosidic bond**.

Disaccharides can be broken down into their component monosaccharides. The reactions shown in Figures 10 and 11 happen in reverse. For this to happen, you can see that a water molecule must be used. For this reason, this kind of reaction is called a **hydrolysis** reaction. 'Hydro' means 'water' and 'lysis' means 'splitting apart'.

QUESTIONS

4. A lactose molecule is made from a β-glucose molecule, joined by a glycosidic bond from its carbon-1 to the carbon-4 of a galactose molecule. Use this information, and Figure 9, to draw a diagram of a lactose molecule.
5. Draw a diagram to show the hydrolysis of the disaccharide, maltose. Refer to the diagrams of condensation in Figures 10 and 11 to help you.

α-Glucose
α-Glucose
H_2O
Condensation reaction
α-1,4 glycosidic bond

Figure 10 The formation of maltose

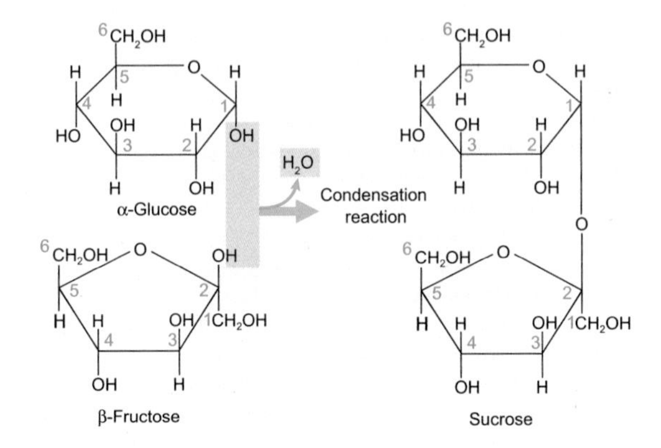

Figure 11 The formation of sucrose

Polysaccharides

Polysaccharides are giant molecules made up from many monosaccharide molecules joined together by condensation reactions. Starch is a polysaccharide. Part of a starch molecule is shown in Figure 2 and Figure 12. Starch molecules are compact, coiled and branched, making them ideal 'energy' stores. Polysaccharides are insoluble, because their molecules are so large and cannot spread out in between water molecules as smaller molecules do.

We have seen that starch is a polymer – a molecule made up of repeating units, rather like links in a chain. The

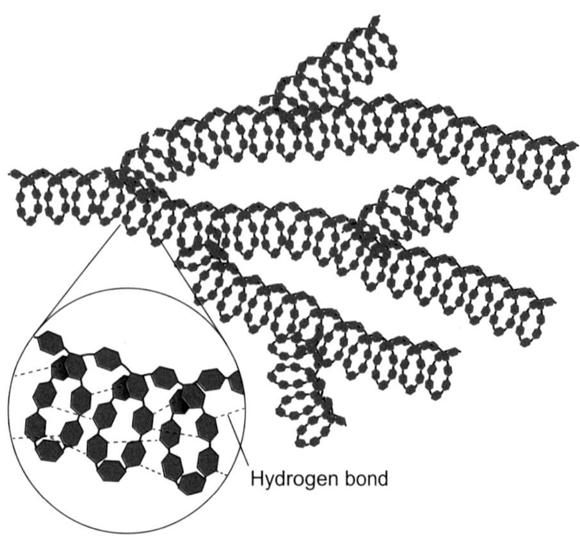

Figure 12 The structure of starch

individual units are called monomers, which in this case are α-glucose molecules. These bonds produce twisted chains of monomers. The coiled and branched chains of starch molecules give them a compact shape which, together with their insolubility in water, makes them ideal 'energy' storage compounds. When required, they can easily be hydrolysed to individual glucose molecules by enzymes that break the α-1,4 glycosidic bonds. The glucose can then be used in respiration to release energy.

Another advantage of using starch as a storage compound is that its large molecules cannot cross membranes. Because of its insolubility, starch does not have any osmotic effects on the cell, as glucose would (it does not affect water potentials). Storing glucose would tend to cause water to move into the cell by osmosis.

Starch is the major storage carbohydrate in plants and thus the major carbohydrate in our diet.

Animals use a very similar substance as their storage carbohydrate, called **glycogen**. (Starch is never found in animals, and glycogen is never found in plants.) Glycogen molecules are very similar to starch molecules – they are made up of α-glucose molecules joined by 1,4 and 1,6 glycosidic bonds – but they have more branches than starch.

Glucose monomers are joined in a very different way to form **cellulose** (Figure 13). Cellulose is the most common polysaccharide in the world, because it is the major substance from which plant cell walls are made. Cellulose molecules are made up of long chains of β-glucose molecules, all joined through 1-4 glycosidic bonds. There are no branches. This produces long, straight molecules that can lie side by side. Cellulose molecules form huge numbers of hydrogen bonds with all the other cellulose molecules lying alongside them, forming groups of parallel molecules called microfibrils. Although an individual hydrogen bond is very weak, the large numbers of them mean that the microbfibrils have great strength. They are also difficult to digest; relatively few organisms have enzymes (celullase) that can break β-4 glycosidic bonds. So cellulose is an excellent structural substance, rather than an energy storage substance like starch and glycogen.

QUESTIONS

6. Construct a table to compare the structure and properties of starch and cellulose. You could include the sugar units from which they are made, the type of glycosidic bonds that join them together, whether or not they branch, the overall shape of the molecules, whether or not the molecules form hydrogen bonds with one another, their solubility, the ease with which they can be digested, where they are found, and their functions.

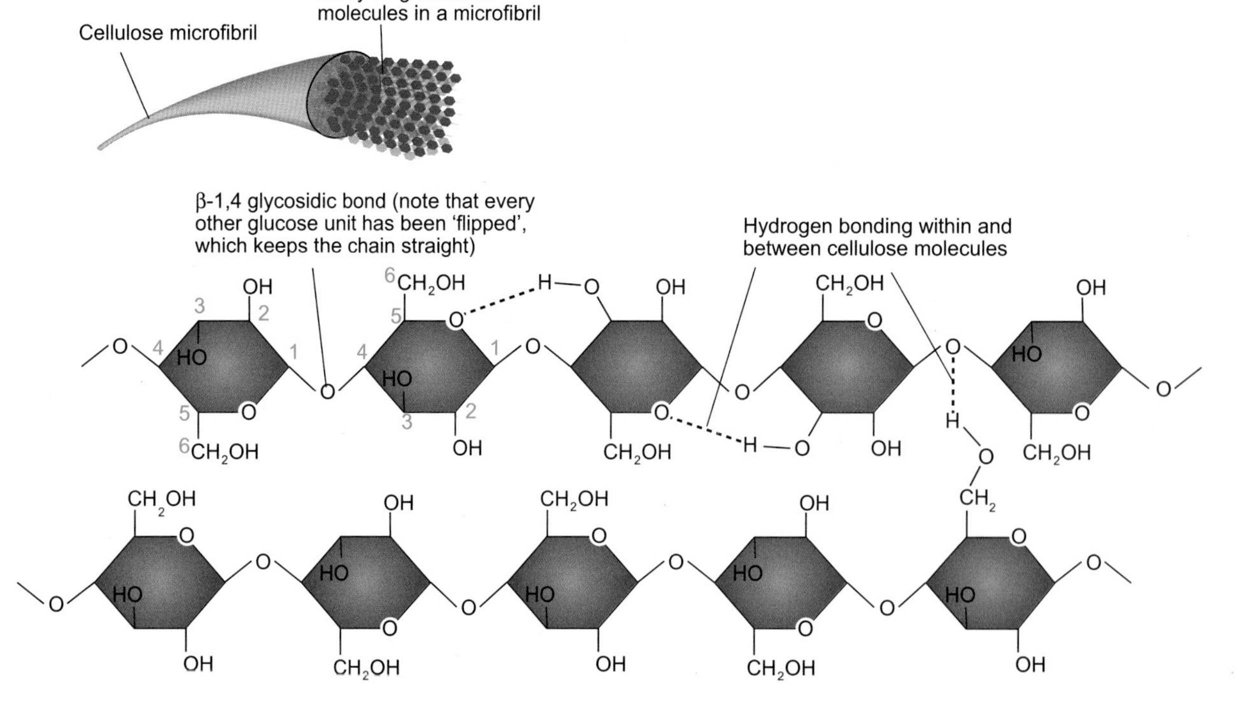

Figure 13 The structure of cellulose

ASSIGNMENT 2: UNDERSTANDING LACTOSE INTOLERANCE

(PS 1.1, PS 1.2)

Lactose is found in milk and dairy products. It is also present in many things you perhaps would not expect – such as crisps, biscuits, crackers, fruit bars, and even some coatings on tablets.

All humans produce lactase in their small intestines when they are babies; this is necessary to digest the lactose present in breast milk or formula milk. However, in most people, lactase production stops as they become adults. There are some exceptions – for example, most people of European descent continue to make lactase throughout their lives.

The lactase in the small intestine hydrolyses lactose into glucose and galactose, which can then be absorbed into the bloodstream. In people who do not produce lactase, the undigested lactose passes into the colon where bacteria ferment it, producing fatty acids and gases such as carbon dioxide, hydrogen and methane. It is these acids and gases that cause the symptoms of lactose intolerance such as wind, pain and bloating. Some of the gases eventually dissolve in the blood and are breathed out from the lungs.

Questions

A1. Use your knowledge of osmosis to explain why water enters the small intestine of a person with lactose intolerance.

A2. What is the result of more water in the large intestine?

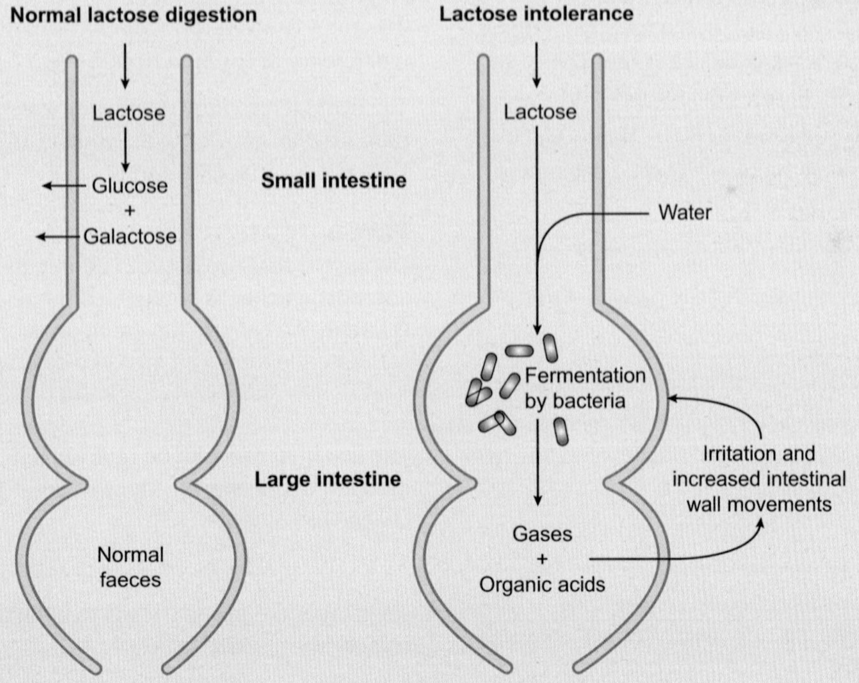

Figure A1 *Lack of the enzyme lactase enables bacteria to ferment the sugar producing acids and gases.*

Worked maths example: The hydrogen breath test

(MS 0.1, MS 1.1, MS 1.11, MS 2.2, MS 2.3, MS 3.3, MS 3.5, PS 3.1, PS 3.2, PS 3.3)

Lactose intolerance can be diagnosed in a variety of ways, but the most reliable is the hydrogen breath test. The subject fasts for about 10 hours before having the test to determine how much hydrogen is present in their breath (in parts per million, ppm). This is their baseline. The subject then drinks 250 cm^3 of a 10% lactose solution and is tested again each hour for three hours. Lactose intolerance is diagnosed if more than 20 ppm of hydrogen is detected, compared with baseline, after three hours.

Use this graph with results of a hydrogen breath test to work out:

a. if Person A is lactose intolerant

b. the rate of change in the concentration of hydrogen, in ppm min^{-1}, in the breath of Person A during the first two hours after they drank the lactose solution, to two significant figures

c. the percentage error of part b. to three significant figures, assuming the breath testing equipment can read to 0.1 ppm.

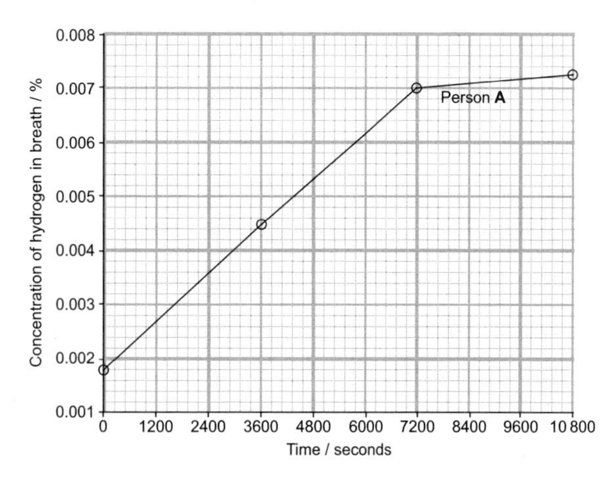

Part a

We need to know the concentration of hydrogen after three hours.

Note that the tester plotted the graph using per cent (y-axis) and seconds (x-axis), rather than ppm and minutes, so we need to ensure we convert the units.

Firstly, we need to convert three hours into seconds; there are 60 minutes in an hour, and 60 seconds in a minute.

So, **3 hours** = 3 × 60 minutes = **180 minutes** = 180 × 60 seconds = **10 800 seconds**.

Next, we must convert per cent to ppm. It may help to think of 'per cent' as 'parts per 100':

$$1\% = \frac{1}{100}$$

$$1\text{ ppm} = \frac{1}{1000000}$$

There are four zeros between these figures, that's a 10 000 (10^4) fold difference. So, to convert percentage to ppm you must multiply by 10 000, and to convert ppm to percentage you divide by 10 000.

1 ppm = 0.0001%

Therefore, 10 ppm = 0.001%

The increments on the y-axis are in 0.001% intervals, so we know that 0.002% = 20 ppm, 0.003% = 30 ppm, and so on.

In the first three hours (between 0 and 10 800 seconds), the concentration of hydrogen rose from 18 ppm (baseline) to 72 ppm (0.0018% to 0.0072%), a difference of **54 ppm** (0.0054%).

The hydrogen concentration after three hours is more than *20 ppm greater* than baseline, so it's likely that Person A is lactose intolerant.

Part b

During the first two hours (120 minutes; 7200 seconds), the line is straight, so the relationship is **linear**. We can use the equation $y = mx + c$ to calculate the rate of change in y compared to x:

$$y = mx + c$$

The equation needs to be re-arranged to find the gradient of the line, m. The gradient is the rate of change (see Chapter 13):

$$m = \frac{y - c}{x}$$

Remember the answer needs to be in ppm, so first convert the y and c values:

0.007% = 70 ppm

0.0018% = 18 ppm

Then put the values into the equation:

rate of change in the concentration of hydrogen

$$= \frac{(70 - 18)}{120} = 0.4333333\text{ ppm min}^{-1}$$

= **0.43** ppm min^{-1} to two significant figures

Part c

The equipment measures to 0.1 ppm, so the uncertainty in measurement is 0.05 ppm. In other words, the equipment is accurate to 0.1 ppm ± 0.05 ppm.

$$\text{percentage error} = \frac{\text{uncertainty}}{\text{value being measured}} \times 100$$

So, in this case:

$$\text{percentage error} = \frac{0.05}{0.43} \times 100 = 11.6279\%$$

= **11.6%** to three significant figures

Biochemical test for starch

To test for starch, add **iodine in potassium iodide solution** to the substance. A blue-black colour indicates the presence of starch.

Biochemical tests for sugars

Tests for sugars can distinguish two groups of sugars: the **reducing sugars** and the **non-reducing sugars**.

A redox reaction is one in which one substance is oxidised while another is reduced:

- oxidation is the loss of electrons or hydrogen, or the gain of oxygen
- reduction involves the gain of electrons or hydrogen, or the loss of oxygen.

Reducing sugars

A reducing sugar readily loses electrons to another substance – the reducing sugar is therefore oxidised, while the other substance is reduced. The monosaccharides glucose and fructose are reducing sugars. The disaccharide maltose is also a reducing sugar.

To test for a reducing sugar, we add Benedict's solution to the substance, then heat the mixture to 80 °C in a water bath. An orange-red precipitate indicates the presence of a reducing sugar.

Non-reducing sugars

Some sugars, though, are not readily oxidised and so do not reduce other substances. These are non-reducing sugars. The disaccharide sucrose is a non-reducing sugar.

A non-reducing sugar will not give an orange-red precipitate when heated with Benedict's solution. To test whether there is a non-reducing sugar present, we boil the substance with dilute acid, and then neutralise by adding hydrogen carbonate. Then we test again with Benedict's solution, and if a non-reducing sugar was present then a precipitate will now form.

Figure 14 shows what happens when the reducing sugar glucose loses electrons in a redox reaction. Carbon-1 forms an exposed –CHO group, which can then accept oxygen so that the glucose is oxidised (to gluconic acid).

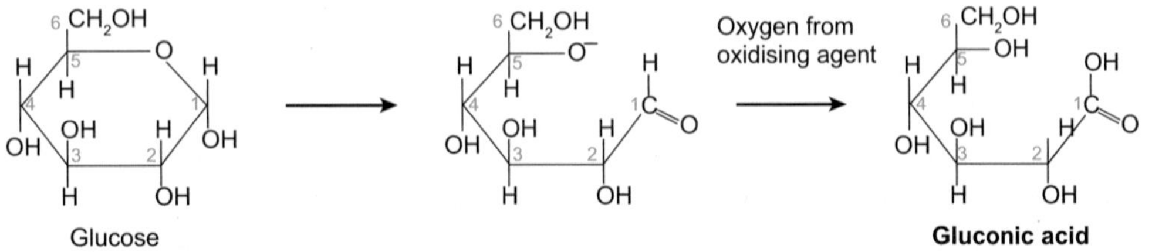

Figure 14 *A redox reaction*

QUESTIONS

7. Explain why Benedict's test works with maltose but not with sucrose.
8. Boiling a disaccharide with acid hydrolyses it into monosaccharides. Explain why Benedict's test gives a positive result after sucrose has been treated with acid.
9. For low concentrations of glucose, a positive Benedict's test ranges from green through yellow to orange and brick red. By comparing the colour of the sample solution with the colour of a standard solution, the glucose concentration of the sample can be estimated. Explain how you would prepare standard solutions for comparison.

ASSIGNMENT 3: IDENTIFYING CARBOHYDRATES FROM BIOCHEMICAL TESTS

(PS 1.2, PS 4.1)

Biochemical tests are very important in both nutritional analysis and clinical diagnosis. Samples of blood, urine and other body fluids can be tested for a range of substances to help in the diagnosis of some disorders, such as diabetes. Different microorganisms are able to ferment different sugars and biochemical tests for carbohydrates can be used on bacterial cultures to determine which carbohydrates they are using and so aid in their identification. This is very important if the sample being tested comes from someone who is ill.

Nutritionists and food scientists use biochemical tests to analyse foodstuffs. It is important, for example, that someone who is lactose intolerant is informed when foodstuffs contain lactose.

Most monosaccharides and disaccharides can be fermented by yeast. The products of fermentation are alcohol and carbon dioxide; the formation of bubbles of carbon dioxide is used to confirm that fermentation is occurring. Although yeast has enzymes for the hydrolysis of most disaccharides, it does not have the enzyme necessary for hydrolysing lactose. So, both lactose and galactose give negative results with this fermentation test. This test can be used to determine whether a substance contains lactose.

Six foodstuffs were tested in a laboratory to determine which carbohydrates were present, to see if they were suitable for lactose-intolerant persons or diabetics. However, the technician had not labelled the test substances, and was relying on the results of biochemical tests. The results of these tests are shown in the table.

Questions

A1. Describe how you would test for:

a. starch

b. reducing sugar

c. non-reducing sugar.

A2. **a.** Which sample contained no carbohydrates? Explain your reasoning.

b. Which samples were likely to contain sucrose? Explain your reasoning.

c. Which sample(s) would be suitable for someone with lactose intolerance? Explain your reasoning.

Sample	Test for starch	Test for reducing sugar	Test for non-reducing sugar	Fermentation test
A	positive	positive	positive	positive
B	positive	negative	positive	positive
C	negative	negative	negative	negative
D	negative	positive	positive	positive
E	positive	negative	negative	negative
F	negative	negative	positive	positive

KEY IDEAS

- Carbohydrates are made up of carbon, hydrogen and oxygen.
- Monosaccharides are single-unit sugars. Glucose, fructose and galactose are monosaccharides. Monosaccharides, such as glucose, can exist in two different forms (for example, α-glucose and β-glucose). These are isomers.
- Disaccharides are made of two monosaccharide molecules joined by a glycosidic bond. Maltose, sucrose and lactose are disaccharides.
- The reaction in which two monosaccharides join together is an example of a condensation reaction. The reaction that breaks them apart is a hydrolysis reaction.
- Polysaccharides are polymers of monosaccharides. Starch, glycogen and cellulose are polysaccharides.
- Starch and glycogen have compact, coiled and branched molecules made of long chains of α-glucose monomers. They are used as energy stores in plants and animals respectively.
- Cellulose has straight, unbranched molecules made of long chains of β-glucose monomers. Cellulose molecules associate with each other to form microfibrils. They form cell walls around plant cells.
- Benedict's solution can be used to test for reducing sugars. Non-reducing sugars do not give a positive result with Benedict's solution.
- Iodine in potassium iodide solution is used to test for starch.

PRACTICE QUESTIONS

1. Figure Q1 shows a glucose molecule and a fructose molecule.

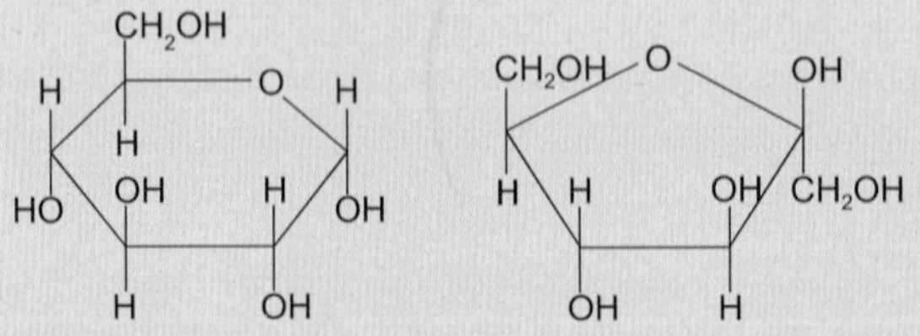

Figure Q1

   a. Is the glucose molecule in the alpha or beta form? Explain your answer.

   b. Write the molecular formula for fructose.

   c. Draw a diagram to show how glucose can join to fructose to form a sucrose molecule.

   d. Name the type of reaction involved in the formation of sucrose.

2. A student was provided with a glucose solution of unknown concentration. He was also given some pure glucose. He had access to Benedict's solution.

   The colour produced when glucose is heated with Benedict's solution is dependent on the concentration of glucose.

   a. Describe how to test a solution for the presence of glucose, and explain what causes the colour change.

   b. Describe how you could make up 1 dm^3 of a 10% solution of glucose.

   c. Describe how you could use this 10% solution to make a range of less concentrated solutions, each of known concentration.

   d. Describe how you would use the Benedict's test to produce a set of colour standards using these solutions.

   e. Explain how you could use these colour standards to find the concentration of the unknown glucose solution.

3. a. Describe the structure of a starch molecule.

   b. Explain how this structure is related to the function of starch molecules in plant cells.

4. Plants lose water from their leaves because liquid water evaporates from the surfaces of the mesophyll cells in the leaf, and then diffuses as water vapour into the air, through the stomata.

(Continued)

Figure Q2 shows the water loss from plants kept for five days at either 22 °C or 28 °C.

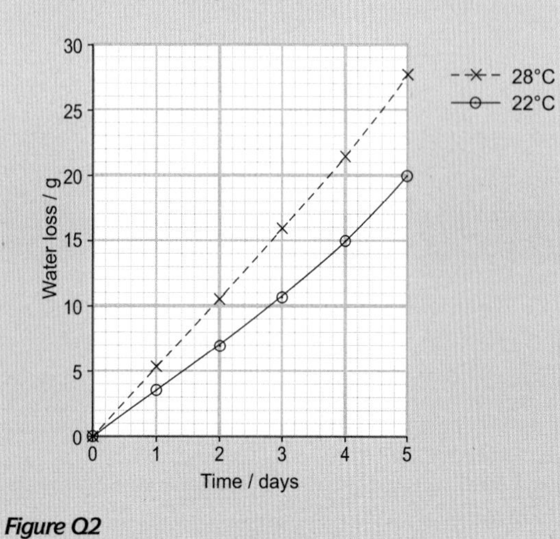

Figure Q2

a. Calculate the mean rates of water loss per day at:

i. 22 °C

ii. 28 °C

b. With reference to the properties of water, explain the reason for the difference in the rates that you have calculated.

c. Explain how the loss of water from the leaves can help to cool the plant.

2 LIPIDS AND PROTEINS

PRIOR KNOWLEDGE

You have probably learned about the importance of fats (lipids) and proteins in the diet. You might want to remind yourself about hydrogen bonds, and also about monomers and polymers, condensation reactions and hydrolysis reactions.

LEARNING OBJECTIVES

In this chapter, you will build on these ideas to understand how the structures and functions of lipid and protein molecules are related to their roles in the bodies of living organisms.

(Specification 3.1.3, 3.1.4)

Everyone knows that they should eat protein as part of their diet; protein is needed to help us to produce new cells, and to repair damaged tissues. But not everyone is aware of the huge number of different kinds of proteins that exist, and the wide range of different functions they have.

Take spider silk, for example. Spider silk is a fibrous protein – one that is made up of long chains of amino acids that lie side by side, forming long, strong fibres. Like many fibrous proteins, spider silk contains repetitive sequences of amino acids, in this case mostly glycine and alanine. It has tremendous strength, and a thread of spider silk is about five times stronger than steel of the same diameter.

Spiders use silk for many different purposes, including catching prey in webs, wrapping prey to immobilise and store, making shelters and ballooning. For ballooning, a spider produces long strands of fine silk thread which blows in the wind and lifts the still-attached spider high into the air. Ballooning spiders have been found at heights of five kilometres above sea level. This is how they are able to disperse to new areas.

Materials chemists are very interested in spider silk, seeing it as a possible starting point for developing technical materials with a very high strength to weight ratio. While people have long kept silkworms (moth larvae) to make large quantities of silk, it is not easy to produce spider silk in large amounts; if you try to keep lots of spiders together, they eat each other. Now, DNA sequences that code for making silk have been isolated from spiders and inserted into silkworms. Spider silk synthesised by these genetically engineered silkworms may form the basis of large-scale production of new fibres and materials.

2.1 LIPIDS

Lipids, like carbohydrates, are substances whose molecules contain carbon, hydrogen and oxygen atoms. However, these atoms are arranged in a completely different way, and the relative number of oxygen atoms compared to hydrogen and carbon is much smaller than in carbohydrates, so lipids have very different properties from carbohydrates.

Lipids are also known as fats and oils. Fats tend to be solid at room temperature, while oils are liquid.

Triglycerides

Triglycerides are a group of lipids that are found in all types of living organisms. Figure 1 shows the formation of a triglyceride molecule. It is made up of one molecule of **glycerol** (a type of alcohol), to which three **fatty acids** are joined through **ester bonds**. The reaction of each fatty acid with a glycerol molecule is a condensation reaction.

a Glycerol

CH_2OH
|
CHOH
|
CH_2OH

b Fatty acid

c Formation of a triglyceride by condensation

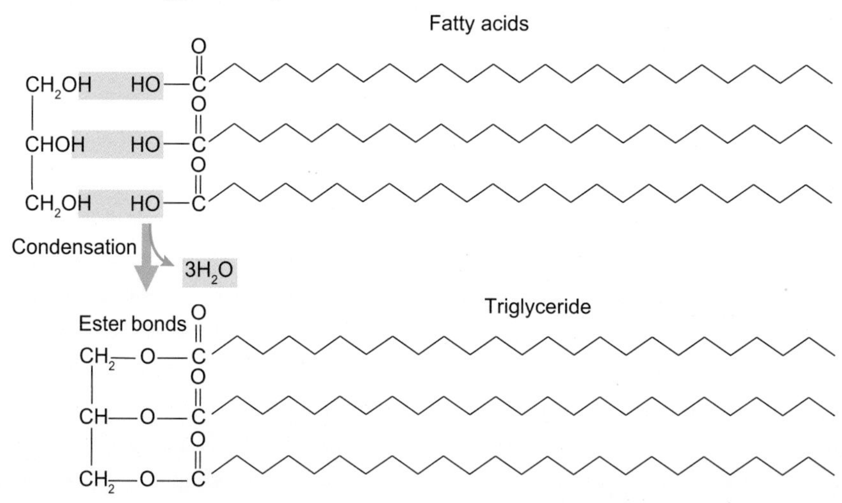

d A simple representation of a triglyceride

Glycerol
Fatty acid
Fatty acid
Fatty acid

Figure 1 The formation of a triglyceride molecule

There are many different types of fatty acids. All of them contain a –COOH group (the acid part), to which a long hydrocarbon chain is attached. The long hydrocarbon chain is sometimes represented by the letter R, so we can represent a fatty acid as RCOOH.

Carbon can bond with up to four atoms, so they can form long chains using two of their bonds to join to the next carbon atoms, and two of their bonds to join to hydrogen atoms. A fatty acid in which all the carbon atoms use up all their four bonds in this way is called a **saturated fatty acid**. There are no spare bonds to which any more hydrogen atoms could be joined. In some fatty acids, however, some of the carbon atoms have a double bond joining them to a neighbouring carbon atom. These are called **unsaturated** fatty acids, because it would be possible to join extra hydrogens to them by changing the double bond to a single bond, leaving a spare one to join to another hydrogen atom. Figure 2 shows saturated and unsaturated fatty acids. Unsaturated fatty acids have a 'kink' in them where there is a double bond.

A triglyceride containing one unsaturated fatty acid is called an unsaturated triglyceride, while a triglyceride containing two or three unsaturated fatty acids is a polyunsaturated triglyceride. In general, unsaturated triglycerides tend to be made by plants, and saturated triglcyerides by animals. The 'kinks' in the unsaturated fatty acids make unsaturated triglycerides behave as liquids at room temperature, and so many of the triglycerides made by plants are oils. Animal fats tend to be saturated, and they are often more solid at room temperature.

Triglycerides are commonly used as energy stores in both plants and animals. Energy is released from triglycerides when they are oxidised in respiration; the high proportion of carbon and hydrogen atoms in their molecules means that they contain more energy per gram than carbohydrates. In animals, they are found in adipose tissue, which often lies just beneath the skin. Each cell contains a droplet of triglyceride, which can almost completely fill the cell. Adipose tissue acts as an excellent heat insulator; animals that live in cold water, such as whales and seals, may have very thick layers of it, called blubber. The density of the triglycerides is low, so the blubber also acts as a buoyancy aid. In plants, triglycerides are often found in seeds, where they serve as lightweight, high-energy stores that the seed can use when it germinates. Many of the oils that we use in cooking, such as olive oil, are extracted from seeds and are mixtures of different unsaturated triglycerides.

Saturated fatty acid

Unsaturated fatty acid

Polyunsaturated triglyceride

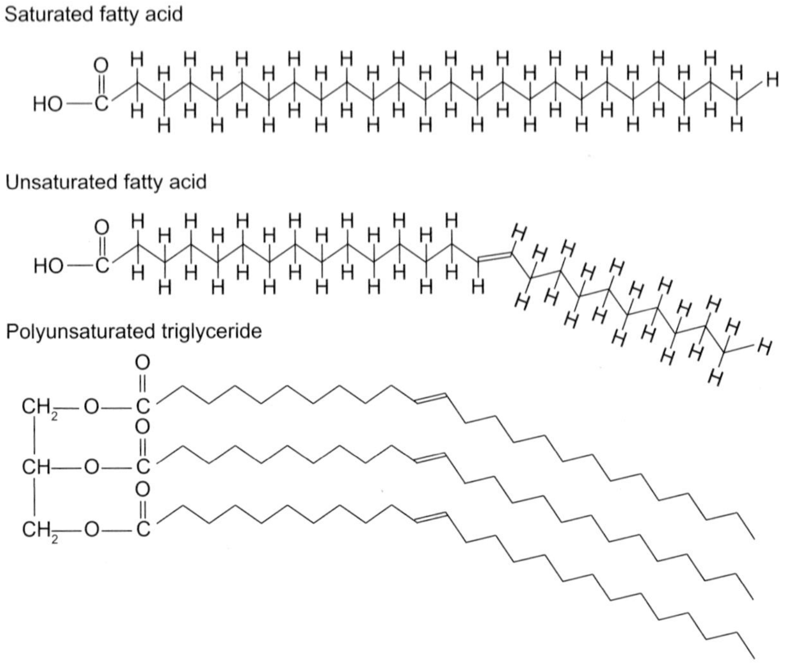

Figure 2 *Saturated and unsaturated fatty acids, and a polyunsaturated triglyceride*

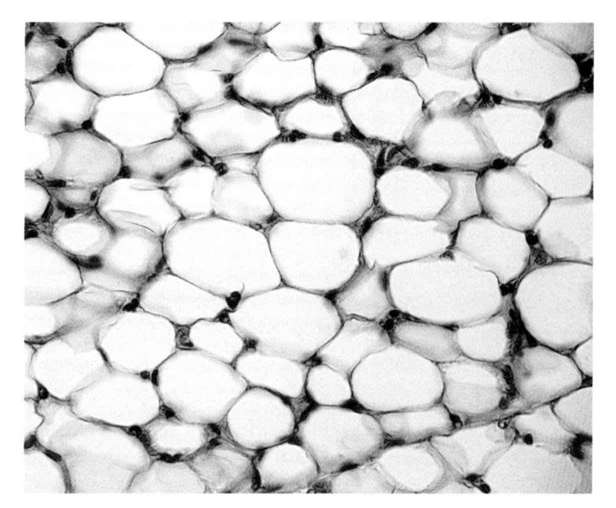

Figure 3 Light micrograph of white adipose tissue stained with hematoxylin and eosin

Triglycerides are insoluble in water. This is because their molecules do not have any uneven distribution of charge, and are said to be **non-polar**. They are therefore not attracted to the charges on water molecules, and are described as **hydrophobic** (water-hating). Most do not even mix with water very well and tend to form a layer on top of it. If stirred into water, triglyceride molecules group together to form droplets. Their insolubility means that they have no osmotic effects on cells. Their large size means that they cannot easily cross membranes.

Phospholipids

Phospholipids, like triglycerides, have a 'backbone' of glycerol to which fatty acids are attached via ester bonds. However, only two fatty acids are present in the molecule. The place of the third one is taken by a phosphate group. Figure 4 shows the structure of a phospholipid molecule.

Like triglycerides, phospholipids do not dissolve in water. The tail of a phospholipid molecule is hydrophobic but the 'head' of the molecule has a negative charge, and so is **hydrophilic** – it 'loves' water. So, when phospholipids are placed in water, they arrange themselves on the surface of the water with their heads down and tails up, or they form spheres in water called micelles. These spheres can be either single-layered or double-layered (Figure 5). In both cases, the hydrophilic heads get to be near the water while the hydrophobic tails stay well away from it.

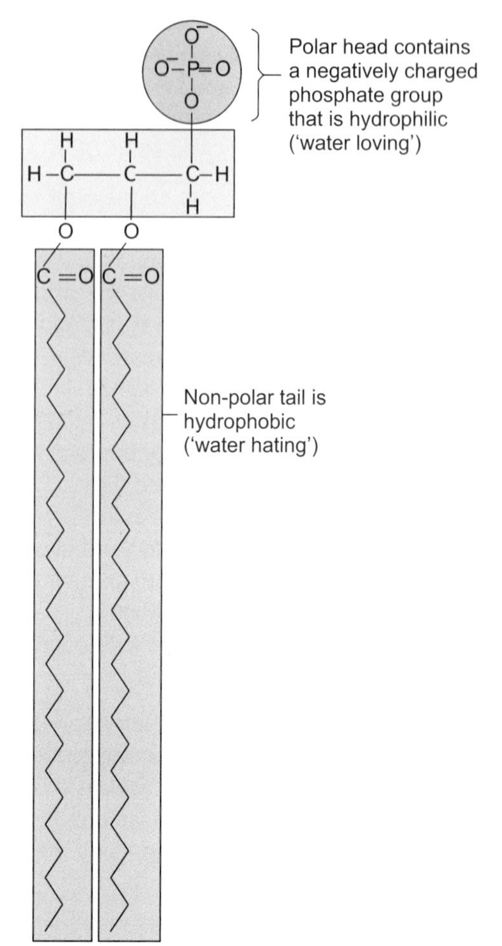

Figure 4 A phospholipid molecule. The phosphate group is polar and hydrophilic, while the fatty acid tails are non-polar and hydrophobic.

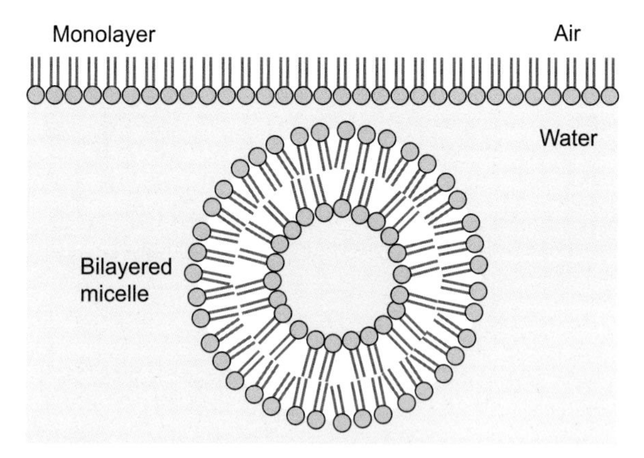

Figure 5 In water, the hydrophilic polar heads of the phospholipids face into the water, while the hydrophobic non-polar tails face away from it. Single or double-layered micelles form in the water, and a monolayer forms on the surface.

The emulsion test for lipids

To test for a lipid, shake the unknown substance with ethanol in a clean test tube. Lipids are soluble in ethanol, so if there is any lipid in your sample, some of it will dissolve.

Now gently pour the ethanol into another test tube containing water. If there is any lipid dissolved in the ethanol, it will form little droplets in the water. The mixture of these microscopic lipid droplets and water is known as an emulsion. The lipid droplets prevent light passing through, so the water becomes cloudy (milky white) rather than being transparent.

QUESTIONS

1. What are the two components of a lipid molecule?
2. What is the difference between a saturated fatty acid and an unsaturated fatty acid?
3. What is the difference between a lipid and a phospholipid?

KEY IDEAS

- Triglycerides and phospholipids are two groups of lipid.
- Triglycerides are formed when three fatty acids form ester bonds with a glycerol molecule, in a condensation reaction.
- Fatty acids in which all the carbon–carbon bonds are single are known as saturated fatty acids. If at least one carbon–carbon bond is a double bond, the fatty acid is unsaturated.
- In phospholipids, a negatively charged phosphate group takes the place of one of the fatty acids.
- Triglycerides are non-polar molecules, and so are hydrophobic and insoluble.
- Phospholipids have charges on their phosphate groups, and so the phosphate heads are polar and hydrophilic, while the fatty acid tails are non-polar and hydrophobic. This causes phospholipids to form monolayers and micelles when mixed with water.

ASSIGNMENT 1: INVESTIGATING THE INSULATION PROPERTIES OF BLUBBER

(MS 1.7, PS 2.1, PS 2.4, PS 3.1)

Cetaceans (whales, seals and porpoises) have a thick layer of adipose tissue beneath their skin, called blubber. This serves as a buoyancy aid, a site of energy storage, and a layer of heat insulation. The latter is necessary because water conducts heat away from a body around 25 times faster than air, and many of these animals live in very cold water.

Scientists investigated the heat insulation properties of blubber from two types of whales. The blubber of short-finned pilot whales, *Globicephala macrorhynchus*, contains fat cells filled with triglycerides. In contrast, the blubber of pygmy sperm whales, *Kogia breviceps*, contains a different type of lipid, in which the backbone of the molecules is a long-chain alcohol rather than glycerol.

The blubber of seven pygmy sperm whales and seven short-finned pilot whales was sampled. The samples were taken from whales that had recently stranded themselves on beaches and died. For each individual, the percentage of lipid in the blubber was measured, and also the thermal conductivity (the rate at which heat could pass through the blubber). The results are shown in the scattergraph.

Questions

A1. Explain how the structure of lipid molecules makes them suitable as energy storage compounds.

A2. Suggest three variables that the researchers kept constant in their investigation.

A3. Suggest why the researchers chose to display their data as a scattergraph.

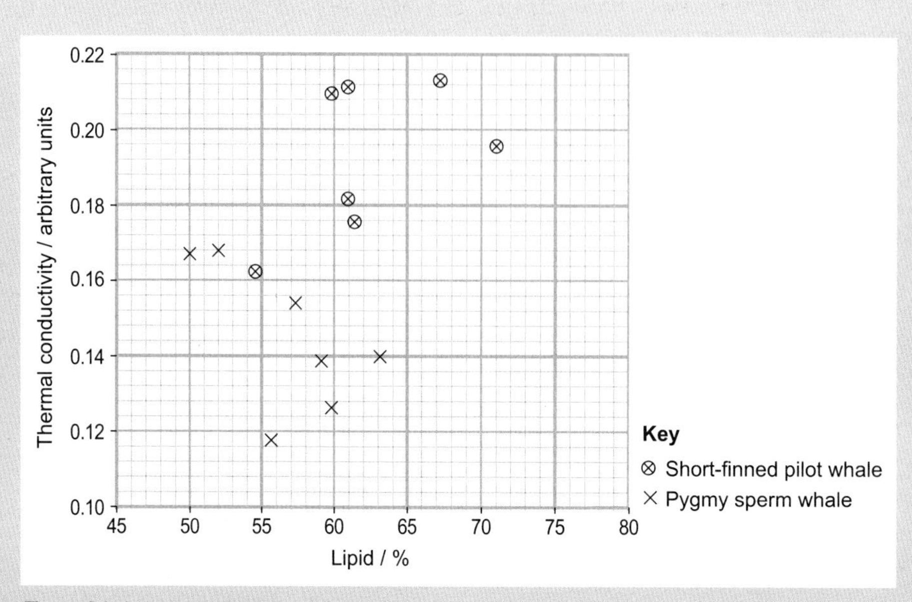

Figure A1 *A scattergraph showing the results*

A4. With reference to the scattergraph, do you consider there is any evidence that:

a. there is a difference between the thermal conductivity of triglycerides and lipids with long-chain alcohol backbones?

b. the percentage of lipid in blubber affects its thermal conductivity?

c. Suggest what could be done to provide more definitive answers to the questions in parts a and b.

Worked maths example: Scattergraphs and correlation coefficients

(MS 1.7, MS 1.9, PS 3.1, PS 3.2)

Scattergraphs allow you to relate one variable to another, in a situation where you can't control either variable. In assignment 1, for example, the variables are thermal conductivity of blubber and the percentage of lipid in blubber.

Imagine you were asked some questions about the scattergraph:

- Does there appear to be any correlation between thermal conductivity and percentage lipid in the blubber of the short-finned pilot whales and the pygmy sperm whales sampled in this experiment?

Just by looking, there seems to be a positive correlation between the two variables. In general, as one increases, so does the other. The trend line would be straight, through the *y*–*x* intercept.

- What statistical test would be appropriate to demonstrate whether these variables correlate?

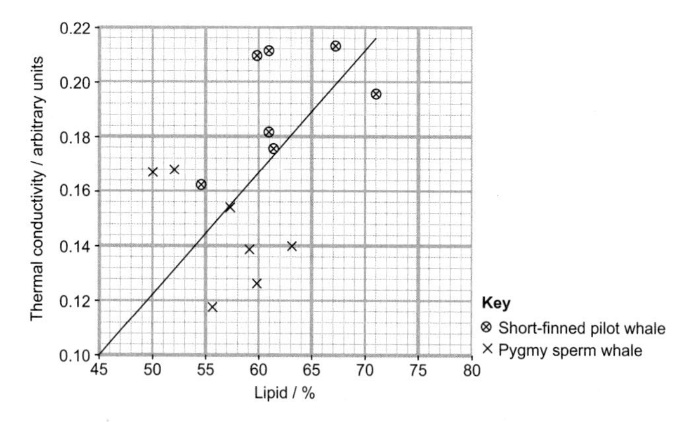

The **correlation coefficient test** is a statistical test for determining correlation between variables. The correlation coefficient is a number on a scale between + 1 and – 1, where + 1 is total **positive correlation**, – 1 is total **negative correlation** and 0 is no correlation. Correlation coefficient calculations can be done in a computer spreadsheet package.

- The correlation coefficient for this data is 0.9. Explain what this suggests about the variables' relationship.

A correlation coefficient of 0.9 suggests that the variables are strongly linked, as predicted. Blubber's thermal conductivity in both types of whale increases as lipid percentage in the blubber increases.

- From the scattergraph, what can you determine about short-finned pilot whales compared with pygmy whales?

The data suggests that short-finned pilot whale blubber has a higher thermal conductivity than pygmy sperm whale blubber. The scattergraph suggests that thermal conductivity and percentage lipid are positively correlated in both blubber types.

2.2 PROTEINS

Proteins are polymers made of many amino acids joined together in long chains. In this respect, they are similar to polysaccharides. However, unlike starch and cellulose, which are made from only one type of monomer (glucose), proteins are made from 20 different naturally occurring amino acid monomer units. These can be assembled in any order, making an endless variety of protein structures possible. It is like having an alphabet of 20 letters to make up words hundreds of letters long.

It is interesting to note that the same 20 amino acids are found in almost all living organisms. This is strong evidence that all types of organism on Earth have evolved from the same common ancestor.

Figure 6 shows an amino acid. All amino acids have an amine group, $-NH_2$, which is chemically basic. This means it reacts with and neutralises acids. It does this by accepting hydrogen ions, to become NH_3^+. Amino acids also have a carboxyl group, $-COOH$, which, like all acids, dissociates (breaks apart) to produce hydrogen ions. H^+, when dissolved in water. This leaves the carboxyl group with a negative charge, $-COO^-$.

Because of these charged groups, amino acids are attracted to water molecules and are therefore soluble in water.

Each amino acid has a different group, called an R group, attached to the central carbon atom (Table 1). The smallest R group is simply a hydrogen atom, and the amino acid with this R group is glycine. All the other R groups contain carbon, and they are sometimes known as side chains. The different side chains give each amino acid different properties and – as you will see – this in turn affects the properties of the proteins they form.

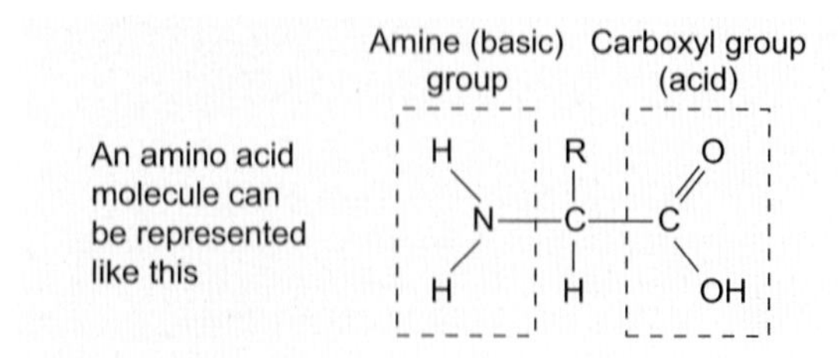

Figure 6 *An amino acid molecule*

Amino acid	R group (side chain)
glycine	H
alanine	CH_3
cysteine	CH_2SH

Table 1 *Some amino acids*

The primary structure of proteins

Figure 7 shows how two amino acid molecules join together. The reaction, like the way in which the component molecules of carbohydrates and lipids join together, is a condensation reaction. The resulting bond between the two amino acids is called a **peptide bond**. The molecule is called a **dipeptide**.

Figure 7 *Two amino acid molecules join by a condensation reaction to form a dipeptide.*

Notice how the acid group joins up with the amine group. The result is that the dipeptide molecule has an acid group at one end and a basic amine group at the other. The long chains that eventually form the protein are made by adding more amino acids. These long chains are **polypeptides**. A protein molecule may be made up of just a single polypeptide, or two or more polypeptides entwined around one another. Each type of polypeptide has its own unique sequence of amino acids, known as its **primary structure**.

QUESTIONS

4. Which four elements are present in every amino acid?
5. Which of these elements is not present in carbohydrates or lipids?
6. What is the formula of the amino group?
7. What is the formula of the carboxyl group?
8. Using Figure 6 as your starting point, draw the structural formula of each of the amino acids listed in Table 1.
9. Write an equation to show the formation of a dipeptide by condensation of alanine and glycine.
10. Draw the backbone chain of carbon and nitrogen atoms formed when three amino acid molecules condense. Label the peptide bonds.

Secondary structure

Polypeptides can form regular coils or pleats. The coiling or pleating of a polypeptide chain is known as the secondary structure of a protein. A very common type of secondary structure is known as an **alpha helix**. The polypeptide chain coils round in a spiral, held in place by hydrogen bonds that form between the hydrogen atom of the amine group of one amino acid, and the oxygen atom of a carboxyl group of another amino acid further along the chain (Figure 8).

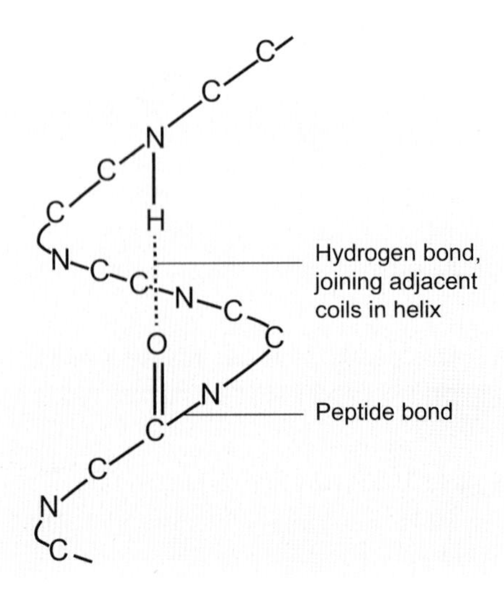

Figure 8 An example of secondary structure: an alpha helix. Note that, for simplicity, the R groups of each amino acid have been omitted from the diagram.

Tertiary structure

The polypeptide chains often fold again, forming all kinds of complex three-dimensional shapes. These shapes are held in place by several different types of bonds between the R groups of the amino acids. Hydrogen bonds (of course) are involved yet again, but also much stronger ionic bonds and disulfide bridges (Figure 9). The three-dimensional shape of the molecule is therefore determined by where these bonds can form, and this in turn is determined by the sequence of amino acids in the chain.

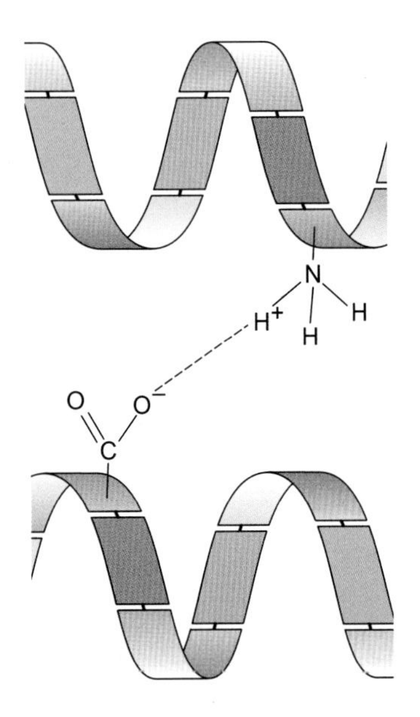

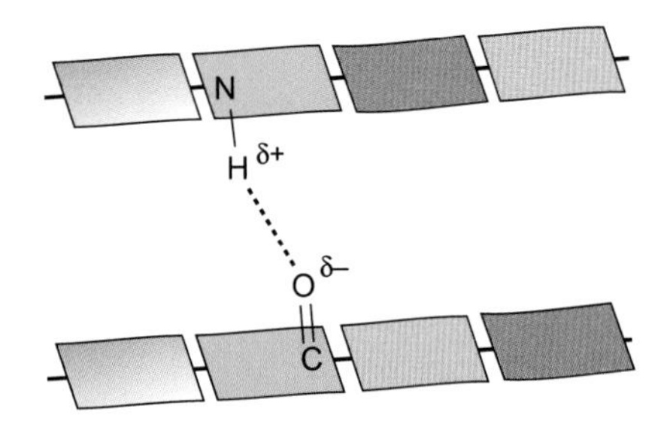

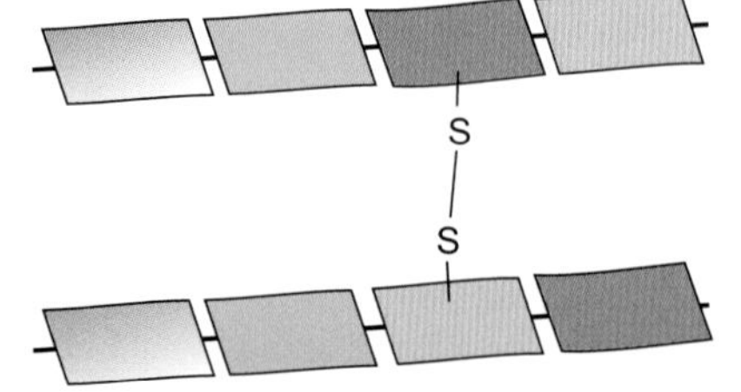

Figure 9 Types of bond that hold the three-dimensional structure (tertiary structure) of a protein in shape

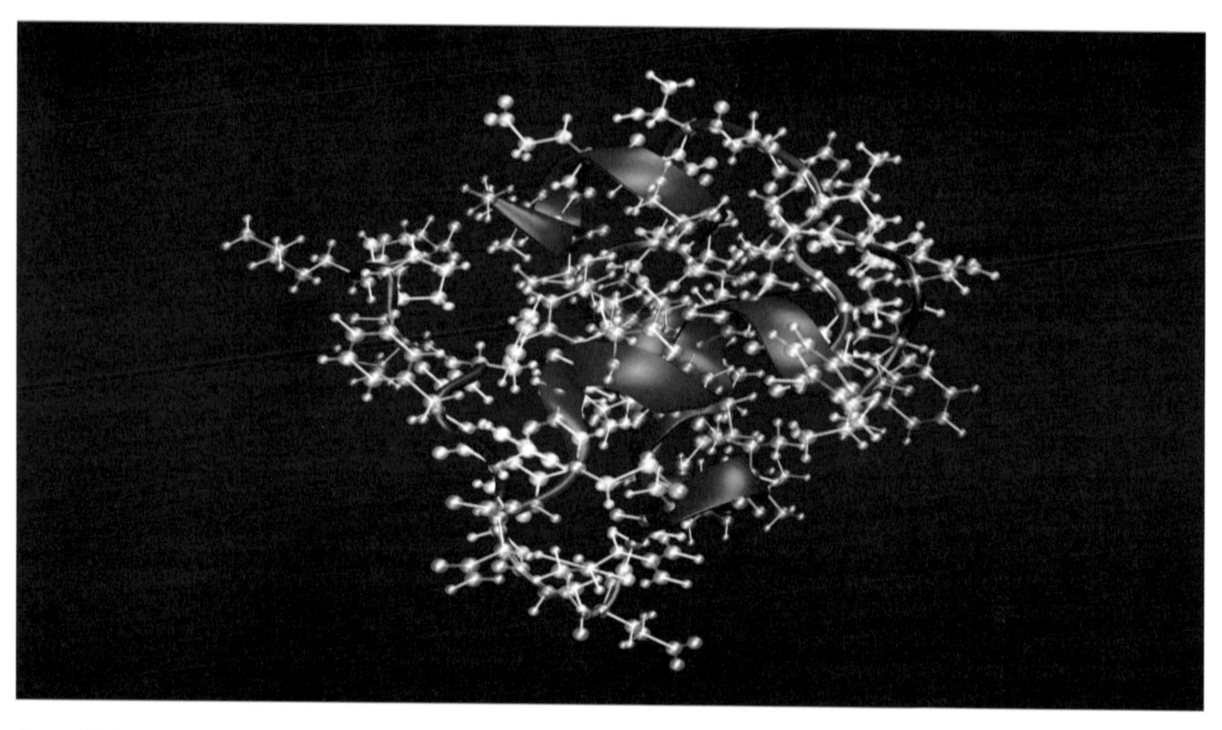

Figure 10 A computer-generated model of an insulin molecule. It has two polypeptide chains, shown in blue and red, which you can pick out as they snake through the structure. The individual atoms of the amino acids are shown as balls joined by sticks. The purple 'sheets' represent disulfide bridges helping to hold the whole molecule in a precise shape.

The three-dimensional shape is the **tertiary structure** of the protein. Each type of protein has its own particular tertiary structure, and this determines its function.

Quaternary structure

As mentioned earlier, some proteins are made up of more than one polypeptide chain. Haemoglobin is a good example. It contains four polypeptide chains, two of one type (called α chains) and two of another (called β chains). Proteins made from more than one polypeptide are said to have **quaternary structure**. (This does not mean they have four chains; it simply means that they contain more than one polypeptide chain.) Haemoglobin is also interesting because it contains 'extra' groups that are not amino acids. These are called haem groups, and they help haemoglobin to carry out its role of transporting oxygen in the blood.

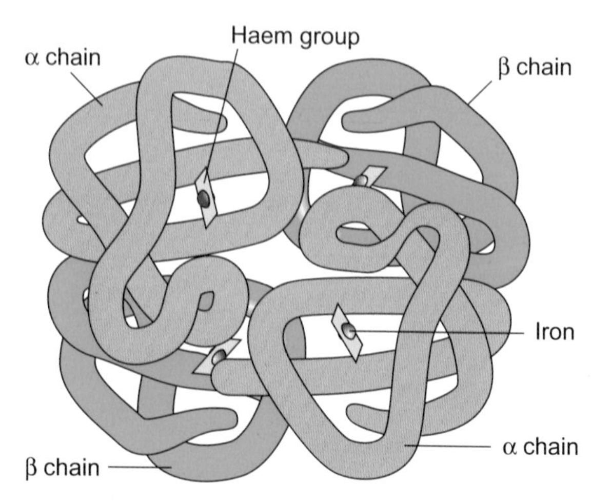

Figure 11 Haemoglobin, a protein with quaternary structure

QUESTIONS

11. Which of the amino acids listed in Table 1 could form disulfide bridges?
12. Explain how the primary structure of a protein determines its three-dimensional shape.

Structure and function in proteins

The function of a protein depends on its three-dimensional shape (tertiary structure). This is determined by its primary structure (the sequence of amino acids), which affects the way in which it folds to form its final three-dimensional shape.

Haemoglobin is an example of a **globular protein**. It forms a ball-shaped molecule. Haemoglobin is soluble in water, because it has many R groups containing hydrophilic side chains on the outside of the 'ball'. These side chains interact with water molecules, enabling haemoglobin to dissolve easily.

Other examples of globular proteins include the hormone insulin and all enzymes. The shape of insulin (Figure 10) allows it to form bonds with other protein molecules in the membranes of liver cells that have a complimentary shape, which then respond by setting off a sequence of events that culminates in joining glucose molecules together to form glycogen. The shapes of enzymes include a depression called an active site, which has a very precise shape that allows the enzyme to bind with other molecules and cause them to react. You can read much more about this in Chapter 3. Many globular proteins are involved in metabolic reactions.

However, many proteins do not curl up into a ball, but form long, thin molecules. These are called **fibrous proteins**. A good example is keratin, the protein that is found in hair, wool, nails and the surface layers of the skin. The chains of amino acids have a regular pattern of hydrogen bonds that cause them to coil into long helices that resemble thin springs. Keratin forms long molecules that coil and associate with each other to form filaments. These are very strong, held together with many disulfide bridges; you can smell the sulfur if you accidentally singe your hair. Unlike most globular proteins, fibrous proteins are not soluble in water (so your hair does not dissolve when you wash it). They therefore tend to have structural roles, rather than being involved in metabolic processes.

Collagen, another fibrous structural protein, gives bone and cartilage their strength. Collagen molecules have three twisted polypeptide chains, even more tightly bound than in keratin. This means that collagen can form more rigid structures. Often the molecules are grouped together to make almost rigid rods (Figure 12).

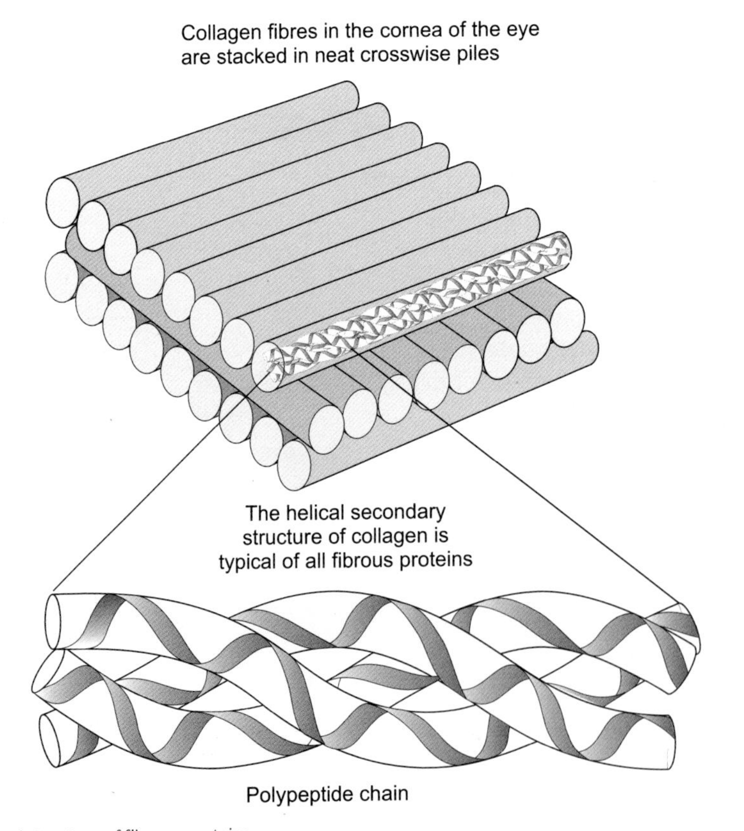

Figure 12 *Three-dimensional structure of fibrous proteins*

QUESTIONS

13. In cartilage the collagen rods are arranged in many different directions. Suggest the advantage of this.

Stretch and challenge

14. In the cornea, collagen rods are stacked in neat piles. Suggest the advantage of this, bearing in mind that light enters the eye through the cornea.

KEY IDEAS

- Proteins are polymers made from amino acids joined by peptide bonds.
- There are 20 different amino acids found in living organisms. Each has a central carbon atom, to which a carboxyl group and an amine group are joined. The fourth bond on the carbon atom holds an R group or side chain, which varies between different amino acids.
- A long chain of amino acids is known as a polypeptide. A protein may be made up of just one or several polypeptides.

The biuret test for proteins

The biuret test detects the presence of peptide bonds. You simply add biuret reagent to the solution that you think might contain protein and mix gently. There is no need to heat it. If protein is present, the mixture will become lilac or purple. It's a good idea to do a control test with pure water, because biuret reagent itself is blue, and you might mistake this blue colour for a positive result.

- Hydrogen bonds, ionic bonds and disulfide bridges help to hold a protein molecule in shape.
- The sequence of amino acids in a polypeptide or protein is known as its primary structure. Regular coiling or folding forms the secondary structure, and further folding forms the tertiary structure. The association of more than one polypeptide forms the quaternary structure.
- The shape of a protein is strongly related to its function. Globular proteins are usually soluble and often have roles in metabolism. Fibrous proteins are insoluble and generally have structural roles.
- The biuret test is used to detect the presence of peptide bonds.

ASSIGNMENT 2: IDENTIFYING AMINO ACIDS USING CHROMATOGRAPHY

(MS 0.3, MS 2.4, PS 2.3, PS 3.3)

Chromatography is a technique that can be used to separate molecules. It can be used to find out which amino acids are present in a protein. Enzymes are used to hydrolyse (digest) the protein into its amino acids and then a concentrated spot of the mixture is placed on a line drawn at the base of a piece of absorbent paper (the origin). The spot is made as concentrated as possible by repeatedly spotting in the same place and allowing it to dry. The paper is then held in a container with the lower edge dipping in a shallow layer of the solvent (butanol/ethanoic acid), ensuring the solvent does not reach the origin line. The container is then covered and left. As the solvent travels up the paper, it carries the different amino acid molecules with it. Different amino acids travel at different speeds and are separated out over a period of about three hours.

Before the solvent front reaches the top of the paper, its position is marked in pencil and the chromatogram is allowed to dry. Amino acids are colourless and so will not show up on the chromatogram until it is sprayed with a suitable dye, such as ninhydrin, inside a fume cupboard. Ninhydrin reacts with the amino acids to give coloured spots, mainly brown, yellow or purple.

R_f values

The distance travelled by a molecule relative to the distance travelled by the solvent is called the R_f value, and is constant for any given molecule, as long as everything else is kept constant.

The R_f value for each amino acid can be worked out using the formula:

$$R_f = \frac{\text{distance travelled by compound}}{\text{distance travelled by solvent}}$$

For example, if an amino acid moved 4.8 cm from the base line while the solvent moved 12.0 cm, then the R_f value for that amino acid is:

$$R_f = \frac{4.8}{12.0} = 0.4$$

This value can then be located on a table of values to identify the amino acid, in this case, cysteine.

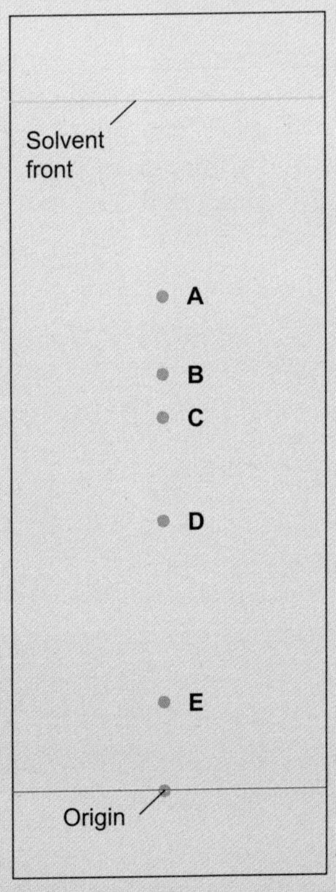

Figure A1 *The chromatogram*

A protein was analysed using chromatography. The chromatogram is shown in Figure A1.

Questions

A1. How many amino acids are present in this protein?

A2. Calculate the R_f values of the spots. Use the table to identify the amino acids present.

Amino acid	R_f value
histidine	0.11
glutamine	0.13
lysine	0.14
aspartic acid	0.24
threonine	0.35
cysteine	0.40
methionine	0.55
valine	0.61
tryptophan	0.66
leucine	0.73

A3. Why might some amino acids be more difficult to identify than others?

A4. Research the other applications of paper chromatography in biology. In what areas of plant science has it been put into use?

PRACTICE QUESTIONS

1. a. Describe a biochemical test to find out if a substance contains a protein.

b. Figure Q1 shows the structural formulae of two amino acids.

i. Name one chemical element found in all amino acids, but not in monosaccharides.

$H_2N — C(H)(H) — C(=O) — OH$

$H_2N — C(H)(CH_2—SH) — C(=O) — OH$

Figure Q1

ii. What type of chemical reaction occurs to form a dipeptide?

iii. Draw the structural formula of the dipeptide formed when these two amino acids combine.

AQA January 2002 Unit 1 Question 1

2. Figure Q2 shows the structure of a molecule of glycerol and a molecule of fatty acid.

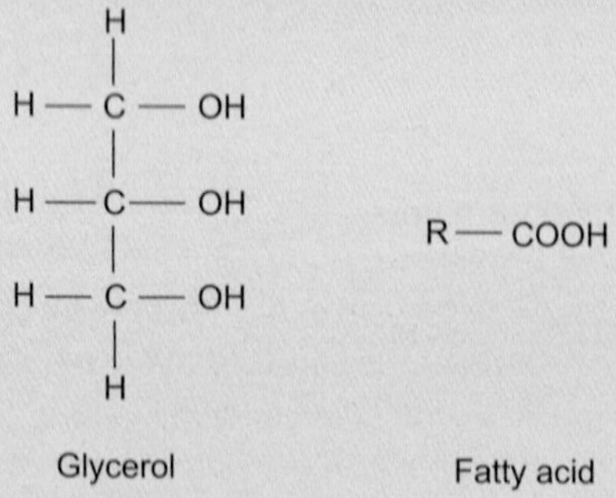

R — COOH

Glycerol

Fatty acid

Figure Q2

a. Draw a diagram to show the structure of a triglyceride molecule.

b. Explain why triglycerides are not considered to be polymers.

AQA January 2004 Unit 1 Question 2

3. a. With reference to named examples, explain the differences between globular proteins and fibrous proteins.

b. Suggest why proteins fulfil a much greater variety of roles in living organisms than carbohydrates.

4. Figure Q3 shows four types of bonds found in protein molecules.

a. Name each bond shown in Figure Q3.

b. Explain the importance of each of these types of bonds in the primary, secondary, tertiary and quaternary structure of proteins.

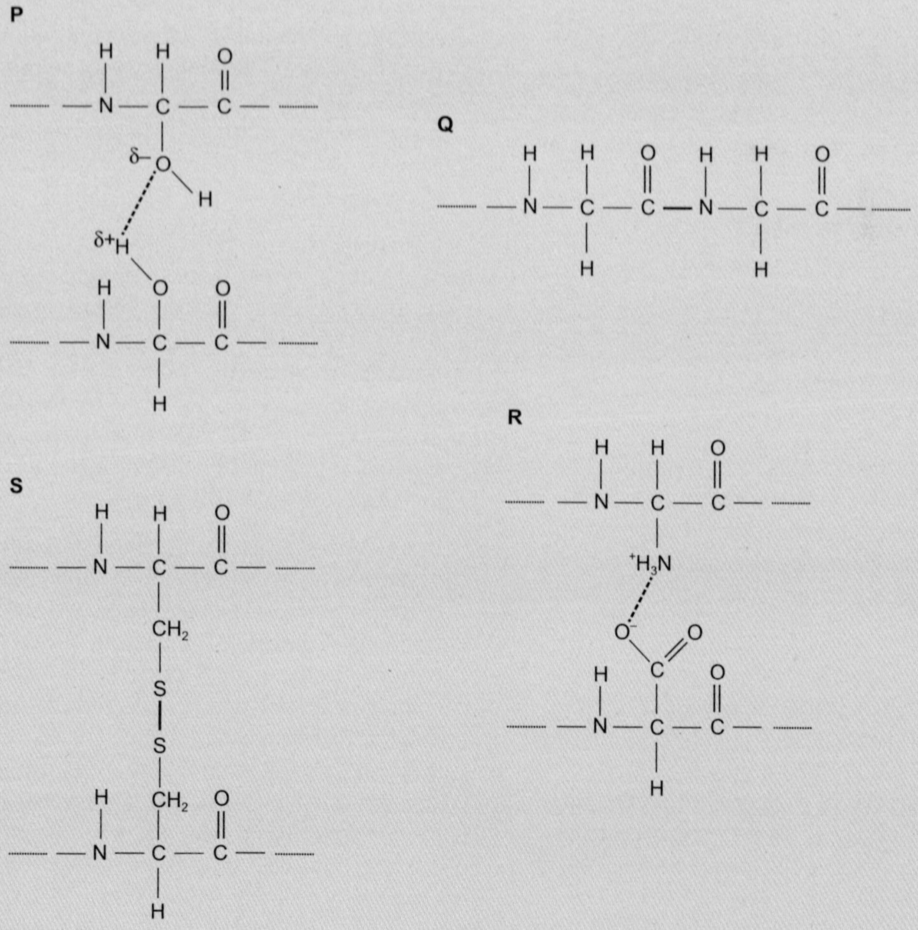

P

Q

R

S

Figure Q3

3 ENZYMES

PRIOR KNOWLEDGE

You may remember, from Key Stage 4, that catalysts are substances that can alter the rate of a chemical reaction without themselves being changed at the end of the reaction. You may also know that increasing temperature can increase the rate of many reactions, and you may recall that pH is a measure of the acidity or basicity of a solution, which is determined by the concentration of hydrogen ions in it. Before you begin this chapter you may want to remind yourself about protein structure (Chapter 2).

LEARNING OBJECTIVES

In this chapter we will look at the essential roles of enzymes – biological catalysts – in controlling the metabolic reactions that take place in living organisms.

(Specification 3.1.4.2)

Enzymes make life possible because without them, the vast majority of chemical reactions that take place inside living organisms would not happen. Enzymes catalyse metabolic reactions, reducing the energy that is needed to allow these reactions to start.

Each enzyme is specific for one type of reaction, so our bodies manufacture hundreds of different enzymes. Cells can control the activity of enzymes by producing inhibitors that attach to enzyme molecules and prevent them from working. This is one of the ways by which cells can prevent a chaotic medley of reactions taking place all at the same time.

Many inhibitors are important for the day-to-day working of an organism's body, while others are used to defend the organism from attack by other species. For example, the seeds of many plants belonging to the pea and bean family contain large quantities of protein, which provides nutrients for the growing young seedling. This high protein content makes the pea and bean seeds a tempting food for many animals. However, the seeds also contain an inhibitor that blocks the action of the protein-digesting enzyme, trypsin, so that animals that eat the seeds are unable to digest and absorb the proteins, and are likely to suffer abdominal pain. Cooking the beans destroys the inhibitor and makes them safe to eat.

Ingesting some enzyme inhibitors can have devastating consequences. Eating just small quantities of the death cap mushroom, *Amanita phalloides,* can be fatal, even when the mushroom has been cooked. This is because the fungus contains an inhibitor of the enzyme that catalyses the production of mRNA from a DNA template, which is the first step in protein synthesis. If this process is inhibited, the organism will die.

Not all enzyme inhibitors produced by other organisms are harmful to us. Penicillin is an inhibitor of the enzyme that helps to build and strengthen the cell walls of bacteria. The enzyme inhibitor is produced naturally by some fungi, presumably to help them to compete with and destroy bacteria that share their habitat. Penicillin is now manufactured on an industrial scale and is used as an antibiotic. Because penicillin specifically inhibits only the enzymes involved in producing bacterial cell walls, it has no direct effect on our own body cells. It is, therefore, a drug that can be used safely inside the human body.

3.1 BIOLOGICAL CATALYSTS

A huge number of chemical reactions – known as metabolic reactions – take place inside every living cell. Some of these reactions release energy, some synthesise new substances and others break down waste products. **Enzymes** enable these reactions to happen and, most importantly, enzymes make it possible for the reactions to happen quickly enough at body temperature.

Many chemical reactions need an input of energy to get them going and this energy is normally supplied as heat. Enzymes are **catalysts**. They reduce the amount of energy that is needed to get a reaction going, but they are not themselves changed at the end of the reaction. This makes them very valuable substances, not only in living organisms, but also for human use. Enzymes from yeast have for centuries been used to brew alcoholic drinks. Enzymes are also commonly added to detergents to digest food stains, and many more industrial processes in the future are likely to make use of enzymes.

Enzymes are not living things. They are proteins. The properties of proteins help to explain how enzymes work and how they lower the energy needed to get a reaction going. The protein structure of enzyme molecules does, however, make them sensitive to environmental conditions both inside and outside the body, which means that changes in temperature or pH can disrupt enzyme action. In humans, severe over-heating or getting so cold that core body temperature falls can upset the delicate balance of enzyme-controlled reactions in the body's cells and can be fatal.

3.2 ENZYMES AND CHEMICAL REACTIONS

Hydrogen peroxide is a colourless liquid (it looks just like water). It has the formula H_2O_2. If you put some hydrogen peroxide into a flask or test tube it will gradually – very, very slowly – decompose to form water and oxygen. The equation for this reaction is:

$$2H_2O_2 \rightarrow 2H_2O + O_2$$

The little bubbles of oxygen that are given off are the only evidence that this reaction is happening.

If you want to liven things up, you can add a tiny piece of fresh, raw liver to the flask. Immediately, the hydrogen peroxide begins to bubble furiously. The bubbles form a rapidly expanding froth, which climbs up the flask and may overflow (Figure 1).

Figure 1 Hydrogen peroxide (H_2O_2) decomposition is accelerated by the catalase enzyme found in raw liver.

What is happening when you add the liver? The liver cells, like all living cells, contain an enzyme called **catalase**. This enzyme catalyses the decomposition reaction shown in the equation above.

Activation energy

We have seen that hydrogen peroxide molecules will break apart to form water and oxygen, but they normally do this only very slowly. This is because the reaction needs energy to make it happen. You could speed the reaction up by heating the hydrogen peroxide. The heat energy (thermal energy) that you put in will give the hydrogen peroxide molecules more energy and make them split apart more readily. However, this is not a good idea, as the reaction can be very violent and the hydrogen peroxide is likely to explode and shatter its container.

The minimum energy needed to make a reaction start is called the **activation energy**. Heating raises the energy of the hydrogen peroxide, and when the activation energy is reached, the reaction takes place very rapidly. Figure 2 shows the energy changes that take place during the reaction. Before the reaction takes place, the energy of the molecules is steady. Then something happens to raise the energy level – perhaps one molecule is hit by another very hard, or perhaps they are heated up. If the molecules reach the activation energy, the reaction then takes place, leaving the product molecules at a lower energy level.

Enzymes and activation energy

Enzymes make reactions happen without having to increase the temperature. They do this by reducing the activation energy required for the reaction to take place.

Figure 3 shows the effect of an enzyme on the activation energy of the same reaction that is shown in Figure 2. You can see that the activation energy for the reaction is much lower with the enzyme than it is without the enzyme. This is just what is needed in living cells.

Cells cannot heat themselves up very much. If they could, the effects would be catastrophic because the high temperatures would damage the proteins, enzymes and other molecules from which they are made. Enzymes allow the cell's metabolic reactions to take place at normal temperatures.

Think about this in the context of catalase. Hydrogen peroxide is produced as a by-product (in this case an unwanted product) of several different metabolic reactions. Hydrogen peroxide is a powerful oxidising agent, and it will quickly damage living cells. But because all cells contain catalase, any hydrogen peroxide is broken down almost as soon as it has been formed, in a relatively steady manner, without having to raise the temperature.

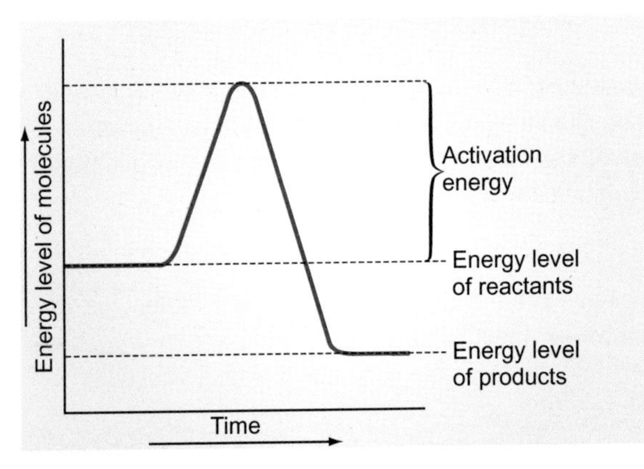

Figure 2 *Changes in energy level during a chemical reaction*

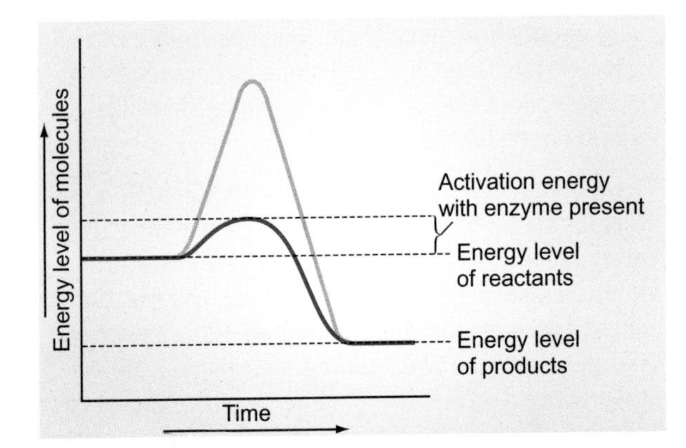

Figure 3 *Enzymes lower activation energy.*

The molecule that an enzyme allows to react is called its **substrate**. The molecules produced at the end of the reaction are the **products**. The substrate for catalase is hydrogen peroxide, while water and oxygen are the products of the reaction that the enzyme catalyses.

KEY IDEAS

- Enzymes are proteins that act as biological catalysts.
- Enzymes work by lowering the activation energy of a reaction, so making it possible for the reaction to happen at normal temperatures in a given organism.

3.3 HOW ENZYMES WORK

We have seen that enzymes are proteins. In fact, they are globular proteins. Their molecules are made of one or more polypeptide chains curled into a very precise three-dimensional shape.

Every enzyme molecule has an **active site**. This is often a 'dent' or 'depression' in the three-dimensional structure of the globular protein. The shape of the active site is very precise. It is lined with R groups (side chains) of particular amino acids that make up the polypeptide chains.

The substrate of the enzyme has a complementary shape to the active site. In a solution containing both enzyme and substrate molecules, both kinds of molecules will be in constant motion. They will often collide with one another. When a substrate molecule hits the active site on an enzyme molecule it will interact with the R groups of the amino acids lining this site. This interaction causes a small shape change in the active site, making it a perfect fit for the substrate molecule. The substrate slots into the active site and forms temporary bonds with the R groups. The combination of enzyme and substrate is called an **enzyme–substrate complex** (Figure 4).

This interaction also changes the shape of the substrate molecule, which stresses chemical bonds,

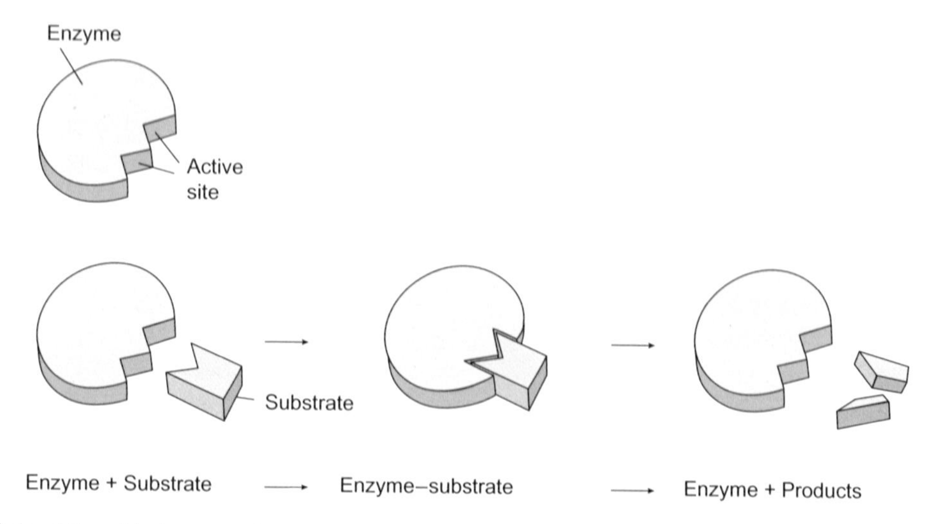

Figure 4 *The induced fit model of enzyme action*

and lowers the activation energy needed for the substrate to break apart into product molecules. The substrate splits into its product molecules, which leave the active site of the enzyme. The enzyme remains completely unchanged, and is now ready to interact with the next substrate molecule that bounces into it.

What we have just described explains how an enzyme can cause a substrate molecule to break apart. Many reactions, however, involve two substrate molecules joining together. In this case, the active site of the enzyme accommodates the two substrates side by side, holding them in exactly the right position for them to react together. The substrates combine to form a product.

This way of explaining how an enzyme works is called the **induced fit model** of enzyme action. 'Induced fit' means that the arrival of the substrate causes (induces) a small shape change in the enzyme's active site, allowing the substrate to bind with it. You may have met the 'lock and key' model of enzyme activity, where the enzyme is thought of as a lock into which the substrate – the key – fits exactly. This is an older model of enzyme action. The induced fit model is a more recent model that is not completely different from the lock and key model, but is a more detailed explanation of it. You simply have to imagine a slightly flexible lock and a slightly flexible key that mould themselves to fit one another once they come into contact.

It has probably taken you several minutes to read this section (longer if you took time to look carefully at the diagram). In that time, one catalase molecule could have caused the breakdown of countless billions of hydrogen peroxide molecules. It is difficult to comprehend, but a single catalase molecule can break down more than 40 million molecules of catalase in one second. So the enzyme–substrate complex shown in Figure 4 exists for only an unimaginably short time.

QUESTIONS

1. Enzymes are specific. A particular enzyme can catalyse only one chemical reaction. For example, catalase will only break down hydrogen peroxide. Some enzymes break only one type of bond – for example, peptidase breaks peptide bonds. Use your knowledge of how enzymes work to explain why enzymes are specific.
2. A small change in the primary structure of an enzyme molecule can prevent it from working. Use your knowledge of protein structure (Chapter 2) to explain why this is so.

KEY IDEAS

- An enzyme molecule has an active site into which its substrate molecule fits.
- The arrival of the substrate molecule slightly changes the shape of the active site, so that the enzyme and substrate can bind together. This is called the induced fit model.
- The temporary combination of the enzyme and substrate is called an enzyme–substrate complex.
- Enzymes can act only on their specific substrate, because only that substrate has the correct shape to bind perfectly with the enzyme's active site.

3.4 FACTORS AFFECTING ENZYME ACTIVITY

Many different factors influence the rate at which an enzyme-catalysed reaction takes place. Temperature can affect the rate of the reaction, while temperature and pH can both affect the shape of the enzyme molecule. The concentrations of both the enzyme and the substrate affect the frequency with which enzyme–substrate complexes form. Other molecules, called inhibitors, can interfere with the reaction.

Temperature and enzyme activity

The rate of a chemical reaction normally increases as the temperature increases. This is because the molecules gain kinetic energy, so they move faster, collide more often and the collisions are more likely to have enough energy to allow the reaction to happen. This is also true for reactions controlled by enzymes, because substrate molecules will collide with the active site more frequently.

However, as the temperature increases, the atoms within the enzyme molecules also gain energy and vibrate so rapidly that the weak hydrogen bonds that help to maintain the tertiary structure of the protein molecule break and the molecule can unravel. Since the shape of the active site in an enzyme molecule is crucial for it to work, any change in shape will inactivate the enzyme. Once broken, the hydrogen bonds generally do not re-form in their original positions. So, even when the temperature is lowered, the enzyme cannot regain its functional shape. We say that the enzyme is **denatured**. Note that you should never say that enzymes are 'killed', since they were never 'alive' in the first place.

The graph in Figure 5 shows these effects. You can see that, as the temperature rises from 0 °C up to the optimum of 38 °C, the rate of reaction increases exponentially. At temperatures higher than this optimum, the rate of reaction drops much more steeply.

Note that this graph shows temperature on the *x*-axis, not time. If you are asked to describe it, take care that you do not fall into the trap of using 'time-related' words inappropriately. For example, it would be wrong to say 'As temperature increases above the optimum, the reaction rate quickly decreases'.

Most enzymes in the human body are denatured at temperatures above about 45 °C. They work fastest at about 40 °C; this is their **optimum temperature**. Human body temperature is about 37 °C, which is just below the optimum for most of our enzymes. It is

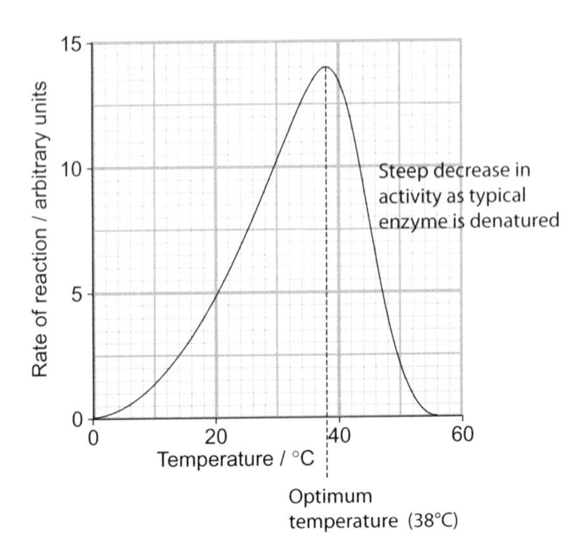

Figure 5 *The effect of temperature on an enzyme-controlled reaction*

thought that this is an adaptation for when our core body temperature does rise above normal – as a result of an illness, for example – and so our body's enzymes can still work and we will remain alive.

Many organisms have enzymes that work perfectly at much higher or lower temperatures. Some fish live in the Antarctic sea, where the water temperature never rises above 2 °C. They are active at this temperature, but die if the water warms by more than a few degrees. Some species of worms and bacteria can live near volcanic vents in the ocean and close to hot springs, where the temperatures are close to boiling point. Many of the enzymes isolated from these specialised organisms are proving useful in manufacturing industry.

QUESTIONS

3. Look at Figure 5. Explain why the rate of reaction is so slow below 10 °C.
4. What are the rates of reaction at 10 °C, 20 °C and 30 °C?
5. What do you notice about the effect of increasing the temperature by 10 °C?
6. What would be the rate of reaction if some of the enzyme and its substrates were frozen and then warmed to 40 °C?
7. The graph in Figure 6 shows the effect of temperature on a protease used in washing powders.

a. What is the optimum temperature of this protease?

b. Suggest the advantage of using a protease with this optimum temperature in a washing powder.

8. Proteases hydrolyse protein molecules into amino acids. Explain how proteases in washing powders help to remove protein stains such as blood or egg from clothes.

Stretch and challenge

9. Suggest why the curve above 40 °C does not drop down directly to 0.

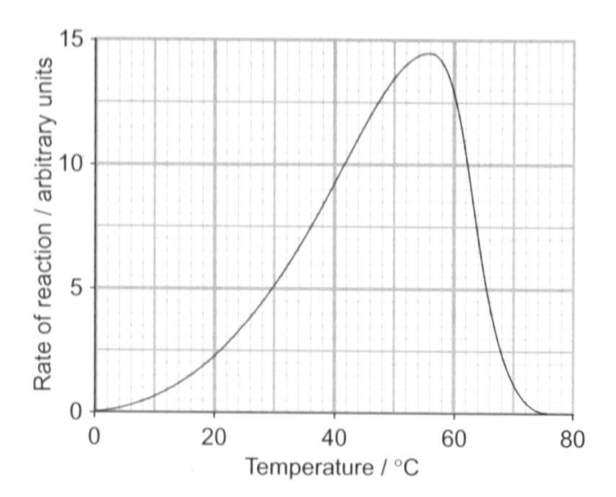

Figure 6 The effect of temperature on a protease used in washing powder

ASSIGNMENT 1: INVESTIGATING BIOLOGICAL WASHING POWDERS

(PS 1.2, PS 2.1, PS 2.4)

Until the late 1960s, washing powders contained a detergent only. These powders worked by emulsifying fats in stains and they required high temperatures and vigorous washing to work. This often resulted in wear on materials and faded colours.

Much work went into harnessing the power of enzymes for cleaning clothes. Although the first patent for using enzymes to clean clothes was granted in 1913, enzymes were not used in commercial washing powders until 1969.

Early attempts to incorporate protease enzymes into washing powders failed because people developed skin irritations, and it was not until the enzymes were enclosed in dust-free granules wrapped in cellulose that they became widely used. Washing powders that contain enzymes are commonly known as biological washing powders.

Stains are made up of different types of molecules, and so different enzymes are needed to hydrolyse proteins (found in blood and egg), starches, and fats and grease. Commercial washing powders contain lipase enzyme to hydrolyse fat, amylase to hydrolyse carbohydrate, and a protease that is produced commercially in bacteria, for hydrolysing peptide bonds. The use of enzymes in biological washing powders enables clothes to be washed at lower temperatures, causing less damage to the fabric and using less energy for heating the water.

Figure A1 Washing clothes used to be hard work using basic detergents.

From December 2013, all new washing machines must be able to wash at temperatures as low as 20 °C, which greatly reduces the running costs of the washing machines. Some washing powders now also contain a cellulase that hydrolyses microfibrils formed on cotton by agitation. This serves to restore and brighten colours.

Questions

A1. Explain why the early use of protease enzymes caused skin irritation.

A2. Explain why different enzymes are needed for different types of stains.

A3. What could be the disadvantage of washing clothes at 20 °C?

A4. Some students wanted to compare the ability of two washing powders – one biological and one non-biological – to hydrolyse protein. They used blocks of gelatine as a protein and put them in beakers of water with either the biological or non-biological powder. The students left the blocks for two hours and then measured them to see how much had been digested.

a. What variables should the students control and why?

b. How could they control each variable you have mentioned?

c. How could the investigation be improved?

A5. Could the method used in this investigation provide enough evidence to indicate how well the two washing powders would clean clothes?

pH and enzyme activity

As well as having an optimum temperature, enzymes also have an optimum pH. Even small changes in pH either side of the optimum will affect enzyme activity, causing small changes in the shape of the active site. Most enzymes are denatured in solutions that are strongly acidic or strongly alkaline. This is because the hydrogen ions (H^+) in an acid or the hydroxyl ions (OH^-) in an alkali are attracted to the charges on the amino acids in the polypeptide chains that make up the enzyme. The hydroxyl and hydrogen ions interact with the amino acids and disrupt the hydrogen bonds and ionic bonds that maintain the enzyme molecule's three-dimensional shape, destroying the active site of the enzyme. Once the original structure is lost it cannot re-form and the enzyme no longer binds to its substrates.

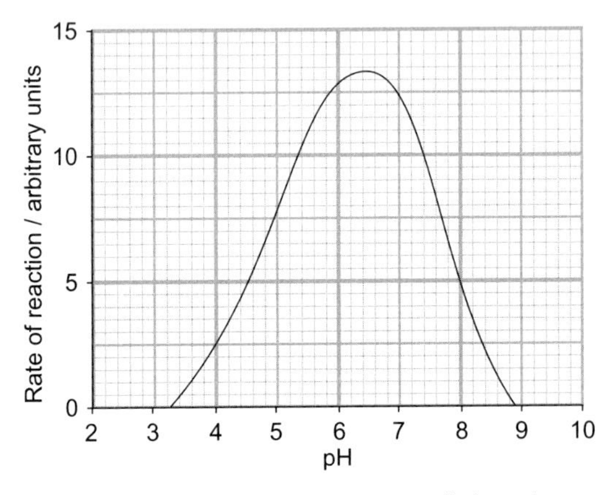

Figure 7 *The effect of pH on an enzyme-controlled reaction*

ASSIGNMENT 2: INVESTIGATING THE EFFECT OF PH ON ENZYMES

(MS 1.3, MS 3.1, MS 3.2, PS 1.2, PS 3.1)

Extremes of pH rarely occur inside the cells of living organisms, and most enzymes have an activity curve similar to the one in Figure 7, with an optimum pH somewhere near neutral. However, some enzymes work best at extreme pH values. For example, proteases in the acid environment of the stomach have an optimum pH of about 2 and are denatured above pH 6.

The marine shipworm is a mollusc that can bore into woodwork. It secretes two digestive enzymes into the wood, using a source of nitrogen obtained from symbiotic bacteria. One of the enzymes is a cellulase that breaks down cellulose in wood. The second is a

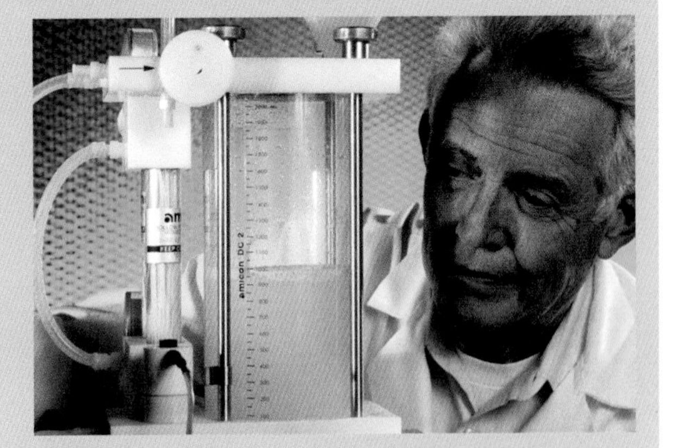

Figure A1 *A chemist extracts an enzyme from a bacterium found in shipworms. The enzyme can be used to help clean contact lenses.*

protease, which has been found to be very good at removing the enzyme lysozyme that accumulates on contact lenses.

The marine shipworm protease has an advantage over other proteases used for cleaning contact lenses because it remains active in the presence of the slightly alkaline solution of hydrogen peroxide that is used to disinfect contact lenses. This means that contact lenses can be cleaned and sterilised at the same time using a single solution, rather than the conventional two solutions.

Questions

A1. Explain why the shipworm needs a source of nitrogen to produce its enzymes.

A2. Why is the production of cellulase a problem for wooden boats?

A3. Table A1 shows the rate of reaction of the marine shipworm protease at different pH values. Plot a graph of the data in Table A1.

pH of solution	Rate of reaction / percentage of maximum
5	25
6	45
7	60
8	70
9	80
10	95
11	85
12	55

Table A1

A4. Use your graph to explain why the protease obtained from the marine shipworm is better for cleaning contact lenses than a protease with a curve like that in Figure 7.

Stretch and challenge

A5. Researchers often analyse the protein content in a cell; they have to break up cells to form a cell lysate. Why might a researcher need to add a protease inhibitor?

REQUIRED PRACTICAL ACTIVITY 3: APPARATUS AND TECHNIQUES

(MS 3.2, MS 3.6, PS 2.1, PS 2.4, PS 3.3, PS 4.1, AT a, AT b, AT c, AT f, AT l)

Investigation into the effect of a named variable on the rate of an enzyme-controlled reaction

This practical activity gives you the opportunity to show that you can:

- use appropriate apparatus to record a range of quantitative measurements
- use laboratory glassware apparatus for a variety of experimental techniques, to include serial dilutions.

Apparatus

There are various ways to investigate enzyme-controlled reactions, using a range of different apparatus.

Change	Examples of apparatus and techniques to measure change
Increase or decrease in mass	› Balance of suitable range and precision
Gas given off	› Displacement of water › Gas syringe (especially if gas produced is soluble in water) › Counting bubbles
Colour change	› Colorimeter › Spectrophotometer (more accurate and precise than a colorimeter, but much more expensive) › Comparison with a colour standard
'Invisible' change in solution	› Sample withdrawn, 'frozen' and mixture analysed; choice of apparatus depends on the quantitative analytical method used
pH	› pH meter and electronic probe › Use of indicator solutions

Table P1 *It is important to select the most appropriate apparatus for taking measurements.*

Data logging allows data gathered electronically to be recorded and graphs of the variable against time generated by a computer. This is particularly useful for reactions that happen very quickly or very slowly.

Temperature

Investigating the effect of temperature on the rate of reaction means that you need apparatus that will maintain the reaction mixture at a fixed temperature. This is not always easy if reactions are strongly exothermic or endothermic. Options include:

- a simple water bath (beaker of water heated on a tripod and gauze or an electrical hotplate); only suitable if a tolerance of ± a few °C is acceptable
- a thermostatically controlled water bath, which can maintain the temperature with much less variation than a manual water bath.

A water bath should be used for all reactions that need a constant temperature. If, for example, you are measuring the effects of pH, you should use a water bath to maintain a constant temperature. This could be a simple container with water at room temperature.

Time

You need a way of measuring time. Stopwatches and stopclocks are the most commonly used apparatus for measuring time manually. Dataloggers will do it automatically.

Colour changes and pH

The digestion of fats is often followed using a pH indicator. As the lipids are hydrolysed into glycerol and fatty acids, the pH of the solution falls and can be followed by noting colour change in the indicator solution. This is a commonly used method if you just wish to time how long the reaction takes to be completed.

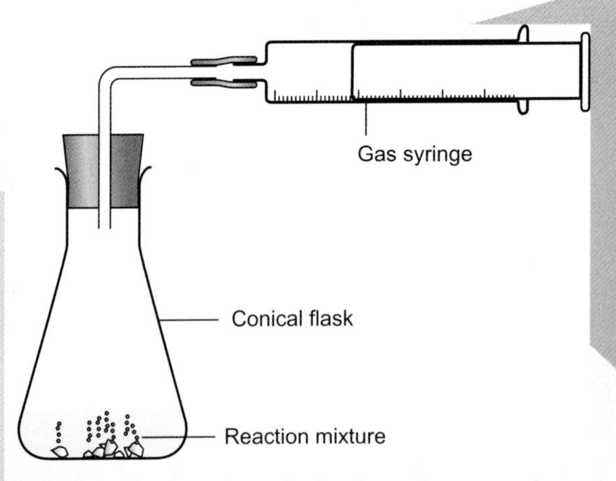

Figure P1 *Measuring the production of gas using a gas syringe*

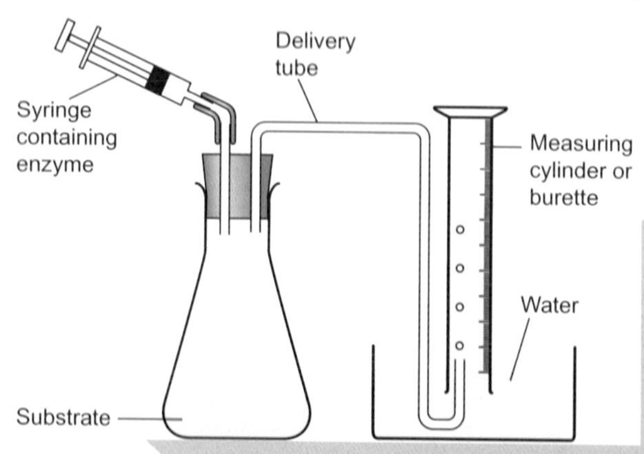

Figure P2 Measuring the evolution of gas by delivery over water

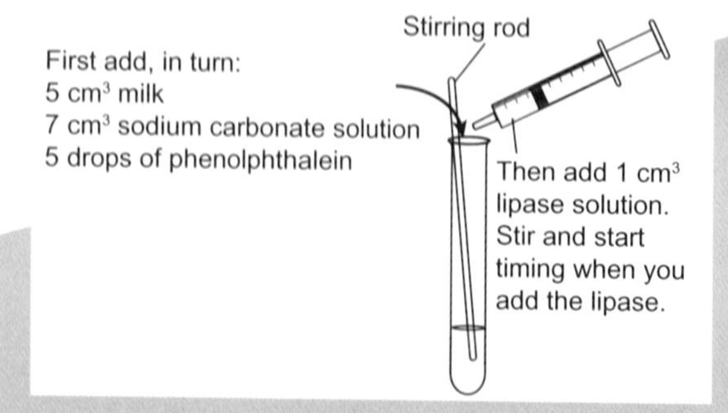

Figure P4 Following a reaction using an indicator

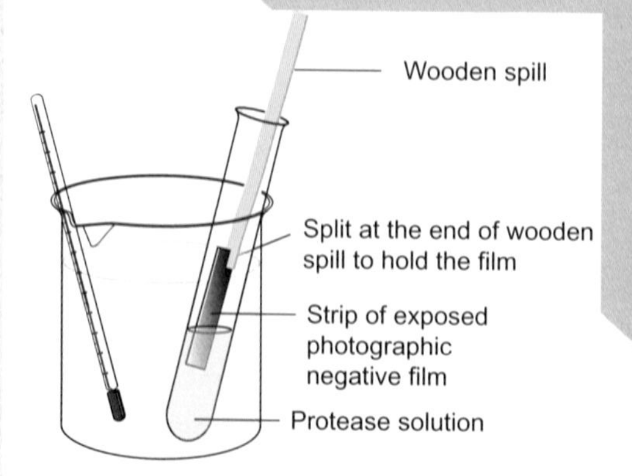

Figure P3 The reaction of a protease is commonly followed using exposed photographic film, which changes from black to transparent as the gelatine (protein) holding the black silver nitrate crystals is digested.

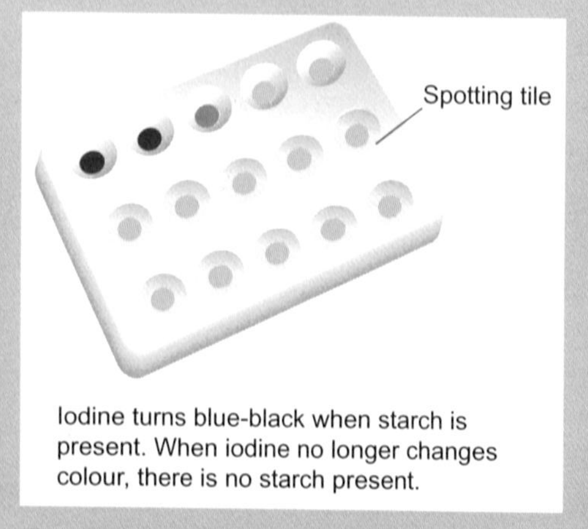

Figure P5 Following a reaction involving a colour change

A pH probe will give a more accurate reading and will enable you to take measurements at frequent intervals to record the rate of reaction more easily.

A reaction involving a colour change can be followed using a spotting tile. This is often used to follow a reaction between starch and amylase. The wells in the spotting tile are filled with potassium iodide solution, which turns blue-black when starch is added. Successive samples are taken from the reaction mixture and the time when a blue-black colour no longer appears is noted.

Techniques

You will usually need to maintain the pH of your reaction within a narrow range. To do this, you would use a buffer solution which maintains the pH at a given value. This is particularly important if the reaction may lead to a pH change.

If you are investigating the effects of pH, a series of buffer solutions of different pH would be used.

If you are changing the concentration of either the enzyme or the substrate, you will need to make up a series of dilutions.

If you are asked to make up a **10% solution**, this means a solution that contains 10 g of the solute in 100 cm^3 of solution.

To make up a 10% solution:

- Use an electronic balance to measure 10 g of solute.
- Put this into a 100 cm^3 volumetric flask.
- Add a small amount of distilled water and swirl until fully dissolved.
- Add more distilled water to make up the solution to precisely the 100 cm^3 mark.

If you need to make up a larger quantity of solution, simply multiply the quantities of solute and solvent. For example, to make 1 dm^3 of a 10% solution, use 100 g of solute and make up to 1 dm^3 (1 dm^3 = 1000 cm^3).

You may also be asked to make up a **1 mol dm^{-3} solution**. For this, you need to know the mass of 1 mole of your solute. For glucose, $C_6H_{12}O_6$, the mass of 1 mole is:

$(6 \times 12) + (12 \times 1) + (6 \times 16) = 180$

To make up a 1 mol dm^{-3} solution, use an electronic balance to measure 180 g of glucose. Make up to 1 dm^3 as described above.

Once you have made up a solution, you can dilute this to make a range of less concentrated solutions.

For example, if you have a 1% solution you can:

- transfer 8 cm^3 of the 1% solution to a clean test tube and add 2 cm^3 of water to make a 0.8% solution
- transfer 2 cm^3 of the 1% solution to a clean test tube and add 8 cm^3 of water to make a 0.2% solution.

You can also use **serial dilutions**, in which each solution is one tenth of the concentration of the previous one. Assuming you start with a 1% solution:

- transfer 1 cm^3 of this to a clean test tube and add 9 cm^3 of water to make a 0.1% solution
- transfer 1 cm^3 of the 0.1% solution to a clean test tube and add 9 cm^3 of water to make a 0.01% solution
- transfer 1 cm^3 of the 0.01% solution to a clean test tube and add 9 cm^3 of water to make a 0.001% solution, and so on.

If you are extracting an enzyme from a solid biological material, you could take a small amount of biological material, grind it in a pestle and mortar with a small amount of distilled water, and then filter it to remove cellular material. This makes your 100% enzyme concentration if you are investigating the effects of this variable.

To investigate the effects of enzyme concentration, you could then prepare a series of dilutions using this as 100% concentration.

Whatever factor you are investigating, you should put both reactants in containers in the water bath and leave them for long enough to come to the required temperature.

You will need to decide how long to leave between measurements. This will depend on the reaction you are investigating and the temperature you have chosen for your water bath.

If you are not investigating the effects of temperature, you should choose a temperature for the water bath that is likely to be close to the optimum for the enzyme you are studying. For human enzymes, 37 °C is a good choice. Plant enzymes often have lower optimum temperatures, and enzymes from bacteria or fungi (the source of many commercially available enzymes used in schools and colleges) are often much higher.

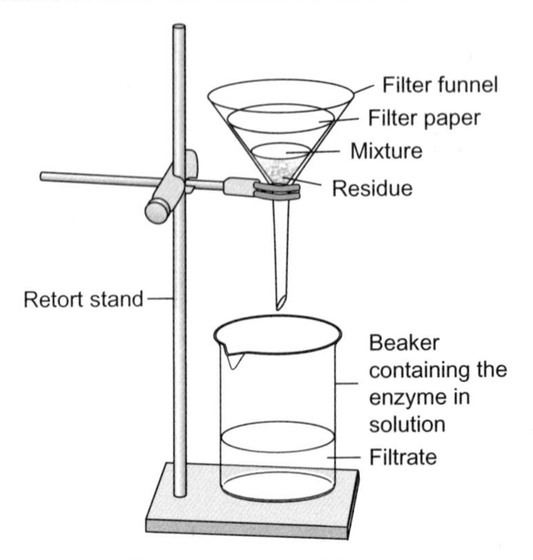

Figure P6 *Filtering an enzyme mixture*

You should aim to take measurements as often as you can and, again, this will depend on the technique you are using. In most investigations, you can take measurements every 30 seconds, but you may need to have a partner to help.

If your reaction is too fast you could:

- decrease the temperature of the water bath
- decrease the concentration of the enzyme
- decrease the concentration of the substrate.

You can do the reverse if your reaction appears too slow.

QUESTIONS

P1. When investigating enzyme or substrate concentrations what factors must be kept constant?

P2. Using a gas syringe or burette is more accurate than using a measuring cylinder. Why?

P3. How would you investigate

a. the effects of pH on the rate of an enzyme-controlled reaction

b. the effects of temperature on the rate of an enzyme-controlled reaction?

P4. Using a spotting tile is not very accurate. Explain why.

P5. Explain which other techniques have similar problems to using a spotting tile.

P6. How can you overcome the problem of maintaining a fixed end point when using an indicator solution or a spotting tile?

Substrate concentration and enzyme activity

Concentration is a measure of the relative proportions of solute and solvent molecules in a solution. A concentrated solution has a large number of solute molecules in a given volume.

Imagine a solution containing enzyme and substrate molecules. All of the molecules are in constant motion. If there are only very few substrate molecules in the solution, then the chance of them hitting an active site is low. If there are a lot of substrate molecules, the chance is much higher. So the rate of reaction is much greater in a concentrated solution containing a high number of substrate molecules than in a dilute solution with a low concentration of substrate molecules (Figure 8).

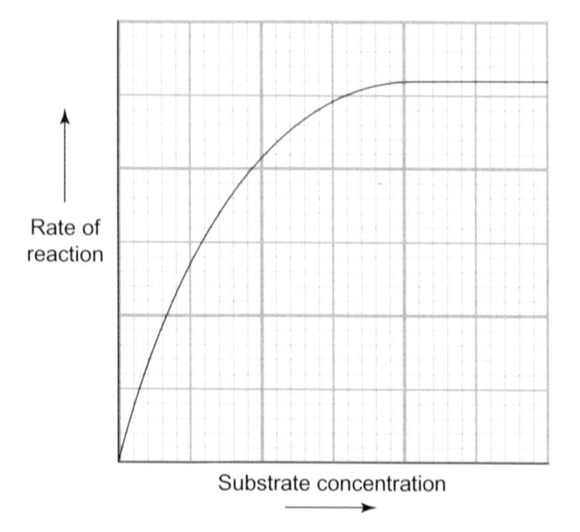

Figure 8 *The effect of substrate concentration on an enzyme-controlled reaction*

Now imagine that we keep on increasing the substrate concentration, while making no change to the concentration of the enzyme. Eventually, we will reach a point where all the active sites are fully occupied at any given time (note: the sites are not all permanently blocked). The enzyme molecules are all working as fast as they can, receiving substrates, turning them into products and releasing them. No matter how much more substrate is available the enzyme cannot work any faster. This explains why the curve in Figure 8 levels off at high substrate concentrations.

Enzyme concentration and enzyme activity

If we keep the substrate concentration constant and increase the enzyme concentration, what effect does this have on the rate of the reaction? This is shown in Figure 9. You can see that the shape of the curve is very similar to that in Figure 8. As we increase the enzyme concentration, the chance of an enzyme molecule colliding with a substrate molecule increases, so the rate of reaction increases. However, at high

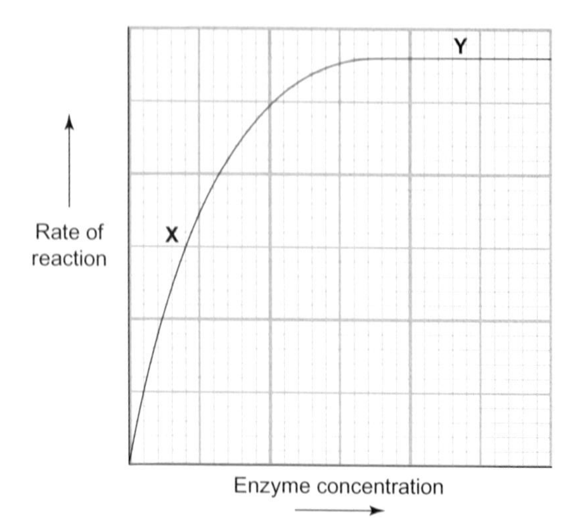

Figure 9 *The effect of enzyme concentration on the rate of reaction*

enzyme concentrations, there may not be enough substrate molecules to keep them all busy all the time, so the rate levels off.

QUESTIONS

10. Explain in terms of active sites why the curve in Figure 9 flattens out.

Stretch and challenge

11. The graph in Figure 9 is labelled with the letters X and Y. Do the following descriptions, a–e, match part X, Y or X and Y on the graph?

a. At this point, there is an excess of substrate in the mixture.

b. At this point, the enzyme's active sites are saturated with substrate.

c. At this point, adding more enzyme could increase the rate of reaction.

d. At this point, adding more substrate could increase the rate of reaction.

e. At this point, increasing the temperature could increase the rate of reaction.

Worked maths example: Rate of reaction (MS 1.3, MS 3.2, MS 3.5, MS 3.6, PS 3.1, PS 3.2)

Your friend carried out an experiment to investigate the rate of a reaction with and without an enzyme. Here are his notes.

Exp 1 – *enzyme*

Time: 0, 30, 60, 90, 120, 150, 180, 210, 240 secs = 0, 2.5, 3.8, 4.5, 5, 5.4, 5.6, 5.7, 5.7

Exp 2 – no enzyme

Time: 0, 30, 60, 90, 120, 150, 180, 210, 240 secs = 0, 0.5, 1, 1.5, 2, 2.6, 3.1, 3.6, 4.2

He needs to display his results and calculate the rate of reaction, in mol min^{-1}, for both experiments at 60 seconds, and has asked for your help.

First, use his notes to create a table, making sure all the data points are recorded to one decimal place, and each column has a clear heading with the units specified.

	Amount of product/moles	
Time/seconds	Experiment 1 (enzyme)	Experiment 2 (no enzyme)
0	0.0	0.0
30	2.5	0.5
60	3.8	1.0
90	4.5	1.5
120	5.0	2.0
150	5.4	2.6
180	5.6	3.1
210	5.7	3.6
240	5.7	4.2

Next, create a graph. Label the y-axis 'Amount of product/moles', and the x-axis 'Time/seconds', and choose suitable scales. Give the graph a title (for example, 'Effect of enzyme on rate of reaction').

Then plot the results, and add lines-of-best-fit to both sets of data. Use different coloured lines and label them 'enzyme' and 'no enzyme'.

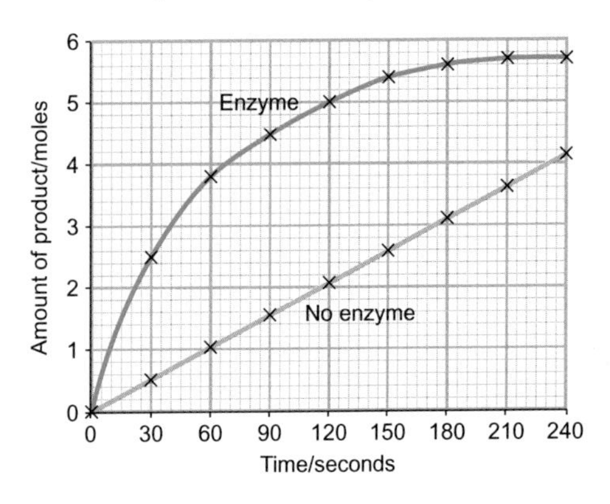

Now use your graph to calculate the rate of reaction for each experiment at 60 seconds.

Experiment 2 (no enzyme)

The orange line is linear, so the rate of reaction is the same at all points in time – it is constant.

$$\text{rate of reaction} = \frac{\text{change in } y}{\text{change in } x}$$

In your friend's experiment, after 120 seconds the amount of product had changed from 0 to 2 moles:

$$\text{rate of reaction} = \frac{2}{120} = 0.01\dot{6} \text{ mol sec}^{-1}$$

The question asks for the rate in mol min^{-1}, so you need to multiply your answer by 60:

$$\text{rate of reaction} = 0.01\dot{6} \times 60 = 1 \text{ mol min}^{-1}$$

We can show that this is the same at other points in a linear graph by choosing another time period. For example: between 150 seconds and 210 seconds (a period of 60 seconds) the change in y is $3.6 - 2.6 = 1$ mole.

$$\text{rate of reaction} = \frac{1}{60} = 0.01\dot{6} \text{ mol sec}^{-1} = 1 \text{ mol min}^{-1}$$

Experiment 1 (enzyme)

The blue line is not a straight line; the rate of reaction slows over time.

You can calculate the rate of reaction at any point in time by drawing a tangent – a straight line which touches the curve at one point only – and then calculating the gradient of the tangent.

So, since your friend wants to know the rate of reaction at 60 seconds, draw a tangent at 60 seconds.

$$\text{rate of reaction} = \frac{\text{change in } y}{\text{change in } x} = \frac{(4.4 - 3.2)}{(80 - 40)}$$

$$= \frac{1.2}{40} = 0.03 \text{ mol sec}^{-1} = 1.8 \text{ mol min}^{-1}$$

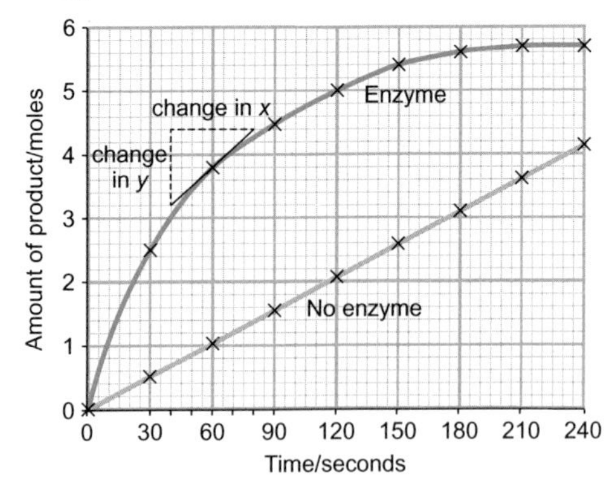

Competitive inhibitors

We have seen that enzymes have high substrate specificity. This means that only molecules of one particular substrate can attach to the active site and a reaction take place. However, sometimes molecules with a very similar shape to the enzyme's specific substrate can also attach to the active site. If such molecules are present in the same solution as the normal substrate, they compete for the active site; they bind to it but no reaction takes place. They are called **competitive inhibitors** (Figure 10a).

The molecules of the competitive inhibitor block the active site, so fewer molecules of the enzyme are available for the normal reaction and the reaction rate is reduced.

The degree of inhibition depends on the relative concentrations of the substrate and the competitive inhibitor. Usually, the inhibitor does not attach permanently to the active site, nor does it damage the site. Therefore, if the concentration of the enzyme's normal substrate is increased, its molecules compete more successfully for a place in the active site and the rate of the reaction increases.

Non-competitive inhibitors

Some substances inhibit enzyme reactions in a different way. Their molecules do not attach to the active site, but to a different part of the enzyme. This alters the shape of the enzyme (protein) and thus the active site does not form. These substances are called **non-competitive inhibitors** (Figure 10b).

The shape of a non-competitive inhibitor molecule may be completely different from that of the usual substrate. When the inhibitor binds to the enzyme – permanently or temporarily – the enzyme is effectively inactivated. Increasing the amount of substrate in the solution does not help and will not increase the reaction rate.

Some non-competitive inhibitors can affect many different enzymes. This is why heavy metal ions such as mercury, lead and arsenic are so poisonous; they are non-competitive inhibitors of many different enzymes, and so prevent many of the body's metabolic reactions taking place. Other non-competitive inhibitors are more specific, and they may even have useful applications, for example, as insecticides.

QUESTIONS

12. Figure 11 (overleaf) shows the structures of succinate and malonate. Succinate dehydrogenase is an enzyme involved in the metabolic pathway of respiration. It removes hydrogen from succinate. Suggest why malonate is a competitive inhibitor of succinate dehydrogenase.

13. Some non-competitive inhibitors can be used as pesticides. What factors would have to be considered before using one as a pesticide for general use?

a Competitive inhibition

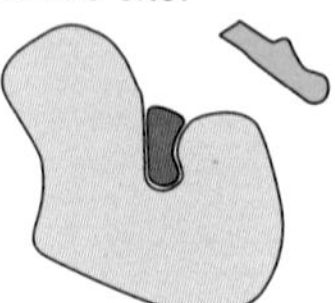

b Non-competitive inhibition

Figure 10 *Competitive and non-competitive inhibition*

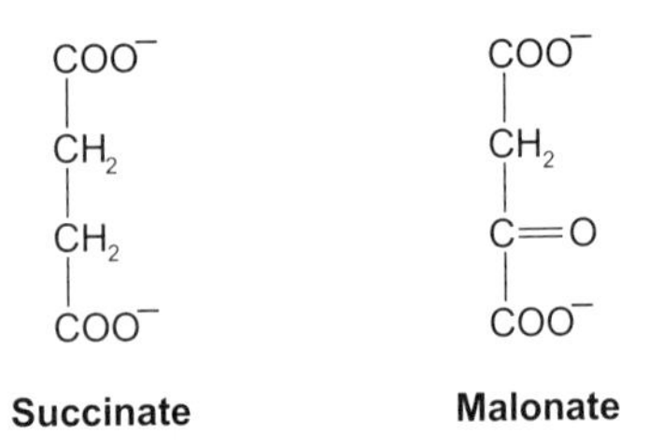

Figure 11 Succinate and malonate

KEY IDEAS

- An enzyme has an optimum temperature at which its activity is greatest. At lower temperatures the enzyme and substrate molecules have lower kinetic energy, and so do not collide as frequently. At higher temperatures, hydrogen bonds within the enzyme molecule break, so the enzyme loses its shape and is said to be denatured.
- An enzyme has an optimum pH at which its activity is greatest. At a higher or lower pH the hydrogen and ionic bonds within the enzyme are disrupted, so the active site loses its shape.
- Increasing either the substrate concentration or the enzyme concentration increases the frequency with which substrate and enzyme molecules collide, and therefore increases the rate of reaction.
- Inhibitors are substances that reduce enzyme activity.
- Competitive inhibitors fit into the active site and prevent the normal substrate from binding. If the concentration of the substrate is increased, then there is a better chance that the substrate molecule will get into the active site before an inhibitor, and the reaction rate increases.
- Non-competitive inhibitors bind with the enzyme at a position other than its active site. This causes the enzyme to change its shape so the active site can no longer bind with the substrate. Increasing the substrate concentration therefore has no effect on the reaction rate.

ASSIGNMENT 3: INVESTIGATING ENZYMES FOR THE FUTURE

(MS 0.1, MS 0.3, PS 1.1, PS 1.2)

The dream of using hydrogen as an economical, pollution-free fuel has moved closer to reality with the discovery of two bacterial enzymes.

Figure A1 A hydrogen-powered bus in London Waterloo; it is 'cleaner and greener' than its diesel-powered brothers.

Working together, the two enzymes use glucose to produce hydrogen gas. During the process, which does not generate any potentially harmful by-products, the enzymes convert glucose to gluconic acid, a substance used in detergents and some pharmaceuticals.

Thermoplasma acidophilum, a type of bacterium that was first found in a coal tip, produces the enzyme glucose dehydrogenase. This enzyme can remove two hydrogen atoms from each glucose molecule it interacts with. One of the hydrogen atoms combines with a carrier molecule while the other goes into solution as a hydrogen ion. The reaction is completed when the second enzyme, hydrogenase, catalyses the reaction between two hydrogen ions to form a molecule of hydrogen gas. Hydrogenase is produced by *Pyrococcus furiosus*, a species of bacterium that flourishes around volcanic vents deep in the Pacific Ocean.

Both the enzymes are unusual in that they can resist quite high temperatures. Glucose dehydrogenase works best at 55 °C, while hydrogenase has an optimum temperature of

85 °C. The ability to operate at these temperatures makes the conversion of glucose very rapid. Contamination by other bacteria is also unlikely because very few bacteria can survive at these temperatures.

Questions

A1. Gluconic acid is made after removing the hydrogen from glucose. Which group in gluconic acid makes it an acid?

A2. Sketch graphs to show the likely rates of reaction for the two enzymes glucose dehydrogenase and hydrogenase at different temperatures.

A3. If both processes go on at the same time in the same container, what would be the optimum temperature to use? Explain your answer.

A4. Why does operating at high temperatures make the reaction very rapid?

A5. One possible source of glucose for this process is the cellulose in waste paper. How could glucose be obtained from the cellulose?

Stretch and challenge

A6. In the USA, it costs about 30 cents to obtain one kilogram of glucose from one kilogram of cellulose. Gluconic acid sells for about $4 per kilogram.

a. Assuming that each kilogram of glucose yields approximately one kilogram of gluconic acid, what would be the percentage profit?

b. About seven million tonnes of cellulose from waste paper could be available. What value of gluconic acid could be produced?

c. What other source of profit would there be from the process?

d. The demand for gluconic acid would be limited and would be less than the amount produced. Explain how this could affect the economics of the process.

e. What other factors would have to be taken into account when assessing profitability?

ASSIGNMENT 4: MEASURING THE ACTIVITY OF AMYLASE

(MS 1.3, MS 3.2, MS 4.1, PS 1.1, PS 2.3, PS 3.1, PS 3.2)

The activity of a digestive enzyme can be determined in two ways:

- by measuring the quantity of products
- by measuring how much substrate is used up in a given time.

One convenient technique for measuring the activity of the starch-digesting enzyme, amylase, is to use starch agar plates. Starch agar is made by adding starch to liquefied agar jelly. The molten starch agar is poured into a Petri dish and allowed to set. Samples to be tested are placed in cavities cut into the agar, or solid samples can be placed on the surface (as in Figure A1a).

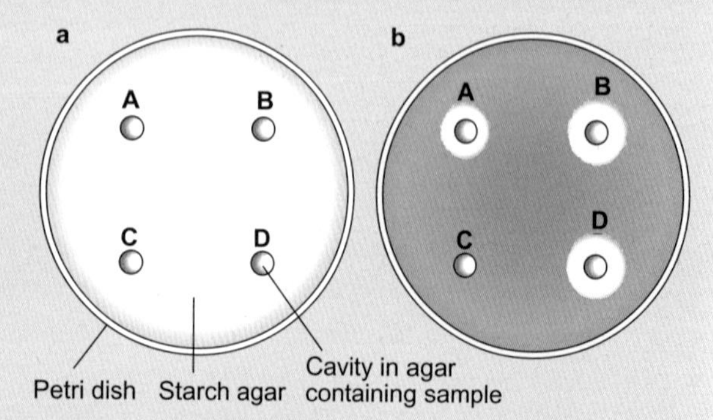

Figure A1 *Using starch agar plates to test for the presence of amylase*

After several hours, the surface of the agar plate is covered with iodine solution. Areas that contain starch stain blue-black, whereas areas in which the starch has been digested remain clear. The size of a clear area can be used as a measure of the concentration of amylase in the sample. The larger the diameter of the clear area, the further the amylase must have diffused from the sample. This is because the higher the concentration in the sample, the steeper the diffusion gradient.

This technique, in which the quantity of a substance is found by comparing its activity with a standard sample, is called an **assay**. A similar technique can be used to measure the activity of other enzymes. For example, white protein powder can be suspended in the agar. Protein-digesting enzymes make the milky-white agar turn clear.

Questions

A1. Look at the clear areas in the starch agar plates in Figure A1b. Well A is the standard concentration.

a. Calculate the area of each zone of inhibition.

b. Present these data graphically.

c. What can you conclude about the samples?

d. What substances would you expect to find in the clear areas of agar?

A2. A manufacturer wants to test three strains of a fungus as possible sources of amylase. Describe how starch agar plates could be used to find the strain that produces the most amylase.

PRACTICE QUESTIONS

1. Liver contains the enzyme catalase. This enzyme catalyses the breakdown of hydrogen peroxide into water and oxygen.

$2H_2O_2 \rightarrow 2H_2O + O_2$

A 1 g piece of liver was dropped into a beaker containing 50 cm^3 of hydrogen peroxide. The loss of mass by the beaker and its contents was measured for the next 15 minutes. The graph in Figure Q1 shows the processed results of this experiment.

a. Why was there a loss in mass during this experiment?

b. i. Describe the most suitable control for this experiment.

ii. Explain why this control is necessary.

c. Describe the relationship between the rate of this reaction and time.

d. Explain the reason for the shape of the curve at:

i. X

ii. Y.

e. Explain, in terms of activation energy, why hydrogen peroxide decomposes slowly at room temperature, but very rapidly if catalase is added.

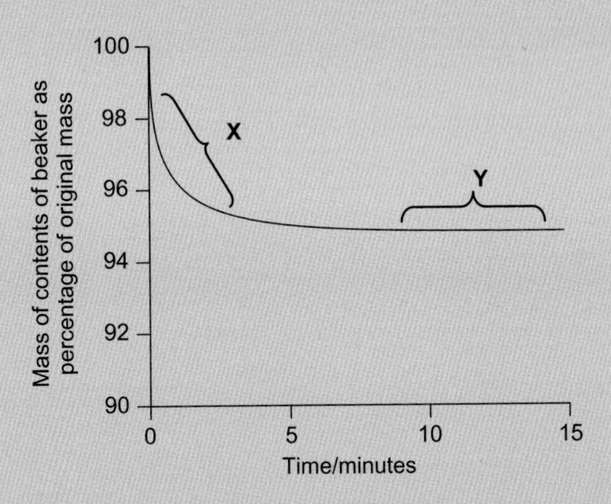

Figure Q1

2. a. Many reactions take place in living cells at temperatures far lower than those required for the same reactions in a laboratory. Explain how enzymes enable this to happen.

b. An amylase enzyme converts starch to maltose syrup which is used in the brewing industry.

Describe a biochemical test to identify:

i. starch

ii. a reducing sugar such as maltose.

c. The graph in Figure Q2 shows the results of tests to determine the optimum temperature for the activity of this amylase.

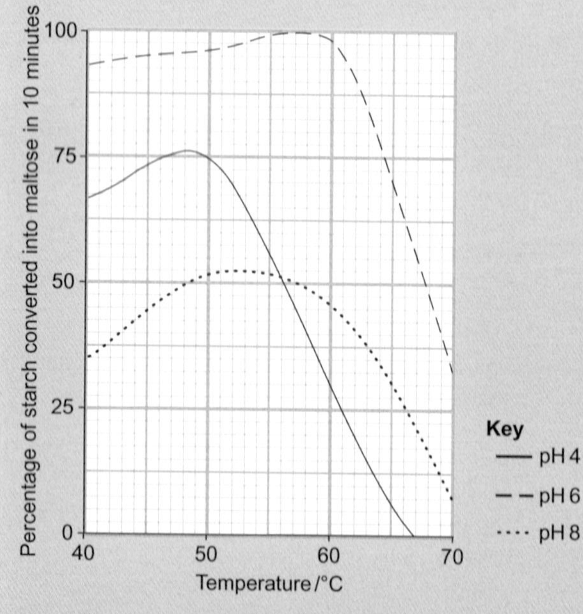

Figure Q2

i. Copy and complete Table Q1 for the optimum temperature for the activity of amylase at each pH value.

	pH		
	4	6	8
Optimum temperature /°C			

Table Q1

ii. Describe and explain the effect of temperature on the rate of reaction of this enzyme at pH 4.

AQA June 2006 Paper 1 Q4

3. a. An enzyme catalyses only one reaction. Explain why.

b. Gout is a disease caused by the build-up of uric acid crystals in joints. Uric acid is produced from xanthine in a reaction catalysed by the enzyme xanthine oxidase.

$$\text{xanthine} \xrightarrow{\text{xanthine oxidase}} \text{uric acid}$$

Xanthine　　Allopurinol

Figure Q3

Allopurinol is a drug used to treat gout. Figure Q3 shows the structures of xanthine and allopurinol.

Use this information to suggest how allopurinol can be used to treat gout.

AQA June 2013 Paper 1 Q2

4. The enzyme tyrosine kinase (TK) is found in human cells. TK can exist in a non-functional and a functional form. The functional form of TK is only produced when a phosphate group is added to TK. This is shown in Figure Q4.

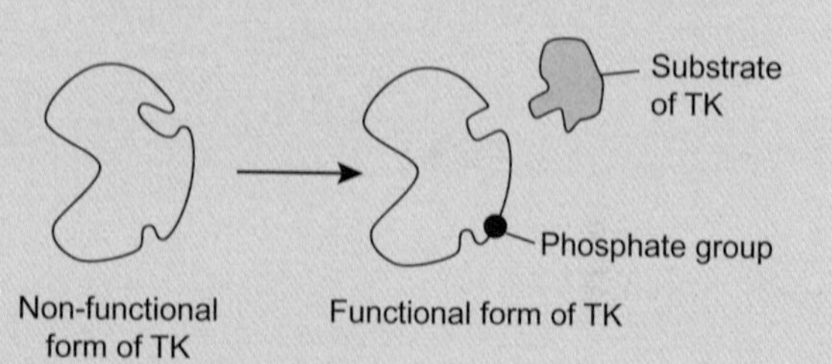

Figure Q4

a. Addition of a phosphate group to the non-functional form of TK leads to production of the functional form of TK. Explain how.

b. The binding of the functional form of TK to its substrate leads to cell division. Chronic myeloid leukaemia is a cancer caused by a faulty form of TK. Cancer involves uncontrolled cell division. Figure Q5 shows the faulty form of TK.

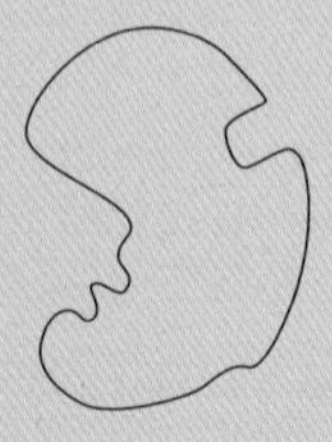

Faulty form of TK

Figure Q5

(Continued)

Suggest how faulty TK leads to chronic myeloid leukaemia.

c. Imatinib is a drug used to treat chronic myeloid leukaemia. Figure Q6 shows how imatinib inhibits faulty TK.

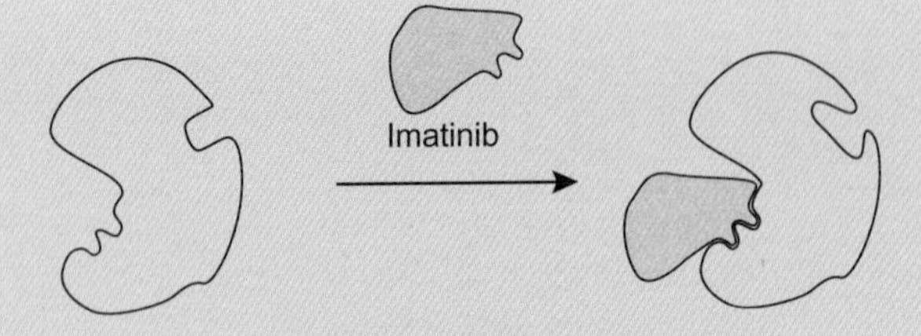

Figure Q6

Using all of the information, describe how imatinib stops the development of chronic myeloid leukaemia.

AQA January 2013 Paper 1 Question 5

5. A student investigated the effect of temperature on the rate of reaction of catalase. Catalase increases the rate of breakdown of hydrogen peroxide to water and oxygen.
   a. The student decided to measure the rate of reaction by measuring the volume of oxygen produced. Explain why, for this reaction, this is better than measuring the rate of disappearance of the substrate.
   b. i. Suggest a suitable range of temperatures that the student should use in her experiment. Justify your suggestion.
      ii. Suggest a suitable interval between the values of temperature used. Justify your suggestion.
      iii. Describe how the student could vary the temperature during her experiment.
   c. i. The student decided that she needed to keep the pH of the reacting mixture constant. Describe how she could do this.
      ii. State two other variables that the student must keep constant during her experiment.

6. Enzymes are proteins. Explain how the structure of protein molecules makes it possible for cells to produce enzymes that are able to catalyse all of the different metabolic reactions that take place in the human body.

7. Read the passage and then answer the questions that follow.

   The enzyme catalase, sometimes known as peroxidase, is found in almost all living organisms. The enzyme catalyses the decomposition of hydrogen peroxide to water and oxygen.

   Catalase is a protein with quaternary structure. It is made up of four polypeptide chains, each of which has a haem group. The precise structure of catalase varies between different species, and the differences between the positioning and types of bonding holding the three-dimensional structure of the molecule together means that catalases from different species can have very different values for their optimum temperature and optimum pH.

   Despite the differences in their structure, all catalases have an active site that allows hydrogen peroxide molecules to form temporary bonds with the side chains of asparagine and histidine amino acids, and this causes an oxygen atom to leave the hydrogen peroxide molecule. The iron ion in the haem group is also involved in catalysing the conversion of the substrate to the two products.

   In humans, lack of catalase activity has been implicated in the process that causes hair to go grey as we age. Normally, catalase in the hair follicles breaks down hydrogen peroxide that is formed there as a by-product of various metabolic reactions. However, as we age, catalase activity in hair follicles decreases and hydrogen peroxide is not broken down as effectively. The hydrogen peroxide bleaches the hair.

   Bombardier beetles make use of catalase to defend themselves from predators. The beetles store two chemicals – hydrogen peroxide and hydroquinone – in two separate compartments towards the rear of their abdomens. When the beetle is threatened, a set of muscles contracts, opening valves between the chambers and allowing the two chemicals to move into a third compartment, where catalase is produced. The result is a rapid exothermic reaction, as the catalase almost instantly breaks down the hydrogen peroxide. This results in the rapid expansion of a very hot liquid with a horrible smell, which is ejected at high speed from the beetle's rear end. (You can see some video clips of this if you search on YouTube.)

a. Explain why catalase is also known as peroxidase.

b. Explain what is meant by the statement that catalase has quaternary structure.

c. State two similarities and one difference between the structures of catalase and haemoglobin.

d. State the types of bond in a catalase molecule that are affected by temperature and by pH.

e. Suggest how, despite the fact that catalases from different species have different structures, all of them are able to catalyse the breakdown of hydrogen peroxide.

f. Since the discovery that catalase may be involved in the greying of hair, several companies have begun marketing pills containing catalase. Explain why these pills are unlikely to stop hair going grey.

g. Explain why the liquid that is ejected from the bombardier beetle is very hot and under high pressure.

4 NUCLEOTIDES

PRIOR KNOWLEDGE

You have probably learned that DNA is the genetic material in our cells, holding information that is transferred from parents to offspring. You might want to remind yourself that many biological molecules are polymers, made from many monomers joined by condensation reactions. These polymers can be broken apart by hydrolysis reactions. You will probably remember that all metabolic reactions are catalysed by enzymes.

LEARNING OBJECTIVES

In this chapter, we will look at the structure of the polynucleotides DNA and RNA, and the nucleotide derivative ATP, and will also see how DNA is able to replicate itself perfectly.

(Specification 3.1.5.1, 3.1.5.2, 3.1.6)

All cells contain DNA, which contains coded information to guide the cell in synthesising proteins. When a cell divides, the DNA is copied so that a full set of information can be transferred to each new cell.

Cell division is normally fully controlled, so that it only happens as and when this is necessary – for example for growth, or to repair damaged tissues. But sometimes this control breaks down, and cells divide over and over again with no curb on the number of new cells that are produced. This can lead to cancer.

There are many different types of cancer, each with its own particular features. Different types of cancer in different people respond to different types of treatment. One of the major weapons in the armoury to fight cancer is chemotherapy. This involves taking drugs that are designed to destroy cancer cells in the body, without causing too much harm to other, normal cells.

Methotrexate is one of these drugs. It is an enzyme inhibitor that prevents the synthesis of nucleotides in cells. Nucleotides are the monomers that join together to form DNA and RNA. Without a good supply of nucleotides, new copies of DNA cannot be made, and so cells cannot divide. Because cancer cells divide much more often than normal cells, methotrexate has a much greater effect on them than on other body cells.

Nevertheless, many people experience side effects when taking this drug. For example, the lack of nucleotides means that cells in the hair follicles, which usually divide regularly to cause hair growth, no longer have the raw materials for making DNA for new cells. Cell division stops and hair loss may occur. Another side effect is feeling nauseous. This happens because cells that line the alimentary canal, which normally divide to replace the ones that get damaged by the passage of food and the activities of digestive enzymes, can no longer do so, affecting the lining of the stomach and intestines. Yet another effect is a reduction in the numbers of red and white blood cells, another group of cells that is regularly replenished and therefore needs constant supplies of nucleotides.

Careful management of the dose of methotrexate can help to make sure that maximum inhibition of division of cancer cells is achieved, while minimising the effects on other, healthy cells. And, once chemotherapy stops, all of these non-cancerous cells will go back to dividing normally again, so the symptoms of the side effects disappear.

4.1 STRUCTURE OF DNA AND RNA

Nucleic acids

Nucleic acids are substances that were first discovered inside the nuclei of cells – hence their name. There are two types: deoxyribonucleic acid (DNA) and ribonucleic acid (RNA). Both of them are polymers of smaller molecules called **nucleotides**. DNA and RNA are therefore **polynucleotides**.

DNA is the nucleic acid that stores genetic information in every living cell. DNA is a remarkable substance. Its properties make it the key to all life on Earth and it has essentially the same structure in bacteria, plants and mammals. It has survived throughout evolution as the one substance that can store blueprints for each of the millions of species that have existed. DNA molecules:

- are huge, and able to store vast amounts of information in a small volume
- have small variations in structure that act as a simple code
- are stable, so that the information is not easily corrupted
- can reproduce themselves and so copy the information.

RNA comes in several different forms, some of which are involved in carrying the information held in the DNA code to the ribosomes, where it is used to make protein molecules.

The structure of DNA

People have selected for favourable features in animals and plants for roughly 10 000 years – really since the start of agriculture. For most of that time they were unaware of the rules of genetics that controlled the inheritance of those features. Genetics as a science began roughly 100 years ago. Molecular genetics, which allows us to explain the reactions of the chemicals that control the inheritance of genes, is even younger. It all started in 1953 when Francis Crick and James Watson at Cambridge University, helped by Rosalind Franklin at Kings College in London, worked out the molecular structure of DNA. Using X-ray diffraction images made by Franklin, plus some creative model-making, Crick and Watson unlocked the puzzle of how the components of DNA fit together into a complex three-dimensional structure (Figure 1).

We have seen that DNA is made up of monomers called nucleotides. A nucleotide molecule has three parts: a five-carbon (pentose) sugar, a phosphate group and an organic base (Figure 2).

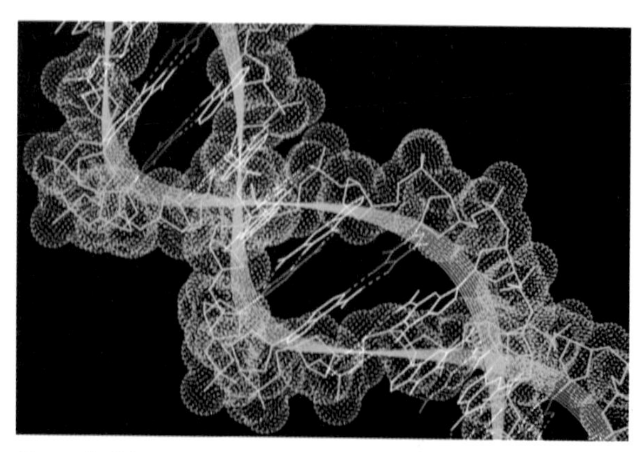

Figure 1 This computer representation of a small piece of DNA may look complicated, but its basic structure is very simple.

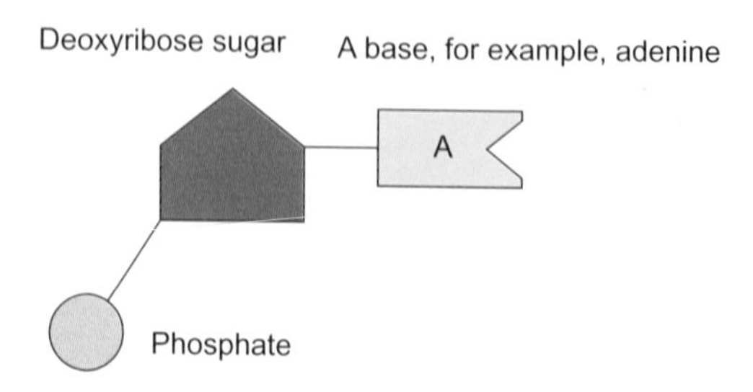

Figure 2 The structure of a DNA nucleotide

In DNA, the pentose sugar in the nucleotides is **deoxyribose**. There are four different bases, all of which contain nitrogen. These four bases are called adenine, thymine, cytosine and guanine, and are often referred to by their initial letters, A, T, C and G. This means that there are four types of nucleotide in a DNA molecule.

The nucleotide monomers combine together by condensation reactions, which form **phosphodiester bonds** between the sugar and phosphate molecules. The resulting molecule is a polymer called a polynucleotide (Figure 3, overleaf).

A DNA molecule consists of two polynucleotide strands joined together to make a structure rather like a twisted ladder. Hydrogen bonds form between the base pairs with complementary shapes (adenine–thymine and cytosine–guanine) to produce the 'rungs' of the ladder. The hydrogen atoms on one base are attracted to oxygen and nitrogen atoms on another. This produces a regular and stable DNA molecule with two sugar–phosphate sides joined by pairs of bases (Figure 4, overleaf).

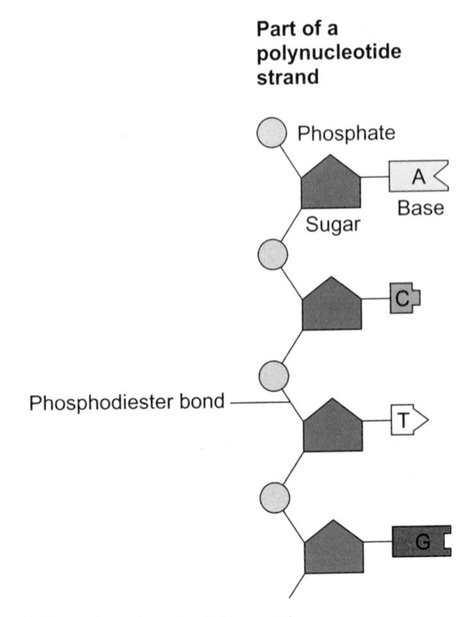

Figure 3 *Part of a polynucleotide strand*

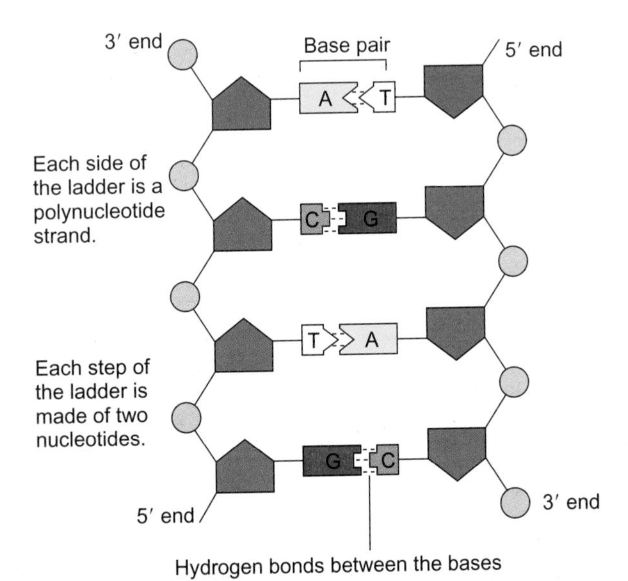

Figure 4 *Part of a DNA molecule, with the two polynucleotide strands joined by hydrogen bonds between the bases*

The pairing between the bases is said to be **complementary** – this means that A always pairs with T, and C always pairs with G.

Notice that the two polynucleotide strands run in opposite directions. They are said to be **antiparallel**.

These two long strands of polynucleotides then twist together to form a double helix (Figure 5). This structure ensures that the hydrogen bonds (which, you may remember, are individually weak bonds, but collectively strong when there are many of them) are protected inside the molecule, which prevents the code being corrupted by other molecules in the nucleus that might interact with them. DNA is a very stable molecule and can withstand quite high and low temperatures. Samples of intact DNA have been found in centuries-old woolly mammoths frozen in the Arctic permafrost, and even from 20-million-year-old fossils of insects preserved in amber – the basis of the book and film *Jurassic Park*.

How DNA holds information

The sequence of bases in a DNA molecule determines what proteins are made in the cell. Essentially, the sequence of bases in DNA determines the sequence of amino acids that are joined together when proteins are made in the cell. A sequence of three bases in the DNA molecule codes for one amino acid. As you have seen, this sequence of amino acids – known as the primary structure – ultimately determines the three-dimensional shape of the protein, and therefore its function. So the sequence of A, T, C and G in the DNA in a cell's nucleus carries information about all the proteins that will be made in that cell, and this in turn determines the structures that are formed in the cell, and the metabolic reactions that it will carry out.

You can read much more about DNA and protein synthesis in Chapter 10.

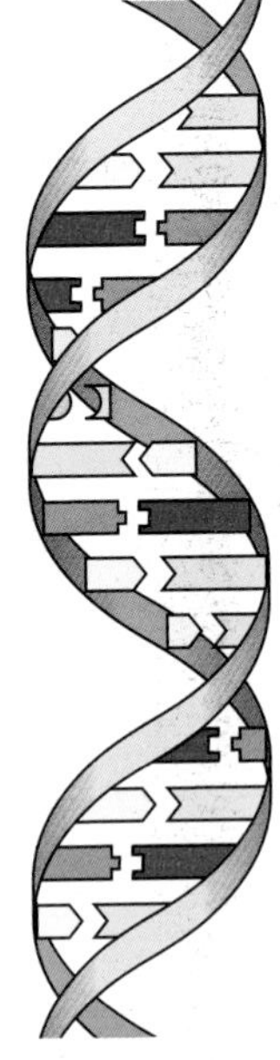

Figure 5 *The double helix of DNA*

ASSIGNMENT 1: DISCOVERING THE STRUCTURE OF DNA

(PS 1.2)

Figure A1 *Watson and Crick with their DNA model*

For many years, it was believed that inheritance was due to a protein in the cell nucleus. This was because only proteins, with their almost infinite variety of complex structures, were thought to be able to store the huge amount of information required to construct and maintain a living organism.

In 1869, a Swiss chemist, Friederich Miescher, was investigating the protein components of leukocytes (white blood cells). He arranged for a local clinic to send him used, pus-coated bandages which he washed, filtered out the leukocytes, and extracted and identified the various proteins.

During his investigations he came across a substance that had chemical properties unlike any protein, including a much higher phosphorus content. Miescher realised that he had discovered a new substance in the nucleus of white blood cells which he called 'nuclein'. This later became known as deoxyribonucleic acid.

In the 20th century, other scientists continued to investigate the chemical nature of this molecule. Russian biochemist Phoebus Levene proposed the structure of nucleotides in 1919. Several scientists suggested how the nucleotides might fit together, but it was Levene's 'polynucleotide' model that proved to be the correct one. Levene suggested that nucleic acids were made up from a series of nucleotides, and that each nucleotide was in turn made up of one of four nitrogen-containing bases, a sugar molecule, and a phosphate group.

In 1944, Oswald Avery showed that the substance responsible for heredity was not a protein by removing proteins from bacteria and showing that the bacteria were still able to reproduce. He showed that genes are composed of DNA. Erwin Chargaff had read Avery's research paper and in 1950 investigated DNA from different species. He discovered that the DNA of different species had different proportions of nucleotides. He also discovered that in all species, the amount of adenine is similar to the amount of thymine, and that the amount of cytosine is similar to the amount of guanine. In other words, the total amount of purines (A + G) and the total amount of pyrimidines (C + T) are usually nearly equal. This is now known as 'Chargaff's rule'.

During this time a biophysicist, Rosalind Franklin, was working on X-ray diffraction images of DNA which demonstrated that it had a helical structure. These images were crucial to James Watson and Francis Crick, who examined Franklin's X-rays without her permission. They finally put together the three-dimensional, double-helical model for the structure of DNA in 1953. Using cardboard cutouts representing the individual chemical components of the four bases and the sugar and phosphates, Watson and Crick arranged the cardboard models as though putting together a puzzle. When they placed the sugar–phosphate backbone on the outside of the model, the complementary bases fitted together perfectly (A with T and C with G), with each pair held together by hydrogen bonds. The structure also fitted in with Chargaff's rule. In 1962, Watson and Crick were awarded the Nobel Prize for their work on DNA structure.

Questions

A1. a. Why did Miescher use white blood cells, rather than red blood cells, for his investigation?

b. What was his evidence that nuclein was not a protein?

A2. Explain the significance of Chargaff's rule.

A3. Use the internet to find out more about how Watson and Crick made use of the work of Rosalind Franklin and others. Discuss whether it was appropriate to award the Nobel Prize only to Watson and Crick, or whether it should have been shared with others.

QUESTIONS

1. Draw a diagram of a polynucleotide strand with the bases in the following order: thymine, thymine, adenine, guanine, cytosine and adenine, using the shapes shown in Figure 3.
2. Complete the other strand of the DNA molecule you have just drawn.
3. Table 1 shows the proportions of the four bases in DNA from four organisms.

Source of DNA	Adenine/%	Guanine/%	Thymine/%	Cytosine/%
human	30	20	30	20
rat	28	22	28	22
yeast	31	19	31	19
turtle	28	22	28	22

Table 1

a. Use your understanding of the structure of DNA to explain the pattern in these proportions.

b. In an organism, 26% of the bases in the DNA are found to be adenine. What percentage would be cytosine? Explain how you worked it out.

The structure of RNA

RNA differs from DNA in four ways.

- The pentose sugar is ribose, not deoxyribose.
- The base thymine is replaced by a different base, called uracil.
- Most forms of RNA are made up of a single strand, rather than two antiparallel strands joined together.
- RNA molecules are shorter than DNA molecules (in other words, they contain fewer nucleotides joined together).

Just as in DNA, these nucleotides can be joined together through phosphodiester bonds to form long chains. Figure 6 shows the structure of an RNA polynucleotide.

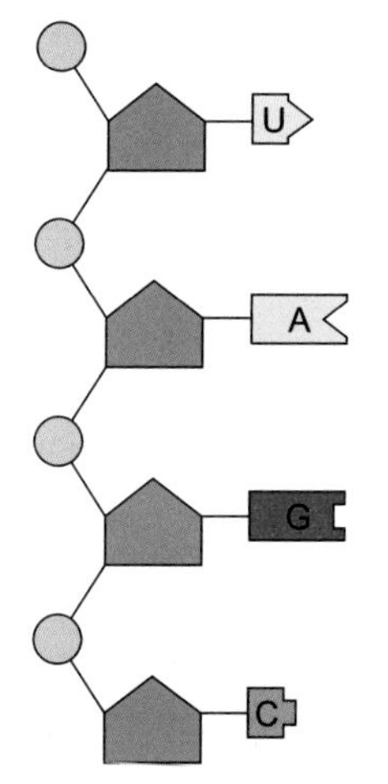

Figure 6 *The structure of RNA*

You will find out about some of the roles of RNA in a cell in Chapter 10. Ribosomes, the tiny organelles where proteins are synthesised in a cell, are made of RNA and protein.

KEY IDEAS

- DNA and RNA are important information-carrying molecules, found in all living cells.
- DNA holds or stores the genetic information, while RNA transfers it to ribosomes, where proteins are synthesised.
- DNA and RNA are polymers of nucleotides. Each nucleotide is formed from a pentose (5 carbon) sugar, a phosphate group and an organic base.
- The organic bases in DNA are adenine, cytosine, guanine and thymine. In RNA, uracil is found instead of thymine.
- Two nucleotides can join together by a condensation reaction, forming a phosphodiester bond between the pentose group of one of them, and the phosphate group of the other.

- Polynucleotides consist of long chains of nucleotide monomers. DNA polynucleotides are usually much longer chains than RNA polynucleotides.
- A DNA molecule consists of two polynucleotide chains running in opposite directions (antiparallel), joined by hydrogen bonds between the bases.
- There is complementary base pairing; A always joins with T, and C always joins with G.

ASSIGNMENT 2: UNDERSTANDING BASE PAIRING

(MS 0.3, PS 1.2)

James Watson and Francis Crick's model of DNA is a double helix with a sugar–phosphate backbone with the organic bases on the inside. The two strands of the DNA double helix run in opposite directions, one in the 5′ to 3′ direction, the other in the 3′ to 5′ direction. (The numbers 5′ and 3′ refer to the carbon atoms in the deoxyribose that are involved in forming the sugar–phosphate backbone.) The term that describes how the two strands relate to each other is **antiparallel**.

DNA samples taken from different cells of the same species have the same proportions of the four bases. For example, in humans the DNA has about 30% each of of adenine and thymine, and 20% each of guanine and cytosine. The figure is different for other organisms, but the amounts of A and T are always the same, as are the amounts of C and G.

The two strands are held together by hydrogen bonds between the bases. There are two hydrogen bonds between adenine and thymine, and three hydrogen bonds between cytosine and guanine.

Hydrogen bonds can only occur between a hydrogen atom on one base and either an oxygen or nitrogen atom on the other base. This explains why only two hydrogen bonds can form between A and T and three can form between G and C, because a hydrogen bond can only form where a hydrogen atom is close to an oxygen or nitrogen atom of a base on the opposite strand.

This type of base pairing is called complementary base pairing.

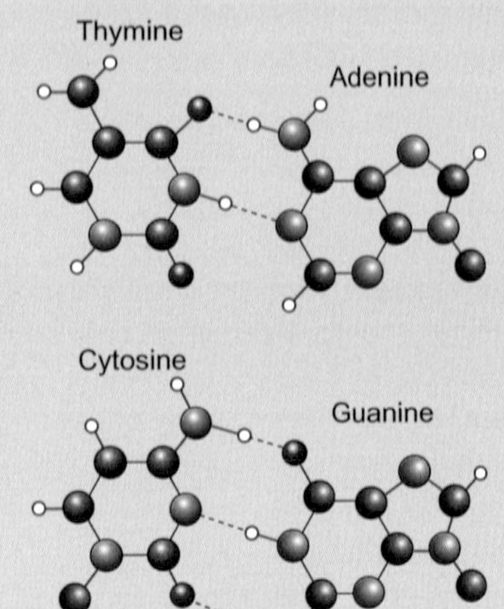

Figure A1 Base pairing

In the formation of RNA, thymine is replaced by uracil, and consequently uracil pairs with adenine.

Questions

A1. State three differences in structure between DNA and RNA.

A2. If a yeast cell contains 31% adenine; what is the percentage of the other bases?

Stretch and challenge

A3. DNA runs in antiparallel directions. What does this mean and why does it happen?

A4. Why would a section of DNA with more C, G pairs require more energy to unwind?

4.2 DNA REPLICATION

You will already be aware that DNA is passed down from parents to offspring, and this is how parental characteristics are passed on from one generation to another. Think, too, about your own body. You began as a single cell (a zygote), which divided to form two cells, which then divided over and over again, eventually forming all of the millions of cells that make up your body. All of these cells contain identical DNA. This can only happen if the DNA in a cell can be perfectly copied before it divides, so that each 'daughter' cell obtains a complete set of uncorrupted information.

The structure of DNA is perfectly adapted to enable exact copying. The way in which the bases pair up enables stored information to be copied quickly and accurately. When a DNA molecule untwists, the hydrogen bonds can be broken so that the strands can be separated like the sides of a zip. Exact copies can then be produced, since specific base pairing means that each exposed base will combine with only one of the four types of nucleotide.

Semi-conservative replication

Every time a cell divides, it first makes a complete copy of its DNA. The copy must be exactly the same as the original in order to preserve the information.

Figure 7 shows what happens. The DNA molecule is unwound and the hydrogen bonds that connect the bases are broken. This is done by an enzyme called **DNA helicase**. The two strands separate easily, exposing the bases. The phosphodiester bonds between the sugar and phosphate groups in the polynucleotide strands are strong, and they keep the separate strands intact.

Free nucleotides can now attach to the exposed bases on each strand. Only the complementary bases will fit together. Another enzyme, called **DNA polymerase**, attaches the new nucleotides to each other, by condensation reactions, forming phosphodiester bonds. A new strand is built on each of the original strands, so that the two new DNA molecules are exactly the same as the original.

This process of making perfect copies of DNA is called **replication**. As you can see from the diagram, each of the two new molecules of DNA has one of the original polynucleotide strands and one new one made from the supply of nucleotides in the cell. The system is therefore called **semi-conservative replication**, because one strand in each molecule is conserved (kept) (Figure 8).

1 DNA helicase unwinds the DNA double helix and breaks the hydrogen bonds holding the two strands together.

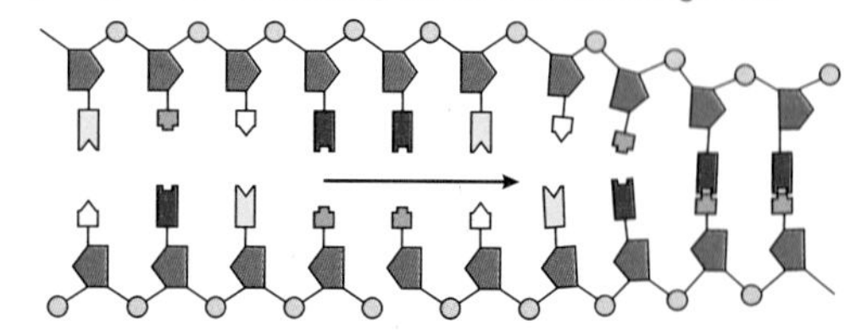

2 Free nucleotides are attracted to their complementary exposed bases, and form hydrogen bonds with them.

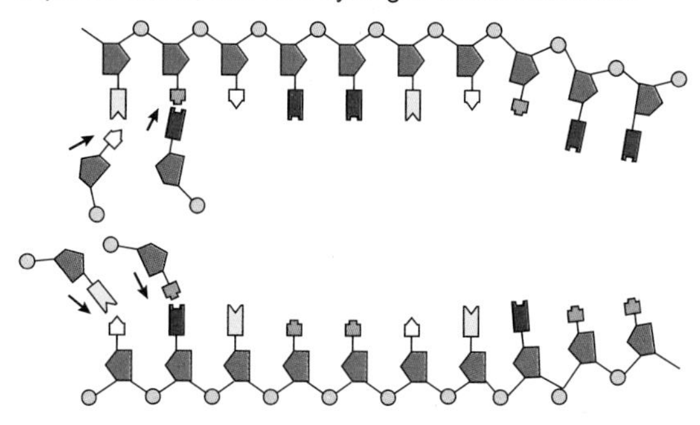

3 DNA polymerase joins the new lines of nucleotides together. Now there are two complete two-stranded DNA molecules, each of which twists to form a double helix.

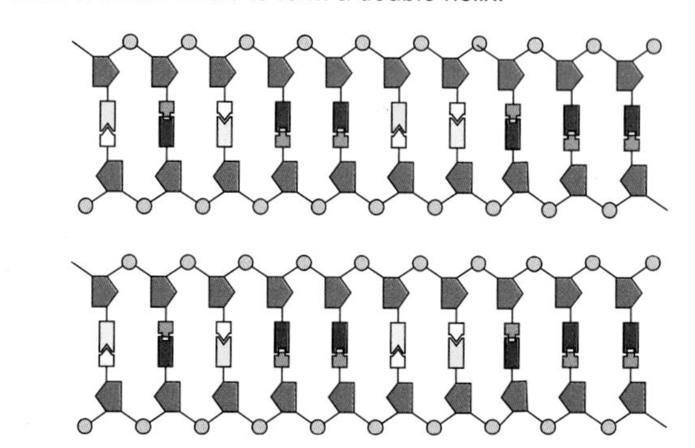

Figure 7 *DNA replication*

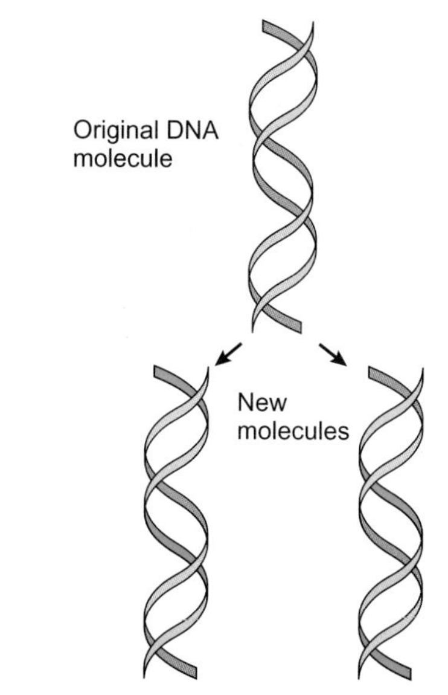

Figure 8 *Semi-conservative replication*

QUESTIONS

Stretch and challenge

4. Draw diagrams to show how the section of DNA in this diagram would be replicated.

Examining the evidence for semi-conservative replication

Shortly after Watson and Crick published their theories and suggested semi-conservative replication, Matthew Meselson and Franklin Stahl set out to investigate whether DNA did replicate like this.

Figure 9 summarises their experiment. Meselson and Stahl grew bacteria in a culture medium containing 'heavy nitrogen' (the isotope ^{15}N). As the bacteria replicated, the nitrogen in the nucleotides of the bacterial DNA was therefore ^{15}N. As the diagram shows, Meselson and Stahl then took bacteria from the ^{15}N medium and grew them on a medium containing the common 'light' isotope ^{14}N.

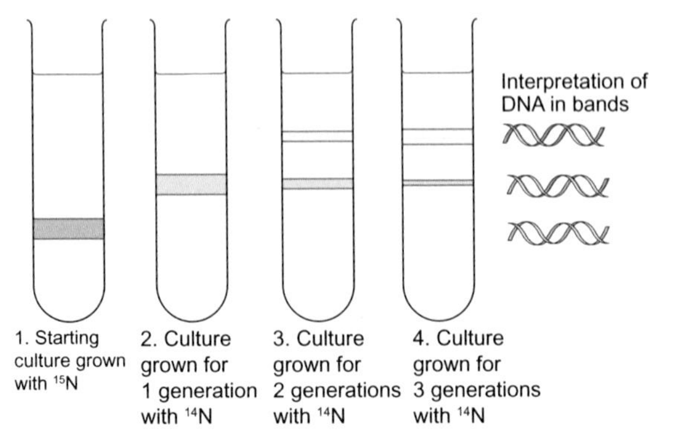

Figure 9 *The Meselson and Stahl experiments*

They extracted the DNA from these bacteria and centrifuged it in a caesium chloride solution.

After centrifuging, the concentration of caesium chloride varies uniformly from the top to the bottom of the tube, with the highest concentration at the bottom. The density changes slowly from the top to the bottom and the DNA extract floats at a particular level (depending on its weight and the density of the liquid it is in).

The DNA molecules float at a point in the centrifuge tube depending on the mass of the molecule. The heaviest molecules settled nearest the bottom of the tube and lighter molecules nearer the top.

QUESTIONS

Stretch and challenge

5. Copy and complete this diagram to show Meselson and Stahl's prediction. Use different colours for the ^{15}N and the ^{14}N strands for the second and third generations.

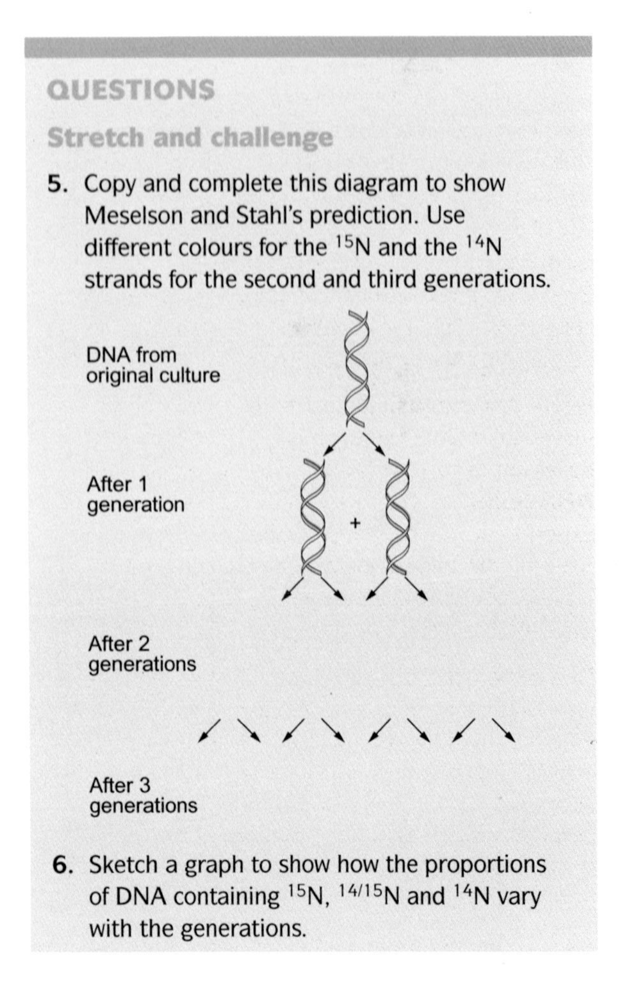

6. Sketch a graph to show how the proportions of DNA containing ^{15}N, $^{14/15}N$ and ^{14}N vary with the generations.

7. Suppose that DNA replicated by producing a new molecule made completely of new nucleotides, as suggested in this diagram. What results would you expect to find in tube 2 after one generation?

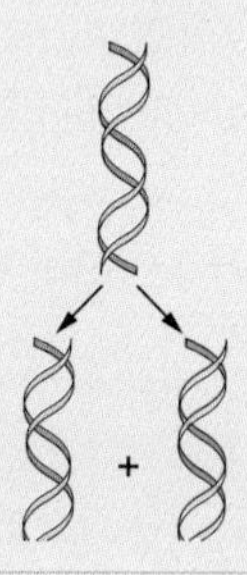

KEY IDEAS

- DNA is copied perfectly by semi-conservative replication.
- DNA helicase unwinds the DNA helix and breaks the hydrogen bonds between the bases.
- Free nucleotides bind to exposed bases on both strands of the unwound DNA by complementary base pairing.
- DNA polymerase joins the new lines of nucleotides by condensation reactions, forming two new polynucleotide strands.
- The two new two-stranded DNA molecules twist to form double helices.
- The semi-conservative replication of DNA ensures genetic continuity between generations of cells.

4.3 ATP

DNA and RNA nucleotides are not the only ones found in cells. All cells contain another vitally important nucleotide derivative called **adenosine triphosphate**, or **ATP**.

Figure 10 shows the structure of an ATP molecule. You can see that, like DNA and RNA nucleotides, it is made up of an organic base (adenine), a pentose sugar (ribose) and phosphate groups. The big difference is that this molecule has *three* phosphate groups. So it is actually a phosphorylated nucleotide. This is why it is called a 'nucleotide derivative' rather than simply a nucleotide.

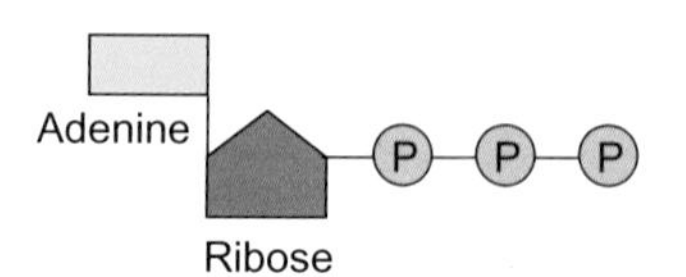

Figure 10 *The structure of ATP*

ATP is an incredibly important molecule. It is the immediate source of energy for almost every process that takes place in a cell. It is often known as the 'energy currency' of cells.

The energy contained in an ATP molecule is released when one of its phosphate groups is removed (Figure 11). This is a hydrolysis reaction, and it is catalysed by an enzyme called **ATP hydrolase**.

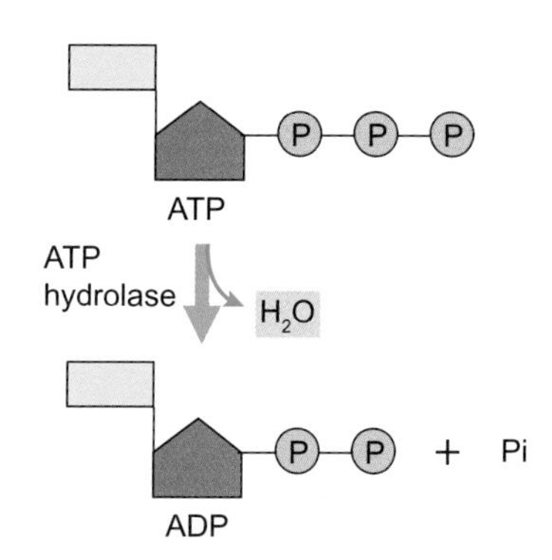

Figure 11 *The hydrolysis of ATP*

The hydrolysis of ATP breaks it down to form adenosine diphosphate, ADP, plus an inorganic phosphate group. This releases energy, which can be used to drive energy-requiring processes in the cell. Usually, the energy from the hydrolysis of the ATP is directly channelled into the process that requires it, so we say that the hydrolysis is 'coupled' to these energy-requiring reactions. We will meet some examples in the next chapter.

What happens to the products of this reaction? The phosphate group can be used to add to other substances, which often makes them more reactive. This lowers the activation energy of any (enzyme-catalysed) reactions they are involved in. For example, before glucose can be used in respiration, it first has to have phosphate added to it. Or the ADP and phosphate can be joined together again to resynthesise ATP. This is done using an enzyme called **ATP synthase**.

This resynthesis of ATP is one of the most important reactions in any cell. Each cell makes its own ATP, and if it cannot make any, it quickly dies, because most of its energy-requiring activities stop.

You can probably see that, if the breakdown of ATP *releases* energy, then energy will have to be *used* to make it join back together again. All cells make ATP through the reactions of respiration, and you can learn about this if you continue your studies to A level. In respiration, the energy used to make the ATP comes from glucose and other nutrient molecules. Some plant cells can make ATP in another way, through photosynthesis. In this case, the energy comes from light. You will learn more about this later in year two of your A-level studies.

QUESTIONS

8. The breakdown of ATP to ADP and P_i is a hydrolysis reaction. Predict the type of reaction that takes place when ATP is formed from ADP and P_i.

KEY IDEAS

- ATP is a nucleotide derivative made of an adenine molecule, a ribose molecule, and three inorganic phosphate groups.
- The hydrolysis of ATP is catalysed by ATP hydrolase. This reaction forms ADP and P_i, and releases energy,
- The hydrolysis of ATP is often coupled to energy-requiring processes in the cell.
- ATP is resynthesised by ATP synthase, which joins a phosphate group to ADP. This happens during respiration and photosynthesis, in which energy from nutrient molecules and light respectively is used to make the ATP.

Worked maths example: Working with large numbers

(MS 0.2, MS 1.1, MS 4.1)

The body contains approximately 37 trillion cells.

Our DNA contains about 24 000 genes, separated by long strands of non-coding DNA.

There are about three billion DNA base pairs in a human genome.

With such large orders of magnitude, **standard form** is commonly used.

- Express 37 trillion cells in standard form.

1 million	1 000 000	six powers of ten	1×10^6
1 billion	1 000 000 000	nine powers of ten	1×10^9
1 trillion	1 000 000 000 000	twelve powers of ten	1×10^{12}

So, 37 trillion = 37 000 000 000 000 = $\mathbf{3.7 \times 10^{13}}$

37 trillion cells is just an estimate – it's difficult to know exactly how many cells we contain, and this number will vary between people.

- The UK population in 2014 was 64.1 million (to three significant figures). Express this in standard form and then calculate the number of cells in the whole population, to two significant figures.

64.1 million = 6.41×10^7

Number of cells in whole population = $(64.1 \times 10^6) \times (37 \times 10^{12}) = 2371.7 \times 10^{18}$

$= 2400 \times 10^{18}$ (to two significant figures) $= 2.4 \times 10^{21}$

- Imagine a subject actually has 38 700 543 000 071 cells. How many cells are there, to three significant figures, expressed in standard form?

38 700 543 000 071 = **38 700 000 000 000** to three significant figures (sf)

38 700 000 000 = $\mathbf{3.87 \times 10^{13}}$

- Viruses are so small they can fit inside cells. Let's assume a virus is a perfect sphere. Calculate the surface area, volume and circumference of a virus with a diameter of 0.6 μm.

The diameter of the virus is 0.6 μm, so the radius is 0.3 μm (radius of a circle is half its diameter).

surface area of a sphere = $4\pi r^2$

$4 \times \pi \times 0.3^2 =$ **1.13 μm²** (to three sf)

volume of a sphere = $4/3\ \pi r^3$

$\frac{4}{3} \times \pi \times 0.3^3 =$ **1.113 μm³** (to three sf)

circumference of a circle (or sphere) = $2\pi r$

$2 \times \pi \times 0.3 =$ **1.88 μm** (to three sf)

› Calculate the surface area (μm²) and volume (μm³) of a cuboidal cell with a length of 10 μm, a width of 12 μm and a height of 6 μm.

surface area of a cuboid = $2\ (lw + wh + hl)$

volume of a cuboid = $w \times l \times h$

where

w = width

l = length

h = height

So, in our example:

surface area = $2\ (12 \times 10\ +\ 12 \times 6\ +\ 10 \times 6)$ = **504 μm²**

volume = $12\ \times\ 10\ \times\ 6 =$ **720 μm³**

› Calculate the surface area and the volume of a cylindrical cell with a diameter of 10 μm and a length of 14 μm.

surface area of a cylinder = $2\pi r^2 + 2\pi rl = 2\pi r(r + l)$

volume of a cylinder = $\pi(r^2)l$

where

r = radius

l = length of cylinder

So, in our example:

surface area = $2 \times \pi \times 5\ (5 + 14) =$ **596.6 μm²**

volume = $\pi \times (5^2) \times 14 =$ **1100 μm³**

PRACTICE QUESTIONS

1. Figure Q1 shows part of a DNA molecule.
   a. Name the structures labelled A, B and C.
   b. The sequence of bases on one strand of a DNA molecule is: ACTGCATCC

      How many amino acids could be coded for by this section of DNA?

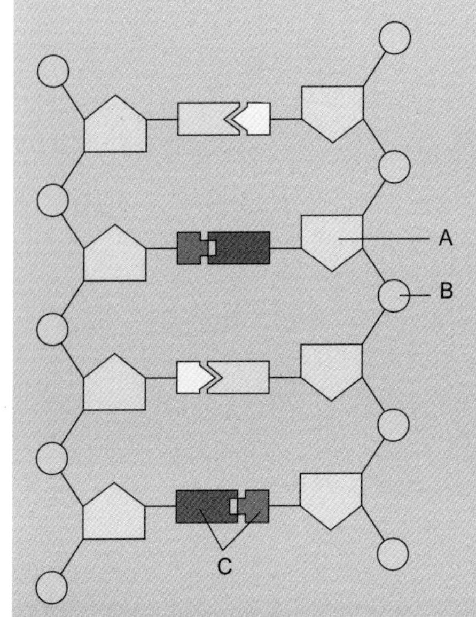

Figure Q1

2. a. A nucleotide of DNA contains a sugar molecule and two other molecules. Name the other two molecules.
   b. Draw a simple diagram to show the structure of a DNA nucleotide.
   c. A piece of DNA contained 16 base pairs. Complete Table Q1 to give the numbers of the bases in the two strands of this piece of DNA.

	Number of bases			
	Adenine	Thymine	Cytosine	Guanine
Strand X	6			
Strand Y	2			4

Table Q1

3. Figure Q2 shows the structure of a DNA molecule.

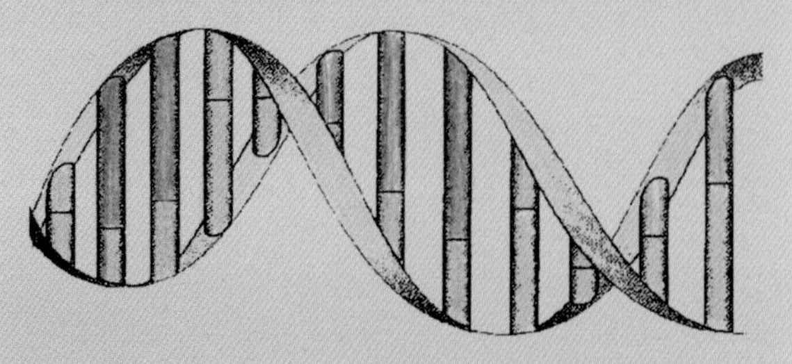

Figure Q2

a. What term describes the shape of a DNA molecule?

b. How are the two nucleotide chains joined to each other?

c. Explain three ways in which the structure of the DNA molecule is related to its function.

4. Figure Q3 shows how DNA replicates. In this case, ^{15}N was supplied and used for the replication in the first generation, but not for the second generation.

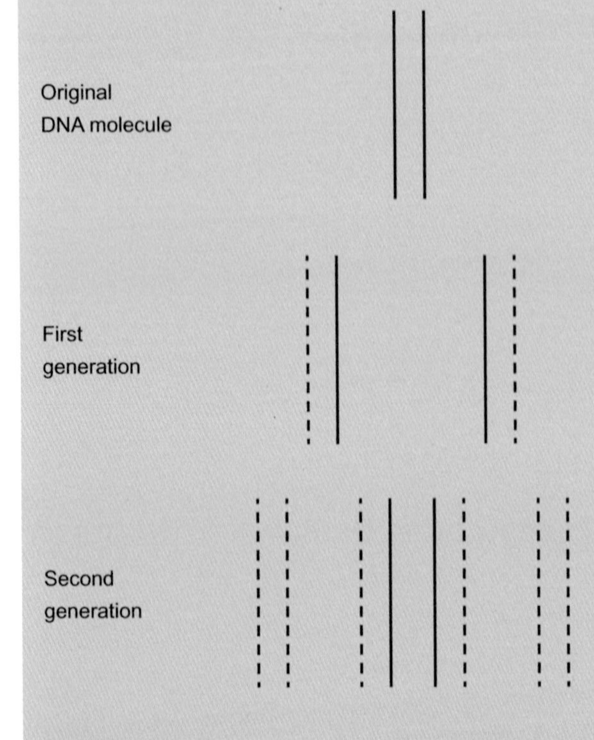

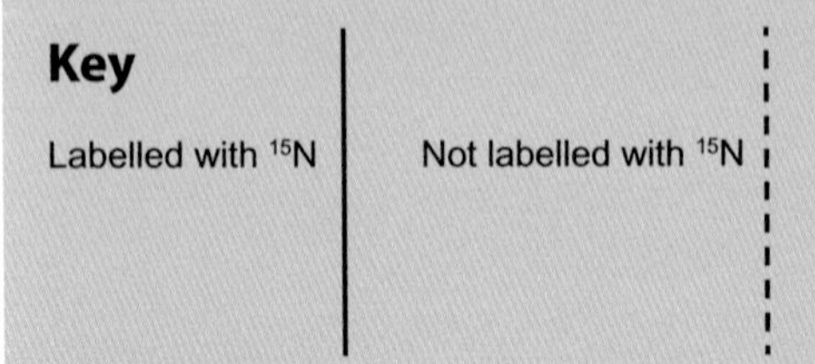

Figure Q3

a. What is ^{15}N?

b. Which part of the DNA molecule is labelled with ^{15}N?

c. Explain how molecules of DNA containing ^{15}N can be distinguished from those which do not contain ^{15}N.

d. Explain as fully as you can:

i. why all the first generation DNA molecules contained ^{15}N

ii. why only half of the second generation DNA molecules contained ^{15}N.

5. a. Figure Q4 shows the replication of a molecule of DNA.

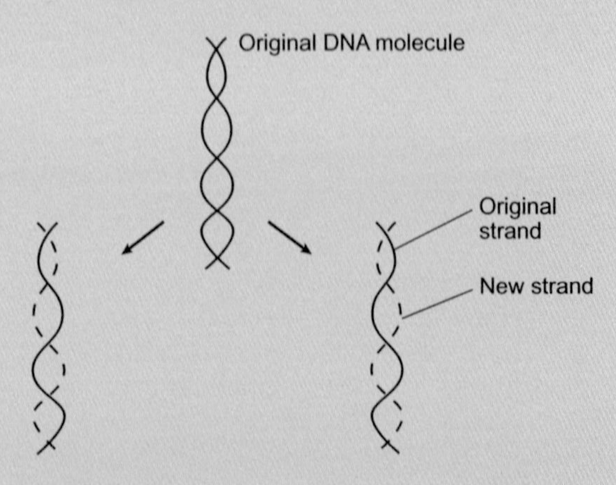

Figure Q4

Explain why DNA replication is described as semiconservative.

b. i. What is meant by specific base pairing?

ii. Explain why specific base pairing is important in DNA replication.

c. Describe two features of DNA which make it a stable molecule.

AQA June 2003 Unit 2 Question 2

6. Figure Q5 shows one nucleotide pair of a DNA molecule.

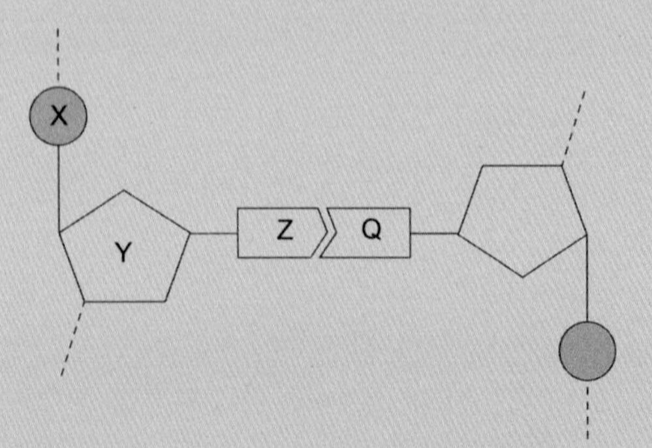

Figure Q5

a. Name the parts of the nucleotide labelled X, Y and Z.

b. What type of bond holds Z and Q together?

c. A sample of DNA was analysed. 28% of the nucleotides contained thymine. Calculate the percentage of nucleotides which contained cytosine. Show your working.

AQA June 2004 Unit 2 Question 1

7. **a.** Explain why the replication of DNA is described as semi-conservative.

b. Bacteria require a source of nitrogen to make the bases needed for DNA replication. In an investigation of DNA replication some bacteria were grown for many cell divisions in a medium containing ^{14}N, a light form of nitrogen. Others were grown in a medium containing ^{15}N, a heavy form of nitrogen. Some of the bacteria grown in a ^{15}N medium were then transferred to a ^{14}N medium and left to divide once. DNA was isolated from the bacteria and centrifuged.

The DNA samples formed bands at different levels, as shown in Figure Q6.

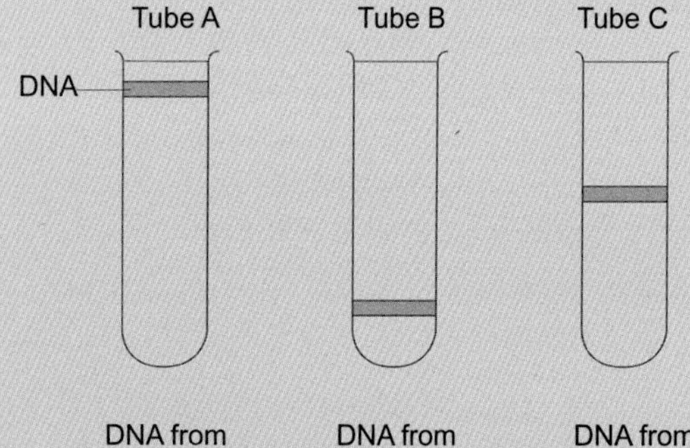

Figure Q6

i. What do tubes A and B show about the density of the DNA formed using the two different forms of nitrogen?

ii. Explain the position of the band in tube C.

c. In a further investigation, the DNA of the bacterium was isolated and separated into single strands. The percentage of each nitrogenous base in each strand was found. The table shows some of the results.

Use your knowledge of base pairing to complete Table Q2.

DNA sample	Percentage of base present			
	Adenine	Cytosine	Guanine	Thymine
Strand 1	26		28	14
Strand 2	14			

Table Q2

AQA June 2006 Unit 2 Question 4

8. Figure Q7 shows a short section of a DNA molecule.

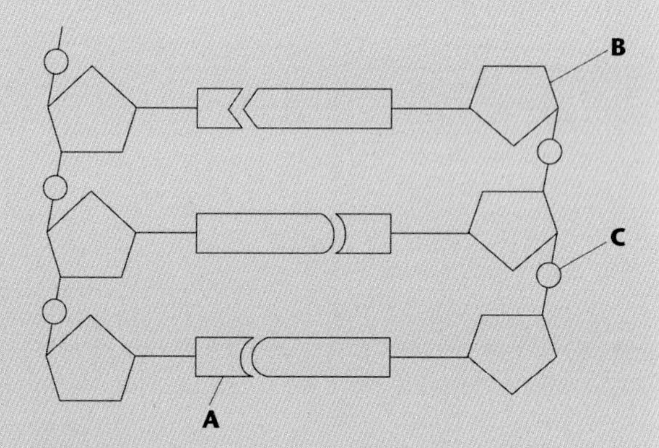

Figure Q7

a. Copy Figure Q7 and draw a box round one nucleotide.

b. **i.** The sequence of bases on one strand of DNA is important for protein synthesis. What is its role?

ii. How are the two strands of the DNA molecule held together?

iii. Give one advantage of DNA molecules having two strands.

AQA January 2005 Unit 2 Question 1

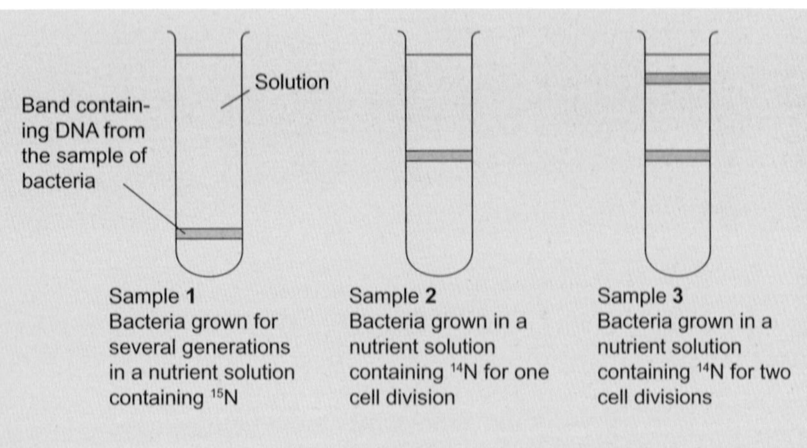

Figure Q8

9. a. DNA helicase is important in DNA replication. Explain why.

   b. Scientists investigating DNA replication grew bacteria for several generations in a nutrient solution containing a heavy form of nitrogen (^{15}N). They obtained DNA from a sample of these bacteria.

The scientists then transferred the bacteria to a nutrient solution containing a light form of nitrogen (^{14}N). The bacteria were allowed to grow and divide twice. After each division, DNA was obtained from a sample of bacteria. The DNA from each sample of bacteria was suspended in a solution in separate tubes. These were spun in a centrifuge at the same speed and for the same time. Figure Q8 shows the scientists' results.

	Type(s) of DNA molecule present in each sample		
Sample	^{15}N ^{15}N	^{15}N ^{14}N	^{14}N ^{14}N
1			
2			
3			

Table Q3

Table Q3 shows the types of DNA molecule that could be present in samples 1 to 3.

Use your knowledge of semi-conservative replication to complete the table with a tick if the DNA molecule is present in the sample.

   c. Cytarabine is a drug used to treat certain cancers. It prevents DNA replication. Figure Q9 shows the structures of cytarabine and the DNA base cytosine.

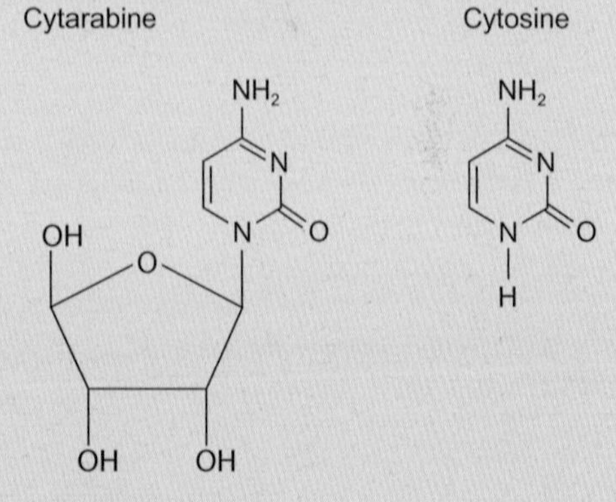

Figure Q9

      i. Use the information in the diagram to suggest how cytarabine prevents DNA replication.

      ii. Cytarabine has a greater effect on cancer cells than on healthy cells. Explain why.

5 CELLS

PRIOR KNOWLEDGE

You have probably learned about the structure of animal and plant cells, and know how to use a simple light microscope. You may remember that, although all cells have particular features in common, cells in multicellular organisms become specialised for different functions, and are organised into tissues, organs and organ systems. You may want to remind yourself about the replication of DNA.

LEARNING OBJECTIVES

In this chapter, you will learn much more detail about the structure and functions of the different components of animal and plant cells, and also about bacteria and viruses, including the ways in which they replicate.

(Specification 3.2.1.1, 3.2.1.2, 3.2.1.3, 3.2.2)

Every living organism is made up of cells. Some, called unicellular organisms, have only one cell. Others – multicellular organisms, such as humans and plants – may have millions and millions of cells. Although different types of cells can look very different, they all have a vast range of features in common with one another, such as the basic structure of their membranes and many of the metabolic reactions that take place inside them. This is indirect evidence that all cells have evolved from a single common ancestral cell.

We think that the earliest cells were probably more like bacterial cells than ours, having no nucleus. Bacterial cells are prokaryotic, meaning 'before a nucleus'. Bacteria not only lack a nucleus, but also lack other membrane-bound organelles. They do not, for example, have either mitochondria or chloroplasts.

In the 1970s, a highly controversial theory was proposed by Dr Lynn Margulis. She suggested that mitochondria may have originated as prokaryotic cells. She thought that, thousands of millions of years ago, a bacterium may have been engulfed by another cell and – instead of being digested – have survived inside it. Gradually, over time, a partnership evolved between these two different organisms, a kind of symbiosis. Symbiosis means 'living together in partnership' and Lynn Margulis's theory became known as the endosymbiotic theory. 'Endo' means 'within'.

Although her theory was initially met with disbelief, today it is widely accepted that this is, indeed, how the mitochondria in the cells of all multicellular organisms arose. Evidence includes the fact that mitochondria reproduce independently of the rest of the cell. They have their own DNA and ribosomes. Their inner membrane is much more like the membrane of a bacterium than the other membranes in the cell. Indeed, it is now also thought that chloroplasts arose in the same way. Today, these mitochondria and chloroplasts are no longer 'cells' in their own right, and are now organelles within a cell.

The partnership between a bacterial cell and another cell would have been very advantageous. Mitochondria provide cells with access to large amounts of energy from glucose, because it is here that the reactions of aerobic respiration take place. Chloroplasts offer the possibility of capturing energy from sunlight and storing it in carbohydrates for future use. Today, all multicellular organisms have cells containing mitochondria. It's odd to think that each of our cells contains many descendants of bacteria that lived long ago.

5.1 CELLS AND LIVING ORGANISMS

Every living organism is made of cells. We think that when life first evolved, perhaps about four billion years ago, it began as a very simple cell – probably just a membrane made of phospholipids, containing a soup of chemicals in which chemical reactions could take place, separated from the outside world. Today, although there are many different kinds of cells in existence, we can find various features that all of them share. This is indirect evidence that all life on Earth today has evolved from those first very simple cells.

Many organisms are made of a single cell (Figure 1), but the organisms with which we are most familiar are multicellular. It is estimated that the number of cells in the human body is 37 trillion (3.7×10^{13}). These are all animal cells, which belong to the type called eukaryotic cells. Plants are also made of eukaryotic cells, but their cells differ from animal cells in several ways – particularly in having a cell wall around them (Figure 2). But much of the teeming life of the microscopic world is made up of a different type of cell that has no nucleus, called prokaryotic cells. It is thought that these were the first kinds of cells to evolve. In fact, as we have seen, we think that eukaryotic cells were originally formed by taking in and 'adopting' prokaryotic cells, which eventually became mitochondria and chloroplasts inside them. So, eukaryotic cells are actually a combination of two very different organisms – a kind of symbiosis.

Cells are very small – even the largest of them is only visible to the human eye as a tiny dot. To study them, we need to use microscopes – both the type that you will use in the laboratory, called light or optical microscopes, and also microscopes that use beams of electrons. Electron microscopes are able to produce images showing much smaller structures than we can see in a light microscope.

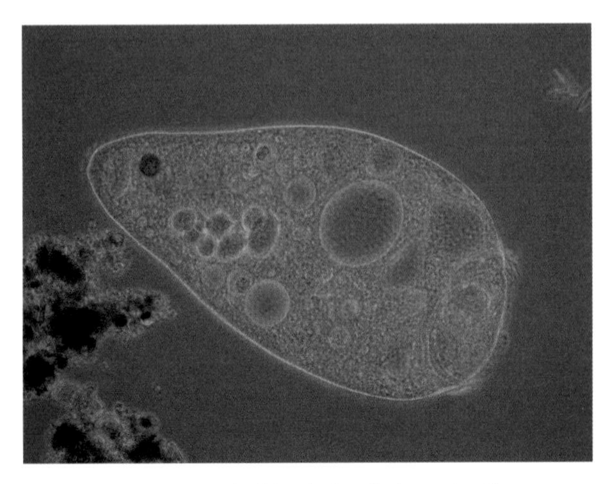

Figure 1 *This photograph of the single-celled organism* Stentor *was taken using a light (optical) microscope. The cell is a eukaryotic cell.*

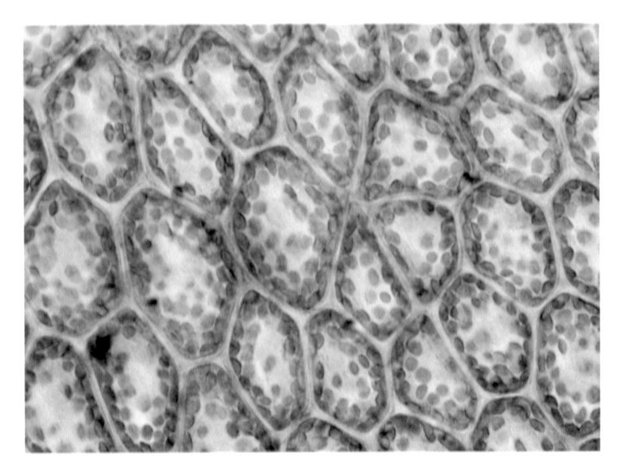

Figure 2 *These are cells from a leaf. Plant cells can be distinguished from animal cells because they are always surrounded by a rigid cell wall, and sometimes contain chloroplasts – the dark green structures in this light micrograph.*

All cells arise from other cells, by cell division. Therefore, all reproduction also involves cell division. One way in which eukaryotic cells can divide is described later in this chapter, and a second method is explained in Chapter 11.

5.2 STRUCTURE OF EUKARYOTIC CELLS

Figure 3 (overleaf) shows the structure of a 'typical' animal cell, showing all of the structures that we can see using an electron microscope. Figure 3 shows part of an animal cell seen using an electron microscope. Images like this are electron micrographs.

Cell-surface membrane

Like all cells, the animal cell in Figure 3 is surrounded by a **cell-surface membrane**, which allows the creation and maintenance of a particular internal environment, different from the external one. The cell membrane has the essential role of separating the contents of the cell from its surroundings, and controlling which substances move in and out. You will find out about how this happens in Chapter 6.

The membrane encloses a semi-solid substance called **cytosol**, coloured pale orange in Figure 3. This jelly-like material is made up of proteins and other substances dissolved in water. Within the cytosol are many small structures called **organelles**. The cytosol plus all the organelles in it, apart from the nucleus, is called the **cytoplasm**.

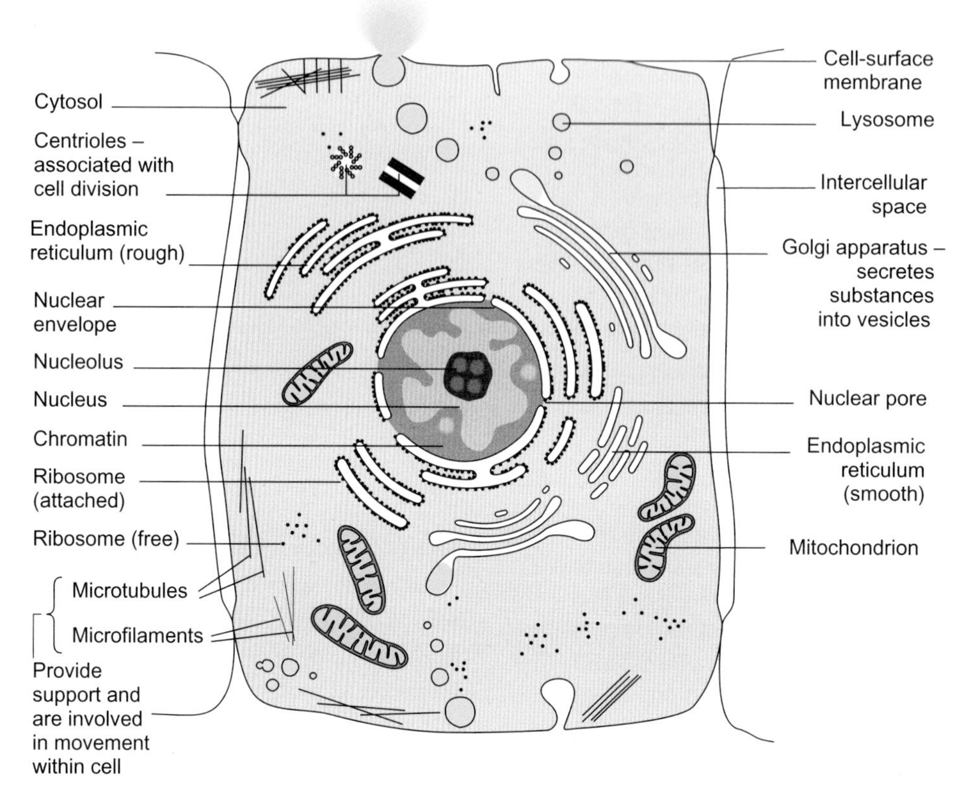

Figure 3 *The structure of a typical animal cell, as seen using an electron microscope*

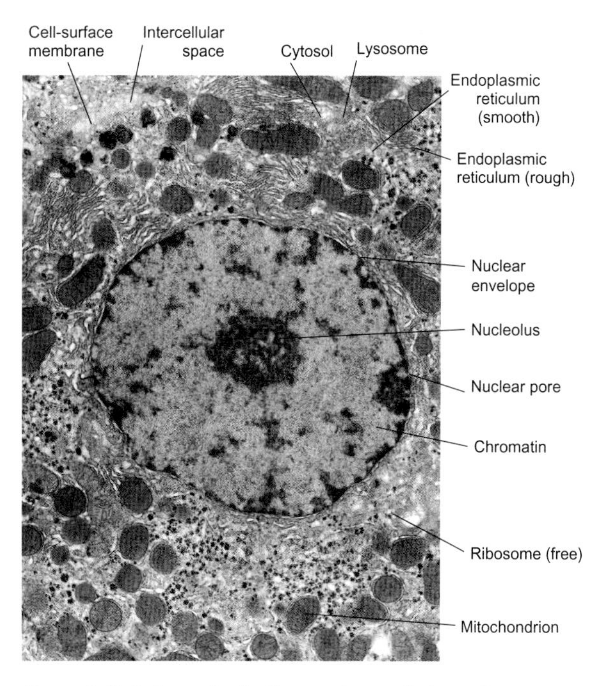

Figure 4 *Part of an animal cell seen using an electron microscope*

Nucleus

In most cells, the largest organelle is the **nucleus** (Figures 3, 5 and 6). It is usually approximately spherical and has a diameter of about 10 μm. Most of the cell's deoxyribonucleic acid (DNA) is in the nucleus. As we have seen, this nucleic acid contains all of the information required to make a new copy of the cell and to control the cell's activities. In a dividing cell, the DNA molecules are condensed to form visible **chromosomes**. At other times the nucleus has a grainy appearance because the chromosomes extend throughout the nucleus as **chromatin**. The chromosomes are made up of linear molecules of DNA, associated with proteins called histones.

Nuclei have one or more **nucleoli** that are visible in electron micrographs as darkly stained spheres. Nucleoli produce the ribonucleic acid (RNA) needed to make ribosomes, the organelles that synthesise proteins using an RNA template. The nucleus is bound by a **nuclear envelope**, which is made up of two membranes. The nuclear envelope contains pores large enough to allow big molecules such as messenger RNA to pass out of the nucleus and into the cytoplasm where they go to the ribosomes to be 'read' to produce proteins (Chapter 10).

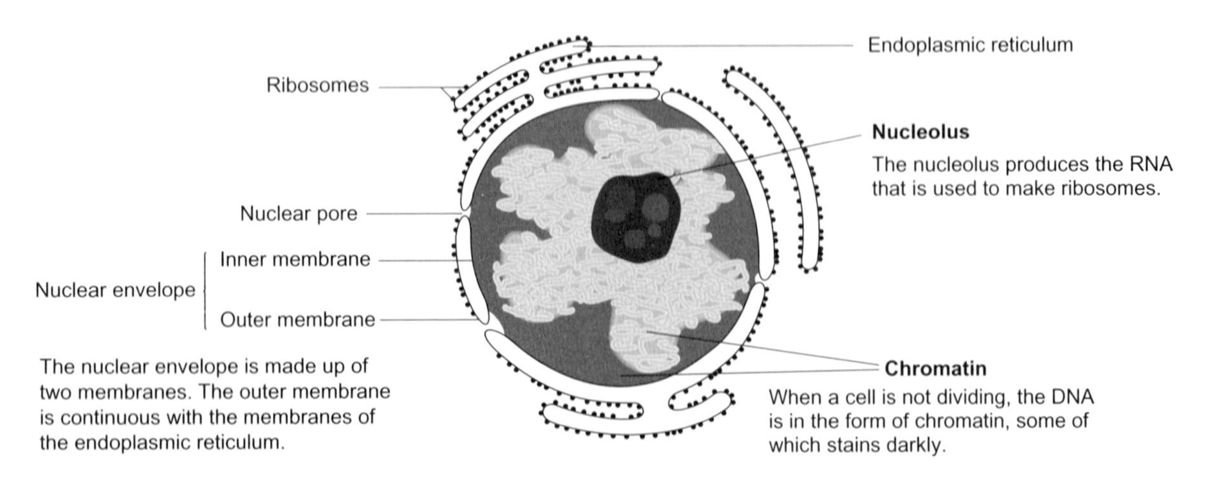

Figure 5 *A nucleus*

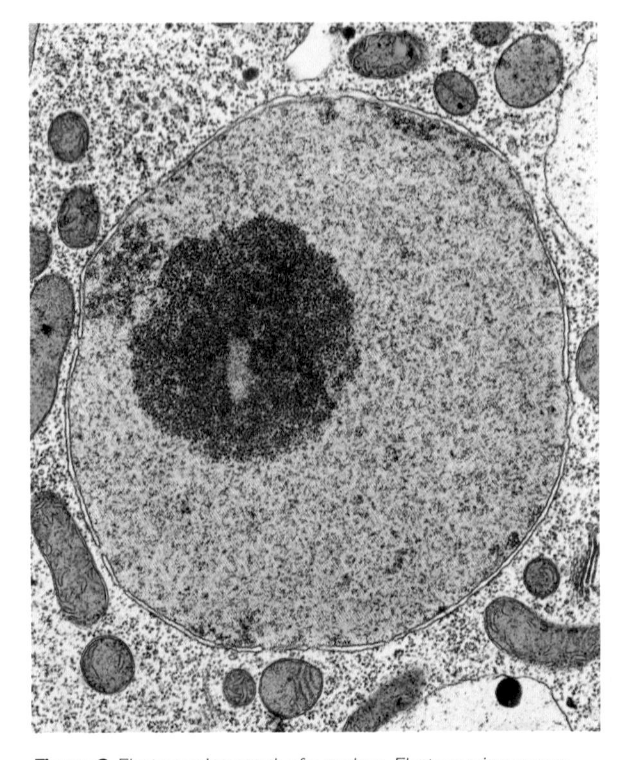

Figure 6 *Electron micrograph of a nucleus. Electron microscopes produce black-and-white images, but this one has had colour added by a computer. The nucleus is shown in light orange, and the nucleolus as a darker brownish area. You can clearly see the two membranes (black lines) making up the nuclear envelope. The green objects are mitochondria.*

Mitochondria

Mitochondria (Figures 7 and 8) are found in almost all living plant and animal cells. (One exception is mature red blood cells, which contain neither mitochondria nor a nucleus.) Mitochondria are the sites of aerobic respiration, the biochemical process that oxidises glucose to release energy and synthesise adenosine triphosphate (ATP).

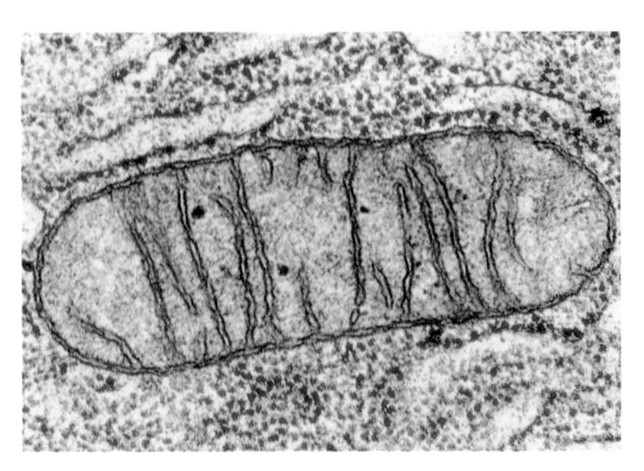

Figure 7 *Electron micrograph of a mitochondrion*

Like nuclei, each mitochondrion is surrounded by an envelope made up of two membranes: an outer membrane that surrounds the entire organelle and a highly folded inner membrane.

The matrix – the central fluid-filled space – contains the free enzymes that catalyse reactions in some of the stages of respiration. The cristae that result from the intricate folds of the inner membrane have a large internal surface area and hold in place many of the enzymes involved in the final stages of aerobic respiration. This is where ATP is synthesised.

Mitochondria contain their own DNA, in the form of a circular molecule. This codes for production of some of the proteins that are required by the mitochondrion, but not all of them.

Endoplasmic reticulum and ribosomes

The **endoplasmic reticulum** (Figures 9 and 10) is a network of membrane-bound channels or lumina (singular: lumen) that run throughout the cytosol of every cell. The channels are called **cisternae**. They

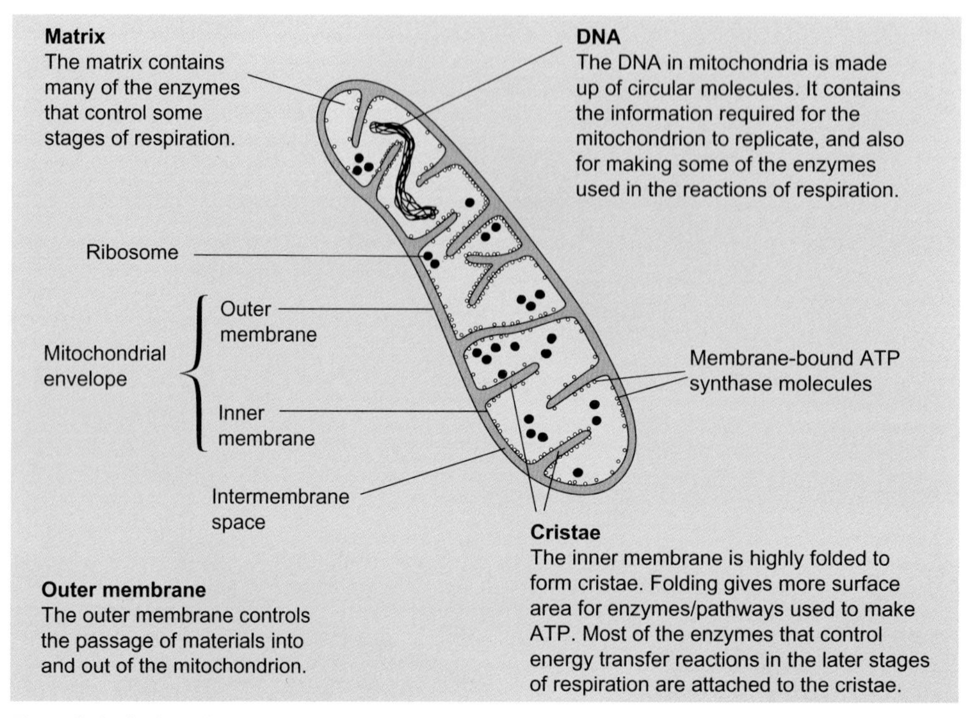

Figure 8 *A mitochondrion*

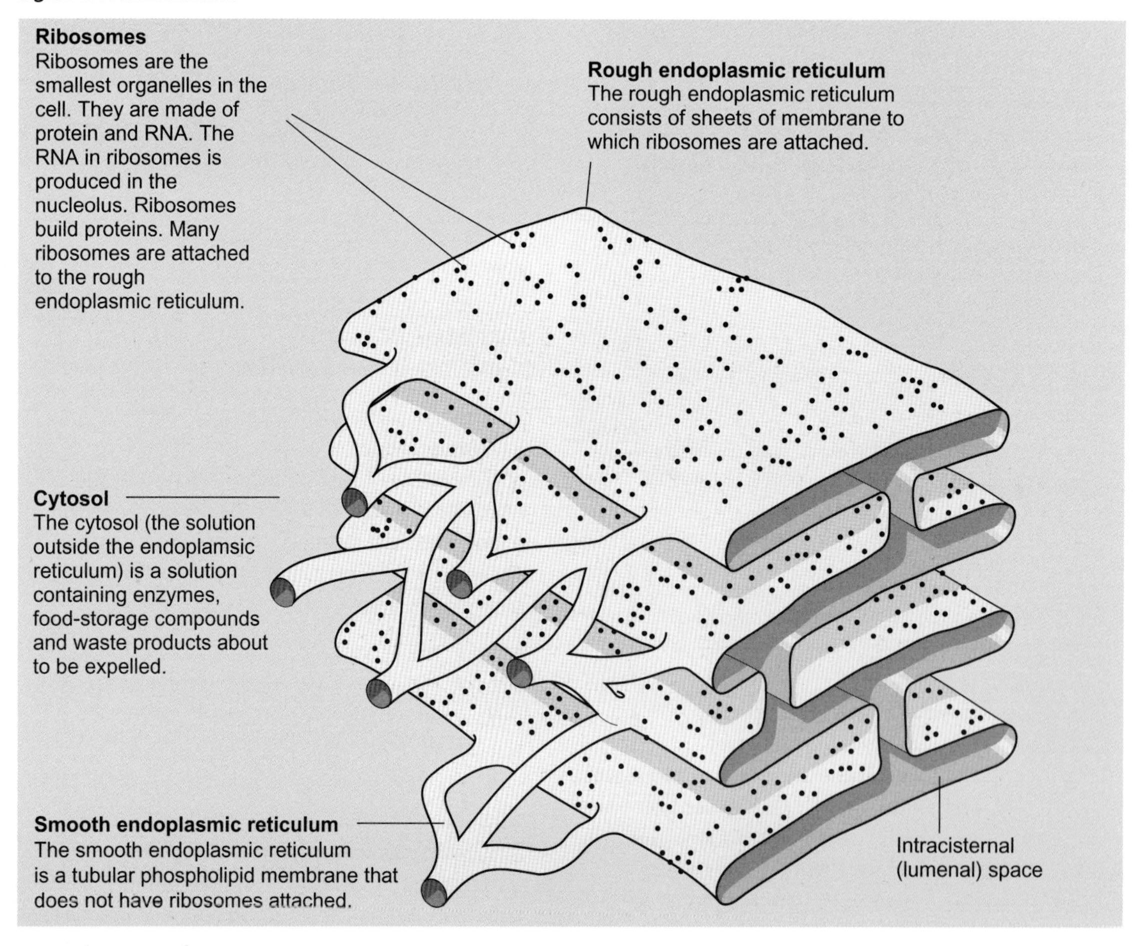

Figure 9 *Endoplasmic reticulum*

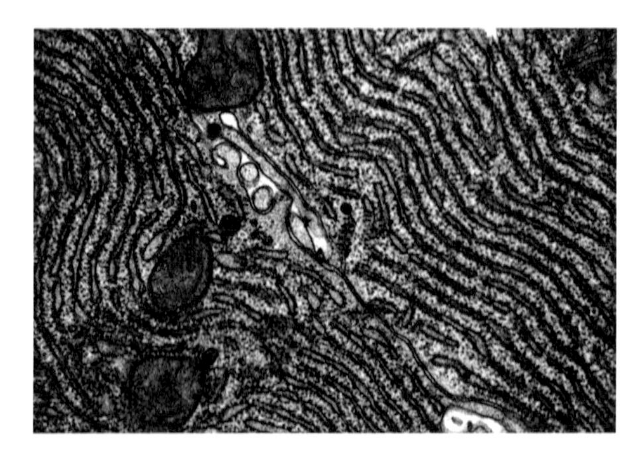

Figure 10 Electron micrograph of endoplasmic reticulum

link up with the nuclear envelope, and also sometimes with the cell-surface membrane.

The membranes of the endoplasmic reticulum often have **ribosomes** on their outer surfaces. This type of endoplasmic reticulum is called **rough endoplasmic reticulum**, or RER for short. Ribosomes are tiny organelles made of RNA and protein, where proteins are synthesised (Chapter 10). Proteins made on the ribosomes move into the lumen between the membranes, from where they can be moved to other parts of the cell.

Other parts of the endoplasmic reticulum have no ribosomes, and are called **smooth endoplasmic reticulum**, or SER. This is where many lipids are synthesised.

Ribosomes are not always attached to endoplasmic reticulum; many are found floating free in the cytoplasm.

Golgi apparatus and lysosomes

The **Golgi apparatus** is a group of flattened membrane-bound cavities (Figures 11 and 12). Its function is to take enzymes and other proteins that have been synthesised in the endoplasmic reticulum (ER) and package them into membrane-bound vesicles. The proteins are also often processed in some way, for example by adding short carbohydrate chains to them to form glycoproteins.

The appearance of the Golgi apparatus is constantly changing as material comes in on one side from the ER and is lost from the other as completed vesicles 'bud off'. Such vesicles transport materials to other parts of the cell, or fuse with the cell-surface membrane, releasing their contents outside of the cell. This process is called secretion. Cells that secrete a lot of enzymes, such as the cells in the pancreas that make pancreatic juice, contain large amounts of RER and Golgi apparatus.

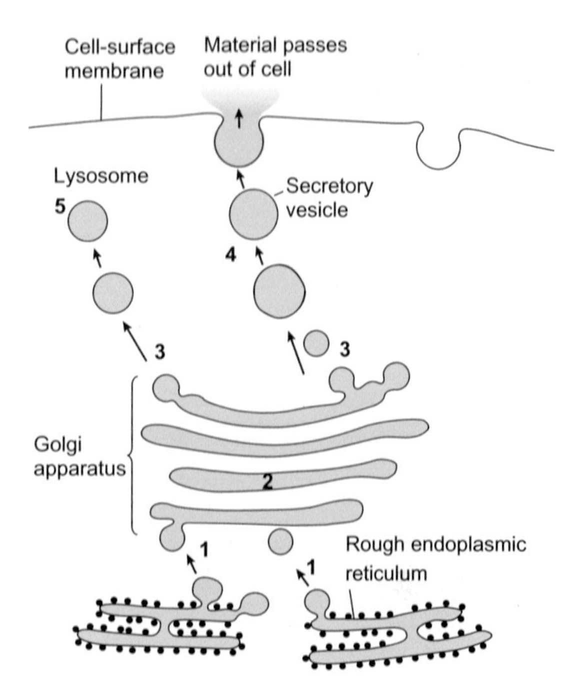

1 Vesicles containing proteins 'bud off' from the RER and fuse with the convex face of the Golgi apparatus.
2 Inside the Golgi, the proteins are modified ready for use elsewhere in the cell, or for export.
3 Vesicles containing the modified proteins 'bud off' from the concave face of the Golgi apparatus.
4 Some of these vesicles move to the cell-surface membrane, where they fuse with it and empty their contents outside of the cell.
5 Some of the vesicles are lysosomes, containing hydrolase enzymes that can hydrolyse material inside the cell.

Figure 11 The Golgi apparatus

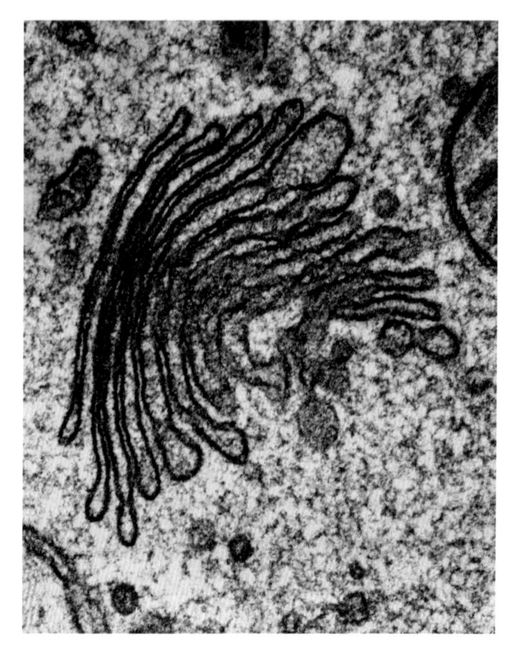

Figure 12 Electron micrograph of Golgi apparatus

Lysosomes are vesicles produced by the Golgi apparatus that contain digestive enzymes called **lysozymes**. These enzymes can destroy, by hydrolysis, old or surplus organelles inside the cell or they can be used to hydrolyse material that has been taken into the cell by the process of endocytosis. Whole cells and tissues that are no longer required can be destroyed if cells nearby allow lysosomes to release their contents at the cell surface. The body uses this process to break down excess muscle in the uterus after birth, and to destroy milk-producing tissue in the breasts after a baby has been weaned.

QUESTIONS

1. Suggest why it is important that lysozymes are produced within vesicles and not released into the cytosol (unless cell death is triggered).

Stretch and challenge

2. There are more than 50 different types of hydrolytic enzymes – but this is a relatively small number compared with the huge diversity of molecules that they digest. Suggest how this relatively small number of available enzymes manages to process such a large array of biomolecules.

Chloroplasts

Chloroplasts are found only in some plant and algal cells, never in animal cells. They are the organelles in which photosynthesis takes place.

Figure 13 shows the structure of a chloroplast. Like a mitochondrion, a chloroplast is surrounded by two membranes, forming an **envelope**. These membranes control what can enter and leave the chloroplast.

The chloroplast membrane encloses a mixture of water, enzymes and other substances, called the **stroma**. Within the stroma is an intricate series of membranes that form interconnecting enclosed spaces called **thylakoids**. In places, these membranes form stacks called **grana**. The membranes have molecules of chlorophyll and other pigments embedded in them.

The first stage of the reactions of photosynthesis, in which light energy is harvested by chlorophyll, takes place in the grana. The arrangement of the thylakoids into grana, and the distribution of the grana, maximises the absorption of light energy. The second stage, in which this energy is used to convert carbon dioxide to carbohydrates, takes place in the stroma.

Like mitochondria, chloroplasts contain their own DNA, in the form of circular molecules, and their own ribosomes (that are smaller than the ones in the cytosol and same size as those in prokaryotic cells). They also contain starch granules (that can sometimes

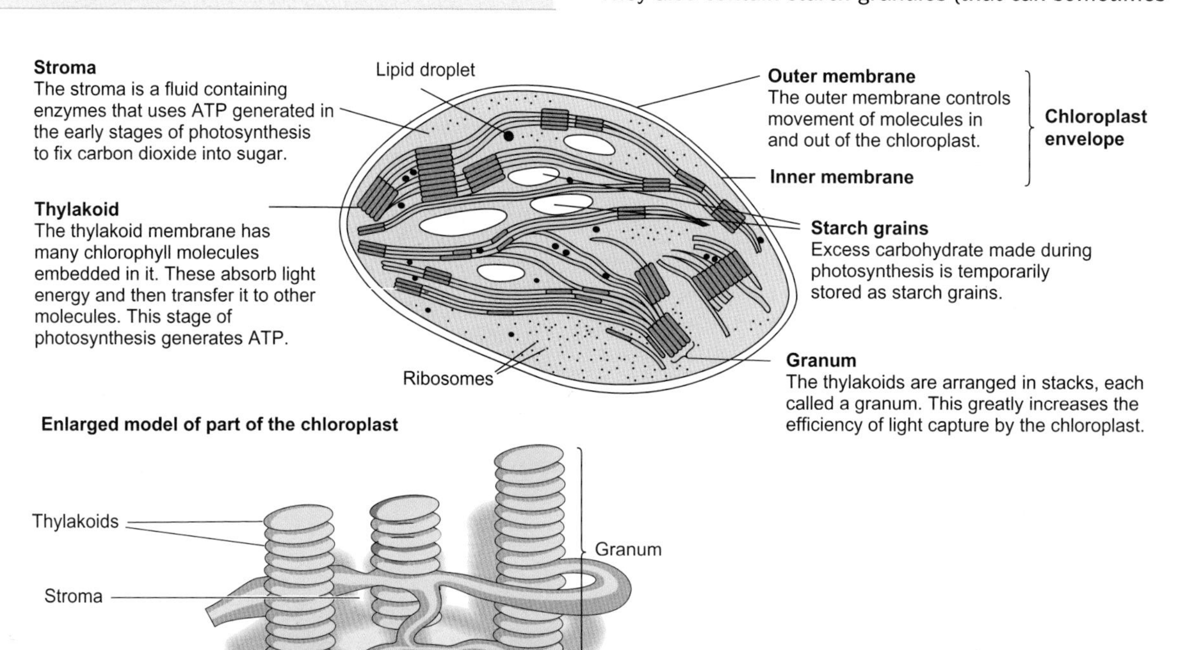

Figure 13 *The structure of a chloroplast*

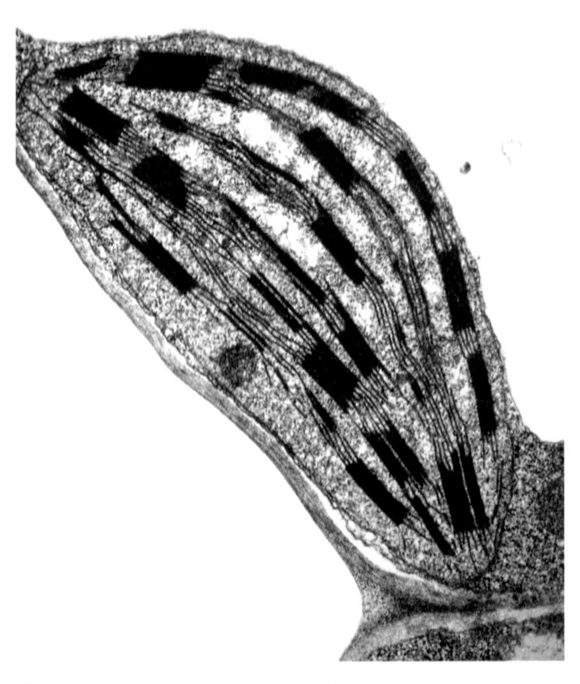

Figure 14 Electron micrograph of a chloroplast

take up a lot of the space inside the chloroplast), which are stores for the carbohydrate that is made inside them. They may also contain droplets of lipid.

QUESTIONS

3. There are three types of organelle in plant cells that are surrounded by an envelope and contain DNA. Name these three organelles.

Stretch and challenge

4. Describe the similarities between mitochondria and chloroplasts, and suggest reasons for these similarities.

Cell wall

Cell walls are structures found on the outside of the cells of plants, algae and fungi. Animal cells never have cell walls.

In algae and plants, the cell walls are made of **cellulose**. As we have seen, cellulose molecules are polysaccharides that lie side by side and link together by hydrogen bonds to form bundles of molecules called microfibrils (Figure 15). In a cell wall, these are embedded in a background material (matrix) made of **pectin**. The cell wall is built up in layers,

Figure 15 Electron micrograph of cellulose microfibrils in the cell wall of an alga. You can see how the microfibrils lie in different directions in different layers. The microfibrils are embedded in, and linked to, a matrix of pectin (not visible here), producing a very strong composite structure.

with the orientation of the cellulose microfibrils being different in each layer. This provides a material that is immensely strong for its weight.

Unlike the cell-surface membrane, the cell wall is fully permeable – that is, it does not control what passes through it, and almost any kind of molecule or ion can easily move through. The main function of the cell wall is to provide support to the cell. For example, when a plant cell absorbs water and its cytoplasm increases in volume, the expanding cell contents push outwards on the cell wall, which resists the expansion. Instead of bursting, the cell simply becomes firm and strong. The cell is said to be **turgid**. It is the turgidity of its cells that helps to hold a leaf out flat; if the cells lose water, they lose their turgidity and the leaf wilts.

Where plant cells lie side by side, their cell walls are physically linked together through a layer of pectin called the middle lamella (Figure 16). There are usually gaps in the wall, lined with cell-surface membrane, through which endoplasmic reticulum from

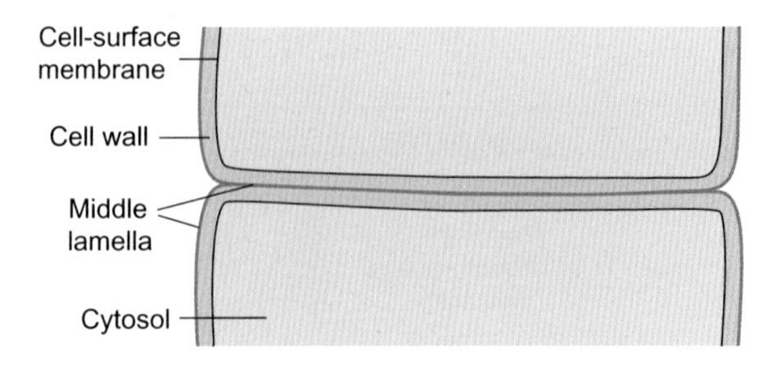

Figure 16 The cell walls of adjacent plant cells are joined via the middle lamella, which is made up of pectin.

one cell connects directly with that of the neighbouring cell. These are called **plasmodesmata**, and they are essential for transferring material between cells in a multicellular plant.

Fungi also have cell walls, but they have a very different structure. They do not contain cellulose. The cell walls of most fungi are made up of fibres formed from a polymer called **chitin** (which is also a major component of the exoskeletons of insects). The monomers of chitin are molecules called acetylglucosamine, which are like monosaccharides with amine groups attached to them. So chitin, unlike cellulose, contains nitrogen.

Cell vacuole

A **vacuole** is a liquid-filled space inside a cell, surrounded by a membrane. All cells have vacuoles. Many of these are very small, and are called **vesicles**. Plant cells, however, often have a very large, central vacuole, which can take up a large proportion of the space inside the cell (Figures 17 and 18). The vacuole contains many different substances in solution, such as sucrose and amino acids. It can be used for storage, to isolate substances that may be harmful to the rest of the cell, and to help to maintain cell turgidity.

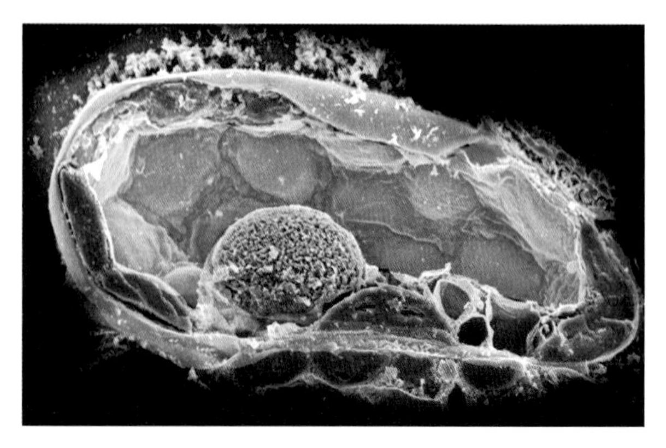

Figure 18 *A scanning electron micrograph of a plant cell. Scanning electron micrographs produce three-dimensional black-and-white images; this one has been coloured by computer. You can see the cell wall around the outside. The membrane is too thin to be visible, but would be pressed tight up against the inner surface of the wall. The large brown sphere is the nucleus, and the dark green objects are chloroplasts. The space inside the cell is largely taken up by the vacuole.*

Specialised cells

The structures that we have described so far in this chapter are effectively a 'toolbox' that individual cells can make use of to help them to carry out their functions. The absolutely essential ones that no cell can do without are the cell-surface membrane and

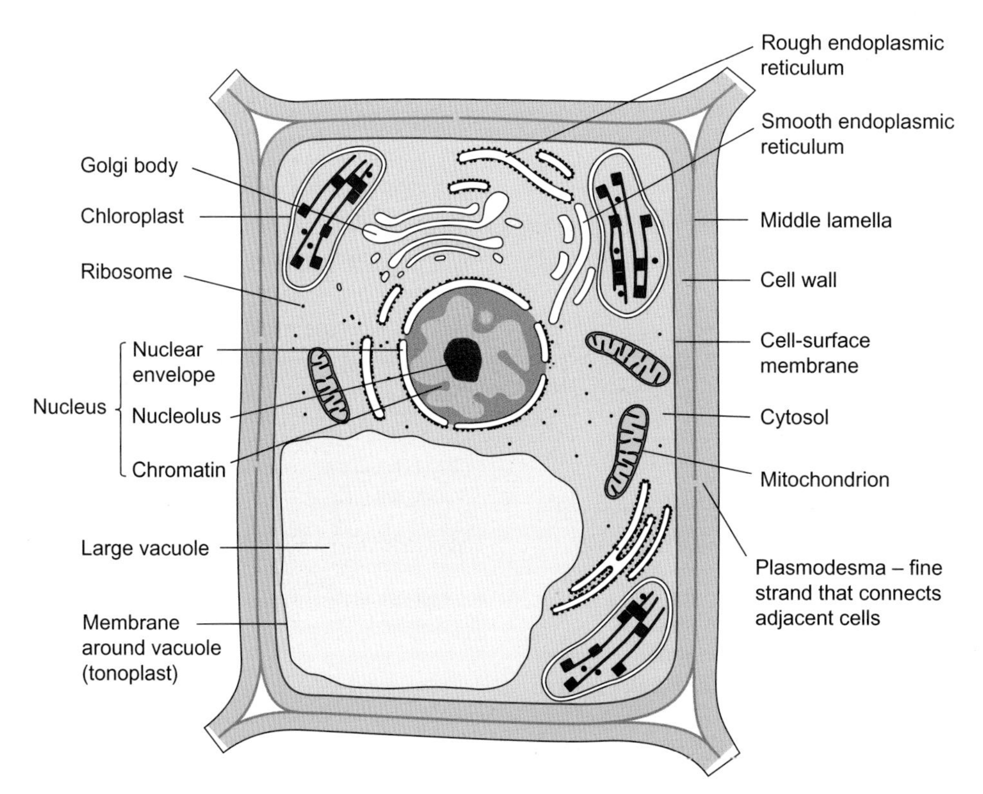

Figure 17 *A typical plant cell, showing all of the structures that can be seen using an electron microscope*

the cytosol. Even a nucleus – that you may imagine would be equally essential – is dispensed with in some cells, such as red blood cells in animals and phloem sieve elements in plants (although dispensing with the nucleus does limit the life expectancy of some cells).

Most cells in a multicellular organism such as yourself or a plant are specialised to carry out particular functions. This happens early in the development of an embryo. Multicellular organisms often begin their lives as a single cell called a zygote, formed from the fusion of a male gamete and a female gamete. This single cell divides repeatedly to form a ball of cells. These cells are **stem cells** – cells that can divide and give rise to more specialised cells. As the embryo grows, by producing more and more cells, the number and variety of specialised cells increases. By the time that the organism reaches adulthood, most cells are specialised for one or other particular functions, and are unable to divide again.

As you continue your biology course, you will meet numerous examples of specialised cells, and you will repeatedly be asked to think about how the structure of the cell is adapted for its particular function. Figures 19 and 20 show two examples of specialised cells in an animal and in a plant. Figure 19 shows cells in the pancreas whose function is to manufacture and secrete digestive enzymes. They have many mitochondria as one of their (evolved) adaptations is to produce the ATP needed to provide energy for this

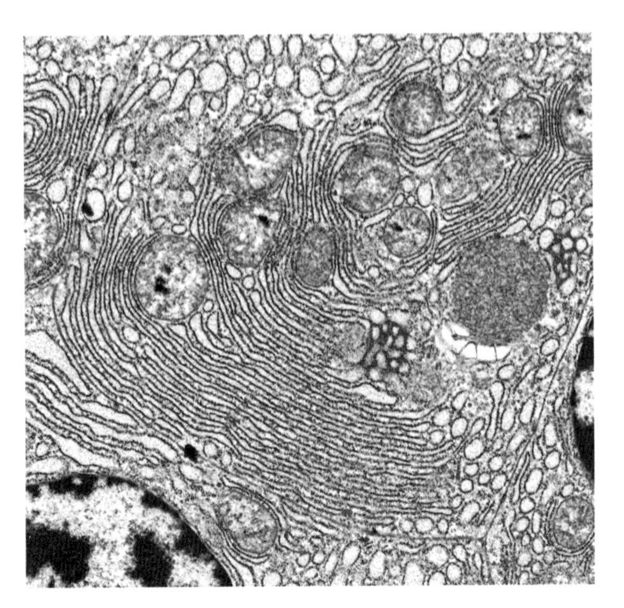

Figure 19 *Enzyme-secreting cells from the pancreas, containing many mitochondria (blue) and extensive rough endoplasmic reticulum (dark orange). You can also see part of the nucleus of one cell at bottom left.*

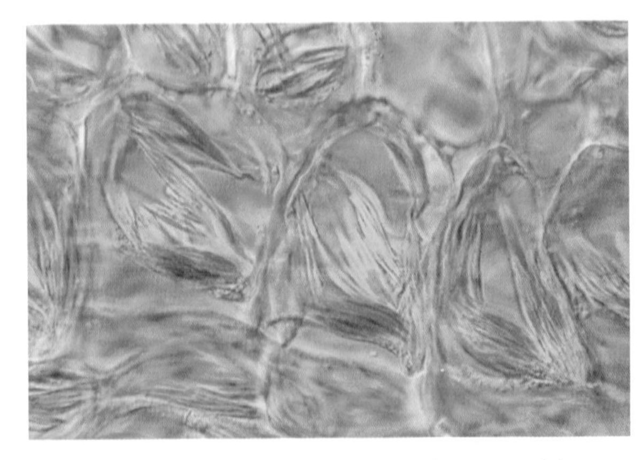

Figure 20 *Cells from the petals of a* Strelitzia *flower, containing chromoplasts whose membranes contain orange pigments*

process, and a lot of rough endoplasmic reticulum on which the enzymes can be synthesised. Figure 20 shows cells from a flower petal; instead of chlorophyll-containing chloroplasts, they have chromoplasts containing coloured pigments that make the petals attractive to insects.

Tissues, organs and organ systems

In a multicellular organism, cells with similar functions are often grouped together to form a **tissue**. For example, the cells lining your mouth are very flat, thin, cells that form squamous epithelial tissue. Your brain is made up of cells specialised for transmitting electrical impulses, forming nervous tissue, and your muscles are made up of cells specialised for contraction, called striated muscle tissue. In plants, leaves contain cells specialised for photosynthesis, forming palisade mesophyll tissue, while stems, leaves and roots contain cells specialised for transporting water, forming xylem tissue.

Tissues are, in turn, grouped to produce larger structures called **organs**. An organ – for example, an eye or a leaf – is a part of an organism that carries out a particular role. On an even larger scale, we can identify **organ systems** made up of many different organs. For example, the digestive system contains the various organs of the alimentary canal (such as the mouth and stomach) and the organs that secrete fluids into the canal to help with digestion (such as the liver and pancreas).

KEY IDEAS

- Eukaryotic cells have nuclei, which are bounded by two membranes (an envelope) and contain DNA in the form of linear molecules associated with proteins.
- The nucleus contains one or more nucleoli, where ribosomes are synthesised.
- Mitochondria are bounded by an envelope of two membranes. The inner membrane is folded to form cristae. The reactions of aerobic respiration, which synthesise ATP, take place inside mitochondria.
- Chloroplasts, like mitochondria and nuclei, are surrounded by an envelope. They are found only in some plant cells. The reactions of photosynthesis take place inside chloroplasts.
- All eukaryotic cells contain rough and smooth endoplasmic reticulum. RER has ribosomes on its outer surface, where proteins are synthesised. SER synthesises lipids.
- The Golgi apparatus is a series of membrane-bound compartments arranged into a stack. It is formed by the coalescence of protein-containing vesicles derived from the RER, and it breaks up into vesicles containing processed protein molecules. These may be used within the cell, or exported (secreted) from it.
- Lysosomes are a type of Golgi vesicle containing hydrolase enzymes called lysozymes.
- The cells of plants, fungi and algae are surrounded by a fully permeable cell wall. In plants and algae this is made of cellulose; the cell walls of fungi often contain chitin. Cell walls provide support.
- Plant cells often contain large vacuoles, surrounded by a membrane, which are filled with a liquid containing sugars and other substances dissolved in water.
- In multicellular organisms, such as animals and plants, cells tend to be specialised for particular functions. Groups of cells with shared functions associate to form tissues. Organs are formed from several different types of tissues, and organ systems are made up of several organs, all working together to perform a function.

5.3 PROKARYOTIC CELLS AND VIRUSES

Prokaryotic cells

'Pro' means 'before', and 'karyotic' means 'with a nucleus'. Prokaryotic cells have no nucleus, and it is thought that this type of cell arose many millions of years before eukaryotic cells originated. Organisms with prokaryotic cells belong to two groups called domains: the Archaea and the Bacteria. Unlike eukaryotic cells, these never form complex multicellular organisms made up of a variety of specialised cells. Most prokaryotic organisms are unicellular (single-celled), although some do form long strands of several cells joined end to end. Prokaryotic cells are much smaller than eukaryotic cells, and it is difficult to see them at all with a light microscope.

Figure 21 shows a typical prokaryotic cell. You can see that, like the animal and plant cells described earlier in this chapter, it is surrounded by a cell-surface membrane and contains cytosol. Another similarity with eukaryotic cells is the presence of ribosomes, where protein synthesis takes place. These ribosomes, however, are smaller than the ones found in the cytosol of eukaryotic cells. Intriguingly, they are much more like the ribosomes found inside mitochondria and chloroplasts, which is evidence for the endosymbiont theory of how eukaryotic cells arose.

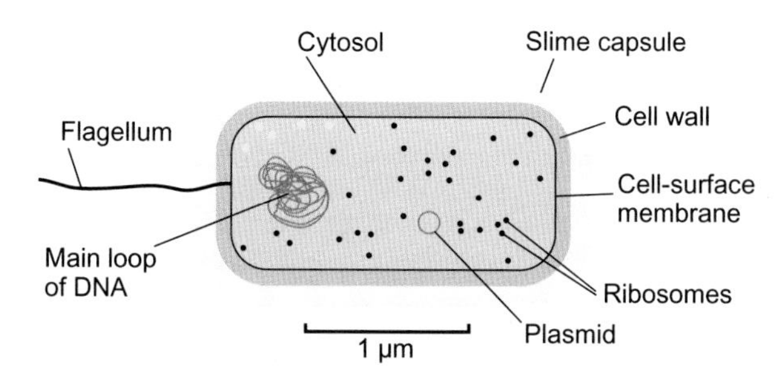

Figure 21 *A bacterium, a single-celled prokaryotic organism*

We have seen that prokaryotic cells do not have a nucleus, but they do have DNA. Unlike the DNA of eukaryotic cells, which is found in the form of linear molecules, the DNA in a prokaryotic cell is a single circular DNA molecule. Another difference between the DNA of prokaryotic and eukaryotic cells is that prokaryotic DNA is not associated with protein. In eukaryotic chromosomes, the DNA is wound around protein molecules, called histones, to form chromosomes. Bacteria do not, therefore, actually have 'real' chromosomes, although the term 'bacterial chromosome' has become quite widely used to refer to their single, circular molecule of DNA.

Many (but not all) prokaryotes also have one or more extra, much smaller, DNA molecules in their cells. These are also circular, and they are known as **plasmids**. Plasmids are relatively easily exchanged between bacteria, and they are often used by genetic engineers as a way of getting genes from one cell into another.

Prokaryotic cells contain no membrane-bound organelles such as mitochondria, lysosomes, Golgi apparatus or endoplasmic reticulum.

You can see in Figure 21 that a cell wall surrounds the cell, as in plant cells and fungal cells. This cell wall, however, is not made of cellulose or chitin, but from a substance called **murein**. Murein (you may also come across its alternative name, peptidoglycan) is a polymer containing both amino acids and sugars, and belongs to a class of substances called glycoproteins.

Some prokaryotic cells have an extra layer, outside their cell wall, called a **capsule**. This is made of polysaccharide, and it provides extra protection, for example against attack by the cells of our immune system. Not surprisingly, capsules are likely to be found around bacteria that are pathogens (in other words, that can cause disease), and are less common in free-living bacteria.

Many prokaryotic cells are not capable of movement, but some have one or more flagella. These are long, whip-like structures made of protein. They are anchored firmly to the cell membrane, where a rotary molecular 'motor' causes the flagellum to move in a twisting motion, propelling the cell along. This is quite different from the flagella found in some eukaryotic cells, which have a very different structure.

QUESTIONS

5. Construct a table comparing the structures of eukaryotic and prokaryotic cells.

Viruses

Viruses are tiny particles of nucleic acid – either DNA or RNA – enclosed in a protein coat called a **capsid** (Figure 22). (Do not confuse a capsid with the capsule found around some bacteria.) They are not made of cells, and so are said to be acellular. (The prefix 'a' means 'not' or 'without'.) According to **Cell Theory**, all life exists as cells – so a virus is not a living organism. They are unable to carry out any of the processes that are characteristic of living things, such as respiration, growth or excretion.

Having invaded a living cell, viruses can be replicated by that cell. First, protein molecules on the capsid

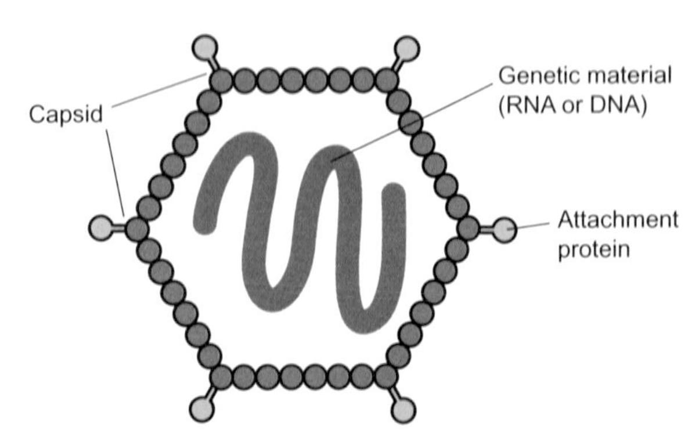

Figure 22 *The structure of a virus particle*

attach to the cell-surface membrane of the cell. The virus then enters the cell, and its genetic material hijacks the machinery of the cell. The cell copies the virus's RNA or DNA, makes new protein coats and assembles new viruses. These eventually burst out of the cell, destroying it.

Viruses are extremely small, only visible using an electron microscope. You could fit a large number of viruses inside one bacterial cell. Indeed, many kinds of viruses do invade bacterial cells; even bacteria can get viral infections.

KEY IDEAS

- Prokaryotic cells are much smaller than eukaryotic cells.
- Prokaryotic cells have no nucleus or any membrane-bound organelles.
- The DNA of prokaryotic cells is in the form of a circular molecule, not associated with proteins. Many prokaryotic cells also contain one or more extra, smaller, circular molecules of DNA, called plasmids.
- Prokaryotic cells have smaller ribosomes than eukaryotic cells.
- Prokaryotic cells have a cell wall made of murein.
- Some prokaryotic cells also have a capsule and/or flagella.
- Viruses are acellular and non-living. They are only replicated inside cells that they have invaded.
- Viruses contain genetic material (RNA or DNA) surrounded by a protein coat called a capsid. Protein molecules on their surfaces help them to attach to and invade living cells.

5.4 METHODS OF STUDYING CELLS

You now know that cells are very small. This makes them difficult to study. We can learn a lot by using microscopes. Another method is to separate the different components of a cell and then study the properties of these components.

Because cells are so small, we have to use units of measurement that are not often used in everyday life.

- The **micrometre** (μm), one thousandth of a millimetre (10^{-3} mm or 10^{-6} m).
- The **nanometre** (nm), one thousandth of a micrometre (10^{-9} m).

Optical microscopes

You have probably used a microscope in the laboratory. The kind of microscope you used would have been one that uses rays of light to pass through the specimen and produce an image on the retina of your eye. This is a light microscope, or an **optical microscope**.

Figure 23 shows a photograph taken using an optical microscope, called a light micrograph. This shows a section across a leaf from a tea plant, *Camellia sinensis*. To make the different tissues show up clearly, stains have been used. The walls of the xylem vessels have taken up a red stain, and many of the other types of cells are stained blue.

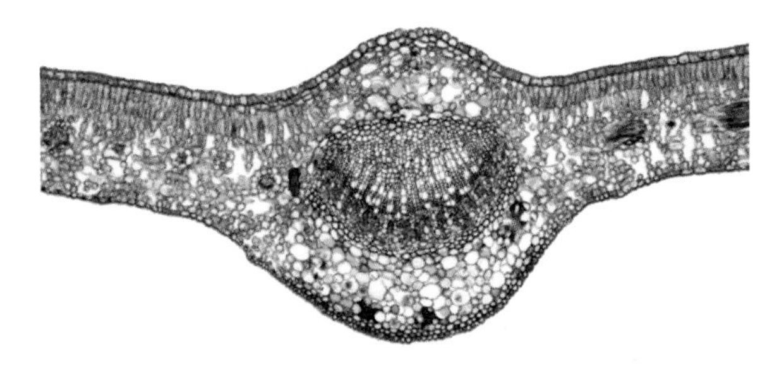

Figure 23 *A light micrograph of a section through a leaf of* Camellia sinensis

Optical microscopes are invaluable tools for the biologist. They are relatively inexpensive and easy to maintain. It is quite easy to prepare specimens to view in them; indeed, we can even view living organisms or living cells through them. The microscopes you will use in the laboratory are compound microscopes – the image is formed by two sets of lenses, called the objective lens and the eyepiece lens (Figure 24). There is also a third lens, the condenser lens, which directs light through the specimen.

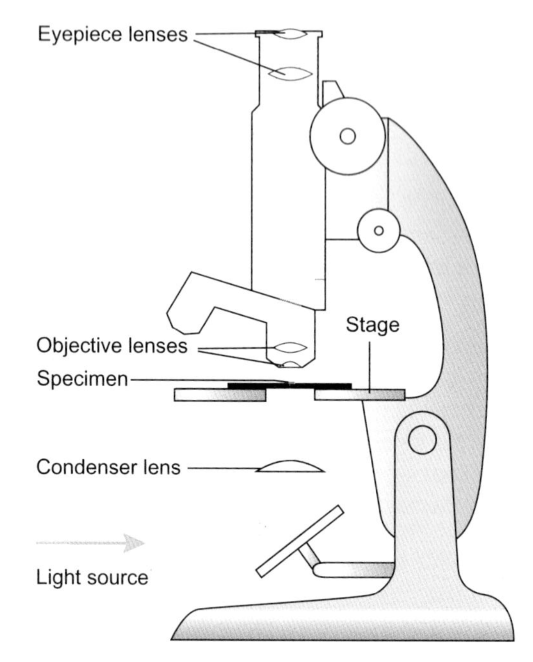

Figure 24 *A compound optical microscope*

Resolving power

The big drawback of an optical microscope is that there is a limit to the smallest things that you can see. You can see only the larger organelles in a cell with an optical microscope because even the larger organelles are near the limit of a light microscope's **resolving power**. Resolving power should not be confused with magnification. Resolving power is the ability to distinguish between two objects. Think of a car at night. When it is far away, the two headlights appear as a single light. As the car gets nearer you can see two headlights. At some point your eyes are able to resolve two lights, not one. With binoculars you could resolve the light into two headlights when the car is much further away.

The limitations of the light microscope are not due to the construction of the equipment, but to the nature of light itself. Even the most powerful lens cannot resolve two dots that are separated by less than 250 nm. This is because no lens system can ever resolve two dots that are closer together than half the wavelength of the light used to view them. The wavelength of visible light is 500–650 nm.

Magnification

Magnification is the number of times larger an image is than the original object. For example, if a beetle is 16 mm long and you make a drawing of it that is 32 mm long, your drawing has a magnification of × 2. If your drawing is 8 mm long, then your drawing has a magnification of × 0.5.

$$\text{magnification} = \frac{\text{size of image}}{\text{real size of object}}$$

Since the lenses in a light microscope will not allow you to distinguish between two objects that are smaller than 200–250 nm (0.20–0.25 µm), the maximum useful magnification of a light microscope is approximately × 1500. This is adequate to see animal cells (diameter 30–50 µm) and individual bacteria (length 5–8 µm). It is also sufficient to detect the changes in a cell that may be the beginning of cancer and to view larger organelles such as the nucleus (diameter 10 µm). But it is not powerful enough to resolve the structure of small organelles such as ribosomes (diameter 20 nm) or cell membranes (thickness 7–10 nm).

The eyepiece lens and the objective lens combine to produce a greater magnification than would be possible with only a single lens. The total magnification is calculated by multiplying the magnifications of the two lenses.

For example, if the eyepiece lens has a magnification of × 10 and the objective lens has a magnification of × 50, the total magnification of the microscope is × 500.

QUESTIONS

6. Why are there no optical microscopes with a magnification of × 2000?
7. Explain the difference between magnification and resolution.
8. If a microscopist uses a light microscope with an eyepiece of ×10 and a set of objective lenses of × 4, ×10 and × 40, what is the total magnification they can view using each objective lens?

Stretch and challenge

9. Research laboratories often have sophisticated types of optical microscope that are able to provide different or higher-quality images than the standard school or college microscope. One example is a **confocal microscope**. Although conceived in the 1950s, it has only recently become widely used. It can image live cells, in 3D. Use the Internet to find out about these microscopes. Explain how they work, and give some examples of how they have opened up new ways for cellular biologists to obtain information about cell structure and activity.

Worked maths example: Calculating size and magnification

(MS 0.1, MS 0.2, MS 1.2, MS 1.8, MS 2.2, MS 2.3, MS 2.5, MS 4.1)

Most biological illustrations have a scale shown – for example, × 60. This means that the object in the picture has been magnified 60 times. In exams you may be asked either to find the actual size of a specimen, or to calculate how many times it has been magnified.

The formula for calculating magnification is:

$$\text{magnification} = \frac{\text{image size}}{\text{actual size}}$$

You need to be able to rearrange this formula – for example, to calculate the real size of an object:

$$\text{actual size} = \frac{\text{image size}}{\text{magnification}}$$

If you find rearranging a formula difficult, you may like to use a formula triangle to help you:

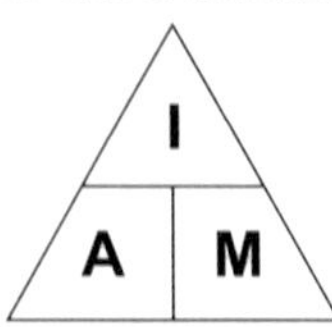

where,

I = image size
A = actual size
M = magnification.

Part a

This photo shows a human egg covered with sperm.

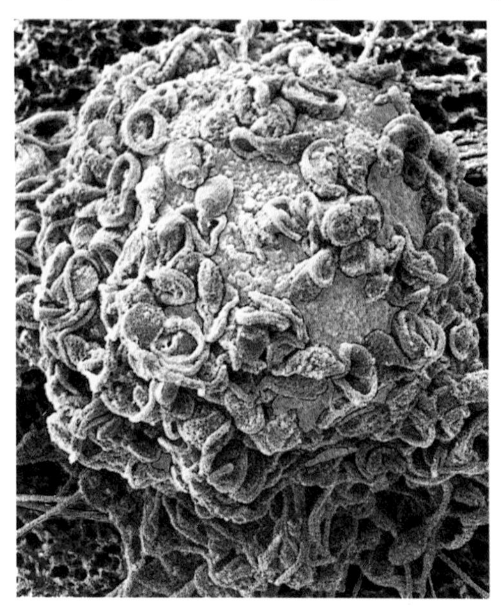

In the photo, the egg has a diameter of 65.6 mm and a single sperm head has a diameter of 5 mm. The actual diameter of the egg is 0.1 mm.

You have been asked to calculate:

- the magnification
- the actual volume of the egg and a single sperm head in cubic micrometres (giving your answers in standard form, and assuming that both are spherical)
- how many sperm heads would fit inside the egg.

First, to work out the magnification we can use the two values given for the egg's diameter. **Note:** It's important to make sure that the values are both the same unit.

$$\text{magnification} = \frac{\text{image size}}{\text{actual size}} = \frac{65.6}{0.1} = 656$$

So, the image has a magnification of × 656 (it has been magnified to 656 times its actual size).

Next, we need to use the formula for the volume of a sphere:

$$\text{volume of sphere} = \frac{4}{3}\pi r^3$$

where r = radius and π = 3.14

Note: When carrying out magnification calculations, it is often useful to convert all measurements to micrometres first. So, if you measure the size of an image in mm, you will need to multiply this measurement by 10^3 (1000) to change your measurement to micrometres.

For the egg:

$r \text{ (actual)} = \frac{0.1}{2} = 0.05 \text{ mm} = 50 \text{ µm}$

$\text{volume} = \frac{4}{3}\pi 50^3 = 523\,333 = 5.23 \times 10^5 \text{ µm}^3$

For the sperm head:

$r \text{ (image)} = \frac{5}{2} = 2.5 \text{ mm}$

$r \text{ (actual)} = \frac{2.5}{656} = 0.0038 \text{ mm} = 3.81 \text{ µm}$

$\text{volume} = \frac{4}{3}\pi 3.81^3 = 231 \text{ µm}^3$

So, to work out how many sperm heads would fit inside an egg, we simply divide the two volumes:

$$\frac{(5.3 \times 10^5)}{231} = 2265.5$$

Part b

This photo shows bacteria magnified 9240 times.

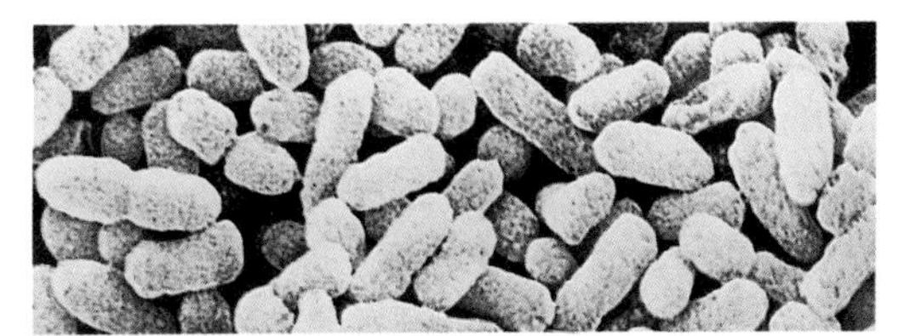

You are asked to calculate the actual length of one bacterium in micrometres.

Using a ruler, we can measure some of the bacteria in the photo and calculate an approximate mean length. We do this by adding up all the values and dividing by the number of bacteria we measured, giving an approximate mean length of 10 mm.

$$\text{actual mean length} = \frac{10}{9240} = 0.00108 \text{ mm} = 1.08 \text{ µm}$$

Part c

This photo shows some pollen grains on the stigma of a flower.

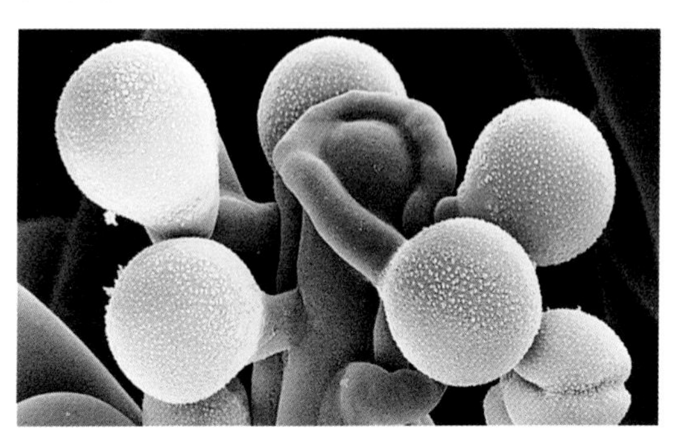

It has been magnified 420 times. Calculate the size of the bottom left grain in micrometres.

image diameter = 16 mm = 16 000 µm

$$\text{actual diameter} = \frac{16\,000}{420} = 38 \text{ µm}$$

Part d

This photo shows an ant magnified 25 times.

Calculate the actual length of its head in millimetres, from the top of its head to the end of its biting mouthparts.

image length = 24 mm

actual length = $\frac{24}{25}$ = 0.96 mm

Part e

This photo shows a section through part of an animal cell magnified 80 000 times.

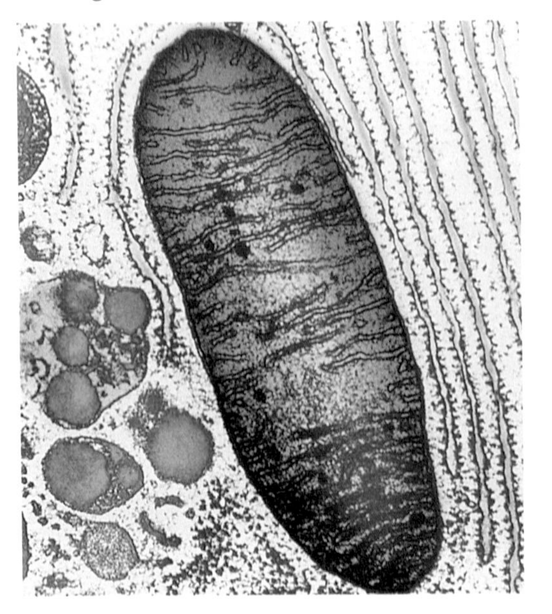

Calculate the length and width of the mitochondrion in nanometres.

Note: 1 mm = 1000 000 nm

image length = 81 mm

actual length = $\frac{81}{80\,000}$ = 0.0010125 mm = 1000 nm

image width = 28 mm

actual width = $\frac{28}{80\,000}$ = 0.00035 mm = 350 nm

Electron microscopes

Electron microscopes use beams of electrons, instead of the beams of light used in optical microscopes. Instead of glass lenses, they use magnets to focus the electron beams.

The detailed ultrastructure of plant and animal cells was revealed in the 1950s when the **transmission electron microscope** (TEM) was first used. A specimen for the electron microscope has to be specially prepared; a simple smear of cells on a microscope slide would not work. Very thin slices of the specimen are fixed and then embedded in resin, and then cut into extremely thin slices. The specimens are then 'stained' using heavy metals. These are taken up by some parts of the specimen, which will then not let electrons pass through and will appear dark on the final image. The specimen is then placed in a chamber inside the electron microscope, which is sealed before the air is sucked out to produce a vacuum. This is necessary because electrons can only pass through air for a very short distance. Electromagnets act as lenses and focus a beam of electrons that passes through the specimen and onto a viewing screen (Figure 25).

The development of the electron microscope has had a huge impact on biology. A TEM has a maximum resolution of 0.5 nm, 400 times better than an optical microscope (Figure 26, overleaf), because electrons have a much shorter wavelength than light. This makes it possible to see the details of cell organelles and led to the discovery of new

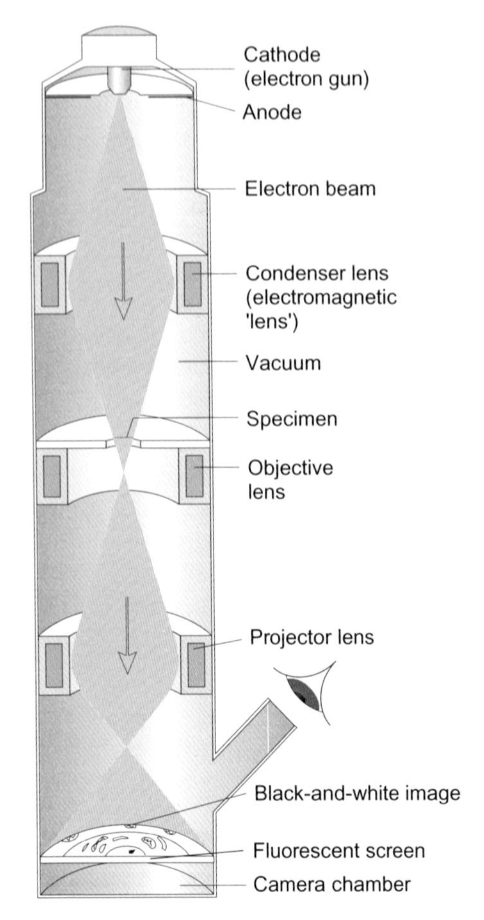

Figure 25 A transmission electron microscope

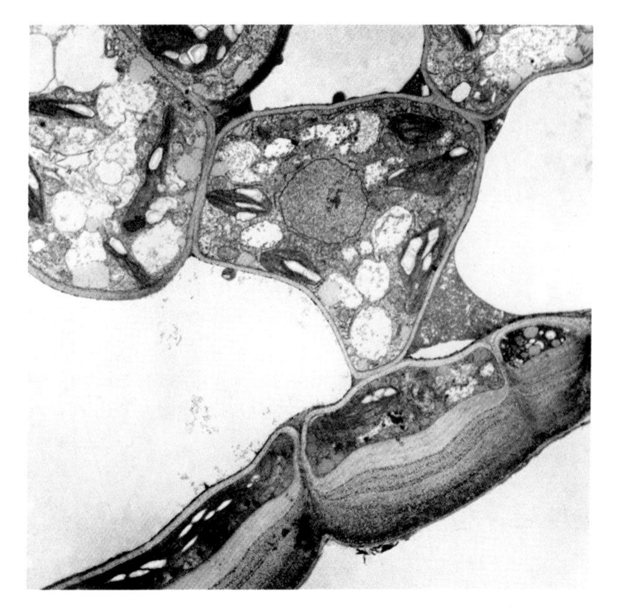

Figure 26 *Transmission electron micrograph of cells from the spongy mesophyll and lower epidermis of a leaf. Compare the detail you can see in this micrograph with Figure 23.*

organelles. The useful magnification of a TEM is up to 500 000 times.

Despite the ability of electron microscopes to make visible much smaller structures than can be seen using light microscopy, they do have their disadvantages. In particular, it is not possible to view living specimens with an electron microscope, because the prepared specimens have to be placed in a vacuum. Also, the very thin sections that must be used in a transmission electron microscope can make it difficult to visualise how the structures are organised in three dimensions.

Transmission electron microscopes do not produce images in colour; they are black and white, as shown in Figure 26. However, it is easy to use a computer to add colour to the images, which makes them more eye-catching and can also make it easier to pick out different structures.

Another type of electron microscope, called the **scanning electron microscope** (SEM) is most often used to visualise the surface features of an organism or a cell. It produces a three-dimensional image (Figure 27). The electron beam is produced in the same way as the beam in the transmission electron microscope. But the beam is then focused by one or two condenser lenses into a beam with a very fine focal spot, sized 1–5 nm. The beam passes through pairs of scanning coils in the objective lens, which deflect the beam horizontally and vertically so that it scans over a rectangular area of the sample surface. The electrons in the scanning beam cause secondary electrons to be emitted from the surface of the sample. The secondary electrons are detected by a device that produces a signal that can be viewed and saved as a digital image.

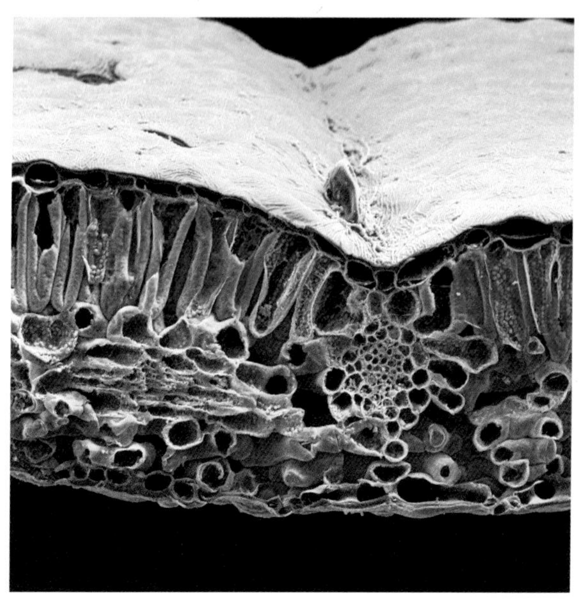

Figure 27 *Scanning electron micrograph of the surface of a section through a leaf of the Christmas rose,* Helleborus niger.

The resolution of the SEM is not as high as the TEM. However, the SEM can:

- image a comparatively large area of the specimen
- image bulk materials (not just thin films or foils).

The resolution of an SEM ranges from less than 1 nm to 20 nm, depending on the instrument. In general, SEM images are much easier to interpret than TEM images.

The preparation of material for any kind of electron microscopy greatly changes its nature. This can result in structures being formed in the specimen that were not there in life. These are called **artefacts**.

Microscopists have always been aware of this problem, but it has not always been easy to distinguish between 'genuine' structures and artefacts. For example, for many years, no-one was sure if the extensive series of membranes seen in cells, which we now know makes up the endoplasmic reticulum, was real or just an artefact. In prokaryotic cells, a frequently visible infolding of the cell-surface membrane, called a mesosome, was not recognised as an artefact until the 1970s.

Cell fractionation and ultracentrifugation

It is easy to see that the availability of the electron microscope was a major help to scientists in working out the structure of cell organelles. But just looking at the organelles in preserved sections of tissue cannot reveal much about their function. In fact, people had found out a great deal about what organelles do, particularly about chloroplasts and mitochondria and the sequence of reactions in photosynthesis and respiration, before they had access to electron micrographs.

They investigated individual organelles by breaking open cells and then separating out organelles such as mitochondria, ribosomes and membranes. This is known as cell fractionation.

To do this successfully, the organelles had to remain intact, with all internal enzymes and structures in place. The technique that made this possible is called ultracentrifugation.

As Figure 28 shows, if a mixture of different sizes and mass is spun at high speed in a centrifuge, the larger, heavier particles tend to accumulate at the bottom of the tube. Since the organelles vary in size and mass, this principle can be used to separate them. Cells are broken up in a homogeniser, a device rather like a kitchen blender that breaks open the outer membrane of the cells but leaves the organelles intact. The liquid used in the homogeniser is ice cold to reduce the rate of enzyme activity and has the same concentration of solutes as the organelles to prevent shrinkage or bursting due to osmosis.

The homogenate is then spun at a relatively slow speed. The nuclei, the largest organelles in the cells, collect at the bottom of the tube. The suspension above is known as the supernatant. This supernatant is then spun at a higher speed. This time the mitochondria separate out. To separate out the ribosomes the process is repeated at a higher centrifuge speed.

At any stage, the pellet that collects at the base of the centrifuge tube can be placed into a fresh solution and mixed to re-suspend the organelles that it contains. These can be identified using microscopy. The structure and function of these organelles can then be studied in detail.

QUESTIONS

10. Why are the cells homogenised:

a. in ice-cold water?

b. in a solution with the same concentration of solute?

11. a. Place these organelles in order of size, with the largest first: ribosomes, mitochondria, nuclei.

b. Draw a centrifuge tube to show the order of settling of these three after they have been centrifuged.

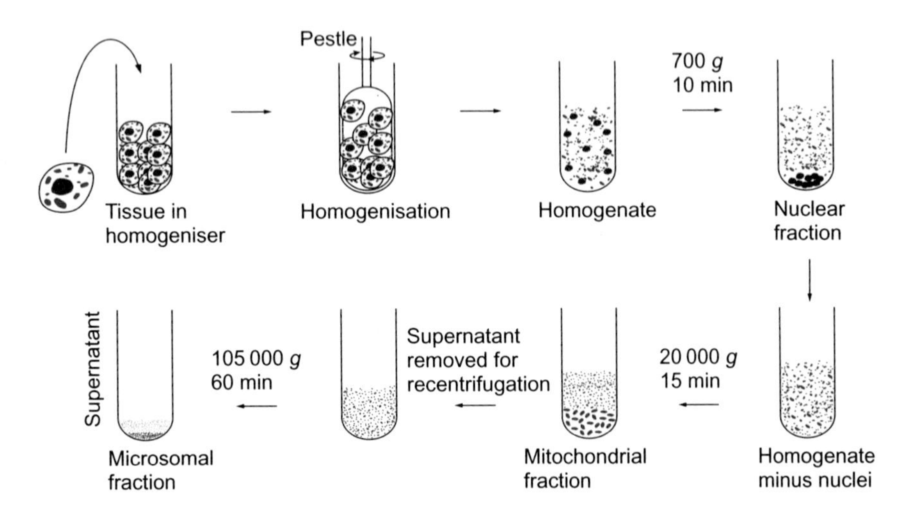

Figure 28 *Separating organelles by differential centrifugation.* ***Note:*** *700 g, 20 000 g and 105 000 g indicate acceleration relative to Earth's gravitational acceleration (g), or 'relative centrifugal force (RCF)'.*

KEY IDEAS

- Optical microscopes use beams of light focused by glass lenses to produce magnified images. Their limit of resolution is between 200 and 250 nm.
- Transmission electron microscopes and scanning electron microscopes use beams of electrons focused by magnetic lenses to produce magnified images. The limit of resolution of a TEM is about 0.5 nm, and that of an SEM is 1 nm.
- Optical microscopes have the advantage of being cheaper and easier to maintain than electron microscopes. They can also be used to view living specimens.
- Electron microscopes have the advantage of being able to view much smaller objects than can be seen in optical microscopes.
- Magnification is the size of the image divided by the size of the real object.
- Resolution is the ability to distinguish between two objects.
- Cell fractionation and ultracentrifugation can be used to separate different types of organelles from cells, so that their structures and activities can be studied.

5.5 MAKING NEW CELLS

Every living cell is produced from another living cell. Cells produce new cells by cell division. The normal method of cell division for eukaryotic cells in a plant or animal is called **mitosis**. Mitosis produces new cells that have exactly the same number and type of chromosomes as the parent cell. Mitosis produces genetically identical cells. Prokaryotic cells divide by binary fission.

The cell cycle

Mitosis is part of a process called the **cell cycle**. This is shown in Figure 29. Not all cells in a multicellular organism are able to divide. Once a cell in your body has become specialised for a particular function, it usually will not be able to divide to produce new cells. Most specialised cells do not show a cell cycle.

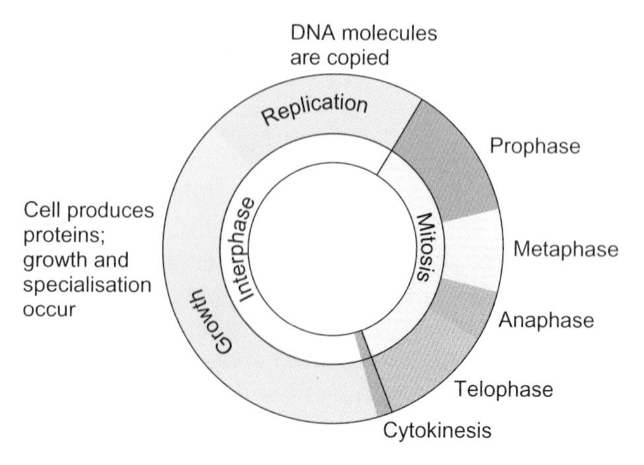

Figure 29 *The cell cycle*

For cells that do have a cell cycle, most of their time is spent in the stage of the cell cycle called **interphase**. This is when DNA replication takes place.

During interphase, the individual chromosomes are not visible. When stained, the whole nucleus appears as a dark mass because the chromosome material, called chromatin, is spread out (see Figure 4 on page 69).

Just before the cell is about to enter mitosis, the long, thin DNA molecules coil up to form much thicker, visible **chromosomes**. These can be seen with an optical microscope. This shrinkage of chromosomes into smaller packages makes it less likely that sections of DNA will get tangled up, break away and be lost during cell division. Each chromosome is made up of two **chromatids** held together at a point called the **centromere**. The two chromatids contain the two identical DNA molecules that were produced by semi-conservative replication during interphase.

The cell now enters the first stage of mitosis, called **prophase** (Figure 30). The two membranes of the nuclear envelope break. Next, spindle fibres, made from microtubules, form a set of radiating strands from each end of the cell. Two spindle fibres, one from each end of the cell, attach to each of the centromeres and pull the chromosomes so that they end up in a line across the centre of the cell. This stage is called **metaphase**.

The spindle fibres now shorten, and pull the two chromatids of each chromosome apart. The spindle fibres continue to shorten, dragging the chromatids to opposite ends of the cell. This stage is called **anaphase**, and is often the easiest to spot when you are looking at stained cells undergoing mitosis. Anaphase ensures one copy of each DNA molecule goes to each daughter cell. The chromosomes look a bit like a tangle of spiders' legs inside the cell.

When the two sets of chromatids arrive at the poles (ends) of the spindle, they start to unravel from their condensed form, so that they become invisible long strands of DNA again. The nuclear envelope reforms. This stage is called **telophase**.

Telophase is the final stage of mitosis, and the nucleus has now completed its division. Usually, the rest of the cell now splits in half, in a process called **cytokinesis**. Two new daughter cells have been produced, each with a perfect copy of all of the DNA that was present in the parent cell. The new cells now enter interphase once more.

You can see that during mitosis the two chromatids that make up each duplicated chromosome split apart. At what point should we start to call these two chromatids 'chromosomes'? You could say that, as soon as they have been split apart, they have become chromosomes in their own right. On the other hand, it is often easier to continue to think of them as chromatids until they have 'settled down' into their new nuclei in the two daughter cells. Either approach is acceptable – the important thing is that you understand the process.

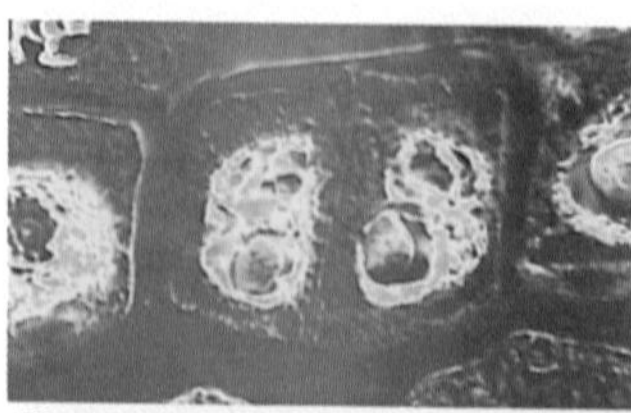

Interphase
For most of the time the chromosomes in a nucleus cannot be seen. The DNA molecules are stretched out and busy synthesising proteins. Just before mitosis begins the DNA replicates. This happens during interphase, before any sign of cell division can be seen.

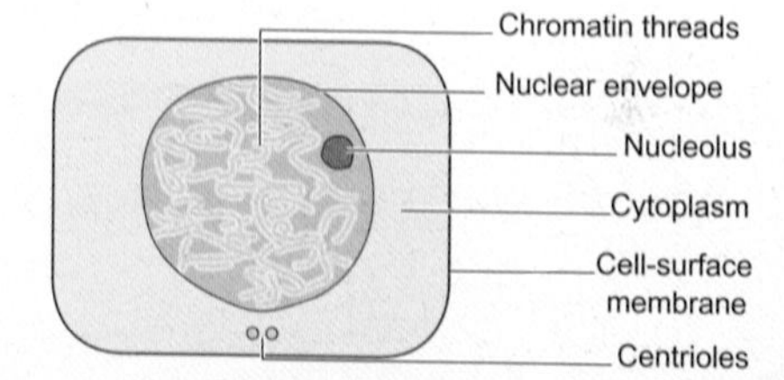

These light micrographs show the stages of mitosis in cells of a hyacinth root.

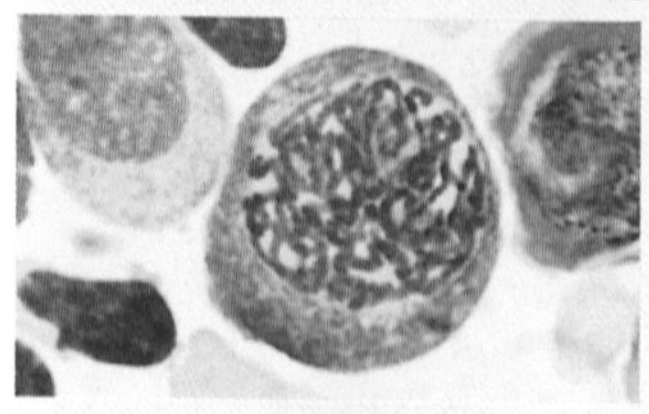

Prophase
After the chromosomes have replicated, they coil up and contract. They then become visible. The replicated chromosomes appear as double strands. In fact they consist of the two new chromosomes, at this stage called chromatids, still firmly joined together at the centromere.

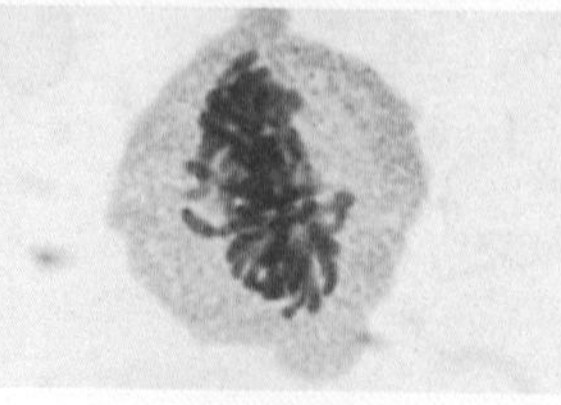

Metaphase
The nuclear envelope breaks down and a web of protein fibres called spindle fibres (microtubules) forms from one end of the cell to the other. The spindle fibres attach to the centromeres and pull the chromosomes to the centre of the cell.

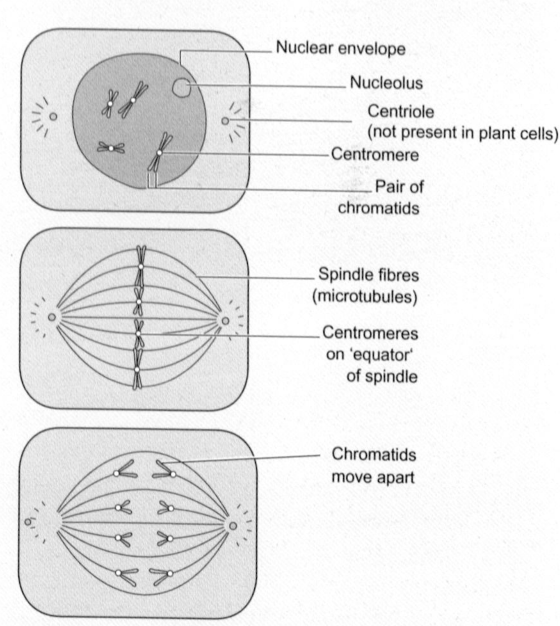

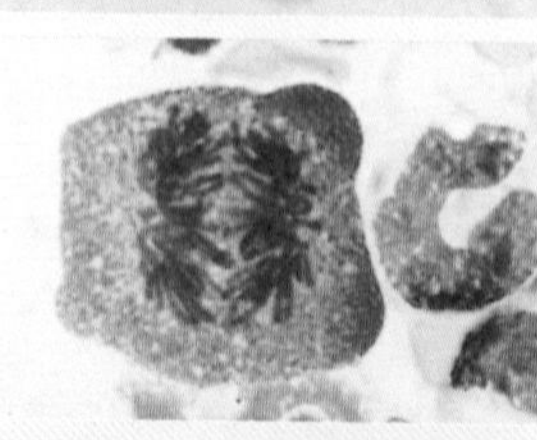

Anaphase
The centromeres now split and the chromatids separate. The chromatids are pulled by the shortening spindle fibres to opposite ends (poles) of the cell.

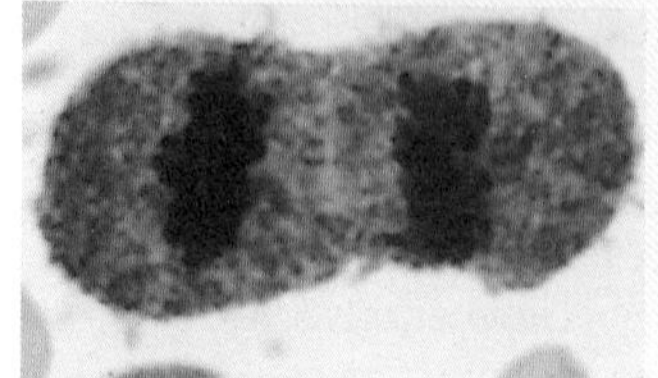

Telophase
The separated chromatids, which are exact copies of the original chromosomes, group together at opposite ends of the cell. After mitosis is complete, the cytoplasm divides (cytokinesis). At this stage, plant cells form new cell walls.

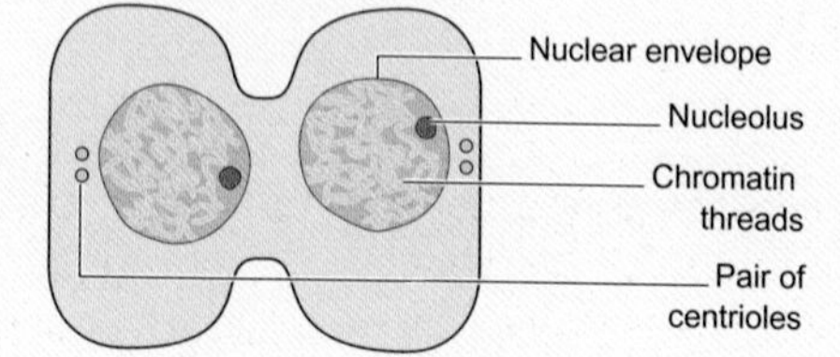

Figure 30 *The stages of the cell cycle*

QUESTIONS

12. a. How does the quanitity of DNA in a nucleus differ between the start and the end of interphase?
    b. There are 46 chromosomes in a human body cell. How many DNA molecules are there in a cell nucleus at the start of prophase?
    c. Suggest why it is important that the chromatids remain attached at the centromere until anaphase.

13. Draw diagrams of metaphase, anaphase and telophase for the cell in Figure 31.

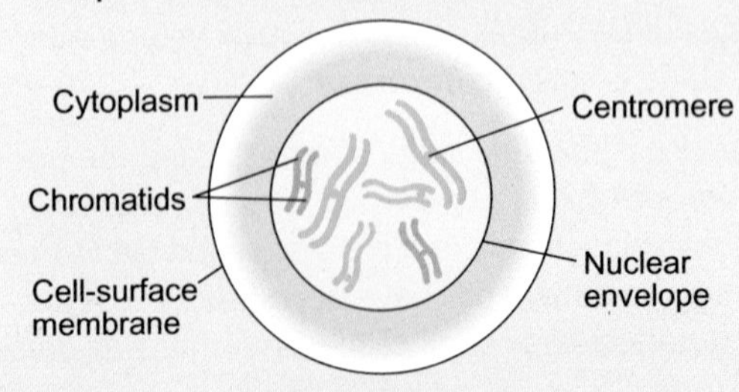

Figure 31 Animal cell with six chromosomes

ASSIGNMENT 1: CONTROLLING THE CELL CYCLE

(MS 1.3, PS 1.2, PS 2.4, PS 3.1)

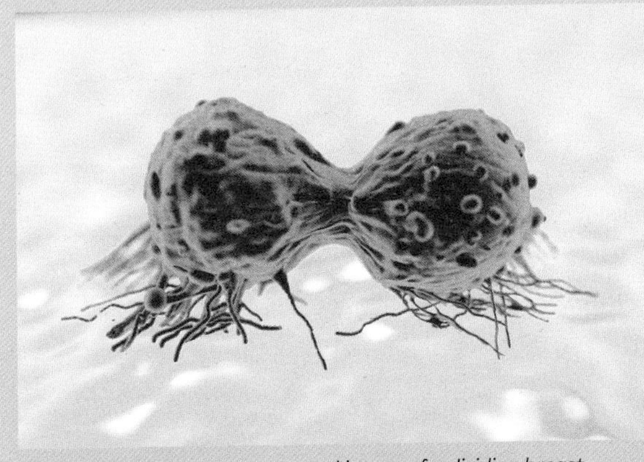

Figure A1 A computer-generated image of a dividing breast cancer cell

What makes your fingers grow to a certain size and then stop? How is it that some cells in your fingers make bone, while others produce skin, nerves, muscles, blood vessels and so on? How do these cells 'know' what to do and when to stop doing it?

Cells produced by mitosis may grow and become specialised, or they may have a relatively short growth period before dividing again. Some, for example most nerve cells, may never divide again. They nevertheless remain active and their genes continue to produce the proteins necessary for their function as nerve cells. Until recently it was thought that no new brain cells are made in humans after about 16 years of age, and that as cells die they are never replaced. However, even with a loss of several thousand per day, there are still plenty to spare. More recently this idea has been challenged, and there is evidence that some cells in nerve tissue retain the ability to divide and form new nerve cells.

Cells in embryos and in tissues that have a high cell turnover – the skin, gut lining and bone marrow, for example – have quite a short interphase. During the first part of interphase the genes are actively involved in protein synthesis and growth occurs and new organelles are formed. After a time a protein called cyclin builds up in the cell. This seems to stimulate production of another protein that in turn stimulates the replication of the DNA and initiates mitosis. This second protein also breaks down cyclin.

Normal concentrations of a form of cyclin called cyclin D1 are known to be critical for normal growth of human mammary (breast) cells. However, the majority of human breast cancer cells have higher than normal concentrations of cyclin D1, which may lead to uncontrolled, cancerous growth.

Scientists have created a new strain of mice that lack the protein cyclin D1. They found that knocking out this important protein causes surprisingly little damage. These results have implications for treating human breast cancer and should lead to a better understanding of cancer.

Cyclin D1 is overproduced in more than half of human breast cancers, but treatment strategies have not targeted this protein because of the damage that they may cause to normal cells in other tissues. The results using the 'knockout' mice suggest that blocking cyclin D action could be a way of preventing the proliferation of tumour cells without harming normal tissues.

Researchers noticed a striking difference in the mutant mice when females delivered pups. The mother mice, whose breast tissue was otherwise normal, were unable to suckle their babies after giving birth. Without cyclin D1 their mammary cells failed to undergo the rapid growth that normally occurs during pregnancy and they produced no milk. The researchers concluded that in adult female mice the cyclin D1 molecule seems critical only for a specialised process in breast tissue – the rapid growth of mammary cells during pregnancy.

The chemical curcumin (found in turmeric) has been shown to inhibit cyclin D1 production in isolated tumour tissues. Figure A2 shows the effects of different concentrations of curcumin on the viability of tumour cells.

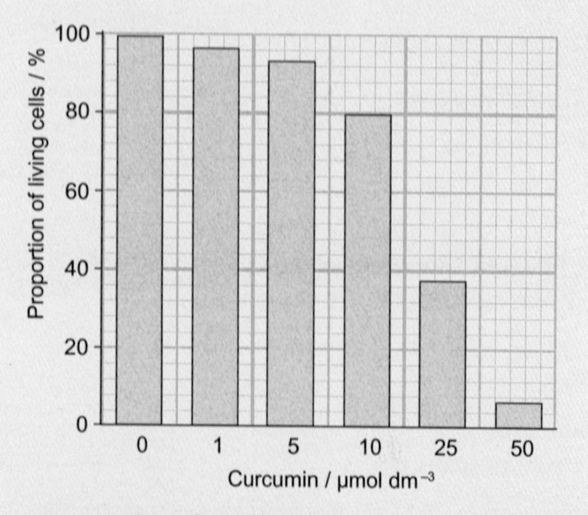

Figure A2 *The effects of different concentrations of curcumin on breast cancer cells*

Questions

A1. Describe the role of cyclin in normal cell division.

A2. Explain what limits the production of cyclin during normal cell division.

A3. Explain what happens to cyclin production in most breast cancers.

A4. Describe the role of cyclin D1 during pregnancy in mice.

A5. Suggest how the research on cyclin D1 in mice may be applicable to treating breast cancer in humans.

A6. Evaluate the ethical issues involved in using mice in these experiments.

A7. a. Describe the effects of curcumin on isolated tumour cells.

b. Describe the experimental control for the research on curcumin.

c. How applicable is the research on isolated tissue to the treatment of breast cancer?

REQUIRED PRACTICAL ACTIVITY 2: APPARATUS AND TECHNIQUES

(MS 0.1, MS 0.3, MS 1.8, MS 2.1, PS 4.1, AT d, AT e, AT f)

Preparation of stained squashes of cells from plant root tips; set up and use of an optical microscope to identify the stages of mitosis in these stained squashes and calculation of a mitotic index.

This practical activity gives you the opportunity to show that you can:

- use a light microscope at high power and low power, including the use of a graticule
- produce scientific drawing from observations with annotations.

Apparatus

The light microscope

There are many different types of light microscope but all of them work in the same basic way. Some have mirrors, and require an external light source (see Figure 24 on page 79), while others have an illuminator.

They generally have two knobs for focusing – a coarse focus and a fine focus.

An eyepiece graticule can be placed within the microscope eyepiece. It can be seen superimposed on your image. It is usually 10 mm in length and each millimetre is divided into 10 parts, so each small division = 0.1 mm = 100 μm.

Techniques

Staining and squashing a root tip

The best place to find cells that are dividing by mitosis is just behind the tip of the growing root. The tip itself is covered by a root cap, which protects the rapidly dividing cells in the growing region just behind it. A recently germinated broad bean seedling, or a head of garlic that has been allowed to stand with its base in water for several days, are good sources of young roots. Cut about 5 mm off the end.

Next, you need to stain the root tip. A good stain to try is 10 parts orcein ethanoic acid (corrosive and flammable) mixed with one part hydrochloric acid (corrosive). This will stain DNA red. Place your root tip in a watch glass and cover it with this stain. Warm it gently on a hotplate for about five minutes. This will begin to break the cells apart a little, and help the stain to bind with the DNA.

Now remove your stained root tip and place it on a clean microscope slide. Add an extra two drops or so of orcein ethanoic acid. Very carefully, trying to keep the tissue in the same general shape, break it apart with a mounted needle. Then put a cover slip over it, wrap it in filter paper, and **very gently** tap it with the blunt end of a pencil. The aim is to spread the cells out into a thin layer on the slide, without breaking them up too much (and without breaking the coverslip).

Finally, warm the slide again by standing in a hotplate for a few seconds, or passing it quickly through a Bunsen flame. Do this while holding it by hand, which will ensure that you do not let it get too hot.

You can now try viewing your slide. If it seems to be stained too darkly, repeat the procedure but leave the root tip in the stain for a shorter time, or heat it less strongly. Use the photographs and diagrams in Figure 30 to help you identify cells in different stages of the cell cycle. Note, however, that the chromosomes in your root tip preparation will be stained red, not blue.

Using a microscope

When using a microscope, you should take the following steps:

- Wind the stage down as far as it will go – this gives you maximum working room.
- Turn the objective lens to the smallest (this is often red and × 4) – this gives you the biggest field of view to locate your specimen.
- Put your slide on the stage and secure it with the clips or fix it to a power driver if your microscope has one.
- Wind the stage up fully while watching it – do not look down the eyepiece at this time or you may break the slide.
- When the stage is as high as it will go, slowly lower it while looking down the eyepiece until your slide is in focus.
- Change to a higher power lens to examine the stages of mitosis, and adjust the fine focus if necessary.

Using a graticule

The specimen on your slide is magnified on to the eyepiece graticule by the magnification of the objective lens. So, a rough approximation of actual amount of specimen seen on each small division of the graticule is:

0.1 mm ÷ objective magnification

For example, when using a ×4 objective lens, each small division on the graticule represents approximately 0.025 mm. With a ×10 objective, each small division is 0.01 mm, and with a ×40 objective each small division is 0.0025 mm.

The graticule can be used to help you count the cells at different stages of the cell cycle and to help you estimate the size of the cells.

Calibrating an eyepiece graticule

Although we can use the magnification labelled on the objective lens to give us a rough approximation of the real sizes represented by the eyepiece graticule scale, this is not at all accurate. To make accurate measurements, we need to calibrate the eyepiece graticule scale when using each objective lens.

To do this, we use a special slide called a **stage micrometer**. This looks like an ordinary microscope slide, but it has a tiny scale engraved on it. The micrometer will be labelled to state the length of the divisions on the scale – it is usually marked off in subdivisions of 0.1 and 0.01 mm.

First, put the stage micrometer on the stage of the microscope, and swing the objective lens you want to use into place. Focus on the scale on the stage micrometer.

Now move the eyepiece around until its scale lines up with the stage micrometer scale. You will see something like this:

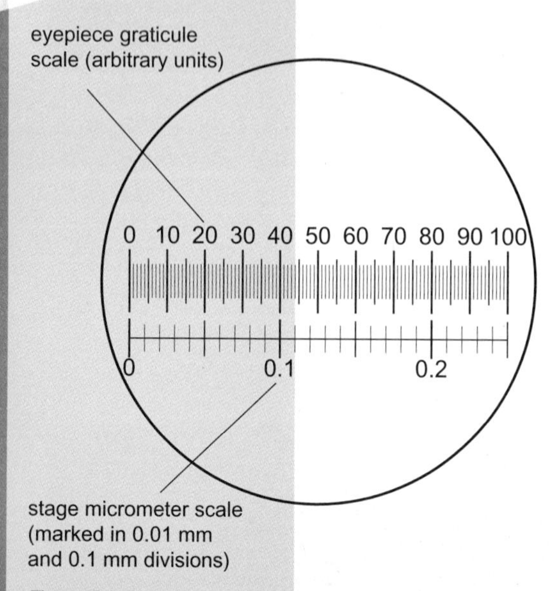

Figure P1 *Calibrating an eyepiece graticule*

Now you can work out the length represented by each division of the eyepiece graticule scale. Here, we can see that 100 divisions on the eyepiece graticule scale equal 0.25 mm on the stage micrometer scale. So each division on the eyepiece graticule is equal to 0.0025 mm, or 2.5 μm.

Next, take away the stage micrometer and replace it with a slide containing the specimen you want to measure. Keeping the same objective lens in place, focus on the specimen. You can now use the calibrated eypiece graticule to measure its length.

You will need to calibrate your eyepiece graticule for each of the different objective lenses on the microscope. Once done, however, you can save your results and refer to them each time you use the same microscope and eyepiece graticule.

Biological drawings

When doing biological drawings you **should**:

- use a sharp pencil
- use plain paper (not lined or squared paper)
- make your drawing large (at least half a page)
- draw a clear, thin outline (no sketching or shading)
- give your drawing a title and a magnification
- label (in pencil) all relevant parts
- draw label lines with a ruler and make sure they do not cross each other
- make sure your label lines touch the part they are labelling.

When doing biological drawings you **should not**:

- use arrows for label lines
- draw a frame or circle showing the field of view.

Calculating the mitotic index

The cell cycle is the time taken for a cell to divide into two, and involves both mitosis and interphase. When examined under the microscope, actively dividing cells can be distinguished from those at interphase. The percentage of cells undergoing mitosis can be used to determine the mitotic index. The mitotic index is much higher in actively dividing tissues such as tumours.

The mitotic index is calculated as:

$$\frac{\text{number of cells undergoing mitosis}}{\text{total number of cells}}$$

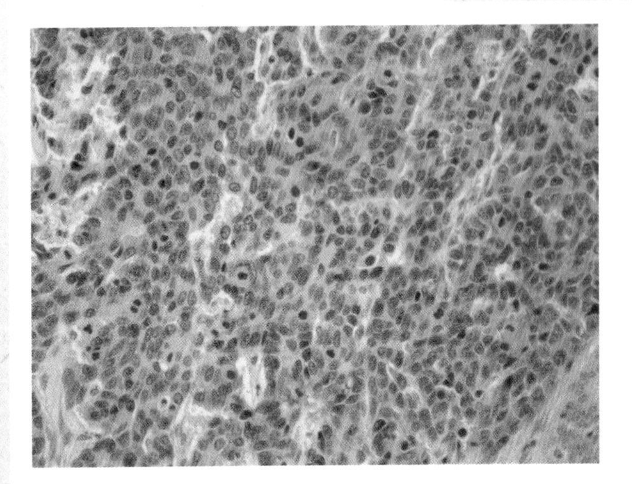

Figure P2 *Cervical cancer. Light micrograph of a section of a large cell neuroendocrine carcinoma of the cervix. Tumours have many more cells undergoing mitosis than normal tissues.*

So, for example, if there are 60 cells in a field of view and 12 of them are in some stage of mitosis then the mitotic index is:

$$\frac{12}{60} = 0.2$$

The percentage of cells in each stage of mitosis can be used to estimate the time spent in each stage if the total time for a complete cycle is known. If we assume that a particular type of cell undergoes a cell cycle that lasts 24 hours (1440 minutes) then the time spent in each stage can be estimated as:

$$\text{percentage cells in a stage} \times \frac{1440}{100}$$

QUESTIONS

P1. Why is it important to treat your root-tip squash gently?

P2. If you are examining cells with a ×10 objective lens and the image spans three small lines on the graticule, what is its approximate size? Give your answer in both mm and μm.

P3. The size of a specimen viewed down a light microscope can only be measured accurately if the eyepiece graticule is calibrated. Describe what an eyepiece graticule is and how it is calibrated.

P4. Explain why you should count the cells in more than one field of view.

P5. You are examining a preparation and it is too dark to see anything clearly. Explain what you could do to improve what you can see.

P6. Look at the drawing. List all of the faults you can see.

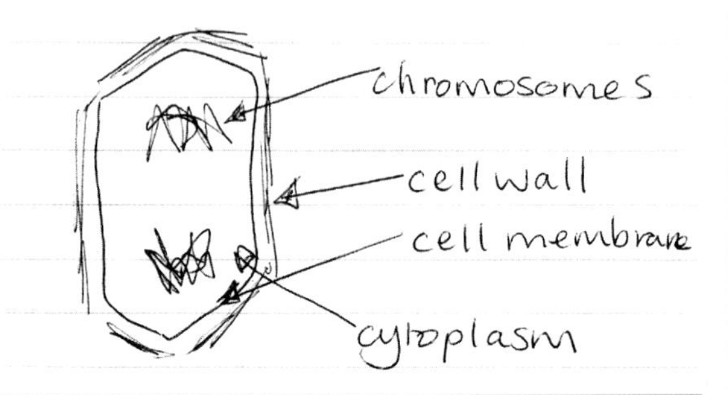

P7. A tissue sample is examined and the number of cells at each stage of the cell cycle are counted and recorded in this table.

Stage of cycle	Number of cells
interphase	270
prophase	35
metaphase	30
anaphase	5
telophase	20

a. Calculate the mitotic index for this tissue.

b. If the cell cycle takes 24 hours, estimate how long each phase lasts.

P8. This table shows the difference in the cell cycle in normal tissue from an ovary and from a cancerous ovary.

Stage of cycle	number of cells in normal ovary	number of cells in cancerous ovary
interphase	19	12
prophase	0	2
metaphase	0	1
anaphase	1	2
telophase	0	3

a. Calculate the mitotic index for both ovaries.

b. Discuss how the mitotic index could be used to determine whether an ovary is cancerous.

Control of the cell cycle

What determines when a cell should divide? This is something that is not yet fully understood, but we know that cells in the human body respond to stimuli which are external factors, for example chemical signals produced by cells in other parts of the body or neighbouring cells, and by internal factors.

The control of the cell cycle is of great interest, because when it goes wrong cells can divide out of control and form tumours. Although many tumours are benign (meaning that they grow relatively slowly, and do not spread into other tissues), cancerous tumours grow aggressively and invade nearby tissues. They may also shed cells that form secondary tumours in other parts of the body. Many cancer treatments attempt to control the rate of cell division of cancerous cells.

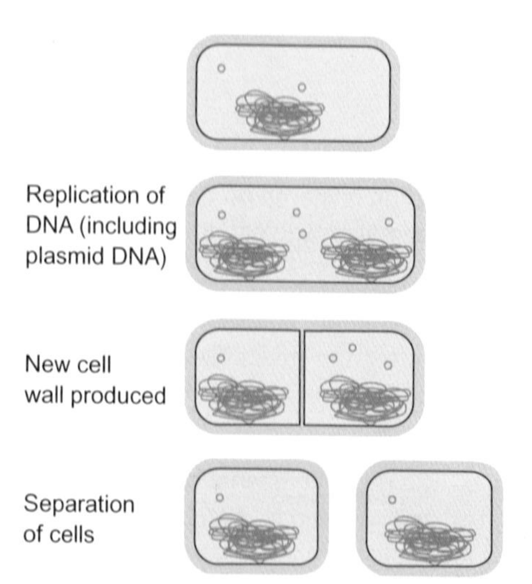

Figure 32 *Binary fission in a prokaryote*

Cell division in prokaryotes

You will remember that prokaryotes do not have true chromosomes, so clearly they cannot divide by mitosis. Prokaryotic cells divide by a process called **binary fission** (Figure 32).

First, the circular DNA molecule replicates. Any plasmids in the cell also replicate. The two identical DNA molecules are moved to opposite ends of the cell, and the cell splits into two. Two new cells have now been formed, each with an identical copy of the original circular DNA molecule. The plasmids also move to opposite ends of the cell, but they may not sort themselves out exactly, so the two new cells may not have the same number of plasmids.

Interestingly, mitochondria and chloroplasts divide in the same way as prokaryotic cells. They divide independently of the rest of the cell. This is further evidence for the endosymbiont theory.

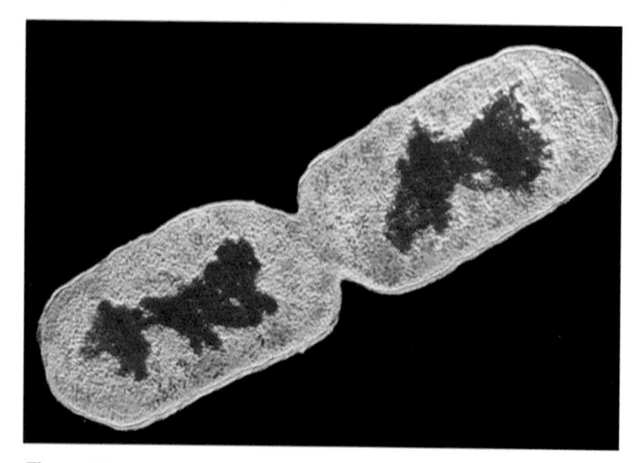

Figure 33 *Transmission electron micrograph of the bacterium* Escherichia coli *dividing by binary fission. The DNA is coloured red.*

KEY IDEAS

- Eukaryotic cells that retain the ability to divide go through a cell cycle.
- Most of the cell cycle is taken up by interphase, when growth and DNA replication take place.
- Mitosis is made up of prophase, metaphase, anaphase and telophase. During mitosis, spindle fibres move the duplicated chromosomes to the centre of the cell and then pull them apart, dragging them to opposite ends of the cell.
- Cytokinesis usually follows mitosis, producing two genetically identical daughter cells.
- The cell cycle is normally controlled to ensure that cells divide as and when required. If this control goes wrong, uncontrolled division can cause tumours and cancer.
- Prokaryotic cells divide by binary fission, in which the circular DNA is copied and the two copies moved to opposite ends of the cell. The cell then splits into two.

Worked maths example: Bacterial growth and logarithms

(MS 0.1, MS 0.5, MS 1.3, MS 2.5)

A single bacterium is placed on a fresh agar Petri dish and kept in optimal growth conditions.

- If this strain of bacteria can divide every 20 minutes, estimate how many bacteria will be present after four hours. Draw a graph to demonstrate bacterial growth over time.

First, we need to ensure that we are dealing in the same units. The rate of bacterial division is every 20 minutes, and we need to calculate how many there will be in four hours.

There are 60 minutes in an hour, so

$4 \times 60 = 240$ minutes

We know that one bacterium replicates every 20 minutes. So, how many new generations will be produced in 240 minutes?

$\frac{240}{20} = 12$ generations

This is an example of exponential growth.

Each new cell will also divide into two, and so will its progeny, and so on. We use this equation:

$$\text{number of bacteria} = 2^x \times n$$

where
x = the number of generations
n = number of bacteria we started with

So, assuming that they all live, the number of bacteria present after 240 minutes = $2^{12} \times 1 = 4096$

We can use this information and the power function on a calculator to make a table of number of bacteria at each 20 minute time point up to four hours (240 minutes):

For example:

- $2^3 \times 1 = 8$ (highlighted in blue in the table below)
- $2^8 \times 1 = 256$ (highlighted in red in the table below)

Plotted on a graph, the data shows a **logarithmic pattern**.

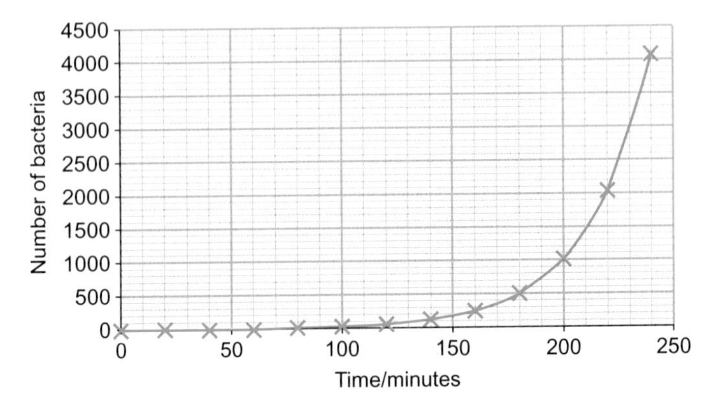

Can you use this graph to estimate how many bacteria are present after 50 minutes?

Because of the huge differences in scale, it's difficult to read bacterial numbers before ~ 140 minutes.

So, we can re-plot the data on a semi-logarithmic scale, making all of the data points clearly visible. It is semi-logarithmic because only the *y*-axis needs to be altered since its data vary significantly in orders of magnitude.

This can be done using software such as Excel®, or by plotting the graph using logarithmic paper.

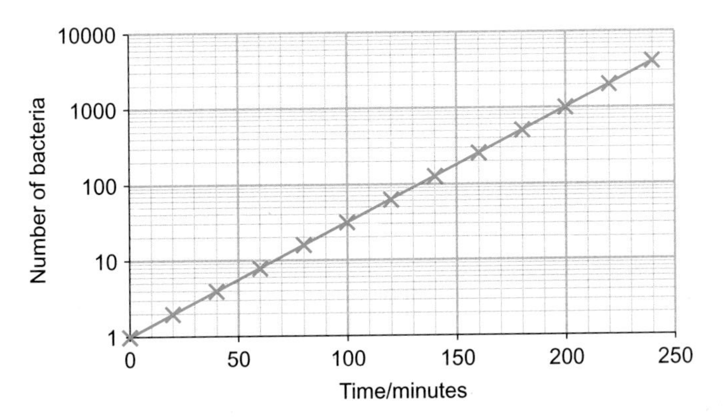

Now we can see that, after 50 minutes, there are approximately 5.5 bacteria.

	Generation												
		1	2	3	4	5	6	7	8	9	10	11	12
Number of bacteria	1	2	4	8	16	32	64	128	256	512	1024	2048	4096
Time/minutes	0	20	40	60	80	100	120	140	160	180	200	220	240

PRACTICE QUESTIONS

1. The drawing in Figure Q1 shows an electron micrograph of part of a plant cell.

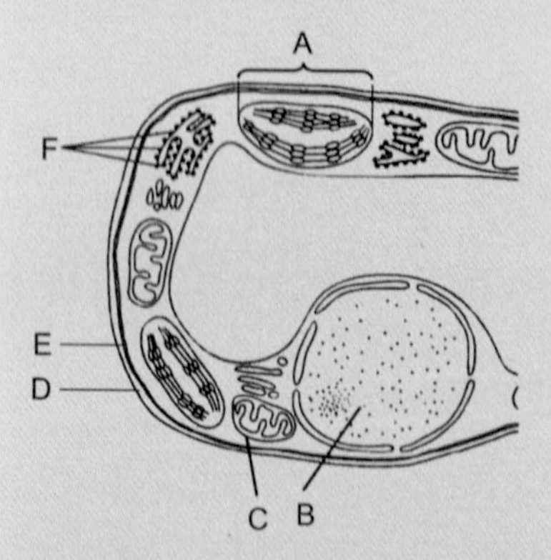

Figure Q1

   a. Name the structures labelled **A–F**.

   b. Name the main carbohydrate found in part **D**.

   c. Give the functions of the parts labelled **C–F**.

   d. Give three differences between the structure of this cell and an animal cell.

2. Figure Q2 shows a mitochondrion.

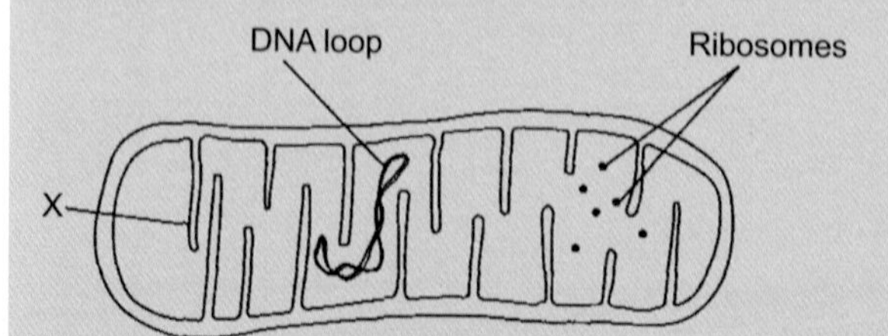

Figure Q2

   a. i. Name the part labelled **X**.

   ii. A human liver cell contains several hundred mitochondria. A cell from a plant root has a much smaller number. Suggest an explanation for this difference.

   iii. Mitochondria contain some DNA and ribosomes. Describe the function of these.

   b. Mitochondria may be separated from homogenised cells by differential centrifugation. During this process, the cells must be kept in an ice-cold solution with a similar concentration to that of the cytoplasm. Explain why.

3. Figure Q3 shows the cell cycle.

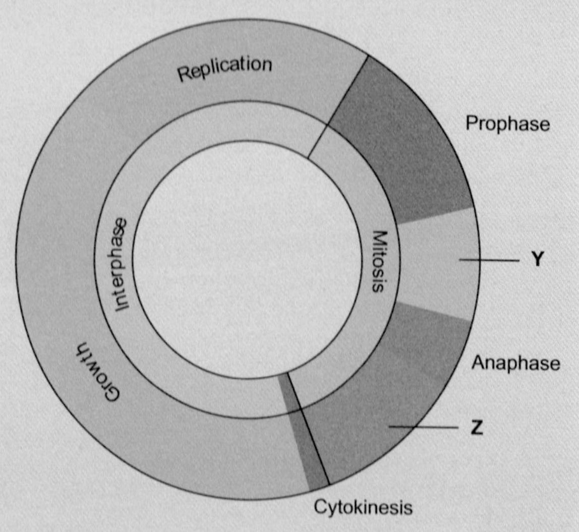

Figure Q3

   a. i. Name the stage of mitosis labelled **Y**.

   ii. Describe what happens during this stage.

   b. i. Name the stage of mitosis labelled **Z**.

   ii. Describe what happens during this stage.

   c. i. Describe what happens during the cytokinesis stage of the cell cycle. Explain the significance of this stage in the development of an organism.

4. a. Boxes **A** to **E** show some of the events of the cell cycle.

A Chromatids separate
B Nuclear envelope disappears
C Cytoplasm divides
D Chromosomes condense and become visible
E Chromosomes on the equator of the spindle

(Continued)

i. List these events in the correct order, starting with **D**.

ii. Name the stage described in box **E**.

b. Name the phase during which DNA replication occurs.

c. Bone marrow cells divide rapidly. As a result of a mutation during DNA replication, a bone marrow cell may become a cancer cell and start to divide in an uncontrolled way. A chemotherapy drug that kills cells when they are dividing was given to a cancer patient. It was given once every three weeks, starting at time 0. The graph in Figure Q4 shows the changes in the number of healthy bone marrow cells and cancer cells during twelve weeks of treatment.

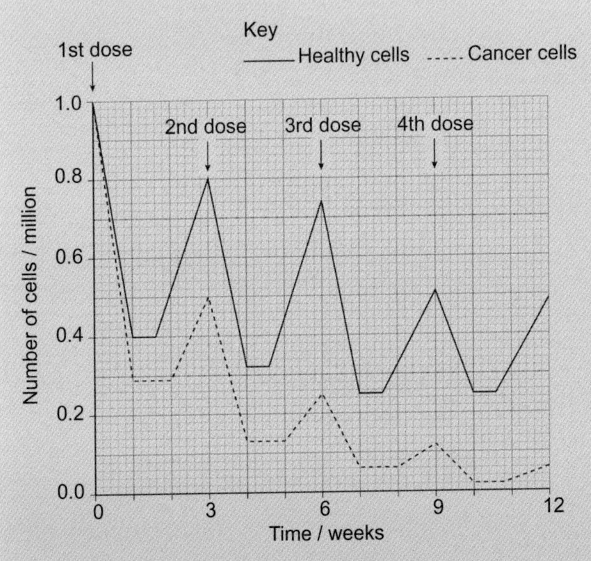

Figure Q4

i. Using the graph, calculate the number of cancer cells present at week 12 as a percentage of the original number of cancer cells. Show your working.

ii. Suggest **one** reason for the lower number of cancer cells compared to healthy cells at the end of the first week.

iii. Describe **two** differences in the effect of the drug on the cancer cells, compared with healthy cells, in the following weeks.

AQA June 2006 Unit 2 Question 2

5. The graph in Figure Q5 shows the changes in the DNA content of cells during the cell cycle.

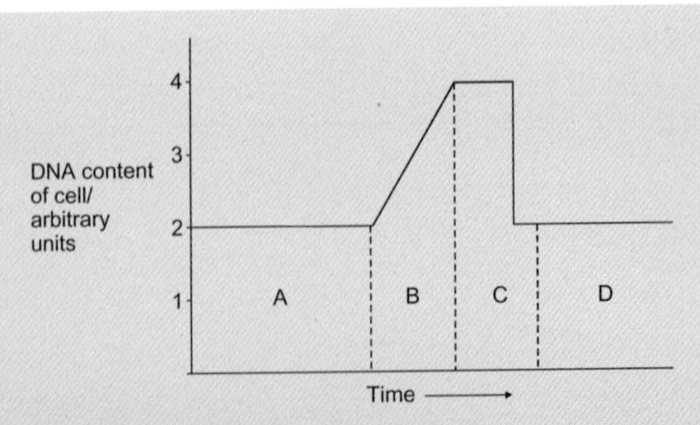

Figure Q5

a. In which of the stages, A to D, does each of the following take place?

i. DNA replicates.

ii. The chromosomes become visible.

b. Describe and explain how the amount of DNA in the cell changes during stage C.

c. i. Cytarabine is a drug used to treat cancer. It inhibits an enzyme needed to synthesise new DNA. Suggest how the graph would be different if cytarabine was present during the cell cycle.

ii. Explain why cytarabine is effective in treating cancer.

6. An amoeba is a single-celled, eukaryotic organism. Scientists used a transmission electron microscope to study an amoeba. Figure Q6 shows its structure.

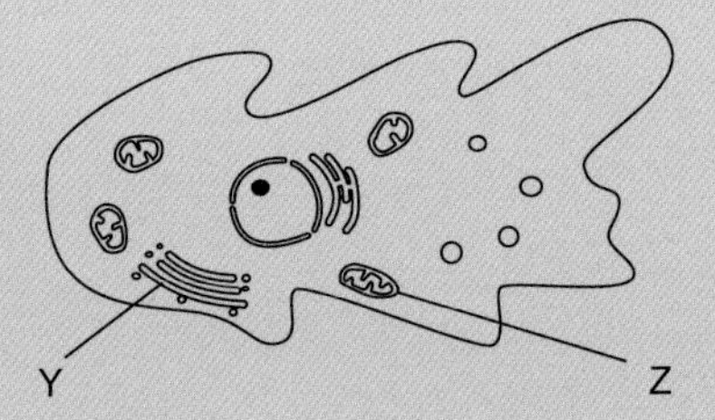

Figure Q6

a. i Name organelle Y.

ii. Name **two** other structures in the diagram which show that the amoeba is a eukaryotic cell.

b. What is the function of organelle Z?

c. The scientists used a transmission electron microscope to study the structure of the amoeba. Explain why.

AQA June 2012 Unit 1 Question 1

7. Figure Q7 shows a bacterium.

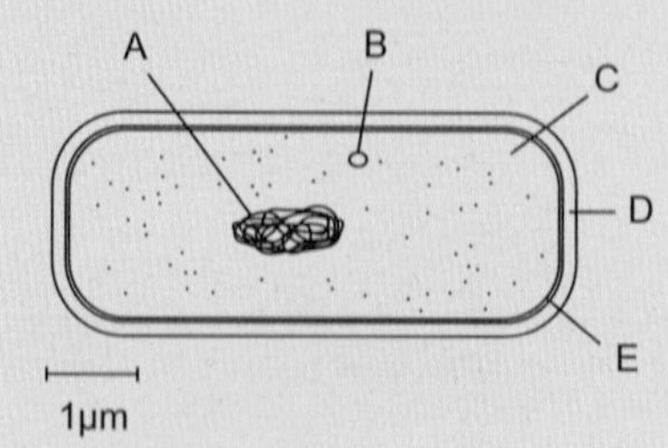

Figure Q7

a. Part **A** is a DNA molecule. Describe how the structure and location of the DNA in a bacterial cell differs from that in a eukaryotic cell.

b. Name parts **B** and **C**, and outline their functions.

c. Describe how the structure of part **D** differs from the equivalent structure in a plant cell.

d. Outline the function of part **E**.

e. Use the scale bar to calculate the actual length of the bacterium. Show your working.

8. a. Describe how DNA is replicated.

b. Figure Q8 shows information about the movement of chromatids in a cell that has just started metaphase of mitosis.

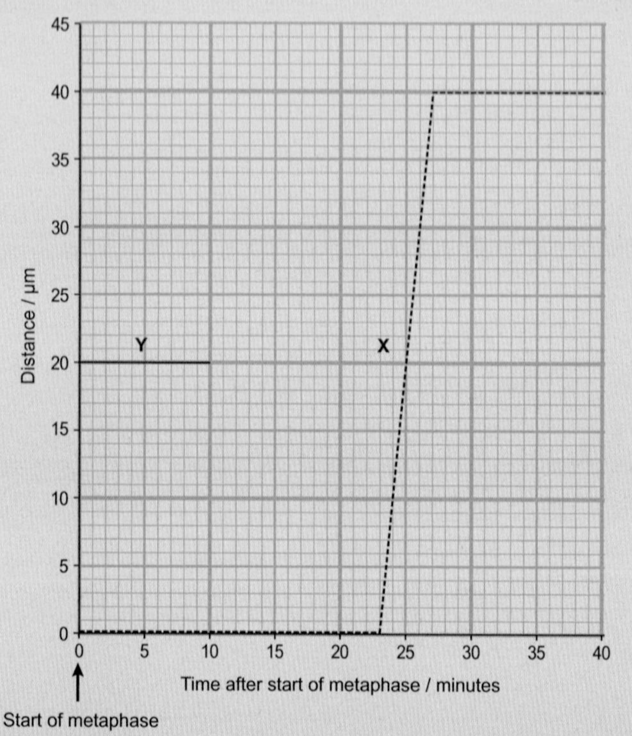

Figure Q8

i. What was the duration of metaphase in this cell?

ii. Use line **X** to calculate the duration of anaphase in this cell.

iii. Describe how you would complete line **Y** on the graph.

c. A doctor investigated the number of cells in different stages of the cell cycle in two tissue samples, **C** and **D**. One tissue sample was taken from a cancerous tumour. The other was taken from non-cancerous tissue. Table Q1 shows his results.

(Continued)

Stage of the cell cycle	Percentage of cells in each stage of the cell cycle	
	Tissue sample C	Tissue sample D
interphase	82	45
prophase	4	16
metaphase	5	18
anaphase	5	12
telophase	4	9

Table Q1

i. In tissue sample **C**, one cell cycle took 24 hours. Use the data in the table to calculate the time in which these cells were in interphase during one cell cycle. Show your working.

ii. Explain how the doctor could have recognised which cells were in interphase when looking at the tissue samples.

iii. Which tissue sample, **C** or **D**, was taken from a cancerous tumour? Use information in the table to explain your answer.

AQA January 2013 Unit 2 Q8

6 CELL MEMBRANES

PRIOR KNOWLEDGE

You have probably learned that all substances consist of particles in constant motion, and that diffusion is the result of their random motion. You may also remember that osmosis is a particular type of diffusion involving water. You may want to remind yourself about the structure and behaviour of phospholipid molecules, and about hydrophilic and hydrophobic (polar and non-polar) substances. You may know that ions are particles that carry relatively large charges.

LEARNING OBJECTIVES

In this chapter, you will learn how phospholipid molecules associate with other molecules to form cell membranes, and how these control the entry and exit of specific molecules and ions into and out of a cell.

(Specification 3.2.3)

About five million children die each year from diarrhoea. Most of these children's lives could be saved by giving them a simple solution containing ions and glucose. This is known as oral rehydration therapy (ORT).

In the normal healthy intestine, there is a continuous exchange of water through the intestinal wall – up to 20 litres of water is secreted and very nearly as much is reabsorbed every 24 hours. When we have diarrhoea this balance is upset and much more water is secreted than is reabsorbed, causing a net loss to the body that can be as high as several litres a day. In addition to water, sodium ions (Na^+) are also lost.

If diarrhoea is not treated, rapid depletion of water and Na^+ occurs. Death occurs if more than 10% of the body's fluid is lost. Simply giving a saline solution (water plus Na^+) by mouth does not help because the normal mechanism by which Na^+ is absorbed by the healthy intestinal wall is impaired when we have diarrhoea. If the Na^+ is not absorbed, neither can the water be absorbed. In fact, excess Na^+ in the lumen of the intestine causes increased secretion of water and the diarrhoea worsens.

If glucose is added to a saline solution, a new mechanism comes into play. The glucose molecules are absorbed through the intestinal wall and Na^+ is carried through in conjunction by co-transport. Water follows by osmosis. The leading medical journal *The Lancet* described the discovery of this mechanism of co-transport of sodium and glucose as 'potentially the most important medical advance this century'.

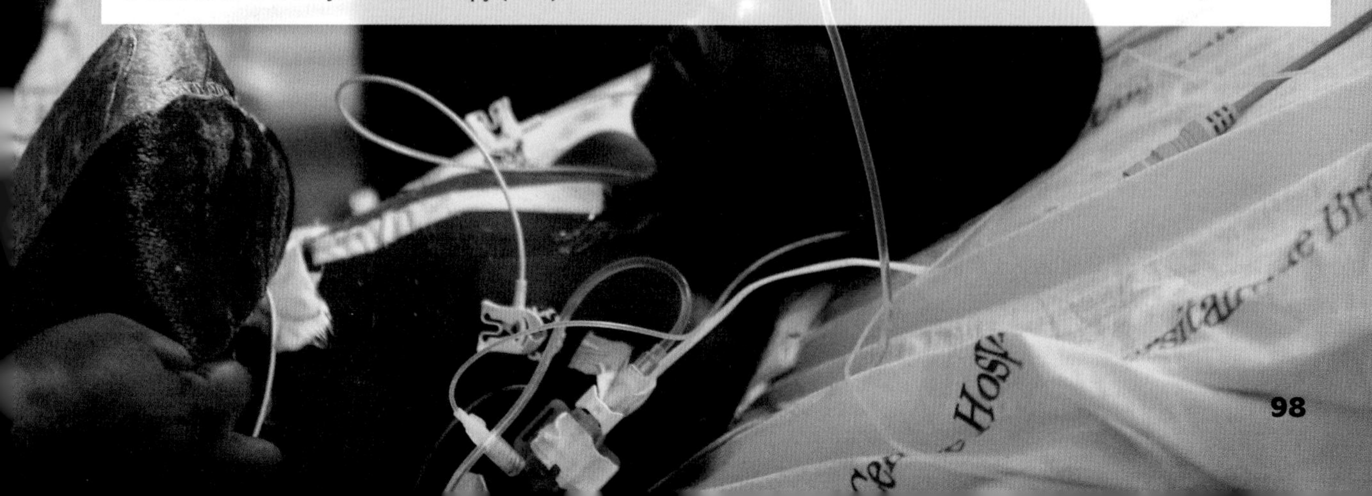

6.1 STRUCTURE OF CELL MEMBRANES

In Chapter 5, we saw that every cell is always surrounded by a **cell-surface membrane**. No cell can ever exist without a membrane surrounding its cytoplasm. Eukaryotic cells also have many membranes inside them, surrounding their organelles and keeping these organelles separate from the cytosol.

The photograph in Figure 1 is a high-power transmission electron micrograph of the cell-surface membrane. You can see the membrane as two dark bands separated by a clear central area. It looks like this because the membrane is not a single layer – it consists of two layers of **phospholipid** molecules.

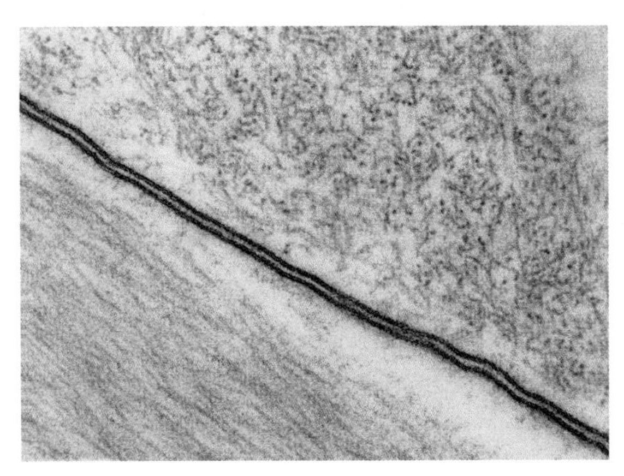

Figure 1 *This transmission electron micrograph was made at the upper limit of the power of the microscope. It shows the cell-surface membrane around a red blood cell. You can see that it is made up of two dark lines, which represent the two layers of phospholipid heads. Magnification × 370 000.*

In Chapter 2, we saw that phospholipid molecules have hydrophilic heads and hydrophobic tails. When they are mixed into a watery liquid, they orientate themselves so that their heads are facing into the water and their tails away from it. They therefore naturally form a bilayer – heads out and tails in. The cytosol of a cell is mostly water, and cells can only survive when there is water immediately outside them, so phospholipids in this situation naturally form a bilayer.

QUESTIONS

1. The image of the cell-surface membrane is quite fuzzy at the magnification of the photograph in Figure 1.
   a. Estimate the thickness of the membrane in nanometres.
   b. How could you make this estimate more reliable?
2. This thickness is approximately twice the length of a phospholipid. Estimate the length, in nanometres, of a phospholipid molecule.

Although phospholipids are by far the most common molecules in a cell membrane, they are not the only ones. The membrane also contains proteins, glycoproteins, glycolipids and cholesterol (Figure 2).

Proteins are embedded in the layers of phospholipids. Some proteins span the membrane from one side to the other. Others appear on one face of the membrane only. Some proteins form passageways through the membrane – you will find out about their functions later. The phospholipid and protein molecules fit together to form a continuous pattern like the tiles in a mosaic. However, unlike mosaic tiles, the molecules that make up a membrane are not all fixed in place; the position of some proteins can change from moment to moment. These proteins can move around within the particular layer or layers of phospholipids that they are in, but they do not normally cross from one layer to the other. The phospholipid molecules are also in constant motion, and so cell membranes are said to have a **fluid-mosaic** structure.

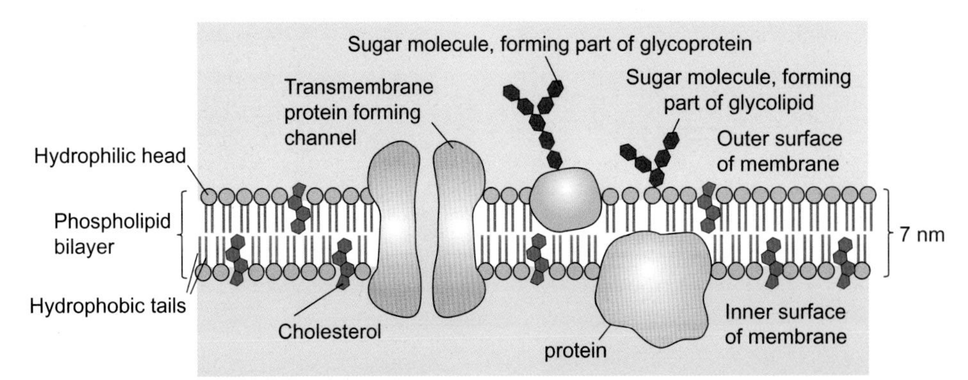

Figure 2 *The fluid-mosaic structure of a cell-surface membrane*

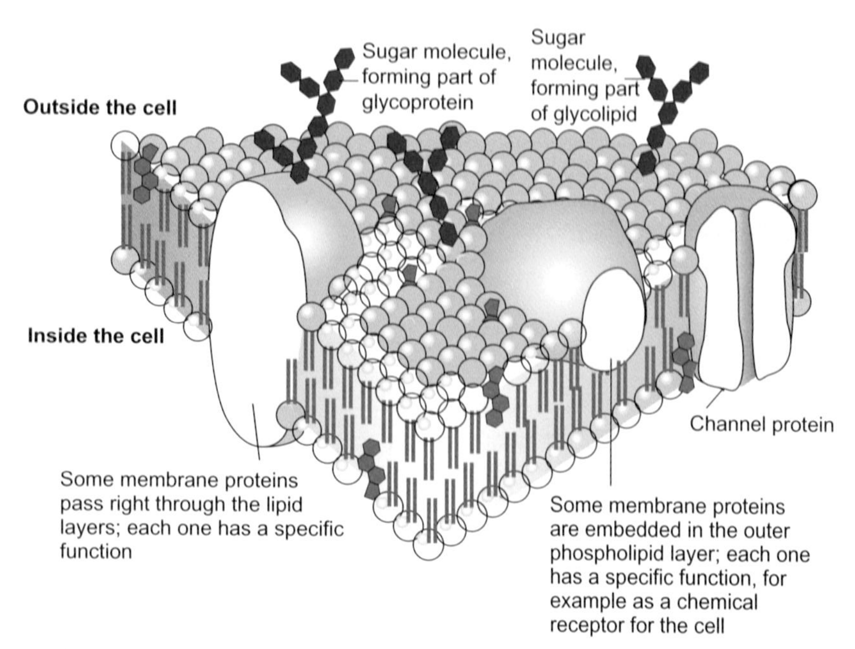

Figure 3 *A 3D representation of part of a cell-surface membrane*

You can see that some of the protein and lipid molecules have 'branches' attached to them. These are little chains of sugar molecules. These are therefore composite molecules – part protein or lipid, and part carbohydrate. They are known as **glycoproteins** and **glycolipids**, respectively. ('Glyco' means 'to do with sugar'.) You will notice that these sugar chains are always on the outer surface of the membrane. They have roles in interacting with molecules that arrive at the cell surface from its environment.

In among the phospholipid molecules there are also molecules of another lipid-like substance – **cholesterol**. Cholesterol is an essential component of cell membranes, where it provides stability, by preventing too much movement of the other molecules in the cell membrane. Cholesterol is not normally found in the cell membranes of prokaryotic cells, and is found in only small quantities in plant cell membranes. Animal cell membranes often contain quite large amounts of cholesterol.

The membranes shown in Figures 1, 2 and 3 are all cell-surface membranes. However, it is important to realise that all cell membranes have the same basic structure – a phospholipid bilayer interspersed with proteins. This includes the endoplasmic reticulum and Golgi apparatus, the two membranes around a nucleus, a mitochondrion or a chloroplast, and the membranes around lyosomes and vacuoles. These membranes can differ in the relative proportions of phospholipids, cholesterol and proteins, and in the quantity and orientation of glycoproteins and glycolipids. They also differ in the specific proteins present, which determine the properties of particular membranes.

KEY IDEAS

- All cell membranes have the same basic structure – a bilayer of phospholipids, arranged with their hydrophilic heads pointing outwards and their hydrophobic tails inwards.
- Proteins can be found in the bilayer. Some of them can move around in their layers, but do not normally swap layers.
- Some proteins and phospholipids have short chains of sugar molecules attached to them, and are called glycoproteins and glycolipids.
- Cholesterol is a type of lipid that is found in most eukaryotic cell membranes. It helps to stabilise the membrane.

6.2 DIFFUSION AND FACILITATED DIFFUSION

All of the particles in liquids and gases are in constant random motion. This motion results in a net movement of particles from a region of high concentration to a region of lower concentration (Figure 4). This process is called **diffusion**.

Figure 4 *Illustrating diffusion with a tea bag; the molecules responsible for the colour and flavour of the tea move randomly in the water, eventually spreading out evenly.*

QUESTIONS

3. Look at the photograph of the tea bag in the beaker (Figure 4). Why is the solution darker near the tea bag?

In a mixture of gases, diffusion causes each gas to spread evenly through the space that the mixture occupies. In the same way, a solute spreads through a liquid until it is evenly dispersed.

It is very important to understand that diffusion is the result of random motion of the particles – they do not move purposefully from one place to another. They just move about randomly, bouncing off in a different direction when they hit another particle. The energy involved is simply their own kinetic energy.

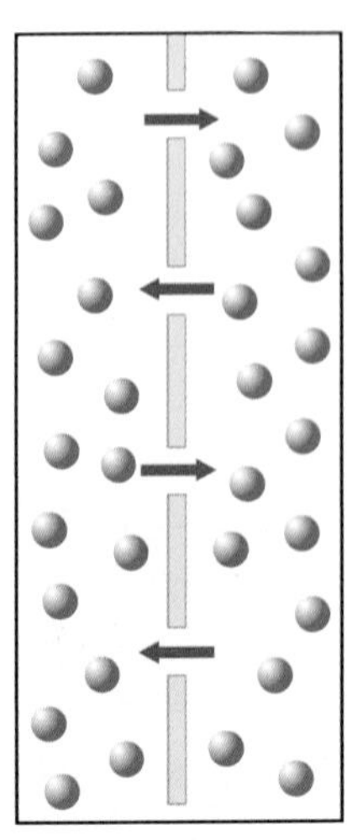

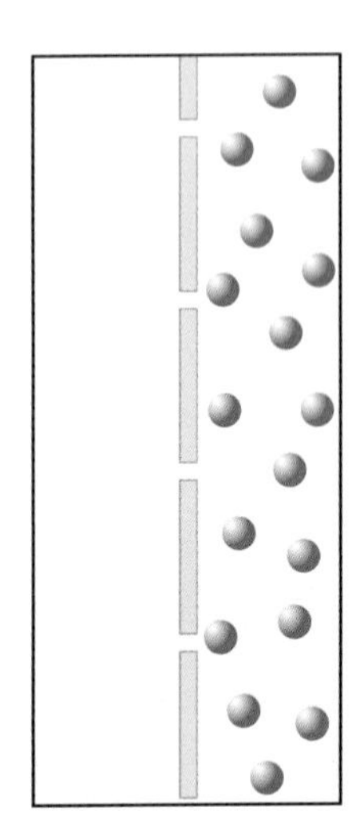

Pores in this membrane are wide enough to allow diffusion

Pores in this membrane are too narrow to allow diffusion

Figure 5 *Diffusion can take place through a membrane so long as there are gaps large enough for the particles to pass through.*

Particles of gas and some solutes can also diffuse through a membrane (Figure 5). This includes cell membranes. Because diffusion does not require the cell to do anything to make it happen, such as an input of energy from ATP, diffusion of a substance across a cell membrane is said to be a **passive** process.

Factors affecting the rate of diffusion

The rate at which diffusion occurs depends on several factors, including:

- the difference in concentration between two volumes, called the **concentration gradient**
- the length of the diffusion pathway between the volumes
- the size of the molecules or ions that are diffusing
- the temperature – the higher the temperature, the greater the kinetic energy of the particles, so the faster they move.

So, the greater the concentration gradient and the smaller the particles, the quicker the net movement of molecules from the area of high concentration to the area of low concentration.

Diffusion between two areas also happens faster when these volumes are only microscopic distances apart, rather than being separated by much larger distances. For example, it would take a small molecule such as oxygen at least four minutes to

ASSIGNMENT 1: MEASURING THE SURFACE AREA OF BLOOD CELLS

(MS 1.2, MS 2.3, MS 3.1, MS 4.1, PS 3.1)

In this assignment, we look at the efficiency of diffusion in a biconcave cell compared with a spherical cell.

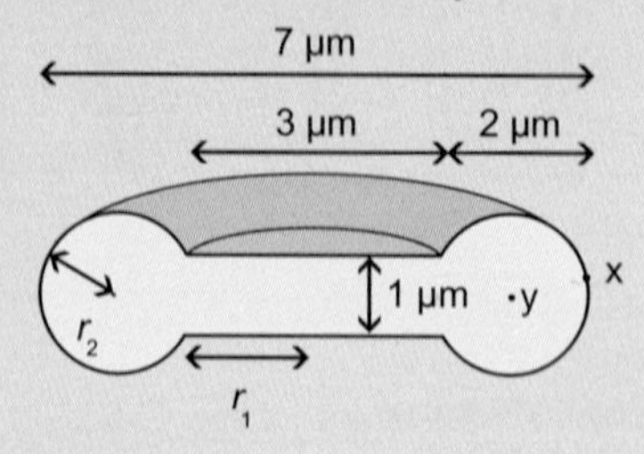

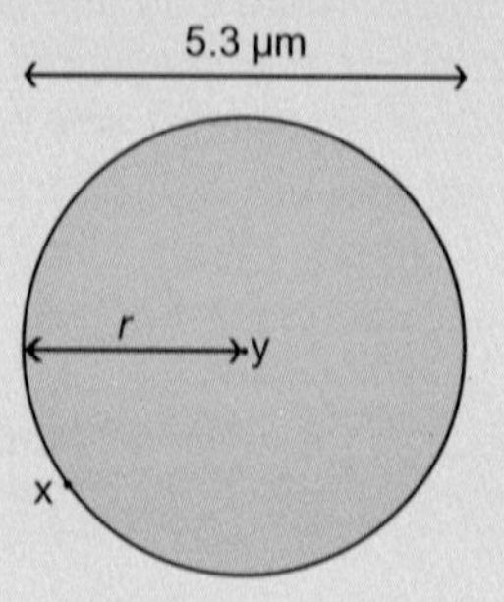

***Figure A1** A cross-section of a red blood cell (left) and a section through a spherical cell with the same volume as the red blood cell (right)*

The surface area of a sphere is:

$$\text{Area} = 4\pi r^2$$

The equation for the surface area of a biconcave disc is very complicated. However, we can come up with a rough value by treating it as a disc:

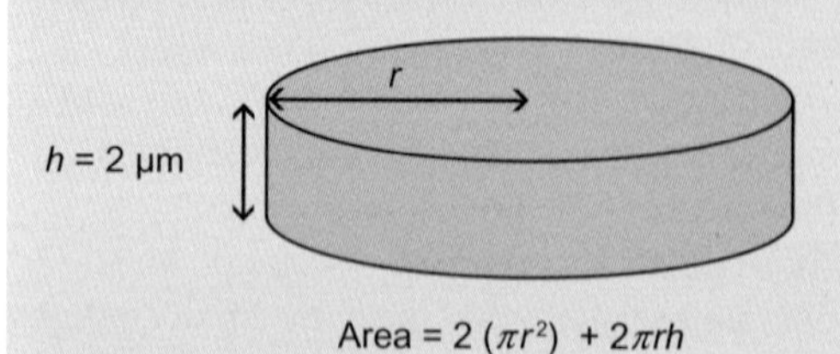

$$\text{Area} = 2(\pi r^2) + 2\pi rh$$

***Figure A2** The blood cell drawn as a disc*

Distance that molecules diffuse	Time required for molecules to diffuse
1 µm	0.4 milliseconds
10 µm	50 milliseconds
100 µm	5 seconds
1000 µm (1 mm)	8.3 minutes
10 000 µm (1 cm)	4 hours

Table A1

Questions

A1. Using the formulae given, estimate the surface area of each of the two cells.

A2. Use the information in Table A1 to calculate the mean speed at which particles diffuse across a distance of:

a. 10 µm

b. 1 nm.

A3. How long would it take for a molecule of oxygen to diffuse from the point marked x to the point marked y on:

a. the red blood cell?

b. the spherical cell?

Stretch and challenge

A4. If red blood cells are put into a dilute salt solution, the cells swell. Look at the change in volume of the red blood cell in Figure A3. Explain how the rate of diffusion of oxygen into the cell would change if the red blood cell swells.

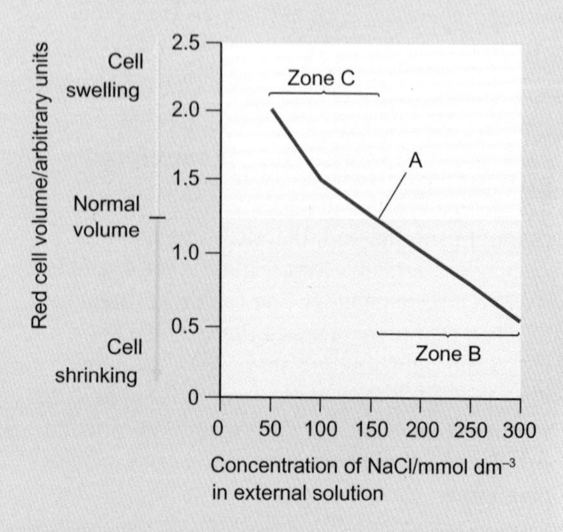

***Figure A3** Red blood cells and salt*

move to the centre of a cell 1 mm in diameter by its random movement.

The overall rate at which a substance diffuses through a membrane also depends on the surface area in contact with the substance. You probably remember that red blood cells have an unusual shape, and that their function is to transport oxygen in the blood. The biconcave-disc shape of a red blood cell gives it a much greater surface area than if it were spherical, allowing large numbers of oxygen molecules to move in and out of it at the same time. Look out for other examples of cells that maximise their surface area for efficient diffusion later in this chapter.

Diffusion through cell membranes

Many different substances diffuse through the cell-surface membrane. Substances consisting of small molecules, or with molecules that are hydrophobic (non-polar) can diffuse through the phospholipid bilayer because they are lipid soluble. For example, oxygen, carbon dioxide and ethanol (alcohol) molecules diffuse freely through cell membranes in this way.

Larger molecules, and substances whose particles carry significant charges, cannot do this. Big molecules cannot diffuse through the phospholipid bilayer. Ions are water soluble and thus unable to diffuse through the hydrophobic phospholipid bilayer. These substances can only diffuse through the membrane if there are special channels that allow them to do so.

Cells can contain special channels for ions to diffuse through. The channels are made from protein molecules and these are called – not surprisingly – **channel proteins**.

The **channel proteins** and, therefore, the channels, are specific for particular ions. Figure 6, for example, shows a sodium ion channel. The channel protein is organised so that it has a space through its centre that is just the right size and shape for a sodium ion to diffuse through it. Any sodium ion that happens to 'hit' an open channel as it is moving around randomly can just slip through the channel to the other side of the membrane.

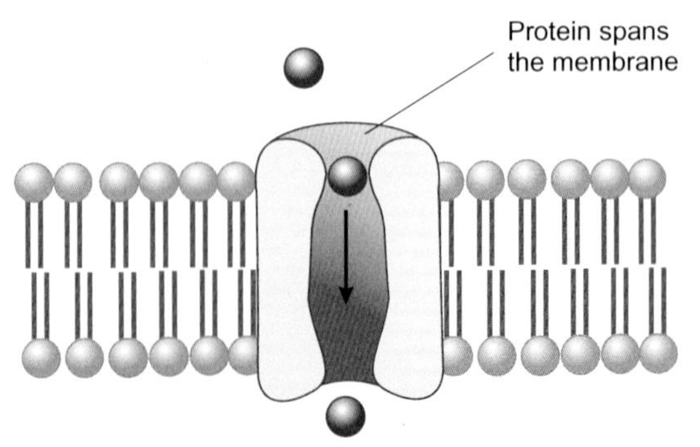

Figure 6 *Facilitated diffusion of sodium ions through a cell membrane*

This is still diffusion. It is still a passive process – the cell is not doing anything to make it happen, other than providing a channel. It is called **facilitated diffusion**. 'Facilitated' means 'made easy', and the channel protein is simply making it easy for the sodium ions to move through.

Nevertheless, the cell does have control over the movement of the ions. This is because it can change the numbers of channel proteins for a particular ion in its membrane. The more channel proteins for sodium ions, the more rapidly sodium can diffuse through the membrane. The cell also controls whether these channels are open or closed, which can be done almost instantaneously by small changes in the shape of the protein. Channels that can be open or closed are called **gated channels**. Various things can make the channels open or close. In nerve cells, for example, changes in electrical charge make sodium ion channels open or close. Channels that respond to voltage changes are called **voltage-gated channels**. Other channels may respond to other signals, such as the arrival of a particular hormone molecule at the cell-surface membrane.

Glucose molecules are quite large, and they have small charges on their –OH groups. This helps them to dissolve in water, but it makes it impossible for them to pass between the phospholipid tails in a cell membrane. Once again, cells provide specific proteins in the membrane for glucose molecules to move through. The glucose molecules bind temporarily with the protein, which then changes shape so that the glucose passes through it to the other side of the membrane (Figure 7). These proteins are called **carrier proteins**.

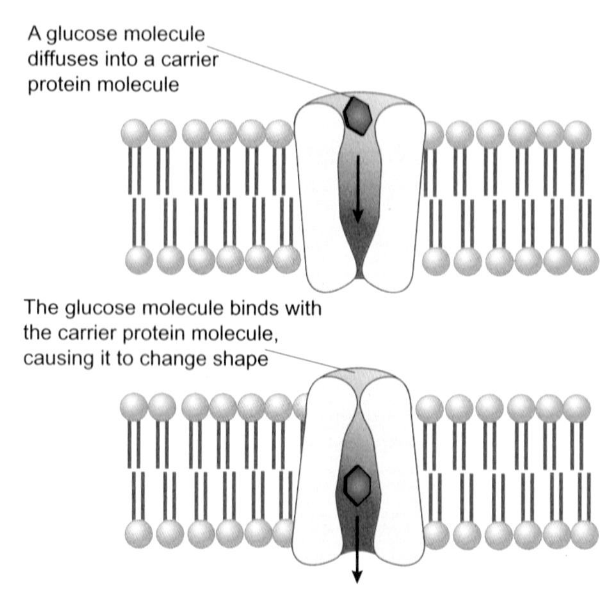

Figure 7 *Facilitated diffusion of glucose molecules through a cell membrane*

Carrier proteins differ from channel proteins only in that they interact a little bit more with the molecule they are transporting. Neither of them provide energy to make the ion or molecule move, so in both cases the movement is still entirely passive.

Both channel and carrier proteins are highly selective – because of their specific tertiary structure, they will only allow one particular type of ion or molecule to pass through or to bind with them. Most cell membranes contain a large variety of different channel and carrier proteins, to allow the diffusion of all kinds of different molecules and ions into and out of the cell. Many of them have their own names; for example, carrier proteins that allow glucose to move into and out of cells are called glucose permeases.

KEY IDEAS

- Diffusion is the result of the random motion of particles of gases, liquids and solutes. The net movement of the particles is from high to low concentration – that is, down their concentration gradient.
- Diffusion is a passive process, relying only on the kinetic energy of the moving particles.
- Small molecules, and larger ones with no charge, are able to diffuse freely through the phospholipid bilayer of cell membranes.
- Other particles, such as large molecules and ions, can only diffuse through protein channels in the cell membrane. This is called facilitated diffusion. Like simple diffusion, it is an entirely passive process.
- Proteins that simply provide a channel through which particles move are called channel proteins. If they interact with the particles and 'help' them through, they are called carrier proteins.
- Many channel proteins are gated – that is, they can open and close, so allowing or preventing a particular substance to pass through.

Worked maths example: Percentage increase in glucose uptake

(MS 0.3, MS 3.2, PS 3.1)

Imagine that you have carried out an experiment to investigate glucose uptake in red blood cells. Using the latest equipment, you have been able to measure glucose concentration in the blood after a meal, both inside and outside the red blood cells.

Time/ seconds	Glucose concentration in red blood cells/ $\mu mol\ dm^{-3}$	Glucose concentration in blood plasma/ $\mu mol\ dm^{-3}$
0	0	56
5	20	55
10	40	50
15	50	40
20	55	20
25	56	0

- Draw a graph of Glucose concentration in red blood cells against Time, and calculate the percentage increase in glucose concentration in red blood cells from 5 to 25 seconds.

First, you need to select the correct data. The question asks for glucose concentration in red blood cells, so we only need the first two columns of the table:

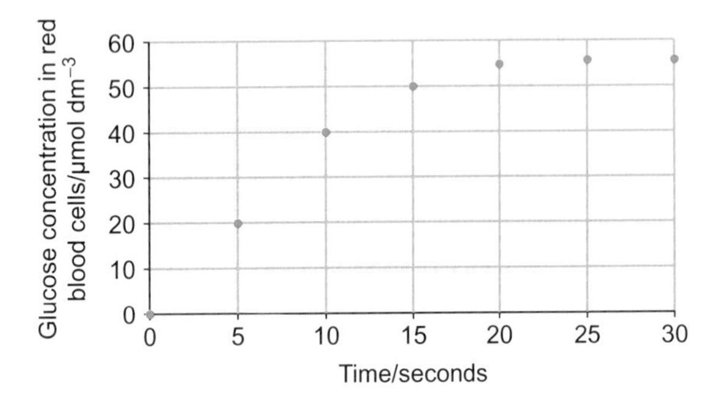

To calculate percentage increase, you use:

$$\text{percentage increase} = \frac{(\text{new value} - \text{original value})}{\text{original value}} \times 100$$

At 5 seconds, the glucose concentration in the red blood cells is 20 μmol dm^{-3}. This is the original value.

At 25 seconds it is 56 μmol dm^{-3}. This is the new value. So:

$$\text{percentage increase} = \frac{(56 - 20)}{20} \times 100$$

percentage increase = 180%

- The experiment continued for a further 30 seconds and ended at one minute. Use your graph of the data to estimate a likely glucose concentration at one minute.

- First, draw a line of best fit on the graph. The rate of glucose concentration begins to slow from about 10 seconds onwards. Following the line of best fit, it is likely that after one minute the glucose concentration would still be 56 μmol dm^{-3}.

6.3 OSMOSIS

The cytosol of every cell is a solution of many different substances in water. Cells are also surrounded by watery solutions. The cell-surface membrane is therefore being continually bombarded by water molecules from both sides, as these molecules move about randomly.

We have seen that water molecules have dipoles – they carry a small charge. This means that they do not associate easily with the hydrophobic (non-polar) tails of the phospholipid molecules that make up cell membranes. However, water molecules are also very small, so it is possible for them to slip through temporary gaps that open up as these tails wiggle around.

Cells can also put channel proteins for water molecules into their membranes, to make it even easier for water molecules to move through. These channel proteins are called **aquaporins**.

So, water can diffuse through cell membranes. This is happening all of the time, in every living cell. Because water makes up so much of the solution on both sides of the membrane, and because so many of the substances dissolved in it cannot diffuse through, this

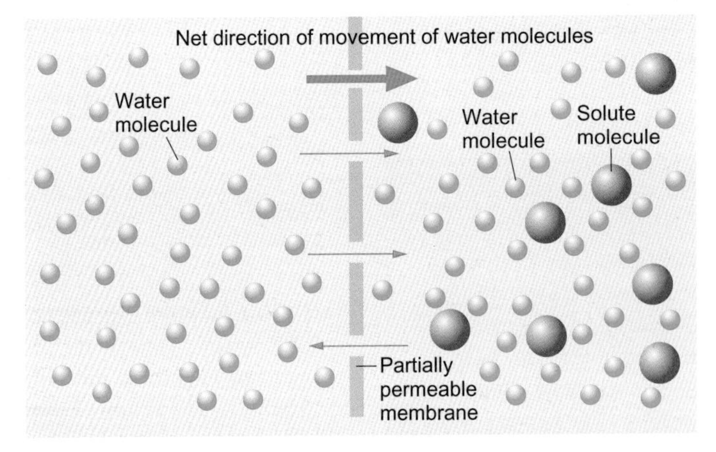

Water molecules pass through the membrane in both directions. The net movement is from the pure water (on the left in the diagram) to the solution. The larger molecules cannot pass through the small pores in the membrane.

***Figure 8** How osmosis happens through a Visking tubing (an artificial, partially permeable membrane)*

diffusion of water is of great importance. It is given its own name – **osmosis**.

Figure 8 shows how osmosis happens through a partially permeable membrane. Two solutions are separated by the membrane that contains pores large enough for water molecules to pass through, but not large enough for the solute molecules to pass through. Water molecules diffuse through the membrane, but the solute molecules have to remain where they are. In cell membranes, osmosis is not quite so simple. Cell membranes are **selectively permeable** – they discriminate on more factors than just the the size of the molecules.

QUESTIONS

4. What is the main difference between osmosis and other types of diffusion?

Water potential

We have seen that diffusion happens down a concentration gradient – it is the net movement of particles from a high concentration to a low concentration. The term 'concentration' describes the number of solute particles in relation to the amount of water in a solution. We therefore need a different term to describe the amount of water. This term is **water potential**.

You can think of water potential as the tendency of water to move out of a solution. Water moves, by osmosis, from a higher water potential to a lower water potential. We say it moves down a **water potential gradient**. In Figure 8, the pure water on the left has a higher water potential than the solution on the right, so the net movement of water is from left to right.

What determines the water potential of a solution? In general, the more water molecules there are in a given volume of the solution, the higher its water potential. Earlier in this book, we saw that water molecules interact with solute molecules, and this reduces their freedom of movement (Figure 9). The fewer solute molecules there are, the more freely the water molecules can move. A dilute solution, therefore, has a higher water potential than a concentrated solution. Water will move, by osmosis, from a dilute solution to a concentrated solution.

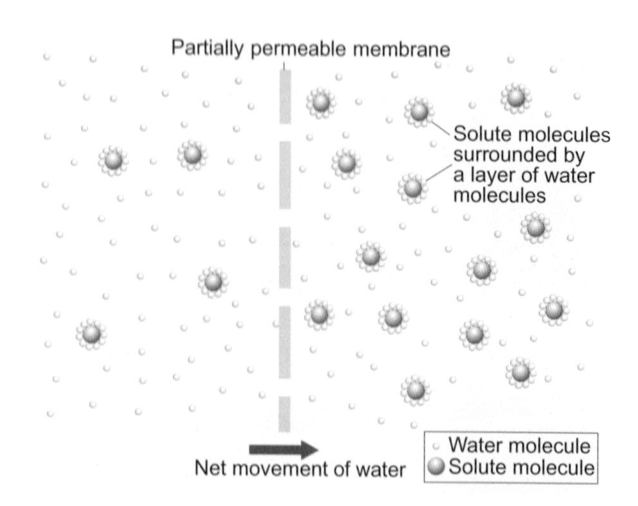

Figure 9 *The presence of solute molecules reduces the freedom of movement of water molecules, and so reduces water potential.*

The importance of osmosis to cells

All cells contain cytoplasm, and most of this cytoplasm consists of a watery solution of many different molecules and ions. The cytoplasm is separated from the environment of the cell by the cell-surface membrane, which is partially permeable. If the water potential of the solution outside the cell differs from the water potential of the solution inside it, then there will be a net movement of water either into or out of the cell by osmosis.

Figure 11 shows what this means for a red blood cell. In the top diagram, the concentration of the solution outside the cell is the same as the concentration of the cytoplasm. As pressure is also the same for both of them, their water potentials are the same. There is, therefore, no net movement of water into or out of the cell (although, of course, individual water molecules will constantly be zipping in and out).

In the middle diagram, the solute concentration of the external solution is less than that of the contents of the cell. The external solution, therefore, has a higher water potential than the cell contents. Water moves into the cell by osmosis.

In the final diagram, the solute concentration of the external solution is greater than that of the contents of the cell. The external solution, therefore, has a lower water potential than the cell contents. Water moves out of the cell by osmosis.

Animal cells have no support around them apart from their thin, flexible cell-surface membranes. If a cell takes up extra water, its volume increases and the cell swells. If water keeps on going in, it may eventually burst. On the other hand, if a lot of water moves out, then the cell shrinks.

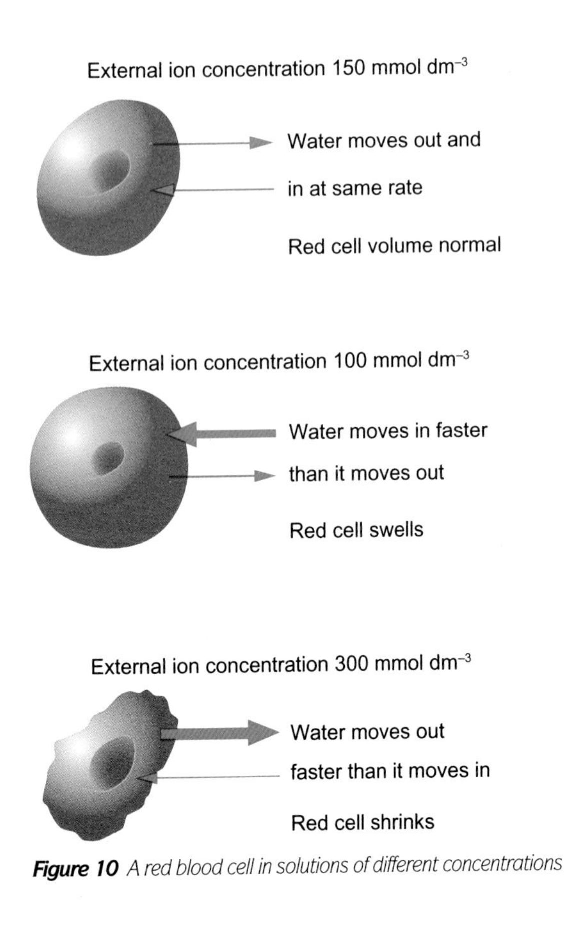

Figure 10 A red blood cell in solutions of different concentrations

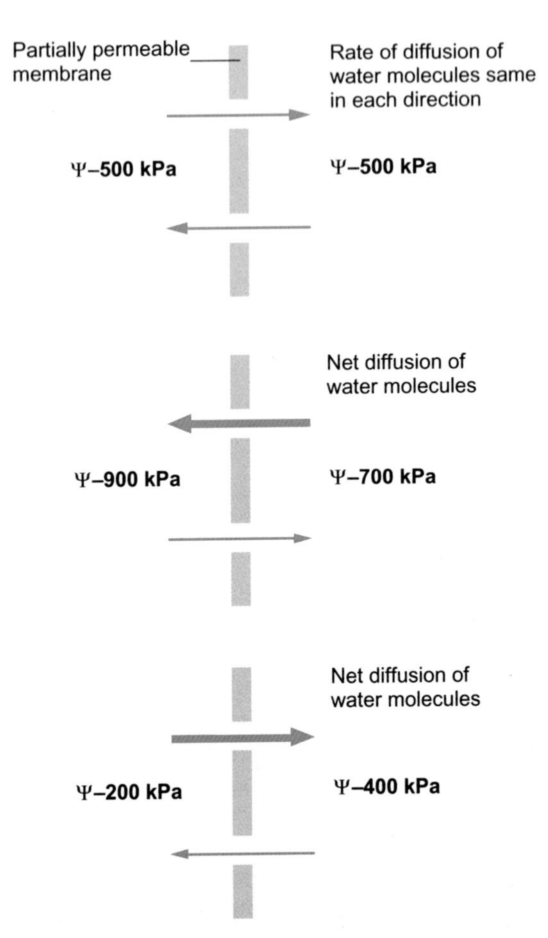

Figure 11 Water always moves down a water potential gradient.

Units for water potential

Water potential is often represented by the Greek letter psi (pronounced 'sigh'). The symbol for this is Ψ. Water potential is measured in pressure units, usually in kilopascals, kPa.

By definition, pure water at atmospheric pressure has a water potential of 0. As we have seen, solutions have lower water potentials than pure water. This means that (rather inconveniently) solutions have negative water potentials.

Solutions that have the lowest water potential have the largest negative values. A solution with a water potential of – 200 kPa, for example, has a lower water potential than a solution with a water potential of –100 kPa. The net movement of water molecules is always towards a region of lower water potential – to where the water potential is relatively more negative, as shown in Figure 11.

QUESTIONS

Stretch and challenge

5. Figure 12 shows the water potential of three cells that are in contact.

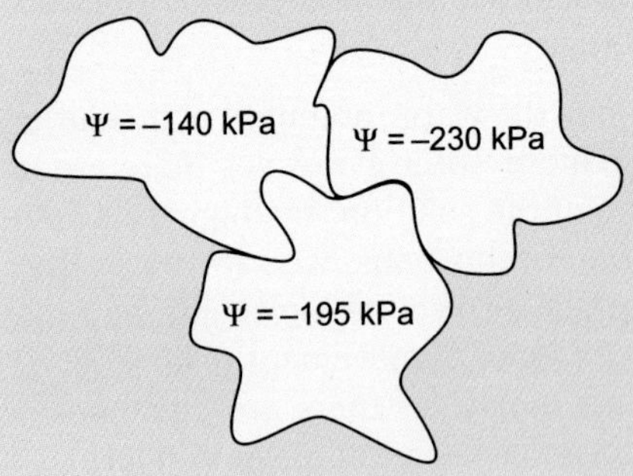

Ψ = –140 kPa | Ψ = –230 kPa | Ψ = –195 kPa

Figure 12 Water potential of three cells in contact

a. Copy the diagram. Use arrows to show the net direction of water movements between the three cells.

b. Which cell has the most negative water potential, and which the least negative?

6. Water molecules can diffuse between the phospholipid molecules in the cell-surface membrane. But they do so at a rate fifteen times slower than the rate that they diffuse in pure water. Suggest why.

ASSIGNMENT 2: APPLYING THE CONCEPT OF WATER POTENTIAL

(PS 1.2)

Energetic dancing in a nightclub is like any other strenuous exercise – it makes you sweat and it can leave you dehydrated. Ecstasy is an illegal drug that is sometimes used by dancers. People have died after taking ecstasy in a club and dancing all night. They did not overdose on the drug; they seem to have overdosed on water. Ecstasy seems to induce repetitive behaviour in some people – this has occasionally led to people drinking 20 litres of water or smoking 100 cigarettes in the space of three hours.

Figure A1 Dancing can leave you dehydrated.

The urge to drink water constantly may also have been encouraged by the popular belief that water is an antidote to ecstasy (it is not), or the drug may affect the way that the brain controls the body's water balance. Whatever the reason, people have collapsed because the large amount of water that they have drunk has diluted their blood so much that they have developed a condition called oedema. This happens when cells and tissues in the body absorb too much water and swell; if this happens in the brain it is really bad news. As brain tissue swells it squeezes against the skull and blood vessels and brain cells become squashed. If the centres in the brain that regulate breathing and the beating of the heart are damaged irreversibly, death is inevitable.

Questions

Stretch and challenge

A1. Use the concept of water potential to explain why drinking large amounts of pure water after sweating may lead to oedema in the brain.

A2. Would it be a good idea for clubs to offer free isotonic drinks instead of water? What would you say to a club manager who asked for your advice?

REQUIRED PRACTICAL ACTIVITY 3: APPARATUS AND TECHNIQUES

(MS 0.3, MS 3.2, MS 3.4, PS 3.1, PS 2.2, PS 3.2, PS 4.1, AT c, AT h, AT j, AT l)

Production of a dilution series of a solute to produce a calibration curve with which to identify the water potential of plant tissue

This practical activity gives you the opportunity to show that you can:

- Use laboratory glassware apparatus for a variety of experimental techniques to include serial dilutions.
- Safely and ethically use organisms.
- Safely use instruments for dissection of an animal or plant organ.
- Use ICT.

Apparatus and techniques

Plant tissue

You could investigate many different plant tissues. Common ones include potato, carrot and apple.

You need to ensure that your pieces of tissue are as uniform as possible. One easy way to do this is to use a cork borer or potato chipper. Cubes or pieces could be cut with a knife or scalpel, but it is much more difficult to ensure consistency. All pieces of plant material should come from the same plant.

If using a cork borer, you will need to remove the cylinder by pushing through the borer with a plastic rod or a piece of wooden dowel.

For safety reasons, when using a cork borer, ensure that you do not hold the plant material in your hand; place it on a white tile.

If your plant material has a skin on it, this must be removed.

Once you have obtained your plant material, you need to take its initial mass using a digital balance to two decimal places. Make sure you pat the material dry (do not squeeze, but pat it gently with a paper towel). Remember to record the mass of your potato cylinders.

If you are leaving the material in solution for more than one hour, the plant material can be left as a cylinder. However, if you are in a hurry, you could cut it into small discs.

Solutions

The usual solutions to investigate are sodium chloride or sucrose. An interesting variation is to use fruit squash, which can be diluted. This has the advantage that the different dilutions can be distinguished visually, so you are much less likely to get them mixed up.

You may be provided with solutions already made up, but if not you will need to produce a series of dilutions for yourself. To get good results you need solutions below 0.5 mol dm^{-3}; so if you are provided with 1.0 mol dm^{-3} solution, you will need to add equal amounts of distilled water to make it 0.5 mol dm^{-3}.

Table P1 shows how to make up 100 cm^3 of some dilutions as an example.

Required solution/ mol dm^{-3}	Volume of 1 mol dm^{-3} solution/cm^3	Volume of distilled water/cm^3
0.5	50	50
0.4	40	60
0.3	30	70
0.2	20	80
0.1	10	90

Table P1

Ensure that you label your solutions and that the volume is sufficient to cover your plant material completely.

When you are ready to immerse the plant material in solution, you will need to make a note of the time, as all pieces need to be in solution for the same amount of time. You will not need to time how long the pieces are in solution to the second, but there should be only a very small difference in time. If you are working with a partner, this will be much easier.

Once the pieces are immersed in the solution (in a beaker or boiling tube) the container should be covered and left for a given amount of time, after which the pieces should all be removed and reweighed as quickly as possible. Again, the piece should be patted dry to remove excess solution prior to weighing. The final mass of each piece should be recorded.

Data analysis

For each plant tissue, you should have an initial and final mass and will need to calculate the percentage change in mass, using:

$$\text{percentage change in mass} = \frac{\text{initial mass-final mass}}{\text{initial mass}} \times 100$$

Some pieces of plant tissue will have gained mass and will have a positive percentage change in mass; some will have lost mass and will have a negative percentage change in mass.

The next step is to plot a graph of percentage change in mass of your plant tissue against the concentration of the solution. Add a calibration curve. The point at which this curve crosses the x-axis is the point at which the plant tissue has zero change in mass. This represents the point at which the external solution has a concentration with a water potential equal to that of the plant tissue.

You can use data tables, such as Table P2, to find the water potential of the solution at various concentrations and, therefore, the water potential of the plant tissue.

Sucrose concentration/ mol dm^{-3}	Water potential/kPa
0.5	– 1320
0.4	– 1060
0.3	– 800
0.2	– 540
0.1	– 260

Table P2

QUESTIONS

P1. Explain why it is important not to squeeze your cylinders when drying them.

P2. Explain why the skin must be removed from your plant material.

P3. Explain why it is helpful to cut your cylinder into discs if you are short of time.

P4. Explain how you would make up 100 cm^3 of a 0.25 mol dm^{-3} solution from a stock solution of 1 mol dm^{-3}.

P5. Explain why it is important to use distilled water in this investigation.

P6. Explain why it is important that all of your plant material is covered with solution.

P7. Explain why you need to calculate percentage change in mass.

REQUIRED PRACTICAL ACTIVITY 4: APPARATUS AND TECHNIQUES

(PS 4.1, AT a, AT b, AT c, AT j, AT l)

Investigation into the effect of a named variable on the permeability of cell-surface membranes

This practical activity gives you the opportunity to show that you can:

- Use appropriate apparatus to record a range of quantitative measurements.
- Use appropriate instrumentation to record quantitative measurements.
- Use laboratory glassware apparatus for a variety of experimental techniques to include serial dilutions.
- Safely use instruments for dissection of an animal or plant organ.
- Use ICT.

Apparatus and techniques

Figure P1 Beetroot is commonly used to investigate cell-surface membrane permeability.

Cell-surface membrane permeability can be investigated using plant material that has a coloured sap – the most commonly used material is beetroot, but red cabbage can also be used.

The dark red and purple pigments in beetroot are in the cell vacuole and belong to a group of chemical compounds called betalains. These pigments cannot pass through cell-surface or vacuolar membranes, but can pass through cellulose cell walls if the membranes are disrupted – by heat (cooking, for example), by surfactants (substances that affect the surface tension of water, such as detergents) or some solvents (ethanol, for example).

As with the investigation into water potential, you need to obtain uniform pieces of plant material and the same methods can be used – a cork borer, a potato chipper or an egg slicer will all work well, or you could cut pieces using a white tile and scalpel or sharp vegetable knife. If you are cutting your own pieces of beetroot, you should rinse them thoroughly in running water until the water runs clear. Any skin must be removed. Your teacher may provide you with already washed pieces to save time. Remember that beetroot will stain fingers temporarily, and may stain your clothes, so a lab coat is essential. The procedure is the same if you are using red cabbage, but preparing uniform pieces is more difficult. As a constant size is important in this investigation you should measure the dimensions of your pieces carefully.

The most common factors to investigate are the effects of temperature or the effects of solvents. Increasing temperature will give the molecules in the cell membrane more kinetic energy and the phospholipids become more fluid, and proteins become denatured. Organic solvents will dissolve the phospholipids. As the membranes become disrupted, the pigment can escape from the vacuole and enter the liquid that the plant material is in. The more damage to the membrane, the darker the fluid will become.

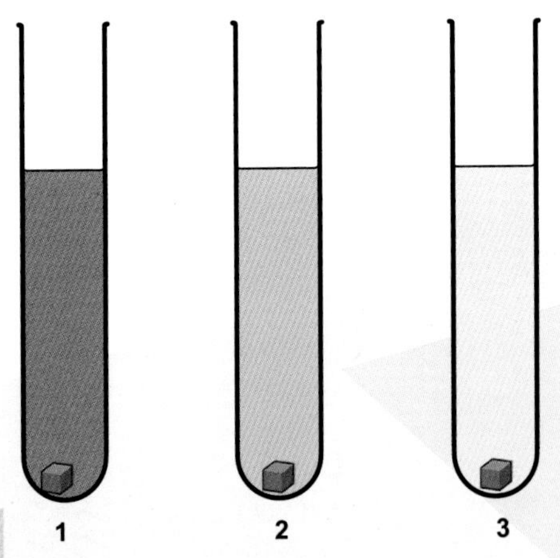

Figure P2 Different amounts of pigment leak from the beetroot as the membranes are disrupted.

The pieces of plant material should be placed in distilled water if you are investigating temperature, or into a solvent if you are investigating that factor. Ensure that volumes are kept constant.

Investigating temperature

You will need access to a range of water baths, which could be thermostatically controlled, or you may need to create your own using a beaker of water and adjusting it as the investigation proceeds.

Ensure that you keep your pieces in each temperature for the same time, and carefully remove them using forceps. You need to keep the liquid, but the beetroot pieces can be discarded.

Investigating solvents

You will need to use a range of solvent concentrations, and ensure that you keep the temperature constant. Again, ensure that the pieces are in the liquid for the same time and, again, remove them carefully.

Using a colorimeter

A colorimeter measures the amount of light absorbed by a liquid; the darker the liquid, the more light will be absorbed (there will be a higher absorbance). The liquid is poured into a special rectangular tube called a cuvette, which should be filled almost to the top but not completely. There are usually ridges on two sides of the cuvette (or sometimes two sides are opaque or have arrows on them) – these are the sides you should hold; the other two sides should be where the light is shone.

Set the colorimeter to respond to a blue/green filter (or wavelength of 530 nm) and check whether you are measuring absorbance or transmission. Check the colorimeter reading with distilled water before every reading.

If you do not have access to a colorimeter you will need a series of colour standards for comparison.

QUESTIONS

P1. Explain why the beetroot must be washed thoroughly before you start your investigation.

P2. Why is it important to have your pieces of plant material as uniform as possible?

P3. All pieces of beetroot should come from the same plant. Explain why.

P4. What would be the best way to analyse your results from this investigation?

KEY IDEAS

- Osmosis is a particular type of diffusion, in which water molecules diffuse through a selectively permeable membrane, such as a cell-surface membrane.
- The tendency of water to leave a solution is known as its water potential. Addition of solute decreases water potential. Applying pressure increases water potential.
- Water moves down a water potential gradient.
- Water potential is measured in kilopascals. Pure water at atmospheric pressure has a water potential of 0. Solutions have negative water potentials.

6.4 ACTIVE TRANSPORT

So far, we have seen how substances can move in and out of cells of their own accord, down their concentration gradients or water potential gradients. All of these types of movement require no input of energy from the cell – they are passive processes.

But cells often move substances into their cells that are only present in small concentrations outside them. Equally, they may move substances from inside them, which are present in smaller concentrations inside than outside. They can make these substances move against their concentration gradients.

In this situation the cell expends some energy, almost always derived from the hydrolysis of ATP. The process is therefore called **active transport**, because the cell is actively making it happen, using its own energy supply.

Some carrier protein molecules allow the cell to use active transport to move ions, or molecules such as glucose, against their concentration gradient (Figure 13). Animal and plant cells that specialise in doing this usually have abundant mitochondria that provide the ATP needed to power active transport.

You can see in Figure 13 that the protein molecules through which active transport takes place are behaving rather like the ones in Figure 7. They are binding with the molecule being transported, and changing shape to make it move through. So they,

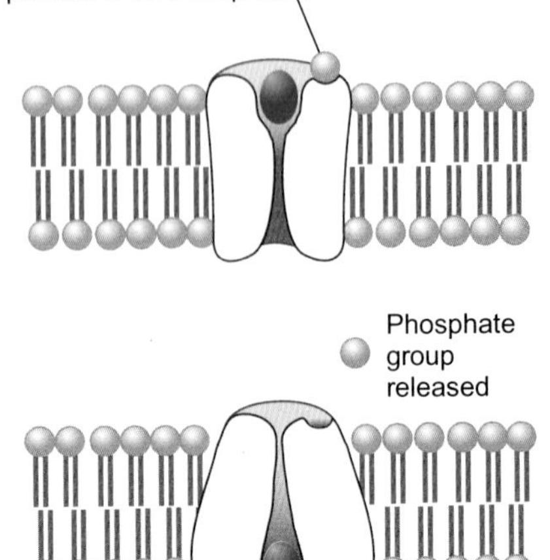

Figure 13 *Active transport involves the use of energy obtained from the hydrolysis of ATP.*

too, are called **carrier proteins**. The big difference here is that the carrier protein has to use energy from ATP to make the molecule move. The **hydrolysis of ATP** to ADP and P_i is essential in order for active transport to take place.

Co-transport

When we exercise and sweat heavily, we lose water and our performance suffers. Sports scientists now recommend that runners should take frequent drinks containing water, mineral ions and glucose during long-distance runs.

These drinks replace the sodium ions and chloride ions lost in sweat, and provide extra glucose for energy. Moreover, the sodium ions stimulate the rate of glucose uptake into the blood – so the athlete gets a quick 'glucose fix'. This happens because of specialised carrier proteins in the cell-surface membranes of the epithelial cells lining the small intestine (Figure 14). The glucose–sodium co-transporter protein transports glucose and sodium into the cell at the same time, but it only works when both substances are present.

At the junction between the epithelial cell and a blood capillary, glucose is transported into the blood by facilitated diffusion and sodium ions by active transport.

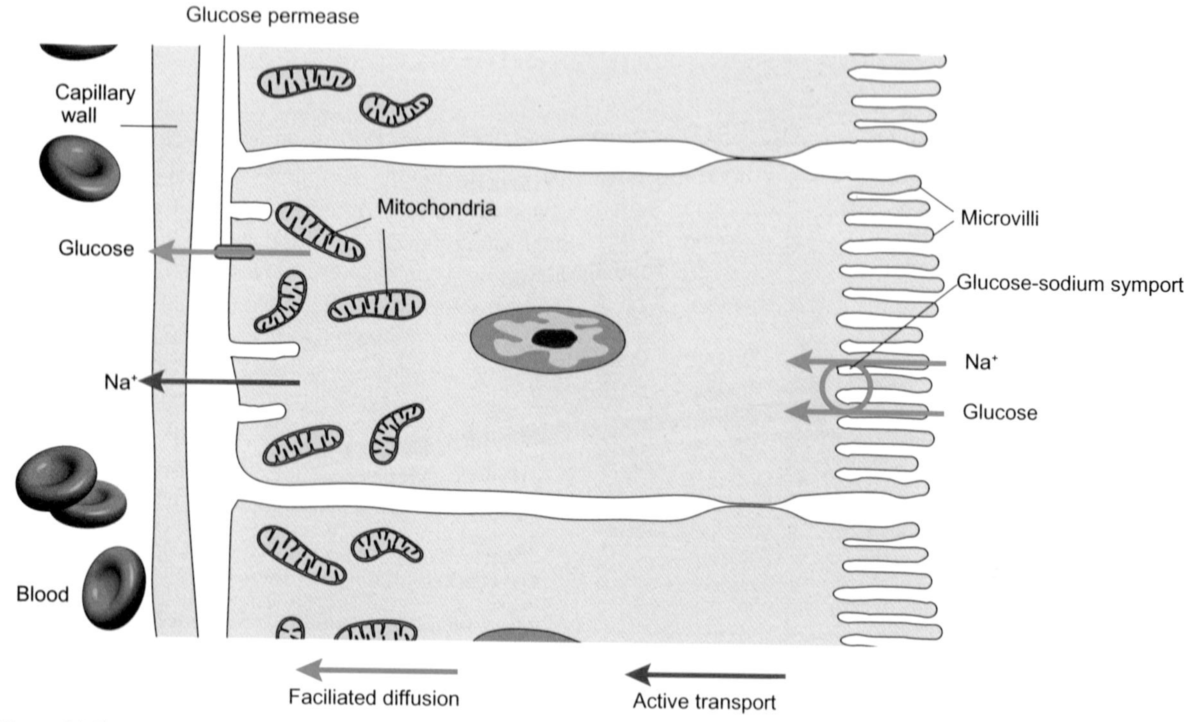

Figure 14 The uptake of sodium ions and glucose happens by a co-transport mechanism.

Figure 14 illustrates how particular cells can be specially adapted to allow substances to move rapidly across their membranes. These adaptations can include:

- having a large surface area (the cells lining the ileum have tiny folds called microvilli, which greatly increases surface area for inclusion of transporter proteins, leading to greater uptake of molecules or ions that can be crossing the membrane at the same time)
- having particularly large numbers of specific channel or carrier proteins in their membranes.
- having a lot of mitochondria (these supply the ATP needed for active transport to take place; the cell is only using ATP to actively transport sodium ions into the blood plasma. This creates and maintains the concentration gradient for sodium ions into the cell from the gut that, in turn, leads to co-transport of glucose).

QUESTIONS

7. Suggest why sports scientists recommend that the maximum concentration of sodium in sports drinks should be 200 mg per dm^3.

Stretch and challenge

8. At the junction between the epithelial cell and a blood capillary, glucose is transported into the blood by facilitated diffusion and sodium ions by active transport. Suggest why different processes are required to transport these two substances from the epithelial cell into the blood.

KEY IDEAS

- Active transport is the movement of molecules or ions against their concentration gradient.
- Active transport requires energy input from the cell. This is provided by the hydrolysis of ATP to ADP and P_i.
- Active transport takes place through carrier proteins in the cell membrane.
- In co-transport, two substances (for example, sodium ions and glucose) are transported together through the same carrier protein.
- Cells may be adapted to allow transport to take place rapidly across their cell membranes. This can involve having a large surface area, having large numbers of mitochondria to supply ATP for active transport, and having large numbers of specific channel proteins or carrier proteins in their membranes.

ASSIGNMENT 3: TREATING KIDNEY DISEASE

(PS 1.2)

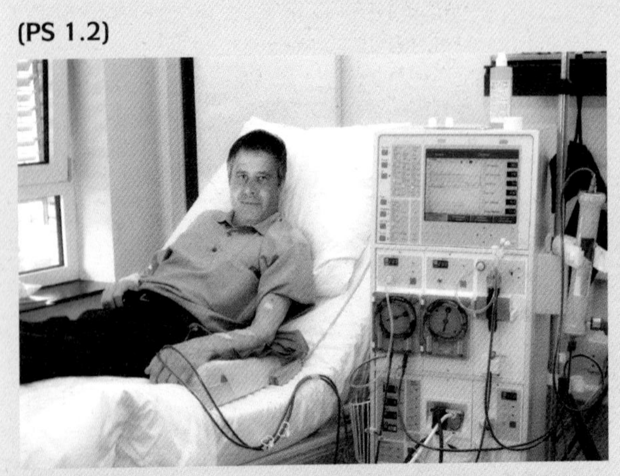

Figure A1 This man is undergoing dialysis.

Normally, the blood is filtered by the kidneys, which remove wastes such as urea, excess minerals and water. When a person has kidney disease, their kidneys do not function properly (or may not function at all). One of the most common treatments is the use of a dialysis machine. The first dialysis machine was built in 1943 from sausage skins, drinks cans and an old washing machine, but it was not until 1945 that a patient had successful treatment.

In a dialysis machine, blood is removed from a vein (usually in the arm) and returned – after filtering – back into the same vein. Within the dialysis machine is a series of plates that are covered with a partially permeable membrane. On one side of the plates, the patient's blood flows. On the other side, dialysis fluid flows in the opposite direction. The dialysis fluid contains water and solutes at the same concentration as normal plasma. Small molecules and ions can pass from the blood to the dialysis fluid, while large molecules and blood cells cannot.

Questions

A1. How do these substances move from the blood into the dialysis fluid?

a. Urea

b. Water

A2. Normal plasma contains 132 mmol dm^{-3} sodium ions and 96 mmol dm^{-3} chloride ions. What will be the concentration of these ions in the dialysis fluid?

A3. The dialysis fluid is often heated to 40 °C. Suggest a reason for this.

Stretch and challenge

A4. Suggest why the blood and dialysis fluid flow in opposite directions.

ASSIGNMENT 4: UNDERSTANDING GRAPHS OF TRANSPORT ACROSS MEMBRANES

(MS 1.3, MS 3.1, PS 3.1)

Substances can be transported across membranes in a variety of ways. Some of these mechanisms are simple and others are more complex.

Simple diffusion will transport lipid-soluble molecules, water and gases but will stop when equilibrium occurs.

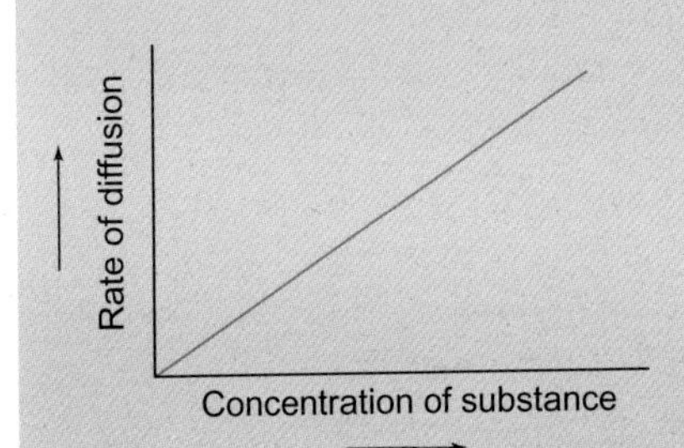

Figure A1 A graph of transport by diffusion against concentration

The rate of transport across the membrane increases as the concentration increases. There is a simple linear relationship.

Facilitated diffusion requires carrier proteins and these may become limiting in the rate of diffusion. Diffusion reaches a maximum rate called V_{max} where the graph levels off.

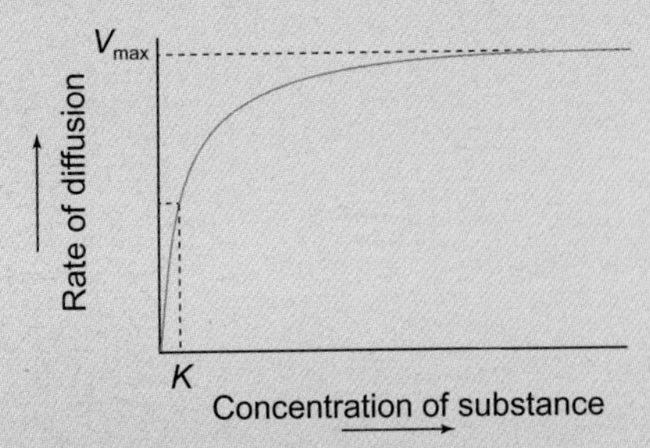

Figure A2 A graph of transport by facilitated diffusion against concentration gradient

Active transport produces a similar graph to facilitated diffusion if the rate of diffusion is plotted against concentration, because the membrane proteins become limiting.

However, if we plot rate of *diffusion of a substance into a cell* against *time*, we can distinguish these two methods of transport.

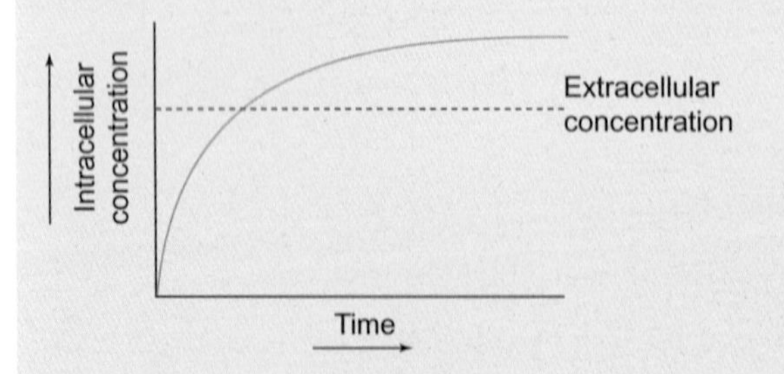

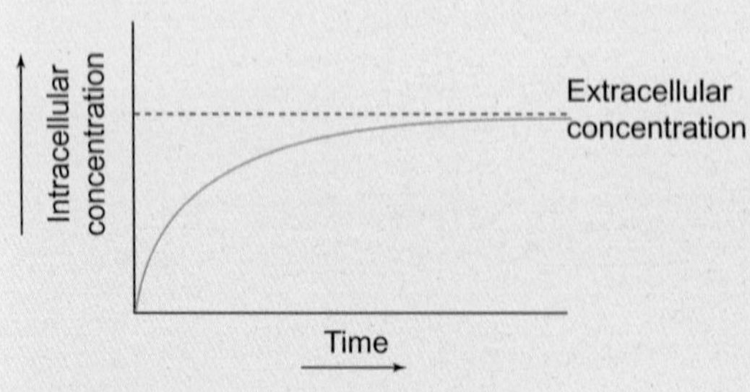

Figure A3 *The top graph is for active transport, whereas the graph on the bottom is for facilitated diffusion.*

Questions

Stretch and challenge

A1. Figure A4 shows the result of an investigation into glucose uptake by red blood cells.

a. Explain the shape of the curve at concentrations below 2 mmol dm^{-3}.

b. Explain the shape of the curve at concentrations above 5 mmol dm^{-3}.

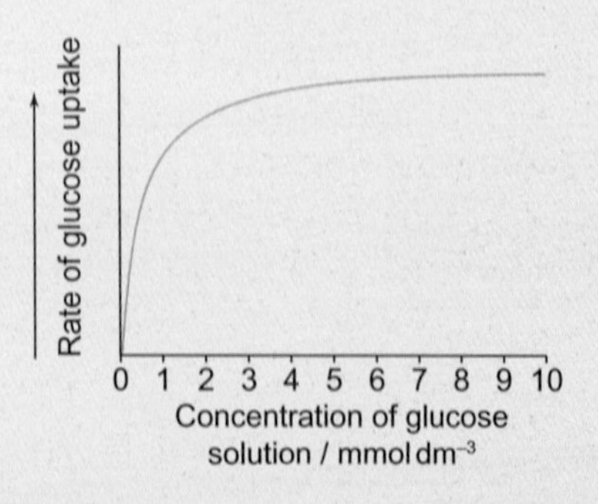

Figure A4

A2. Figure A10 shows the active uptake of magnesium ions by root hair cells.

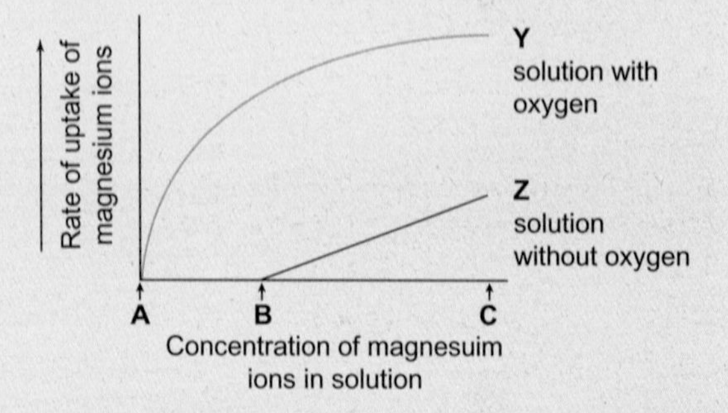

Figure A5

a. Explain the shape of curve Y.

b. Explain why there is no uptake of magnesium in the solution without oxygen between points A and B.

c. Explain the shape of curve Z between points B and C.

PRACTICE QUESTIONS

1. A student placed cylinders cut from a potato into sucrose solutions of different concentrations. He put 10 cylinders in each solution. He left them for one hour. The length of each cylinder was measured before and after immersion in sucrose solution. The graph in Figure Q1 shows the results of the experiment. To make the graph easier to plot, ratios are plotted as single numbers (for example, 1.1 instead of 1.1:1).

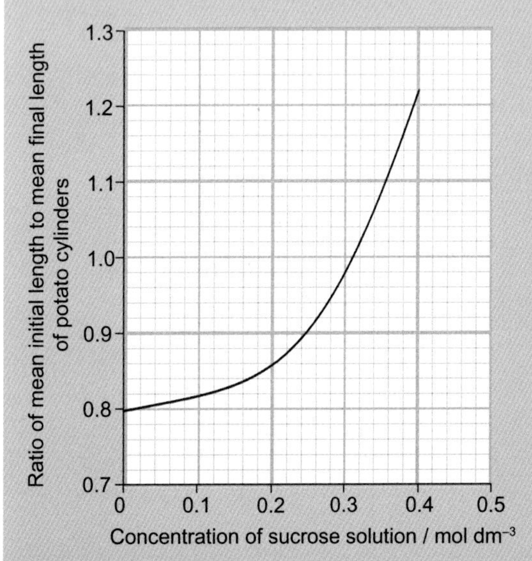

Figure Q1

a. For this experiment:

i. name the independent variable

ii. name the dependent variable

iii. name two variables that should be controlled.

b. Why were 10 potato cylinders placed in each solution?

c. In which concentration of sucrose solution did the potato cylinders remain the same length?

d. Explain in terms of water potential the result for

i. the potato cylinders in 0.1 mol dm^{-3} sucrose solution

ii. the potato cylinders in 0.4 mol dm^{-3} sucrose solution.

2. Figure Q2 shows part of a cell-surface membrane.

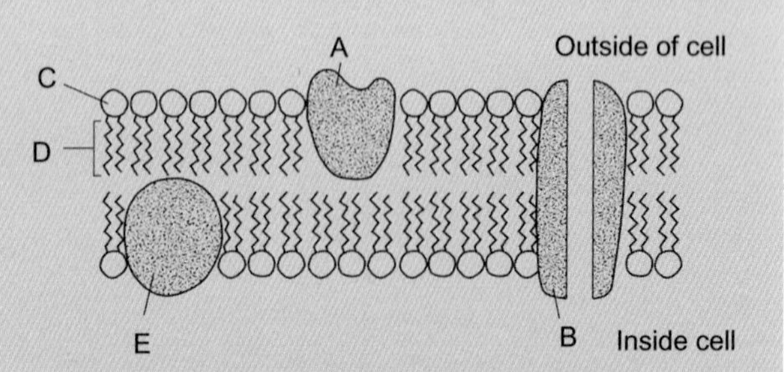

Figure Q2

a. Copy and complete Table Q1 by writing the letter from the diagram that refers to the part of the membrane described.

Part of membrane	Letter
Channel protein	
Contains only the elements carbon and hydrogen	

Table Q1

b. Explain why the structure of a membrane is described as fluid-mosaic.

c. When pieces of carrot are placed in water, chloride ions are released from the cell vacuoles. Identical pieces of carrot were placed in water at different temperatures. The concentration of chloride ions in the water was measured after a set period of time. The graph in Figure Q3 shows the results.

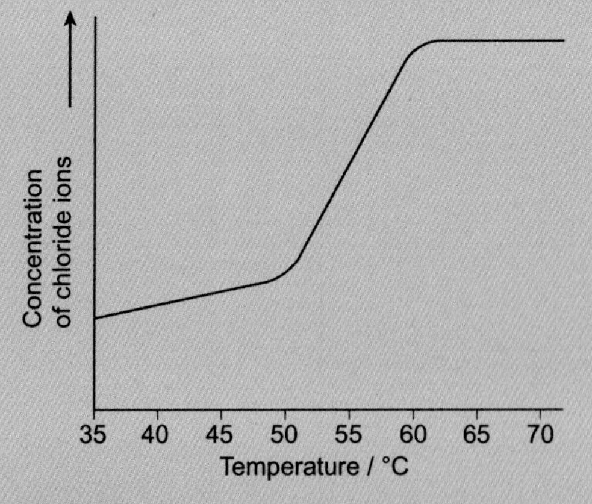

Figure Q3

Describe and explain the shape of the curve.

3. A scientist investigated the effect of cyanide on the uptake of sodium ions by animal tissue. He set up two beakers, **J** and **K**. He put equal volumes of a solution containing sodium ions and equal masses of an animal tissue in each beaker.

   - He added cyanide to beaker **J**.
   - He did not add cyanide to beaker **K**.

   He measured the concentration of sodium ions remaining in the solution in each beaker, for 80 minutes. The graph in Figure Q4 shows his results.

   a. Calculate the rate of uptake of sodium ions by the tissue in beaker **K** for the first 30 minutes. Show your working.

   b. Adding cyanide affects the uptake of sodium ions by the tissue. Use the graph to describe how.

   c. Cyanide is a substance that affects respiration. Use information in the question to explain the effect of cyanide on the uptake of sodium ions by the tissue.

   AQA June 2012 Paper 1 Question 4

4. Imatinib is a drug used to treat a type of cancer that affects white blood cells.

   Scientists investigated the rate of uptake of imatinib by white blood cells. They measured the rate of uptake at 4 °C and at 37 °C.

   Their results are shown in Table Q2.

Concentration of imatinib outside cells/ μmol dm^{-3}	Mean rate of uptake of imatinib into cells/ μg per million cells per hour	
	4 °C	37 °C
0.5	4.0	10.5
1.0	10.7	32.5
5.0	40.4	420.5
10.0	51.9	794.6
50.0	249.9	3156.1
100.0	606.9	3173.0

Table Q2

   a. The scientists measured the rate of uptake of imatinib in μg per million cells per hour. Explain the advantage of using this unit of rate in this investigation.

   b. Calculate the percentage increase in the mean rate of uptake of imatinib when the temperature is increased from 4 °C to 37 °C at a concentration of imatinib outside the cells of 1.0 μmol dm^{-3}. Give your answer to one decimal place.

   c. Imatinib is taken up by blood cells by active transport.

      i. Explain how the data for the two different temperatures support this statement.

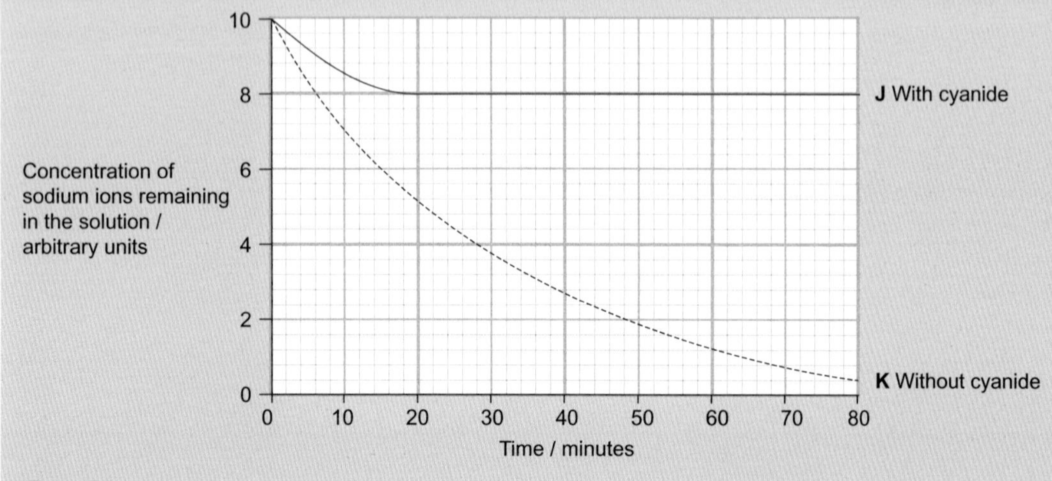

Figure Q4

(Continued)

ii. Explain how the data for concentrations of imatinib outside the blood cells at 50 and 100 μmol dm^{-3} at 37 °C support the statement that imatinib is taken up by active transport.

AQA June 2013 Paper 1 Question 5

5. a. Give two ways in which active transport is different from facilitated diffusion.

Scientists investigated the effect of a drug called a proton pump inhibitor. The drug is given as a tablet to people who produce too much acid in their stomach. It binds to a carrier protein in the surface membrane of cells lining the stomach. This carrier protein usually moves hydrogen ions into the stomach by active transport. The scientists used two groups of people in their investigation. All the people produced too much acid in their stomach. People in group **P** were given the drug. Group **Q** was the control group. The graph in Figure Q5 shows the results.

b. i. The scientists used a control group in this trial. Explain why.

ii. Suggest how the control group would have been treated.

c. Describe the effect of taking the drug on acid secretion.

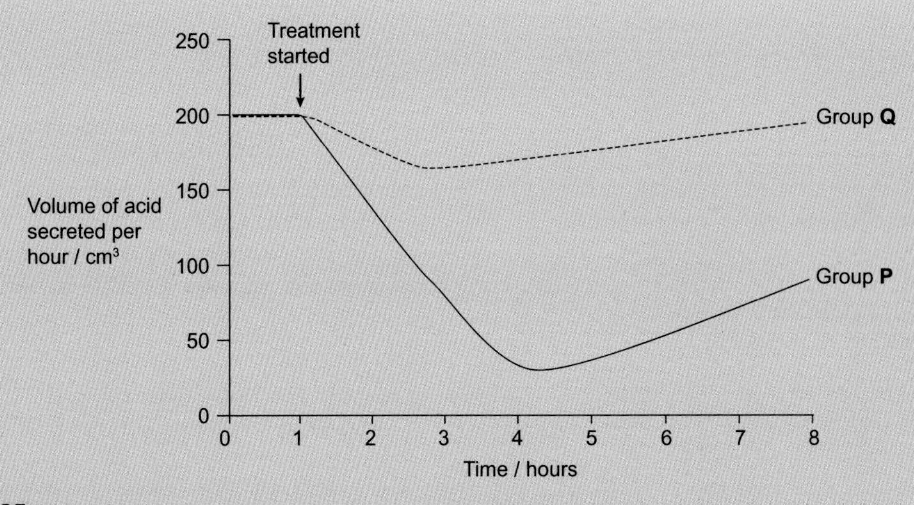

Figure Q5

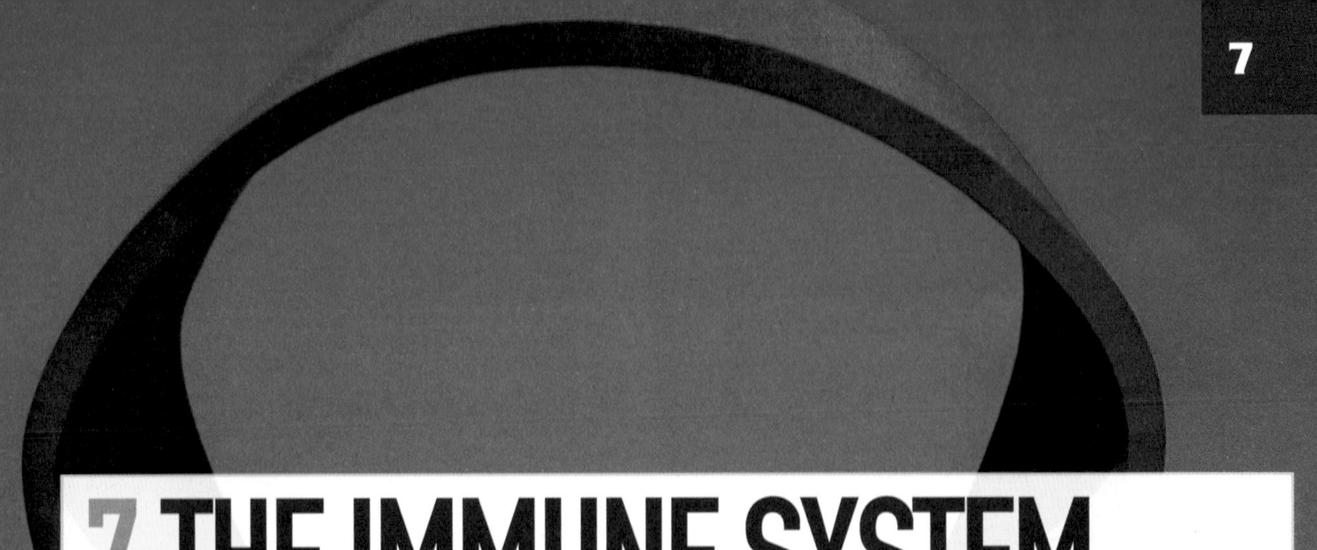

7 THE IMMUNE SYSTEM

PRIOR KNOWLEDGE

You may remember that all cells are surrounded by a cell-surface membrane, which contains numerous types of proteins and glycoproteins. You should also remember that the primary structure of these proteins is determined by the sequence of bases in the DNA of the cell, and that the primary structure in turn determines the three-dimensional (3D) shape of the protein.

LEARNING OBJECTIVES

In this chapter, we will look at how the proteins and glycoproteins in the cell-surface membranes of your body cells identify them as belonging to you ('self'), while cells or viruses that contain non-self antigens are attacked and destroyed by the immune system.

(Specification 3.2.4)

AIDS emerged as a new and serious infectious disease in the 1980s. Research soon established that it was caused by a retrovirus, HIV.

The first AIDS cases in South Africa were all among homosexual men. By 1989, almost 13% of homosexual men in South Africa's largest city, Johannesburg, were HIV positive. Yet the government did nothing, blaming the men's immoral activities for bringing the epidemic upon themselves.

By 1991, just as many people were acquiring AIDS through heterosexual activity. But, with a very turbulent political climate in the later 1980s and 1990s, as South Africa lurched out of apartheid into democracy, still nothing was done about the AIDS epidemic. Many people, including those with strong influence on political parties, believed that AIDS was deliberately introduced from a laboratory, or was caused by police tear gas, or spread because of the newly allowed association between people with skins of different colours. There was widespread disbelief that HIV caused AIDS.

After the first democratic elections in 1994, a rational, scientifically based plan to tackle AIDS was proposed. But no coherent action was taken and politicians continued to put forward non-scientific arguments to justify their reluctance to support the use of antiretroviral drugs, which were by now widely available and proven to reduce HIV transmission and prolong the lives of people with AIDS. Instead, a health minster advocated eating a good diet, with plenty of garlic, beetroot and lemons.

Not until 2009, when Jacob Zuma became president, was a widespread and effective plan to combat the AIDS epidemic begun. Today, South Africa still has the biggest HIV epidemic in the world, but it also has the largest antiretroviral treatment programme. Hopes are high that, despite all the wasted years in which almost nothing was done, people in South Africa are moving towards a future in which HIV/AIDS can be brought under control.

7.1 CELL-SURFACE ANTIGENS

In Chapter 6, we saw that every cell is surrounded by a cell-surface membrane, made of a bilayer of phospholipids. This membrane also contains protein molecules, many of which have short carbohydrate chains attached to them. These composite molecules are known as glycoproteins.

The proteins and glycoproteins in a cell-surface membrane vary in different types of cell in an organism, and between different organisms. We say that they are specific to particular cell types, and particular organisms. They effectively identify each type of cell. These cell-specific proteins and glycoproteins are known as **cell-surface antigens**.

Within your body, each cell carries a set of cell-surface antigens that identify the cell as 'self' – that is, it belongs in your body. However, if a foreign cell finds its way into your body, its antigens will be recognised as being different, and it will be attacked and destroyed. The destruction of the foreign cell is done by your immune system, and the response of the immune system to this invasion is called the **immune response**.

We can therefore define an antigen as a substance that causes an immune response.

How can a cell carrying 'foreign' antigens end up inside your body? There are several different ways, some of which happen very frequently, and some much more rarely. Four of these are outlined here.

Pathogens

A pathogen is an organism that causes disease. Pathogens are usually bacteria or viruses, but there are also some fungi and protoctists that are pathogenic. Bacteria carry antigens in their cell-surface membranes and cell walls; viruses have antigens in their protein coats. Pathogens can enter the body in various ways, such as in the air that you breathe in, in your food or drink, through the transfer of body fluids (for example, through an injection using a needle contaminated with someone else's blood) or directly through the skin.

Cells from other organisms of the same species

If a person suffers a failure of a particular organ or tissue, it may be possible for a healthy organ or tissue to be transplanted into their body. The healthy organ or tissue is sometimes taken from a relative who is willing to donate it (for example, a kidney or bone marrow) or from a person who has died suddenly (for example, a heart or a liver).

The transplanted cells will have antigens on their surfaces that the recipient's immune system will recognise as being foreign. The transplant will therefore be attacked and destroyed, a process called rejection. To avoid this happening, great care is taken to use an organ whose antigens are as similar as possible to those of the recipient. Sometimes, a close relative is found to have similar antigens, but for organs such as a heart, which has to come from someone who has died, a potential recipient may have to wait a long time before an organ whose cells carry antigens that closely match their own becomes available.

Even if the match is close, it will never be exact, unless the organ has been donated by an identical twin. The recipient is therefore given drugs, called immunosupressants, that 'tone down' their immune response. They will have to take these for the rest of their life. Of course, this also makes the immune system less effective at destroying invading pathogens, so there is an increased risk of suffering bacterial or viral infections.

Abnormal body cells

We have seen that some types of cell undergo a cell cycle, in which they repeatedly grow and divide. This cycle is usually tightly controlled, ensuring that each cell divides only as and when it should. However, if the DNA in the cell becomes damaged, then the genes that control the cell cycle may be altered. The cell cycle may take place too frequently and too rapidly, producing a tumour.

Cells in which the DNA has been altered in this way will have often different antigens in the cell-surface membrane than normal body cells. The immune system therefore recognises them as being 'foreign', and will usually destroy them. It is thought that this is a fairly common occurrence, and that most of the time the immune system deals with these potentially cancerous cells very effectively, eliminating them before they can develop into cancer.

Toxins

A toxin is a poison. For example, many pathogenic bacteria produce toxins that are released into the blood and cause damage to cells and metabolic processes, making us feel ill. The toxin molecules act as antigens, and cause an immune response, in which the immune system destroys the toxins.

KEY IDEAS

- Proteins and glycoproteins in the cell-surface membrane of all body cells identify them as being 'self'.
- Any cell or object carrying other, non-self antigens, is attacked and destroyed by the immune system. This includes pathogens, cells from other organisms of the same species, cancer cells and toxins.

7.2 PHAGOCYTOSIS

The cells in our bodies that help to destroy 'foreign' cells are the white blood cells. There are many different types of these, some of which are visible in Figure 1.

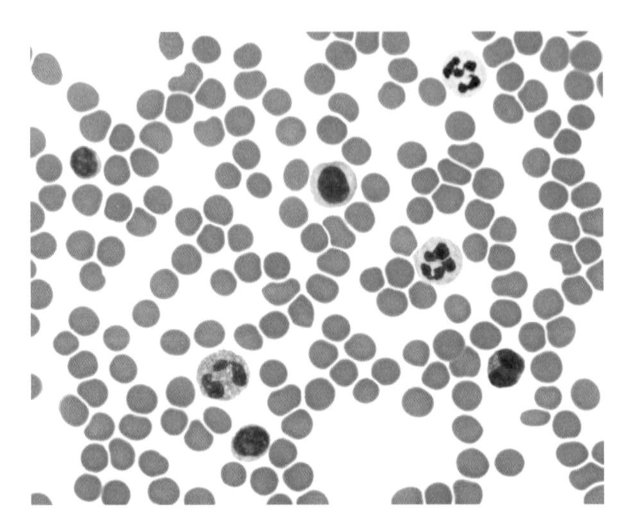

Figure 1 *A photomicrograph of a very thin film of blood. The red circles are red blood cells, and you can see that these are by far the most common type of cell in blood. The nuclei of the white blood cells have taken up a purple stain.*

Several types of white blood cells (principally neutrophils and macrophages), can take in bacteria by protruding pseudopodia (cytoplasmic 'arms') that flow around the pathogen (Figure 2 and Figure 3 (overleaf)). This process is called phagocytosis, and the white blood cells that do this are phagocytes.

The membranes of the 'arms' then fuse together, so that the pathogen is completely enclosed by a membrane. This produces a small, membrane-bound vesicle containing the pathogen, called a phagosome.

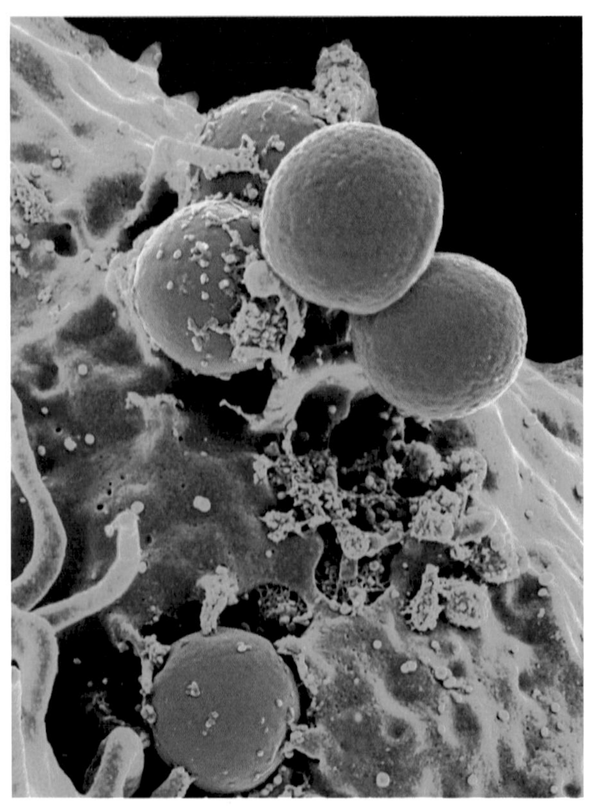

Figure 2 *This is a transmission electron micrograph of a neutrophil (green) engulfing bacteria (pink). The 'arms' being put out by the neutrophil will totally surround each bacterium, enclosing each one in a membrane-bound vacuole.*

The phagosome moves deeper into the cell, and fuses with one or more lysosomes, forming a phagolysosome. As we have seen, lysosomes contain enzymes, principally lysozyme and other hydrolytic enzymes. Lysozyme destroys bacterial cell walls by hydrolysing murein (peptidoglycan), allowing other hydrolytic enzymes to digest the rest of the pathogen.

QUESTIONS

1. Explain what is meant by a hydrolytic enzyme.

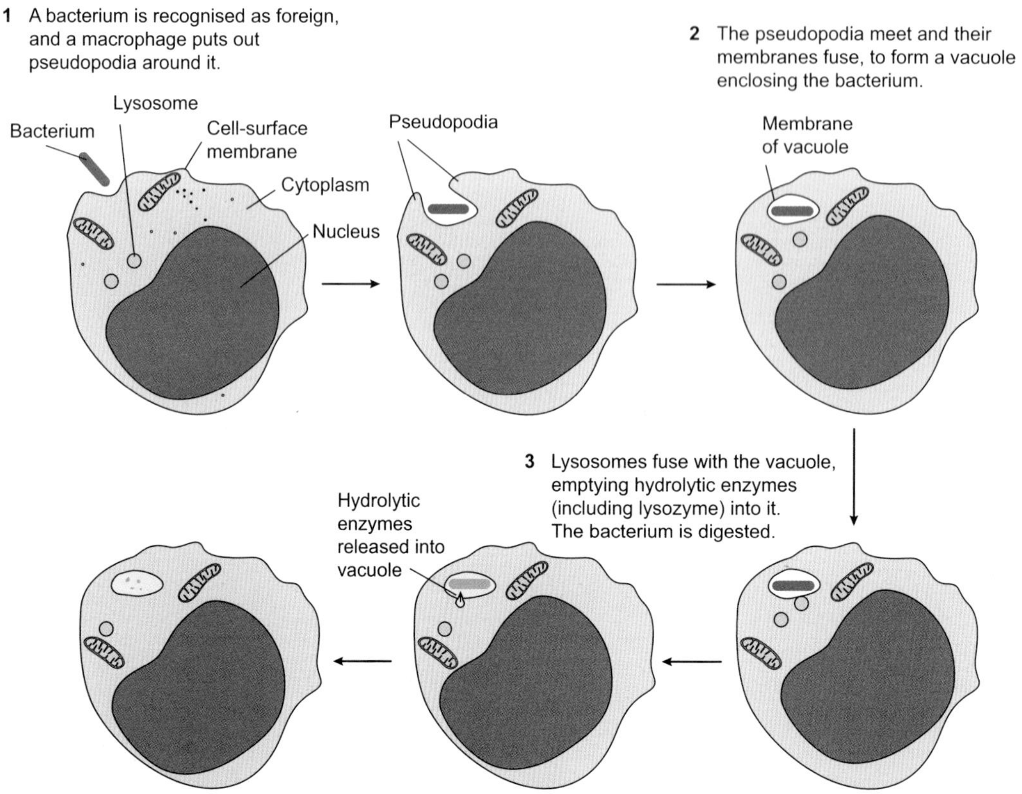

1 A bacterium is recognised as foreign, and a macrophage puts out pseudopodia around it.

Lysosome
Bacterium
Cell-surface membrane
Cytoplasm
Nucleus
Pseudopodia

2 The pseudopodia meet and their membranes fuse, to form a vacuole enclosing the bacterium.

Membrane of vacuole

3 Lysosomes fuse with the vacuole, emptying hydrolytic enzymes (including lysozyme) into it. The bacterium is digested.

Hydrolytic enzymes released into vacuole

Figure 3 *Phagocytosis*

KEY IDEAS

- Neutrophils and macrophages can carry out phagocytosis.
- The phagocytic cell puts out pseudopodia, which surround the structure to be ingested. The membranes of the pseudopodia fuse together, enclosing the structure in a vacuole called a phagosome.
- Lysosomes fuse with the phagosome and empty hydrolytic enzymes into it. These include lysozyme, which hydrolyses bacterial cell walls.

7.3 THE IMMUNE RESPONSE

Phagocytes are able to attack and destroy almost any foreign cell or material that enters the body. Their response is non-specific. However, the immune system also includes other types of cell that respond to specific antigens. These are the **lymphocytes**. Lymphocytes are white blood cells with a large nucleus that almost fills the cell (Figure 1).

There are two types of lymphocytes, which look identical and can only be told apart by their actions. **T lymphocytes**, or T cells, are involved in a defence mechanism called the **cellular response**. **B lymphocytes**, or B cells, are involved in the **humoral response**.

Both of these types of cell exist in a huge number of varieties. They are formed very early in life. Each cell carries specific receptors on its cell-surface membrane. Each specific receptor can bind to a specific antigen.

T lymphocytes and the cellular response

When some phagocytes (neutrophils) engulf and destroy a bacterium, they often use up so much energy that they die shortly afterwards. Some phagocytes (macrophages) are long-lived cells. When a macrophage has engulfed a pathogen, it

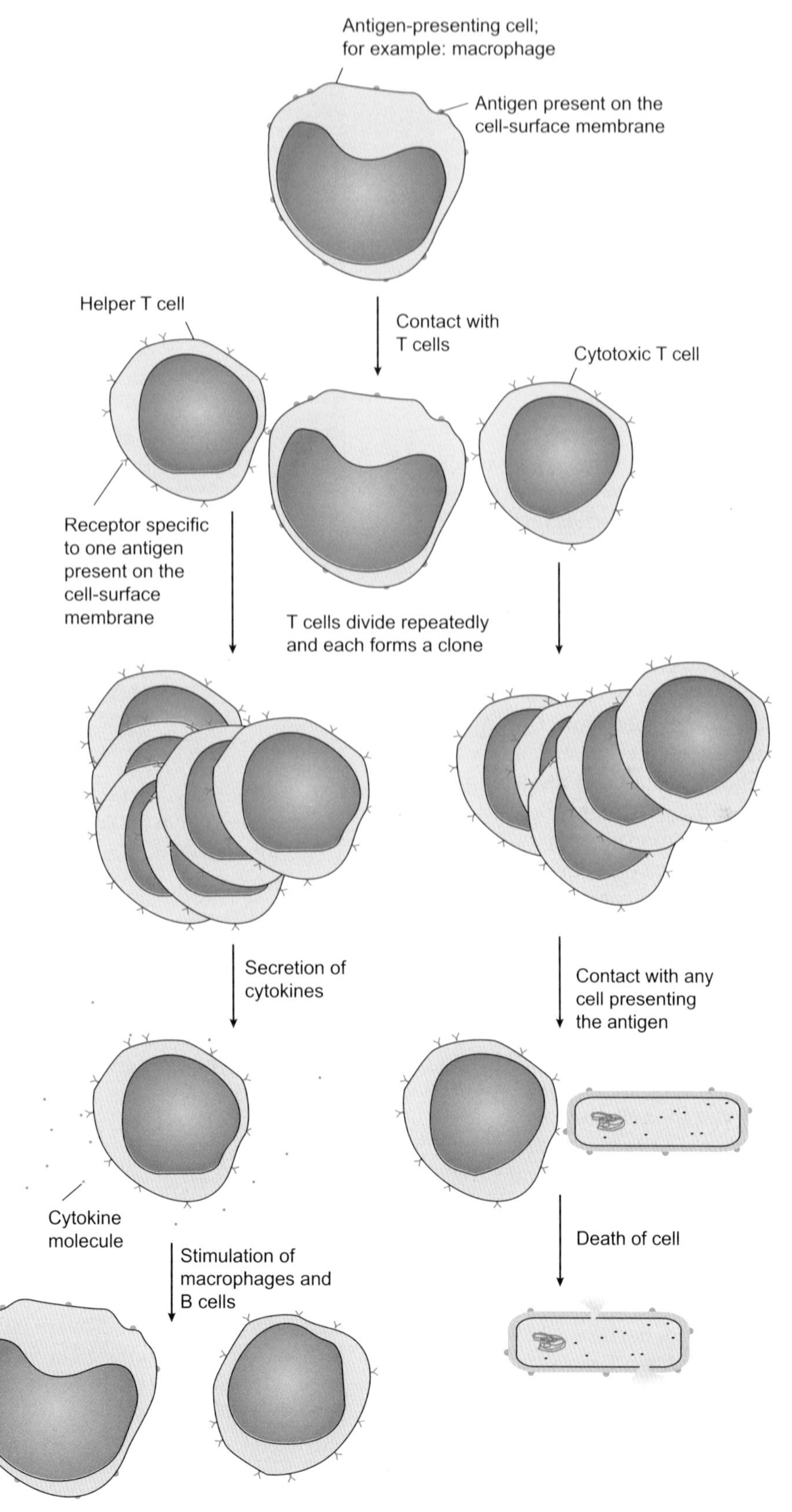

Figure 4 *The cell-mediated response to an antigen*

breaks it apart and places some of the pathogen's antigen molecules in its own cell-surface membrane, where they can be encountered by T lymphocytes. This display of antigens by the macrophage is called antigen presentation, and the macrophage is said to be an **antigen-presenting cell**. This is sometimes abbreviated to APC.

Other cells can also function as antigen-presenting cells. In particular, any cell that has been invaded by a virus will have antigens from the virus in its cell-surface membrane. This effectively 'labels' the cell as being infected, sending out a signal that it should be destroyed.

As we have seen, there are many types of T lymphocyte in the blood, each with its own unique set of protein and glycoprotein molecules in its cell-surface membrane, which act as receptors. If a T lymphocyte encounters an antigen-presenting cell displaying the antigen that has a complementary shape to the specific receptor on its surface, it is stimulated into action (Figure 4 on page 124). First, the T cell responds by dividing by mitosis, over and over again, to form a clone of identical T cells. This process is called **clonal selection**. Only a T cell carrying a receptor that specifically binds to the detected antigen is stimulated to divide and form a clone. All of the daughter cells produced by this division carry the same receptor in their cell-surface membranes, so all of them can detect and respond to the specific antigen that stimulated the response.

If the original T cell was of a type called a **cytotoxic T cell** (T_C cell), each of its 'daughter' cells will attach itself to any cell displaying copies of the specific antigen the T cell has the receptor for, and then destroy it. It may do this by secreting toxic substances into the cell, such as hydrogen peroxide, which pierce holes in its cell-surface membrane. Cytotoxic T cells therefore destroy cells that have been infected by the pathogen that carried the specific antigen to which it can respond.

If the original T cell was a **helper T cell** (T_H cell), the clone of daughter cells does not attack and destroy the infected cells. Instead, the helper T cells release substances called **cytokines**. These cells stimulate macrophages to carry out phagocytosis of infected cells, and they also stimulate B cells and other T cells to respond to the invasion.

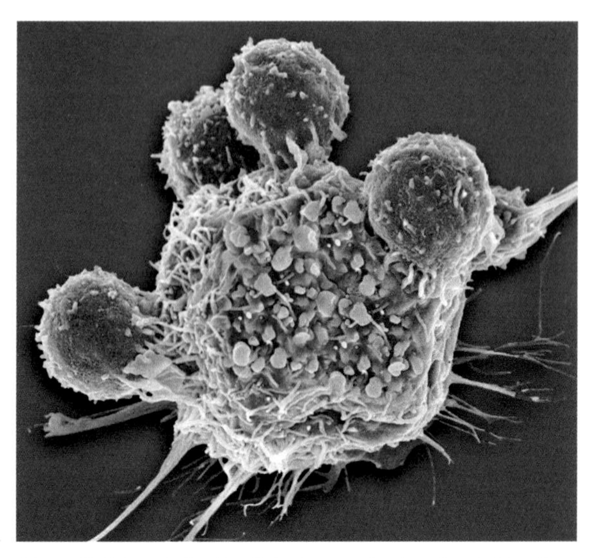

Figure 5 *Cytotoxic T cells will destroy any body cell containing the antigen that is complementary to its specific receptors. These do not have to be antigens from pathogens, but could be antigens specific to a cancer cell, which are different from the normal 'self' antigens. Here, four cytotoxic T cells (red) are attacking a cancer cell.*

QUESTIONS

2. Explain why all of the T cells produced by the division of an activated T cell carry the same receptors in their cell-surface membranes.
3. Suggest why piercing holes in the cell-surface membrane of a cell causes it to die.

Stretch and challenge

4. We have seen that cytotoxic T cells are able to destroy body cells that carry antigens that are not 'self'. Suggest why these cells do not always kill all types of cancer cell.

B lymphocytes and the humoral response

Just as we have a very large number of T cells, each with a different, specific receptor, so we have an equally large number of different types of B cells. Most of the time they do nothing but circulate around the body in the blood or the lymph. But if one encounters an antigen that fits its receptor, then it is stimulated into urgent action.

As with T cells, clonal selection occurs. The binding of the antigen with the B cell receptor stimulates it into action. Helper T cells are also involved in this activation. The first response of the activated B cell is to divide repeatedly by mitosis, forming a clone of cells all carrying the same receptor. The daughter cells are of two types – **plasma cells** and **memory cells**.

Plasma cells (Figure 6) rapidly become filled with large numbers of ribosomes attached to extensive endoplasmic reticulum, with many mitochondria to provide ATP. These allow the cell to manufacture and secrete huge quantities of a specific protein, called an antibody. All of the antibody molecules produced from/by the same genetically identical clone of plasma cells are identical in structure, and so are called **monoclonal antibodies**.

Antibodies are also known as **immunoglobulins**. We can define an antibody as an immunoglobulin secreted by a plasma cell, which interacts with a specific antigen.

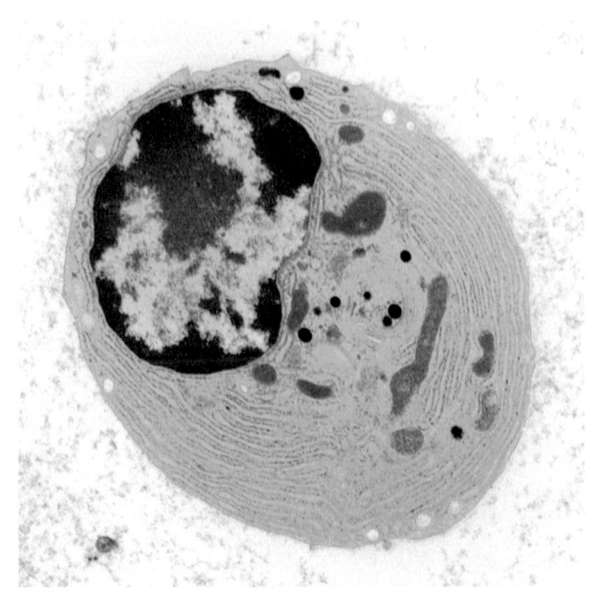

Figure 6 *A plasma cell. You can see the large quantities of rough endoplasmic reticulum, and also some mitochondria (blue) that supply ATP for use in protein synthesis.*

Figure 7 shows the structure of an antibody molecule. The variable region of the antibody is the part that binds to an antigen. It is known as the antigen-binding site. (Note – it is not an 'active site', because antibodies are not enzymes.) The shape of this region, and the side chains on the amino acids that are present here, is different in different antibodies. Here, bonds can form with a specific antigen with a complementary shape to the antibody, forming an antigen–antibody complex. This process can have various effects on the antigen, or the cell to which the antigen is attached. For example, the attachment of many antibodies to the antigens on the surfaces of bacteria can cause the bacteria to clump together, effectively stuck to one another and unable to move or reproduce. This is called **agglutination**. Agglutinated bacteria attract macrophages and neutrophils, which destroy them by phagocytosis.

The response of B cells to an antigen is called the humoral response. 'Humour' is an old name for a liquid, and this refers to the fact that the antibodies secreted by B cells are just molecules that dissolve in the body fluid, rather than the whole cells that are involved in the cellular response (Figure 8).

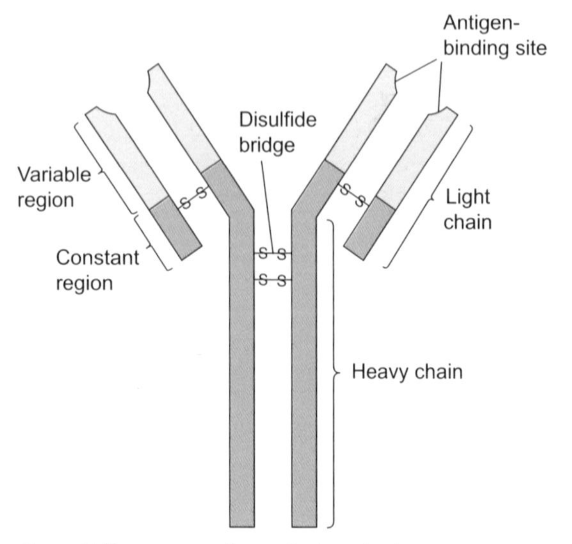

Figure 7 *The structure of an antibody molecule*

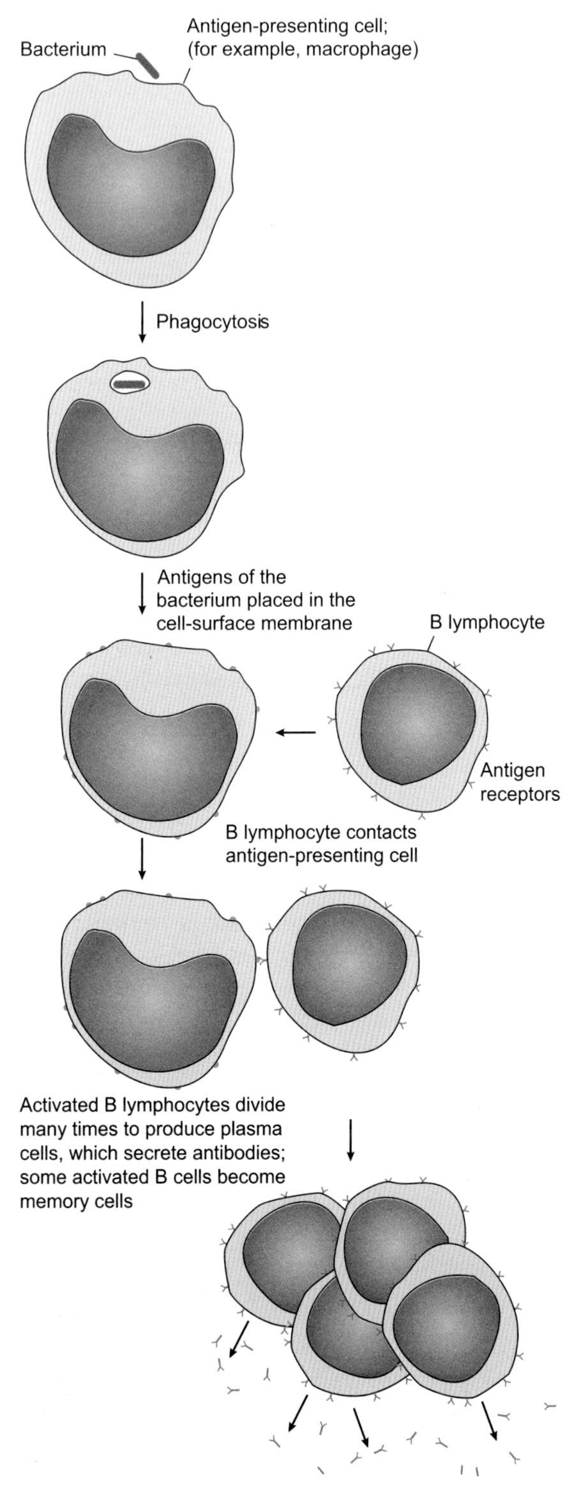

Figure 8 *The humoral response to an antigen*

QUESTIONS

5. Summarise the differences between the cellular response and the humoral response.

KEY IDEAS

- T lymphocytes (T cells) are responsible for the cellular response to an antigen.
- Macrophages place antigens from structures that they have ingested in their cell-surface membranes; this is known as antigen presentation.
- If a T cell encounters a presented antigen that has a complementary shape to the specific receptor on its surface, it is activated. The activated T cell divides repeatedly, forming a clone.
- Some T cells are cytotoxic. They fuse with cells carrying their antigen and kill them.
- Some T cells are helper cells. They secrete cytokines that stimulate macrophages and B cells to act.
- B lymphocytes (B cells) are responsible for the humoral response to an antigen.
- If a B cell encounters a presented antigen that has a complementary shape to the specific receptor on its surface, it is activated. The activated B cell divides repeatedly, forming a clone.
- Some of the clone of B cell become plasma cells, and secrete specific antibodies. Some of them remain as memory cells.

Primary and secondary responses

Plasma cells do not live for long. They wear themselves out through the production of large quantities of antibodies over a relatively short period of time. However, the other type of cell produced by the division of the original activated B cell – memory cells – can live for a very long time. After an infection by a pathogen, the memory cells remain in the circulation. Imagine that the same pathogen enters the body again. As there are already many memory cells carrying the receptor type specific to this pathogen, this increases the chance that one of the memory cells will meet quickly with the pathogen and be activated to produce plasma cells. These plasma cells usually secrete specific antibody fast enough to destroy the pathogen before it has time to reproduce – or replicate, if a virus – and cause illness.

This explains why we often get ill when a particular type of pathogen first invades our body, but do not get ill if we are invaded by the same pathogen on a subsequent occasion. Figure 9 illustrates what happens. On the first occurrence, it takes time for the appropriate B cell to meet the pathogen and respond by producing a clone of plasma cells. This is sometimes called the latent period. It takes more time for these plasma cells to gear up to producing large quantities of the antibody. Meanwhile, the pathogen has been reproducing and making us feel ill. Eventually, the antibodies are being produced in large enough amounts to lead to the destruction of the bacteria, and we recover.

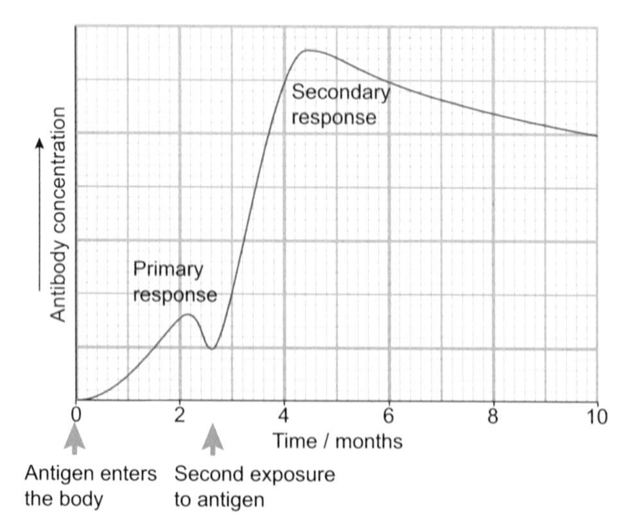

Figure 9 *Primary and secondary immune responses*

On the second occasion, however, we already have large numbers of memory cells in the body, each with receptors for this particular antigen. Now the response is much greater and much faster, often producing antibodies in such large quantities, and so quickly, that the invaders are destroyed before they have time to divide and make us ill. We are said to be **immune** to the disease caused by that pathogen. However, these antibodies are only made in response to a specific antigen. Mutations can cause **antigen variability**. For example, the influenza virus ('flu) during one winter may mutate into a new strain the following winter. The antibodies created against the previous virus will not necessarily work against the new one. This is why a new 'flu jab is available each year.

QUESTIONS

6. Compare the primary and secondary immune responses.
7. Explain the reasons for the differences you have described in your answer to question 6.

Vaccination

When the body is exposed to a pathogen, we have seen that an impressive array of defence mechanisms contributes to maintaining health. However, some bacteria and viruses cause significant harm while the primary response is happening, before the secondary response kicks in and antibody concentration increases. Some of these pathogens may kill or permanently disable the person before the immune response can kill the pathogen. Examples of such diseases are cholera, smallpox and diphtheria. Diseases such as these were the targets of public health programmes for proper sewage treatment and the development of safe water supplies, and for the many vaccinations now available.

The term 'vaccination' comes from the Latin word '*vaccinia*' ('cowpox'), which is derived from '*vacca*' ('cow'). The first vaccinations carried out used material taken from cowpox pustules to vaccinate children against smallpox, a much more serious disease. At the time, no-one knew how this worked, but today's vaccines are thoroughly researched and trialled before use.

A vaccine contains antigen derived from pathogenic organisms. Some vaccines contain just the antigen, while some contain weakened forms of the virus or bacterium that causes the disease. When injected into an individual, the antigen stimulates a primary response that leaves memory cells to generate the secondary response if the individual is subsequently infected by the relevant pathogen.

Most adults have immunity to many diseases, apart from the common cold and influenza. This immunity may be acquired through contact with the pathogen during childhood. For example, many children get chickenpox, and become immune to it once they have recovered from the illness. We can also become immune through vaccination. Most children are vaccinated against a range of diseases during childhood, including diphtheria, tetanus, whooping cough, poliomyelitis, haemophilus influenza (Hib), rotavirus, meningitis, measles, mumps and rubella (Figure 10, overleaf). Both infection and vaccination for these diseases give **active immunity** – the person is exposed to the antigen, and makes their own antibodies against it, retaining memory cells for long-lasting immunity.

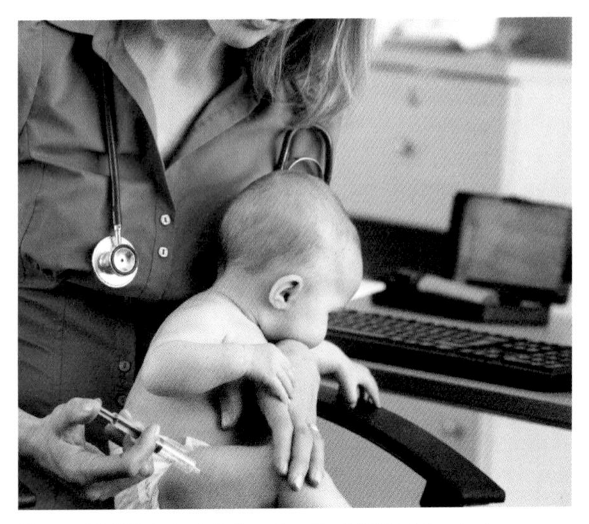

Figure 10 This baby is being given the MMR jab to give her immunity against measles, mumps and rubella.

QUESTIONS

8. Many childhood vaccinations are given in two doses, separated by several months or years. Suggest why this is done.

A newborn baby's immune system is not fully developed. During the first few days of breastfeeding, the mother's breasts produce a high-protein low-fat liquid called colostrum. This contains many antibodies, produced by the mother as a result of infections and vaccinations that she has had during her lifetime. These antibodies provide the infant with immunity against the same diseases to which its mother is immune. Antibodies are also present

ASSIGNMENT 1: UNDERSTANDING VACCINATIONS

(MS 1.3, MS 3.1, PS 1.2)

Figure A1 Edward Jenner

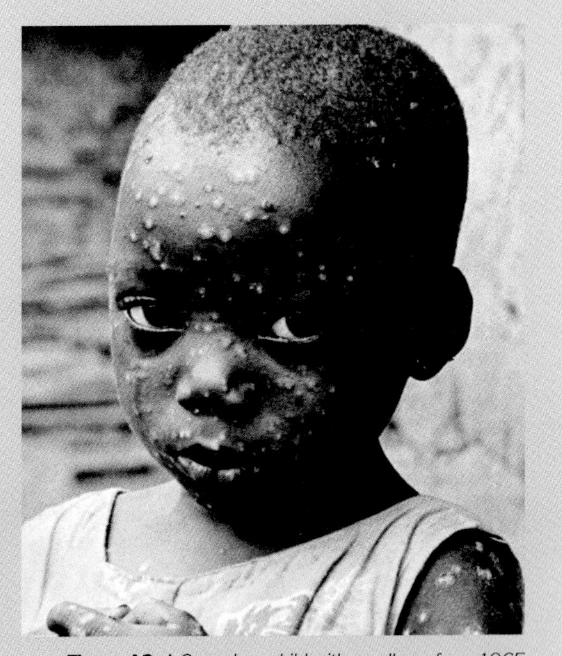

Figure A2 A Congolese child with smallpox, from 1965

The first vaccinations were carried out by Edward Jenner on an eight-year-old boy called James Phipps. Jenner noticed that milk maids rarely got smallpox, but did get a much milder disease called cowpox. Jenner thought that the cowpox gave the milkmaids immunity to smallpox, so he set about testing his hypothesis. He scratched James's arm and rubbed pus from a cowpox pustule into the wound. James developed a mild fever but recovered after a week. Two months later Jenner concluded his investigation by inoculating James with pus from a smallpox lesion. James did not develop smallpox.

The effectiveness of vaccinations is due to the ability of the immune system to develop memory cells to antigens that it has previously encountered. This allows a much bigger and faster response to an antigen if it is met the second time. In some vaccination programmes two injections are given a few weeks (or longer) apart, which increases the number of memory cells.

Questions

The graph shows the level of antibodies in a person's blood during a vaccination programme.

A1. How long after the second injection did it take for the level of antibodies to reach the immune level?

A2. By how many arbitrary units did the antibody level rise after the second injection?

Stretch and challenge

A3. After several weeks, the concentration of antibodies may drop below the concentration required for immunity. However, the person may still be immune to the disease. Explain why.

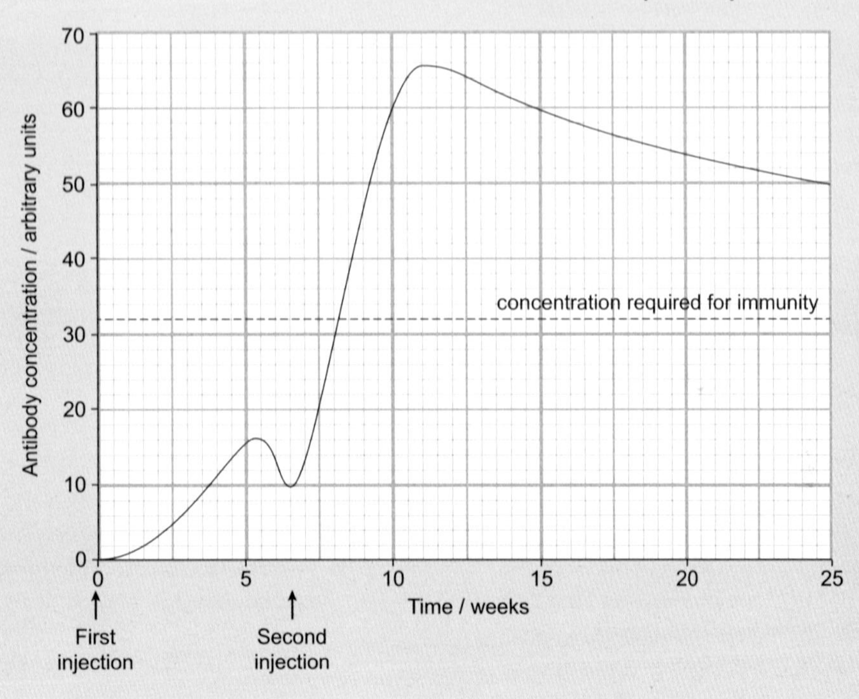

in the milk that is produced after the colostrum. The type of immunity given to the baby through breast milk is called **passive immunity**. The baby is simply provided with ready-made antibodies. It does not produce its own antibodies, nor does it produce memory cells. Passive immunity, therefore, does not last very long – only as long as the antibodies remain in its body.

Sometimes, we may be given passive immunity through injections. For example, an aid worker travelling out to help to fight an epidemic of an infectious disease in a foreign country may be given a vaccination of antibodies against that disease, and also others that they could be expected to encounter as they work. This gives them instant immunity, whereas a vaccination with the antigen would require them to wait for some weeks before travelling, while their body completed the primary response and produced enough antibodies to protect them.

QUESTIONS

9. Passive immunity lasts for only a relatively short time, whereas active immunity lasts much longer. Explain why this is so.

Herd immunity

In the UK, a carefully thought-out programme of vaccinations is available to children through the National Health Service (NHS). However, not every child receives all of the vaccinations. This is almost always because the parents decide not to have the child vaccinated. For example, some parents have been confused by faulty information provided in the media about potentially harmful effects of the MMR (measles, mumps and rubella) vaccine. Although great efforts have been made to explain that this vaccination is very safe, a few parents are still sufficiently worried to decide that their child should not have this vaccination.

In fact, as long as most children are vaccinated, it does not usually matter if a few are not. This is because it would be difficult for a virus or bacterium to move from one non-vaccinated child to another, if there are not many of these non-vaccinated individuals in the population. This is sometimes called **herd immunity**. For example, it is estimated that as long as 94–95% of children are vaccinated against measles, it is very unlikely that anyone – even the non-vaccinated children – will get the disease. Herd immunity also protects very young children, before they have reached an age at which they can be protected. It protects anyone with an immune system that is not working effectively, such as those who are elderly, are taking immunosuppressant drugs to prevent rejection of a transplant or have an illness affecting the immune system.

Unfortunately, worries about the MMR vaccine have meant that, in some parts of the country, vaccination rates have fallen below this level. In Swansea, in South Wales, for example, it fell to 67.5% in 2012–2013. GPs did their best to get more parents to let their children be vaccinated, but with little success; one GP said that some parents had been invited more than 15 times, but still refused. As a result, a measles outbreak occurred in this region in 2013–2014, in which 1200 people became ill, 88 were hospitalised and one died.

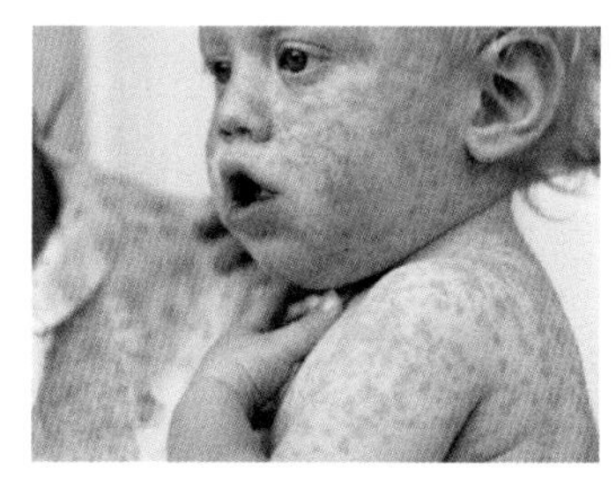

Figure 11 Measles is often seen as a mild disease that just gives you spots and makes you feel ill. However, it may sometimes have serious and long-lasting complications, including ear infections, pneumonia, deafness and brain damage.

QUESTIONS

10. The graph in Figure 12 shows the global number of cases of diphtheria and the global percentage of children immunised against diphtheria.

a. Describe the relationship between the number of cases of diphtheria and the percentage of children immunised.

b. Suggest an explanation for the number of cases of diphtheria in 1994 and 1995.

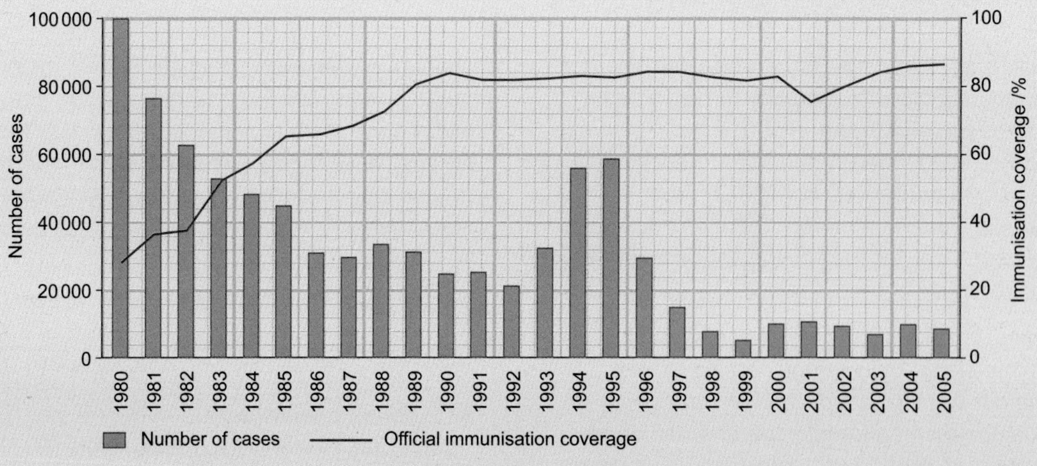

Figure 12 Global cases of diphtheria and immunisation coverage

ASSIGNMENT 2: PREVENTING CERVICAL CANCER

(MS 0.3, MS 1.3, MS 3.1, MS 3.2, PS 1.2, PS 3.1, PS 3.2)

According to Cancer Research UK, cervical cancer is the second most common cancer in women under the age of 35. Unlike most cancers, cervical cancer is caused by a virus, which is passed from one person to another during sexual activity. In the UK, 2900 women are diagnosed with cervical cancer every year. In 2011, just under 1000 women died of cervical cancer, and it is thought that up to 400 lives a year could be saved by protecting girls against infection with HPV (Human Papilloma Virus) before they become sexually active.

HPV vaccination with Cervarix was introduced into the national immunisation programme in September 2008 across the UK for girls aged 12–13, and consists of three injections over a period of 12 months. From September 2012, Gardasil replaced Cervarix as the HPV vaccination of choice due to its added protection against genital warts.

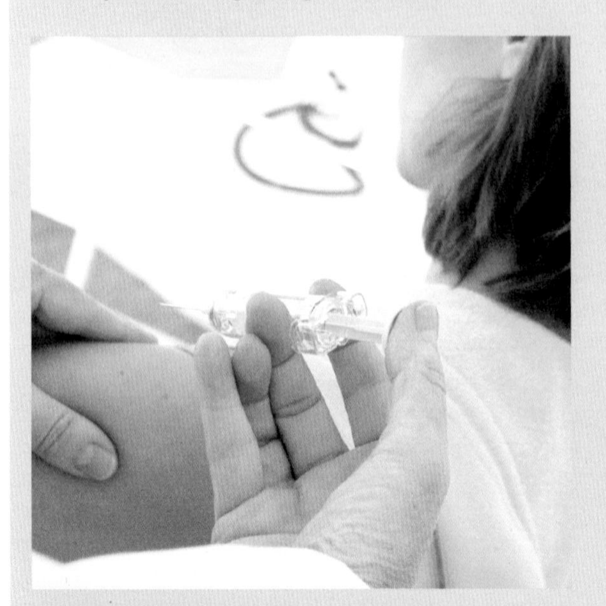

Figure A1 *Girl receiving the HPV vaccination*

The HPV vaccine is mainly given to girls in secondary schools. From September 2014, the number of doses of vaccine that girls aged 12 to 13 years received was reduced from three to two, given at least six, and not more than 24, months apart. In the UK, boys are not currently vaccinated and parents who want their boys protected have to pay.

Research has indicated that the HPV vaccine provides effective protection for at least 20 years, but it will be many years before the vaccination programme has an effect upon cervical cancer incidence, so women are advised to continue accepting their invitations for cervical screening.

Table A1 shows the incidence of some cancers and the role of HPV in the USA.

Type of cancer	Number of cases per year (average)	Number of cases attributed to HPV (estimated)
cancer of the cervix	11 967	11 500
cancer of the vulva	3136	1600
cancer of the vagina	729	500
cancer of the penis	1046	400
TOTAL	**16 878**	**14 000**

Table A1

Questions

A1. Explain why the HPV vaccination programme is delivered through secondary schools.

A2. a. Plot the data in the table on a suitable graph.

b. Explain why the number of cases of these cancers attributed to HPV can only be estimated.

c. Calculate the percentage of cases of all four cancers due to HPV infection.

A3. Why will it be many years before the vaccination programme affects the incidence of cervical cancer?

A4. Vaccinating boys could help to reduce the number of girls that become infected with HPV. Discuss the ethical issues that need to be considered when deciding whether or not all boys should be vaccinated against HPV.

KEY IDEAS

- The first time an antigen is encountered, a primary response is elicited. On the second encounter with the antigen, a secondary response is elicited.
- The secondary response is faster than the primary response, and produces greater quantities of antibodies. This is because memory cells specific to that antigen are already present in the body.
- Although we develop immunity to each strain of the cold or influenza virus that we are exposed to, new strains are constantly being produced, and their antigens are not the same.
- Vaccination introduces harmless antigens into the body, causing a primary response and the production of memory cells, without causing illness.
- Active immunity is the result of the response of the immune system to antigens, and involves the production of antibody and memory cells.
- Passive immunity is the result of acquiring antibodies from another organism (for example, a baby's mother), and does not involve production of antibody or memory cells.
- Herd immunity results if enough members of a population have immunity to a particular pathogen (for example, because they have been vaccinated against it) to make it difficult for the pathogen to pass from one non-immune person to another.

7.4 HIV/AIDS

HIV stands for human immunodeficiency virus. This virus was first recognised in the early 1980s, and is now known to have been introduced to humans from contact with chimpanzees in West Africa, probably early in the 20th century.

HIV is a retrovirus (Figure 13). This means that its genetic material is RNA, rather than DNA. It also contains enzymes that are needed to help its replication once it enters a cell, such as reverse transcriptase. The genetic material and enzymes are surrounded by two layers of protein molecules.

The virus also has a layer of phospholipid around itself, called the viral envelope, which is actually part of the host cell cell-surface membrane that is picked up as the virus leaves the host cell in which it was produced. This membrane contains two types of glycoproteins called gp41 and gp120. These enable the virus particle to identify and then bind to and fuse with potential host cells and infect them.

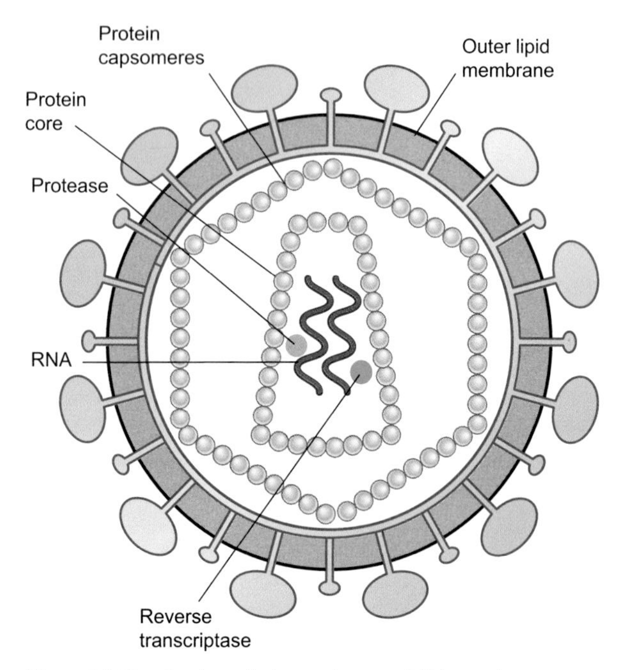

Figure 13 *The structure of a human immunodeficiency virus*

The replication of HIV

HIV infects a particular type of helper T cell. Once it has entered a cell, its RNA is converted to DNA, using the enzyme reverse transcriptase. This DNA has base sequences that code for the production of new viruses. The DNA is incorporated into the chromosomes of the host cell, being replicated if the cell divides. It may remain in the chromosomes of the cells for a long period of time, having no effects at all on the person's health. However, the immune system will produce antibodies against the virus, and these can be detected in the blood – only virus particles in the blood are attacked because the antibody cannot enter the infected cells. A person who has anti-HIV antibodies is said to be HIV-positive.

Eventually, perhaps after many years, the viral DNA is activated, and begins to be transcribed and translated using the host's cells own enzymes and ribosomes. Large numbers of new viruses are formed and assembled, and burst out of the cell, ready to infect more helper T cells.

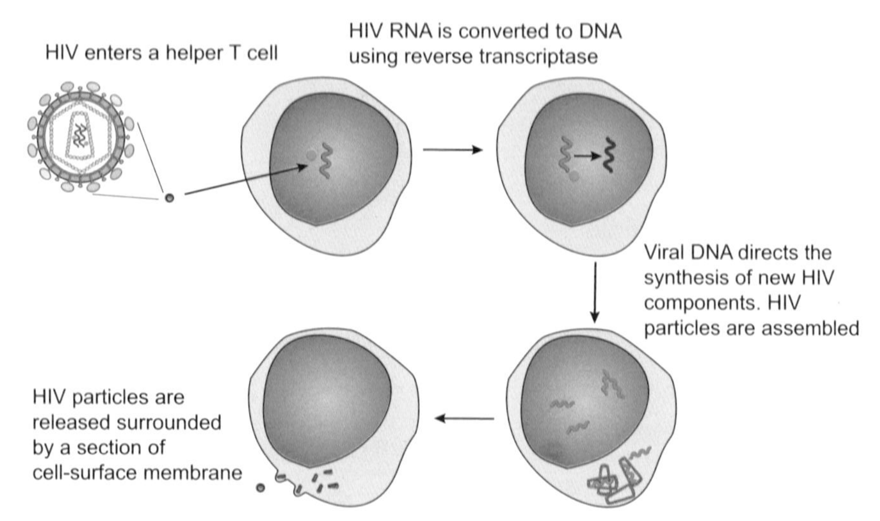

Figure 14 *HIV replication*

Gradually, the number of functioning helper T cells in the person's body decreases. You will remember that helper T cells are needed to help to activate B cells as well as T cells, so the ability to produce antibodies decreases. Eventually, there are so few helper T cells left that the person is unable to fight off other infections, or to destroy cancer cells that may have formed in their body. This is when they are said to have AIDS – acquired immunodeficiency syndrome. Two common causes of death from AIDS are a form of cancer called Kaposi's sarcoma, and degenerative diseases of the brain. AIDS also makes people much more susceptible to diseases such as tuberculosis (TB) and malaria, and many die from pneumonia and various gut infections.

Treating HIV

When HIV was first discovered, there was no cure. Almost everyone infected with HIV eventually developed AIDS and died. Antibiotics, widely used to control bacterial diseases caused by bacteria, are totally ineffective against viruses. This is because most of them work by destroying bacterial cell walls or interfering with the metabolic pathways in their cells. Viruses are not cells (acellular), so do not have cell walls, or any metabolism for antibiotics to act against.

Today, however, drug therapy is available that can greatly slow down the development of AIDS. Many HIV-positive people now have life expectancies almost as long as if they had never been infected. The drugs that are used are called antiretrovirals, or ARVs. They work by binding to the viral reverse transcriptase and stopping it from copying the virus's RNA to make DNA. These drugs cannot cure AIDS, but they do keep the quantity of the virus in the body to a low level, so that the person retains enough healthy helper T cells to have a well-functioning immune system.

ASSIGNMENT 3: UNEARTHING THE ORIGINS OF AIDS

(PS 1.2)

Figure A1 *Did HIV spread to humans because a hunter ate chimpanzee?*

The origin of HIV has been the subject of debate for more than 30 years since it first became a problem in the 1980s. The first recognised case of AIDS was in the USA in the early 1980s when a number of homosexual men began to develop rare opportunistic infections and cancers. The discovery of HIV was made soon afterwards.

HIV is a lentivirus that attacks the immune system; lentivirus literally means 'slow virus', so named because of the long time they take to produce any effects on the body. Lentiviruses have been found in many animals, but the most closely related to HIV is SIV (Simian Immunodeficiency Virus) that affects monkeys and has been around for more than 32 000 years. It is generally believed that SIV gave rise to HIV by a mutation that enabled the virus to transfer between monkey species, to chimpanzees and eventually to humans.

There are several theories about how HIV first infected humans. The most common theory is that a West African hunter caught a chimpanzee for food and became infected while butchering it through an open wound. Meat obtained from wild animals such as this is known as bush meat. Viral transfer from primates to humans occurs with other viruses; in Cameroon 1% of the population is thought to be infected with Simian Foamy Virus (SFV), which has been one of the reasons for the call to ban bush meat hunting.

A second – somewhat controversial – theory is that the virus was transferred by vaccinations for polio during the 1970s, which were given to people in West Africa. It is suggested that these vaccines were grown in chimpanzee tissues that may have been contaminated. However, polio vaccine at that time was administered orally, and recent analysis of a vial of vaccine has shown no traces of SIV or HIV.

In 2014, a group of scientists published the result of research that analysed the base sequences in archived samples of DNA from HIV going back to the early 20th century. They looked for mutations in the HIV genetic code, and used these to trace the probable history of the development of the virus. Their evidence, which they refer to as 'viral archaeology', suggests that the explosion of HIV infections began in Kinshasa, in what is now the Democratic Republic of the Congo. It was fuelled by a 'perfect storm' of rapid population growth, sex trade in the city, unsterilised needles used in health clinics and the rapid development of a railway system that meant people carried the virus to other regions nearby.

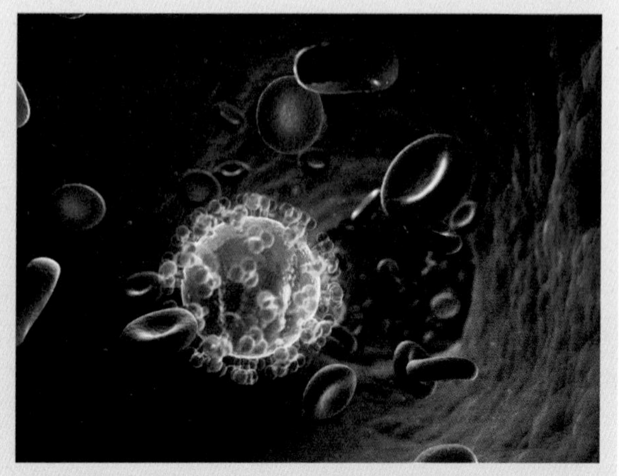

Figure A2 *Computer artwork illustrating HIV (human immunodeficiency virus) infection*

Questions

A1. Suggest why HIV may have been present in West Africa for many years, but was first recognised in the USA in the 1970s.

A2. Several different viruses, such as ebola and Lassa fever, seem to have first emerged in West Africa. Suggest how the consumption of bush meat in this region may have contributed to the emergence of these viruses.

Stretch and challenge

A3. Use the Internet to find out about the most recent theories for the emergence of HIV. How certain can we be about the validity of these theories?

KEY IDEAS

- The human immunodeficiency virus, HIV, is a retrovirus. It contains RNA and is surrounded by protein coats (capsid) and a phospholipid membrane derived from its host cell.
- Once in a host cell, the RNA of the virus is used to make a complementary DNA molecule, which inserts itself into one of the cell's own chromosomes. Eventually, the code on the DNA is used by the cell to make new viruses.
- HIV infects a particular type of T cell and destroys them. This weakens the immune response, so that a person is not able to defend themselves against pathogens. This causes the disease AIDS.
- AIDS cannot be cured, but the replication of HIV in T cells can be greatly reduced by anti-retroviral drugs.

7.5 MONOCLONAL ANTIBODIES

We have seen that, once a particular B cell has met the specific antigen to which its receptors can bind, it is activated and divides by mitosis to form a clone of genetically identical cells. Some of these are plasma cells, which secrete many copies of the antibody that can bind specifically with the antigen.

It is now possible to produce a large clone of plasma cells in the laboratory. Since the clone is produced from a single plasma cell, all the cells of the clone secrete the same, specific antibody. The antibody that they produce is called a **monoclonal antibody**, or Mab for short. They have opened up many new techniques for research, treatment and diagnosis of medical conditions.

For example, monoclonal antibodies can be produced that have complementary shapes to abnormal proteins (that act as an antigen) on particular cells – perhaps the cells in a particular type of cancer. The monoclonal antibody molecules can be attached to a drug that can kill the cancer cells. These combination molecules are then injected into the patient's body. The monoclonal antibody molecules will bind to the cancer cells and deliver the drug to them only – avoiding potential harm to healthy body cells that do not have the same receptor.

Rheumatoid arthritis and Crohn's disease can also be treated using monoclonal antibodies. Rheumatoid arthritis is caused when the immune system fails to recognise some of the cells in the joints as being 'self', and attacks them as though they were invading cells. Crohn's disease results from an immune system attack on the lining of the alimentary canal. Such diseases are known as autoimmune diseases. A substance called TNF-α is involved in this response, and if TNF-α can be disabled, then the immune system's attack is calmed down, and the person's symptoms are reduced. Monoclonal antibodies have been developed that specifically bind to TNF-α and put it out of action (Figure 15).

Monoclonal antibodies can also be used in diagnosis. For example, pregnancy testing kits contain a monoclonal antibody that binds specifically to a hormone called HCG (human chorionic gonadotrophin) which is secreted in the early weeks of pregnancy. Very small quantities of this hormone are present in the pregnant woman's urine. The monoclonal antibody is attached to a dipstick, which is used to test a sample of the woman's urine. Any HCG in the urine binds with the monoclonal antibodies and causes a colour change on the dipstick (Figure 16).

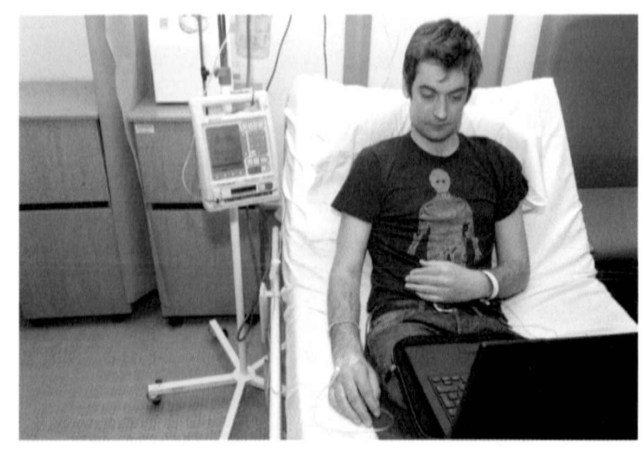

Figure 15 *A monoclonal antibody called infliximab reduces the activity of TNF-α. This person with Crohn's disease is being given an infusion of infliximab as part of his treatment.*

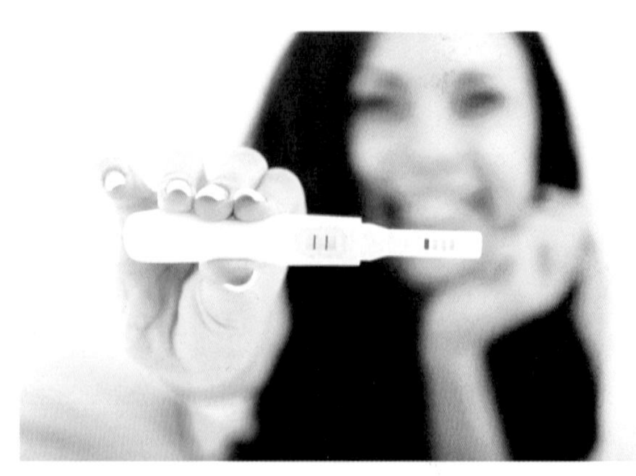

Figure 16 *Monoclonal antibodies that bind with HCG are used in pregnancy testing kits. This one is showing two pink stripes, meaning that the test is positive.*

ELISA tests

ELISA stands for enzyme-linked immunosorbant assay. ELISA tests are used to detect a particular substance (such as an antigen) in a sample of liquid, using monoclonal antibodies. For example, they could be used to test whether a particular virus is present in a sample of blood plasma.

First, a monoclonal antibody is produced that binds specifically to an antigen found on the virus. An enzyme is then attached to this antibody.

In one type of ELISA test, the liquid that is thought to contain the suspected antigen is placed into one of the wells in a plastic plate (Figure 17, overleaf). Further samples of the same liquid can be added to the other wells, or different liquids may be put into each one. Usually, a sample that is known to contain the antigen is added to some of the wells, and a sample that is known not to contain it is added to others.

The liquid samples are left for a short while, to give the antigen time to adhere to the plastic (it is immobilised). Then the monoclonal antibody–enzyme combination is added to the wells. If the suspected antigen is present in any of the wells, the monoclonal antibodies will bind to it. The plate is then washed, so that any unbound antibody–enzyme molecules are washed away. Only those bound to the suspected antigen remain in place.

Finally, the substrate of the enzyme is added. This is often a substance that changes colour when the enzyme acts on it. If the colour of the liquid in any of the wells alters, then this shows that the antibody–enzyme complex is present, meaning that it has bound to the suspected antigen.

Virus particles are attached to a solid surface

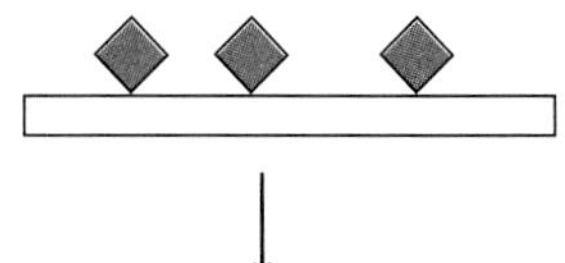

Antibody for the virus is attached to an enzyme and the antibody binds to the virus

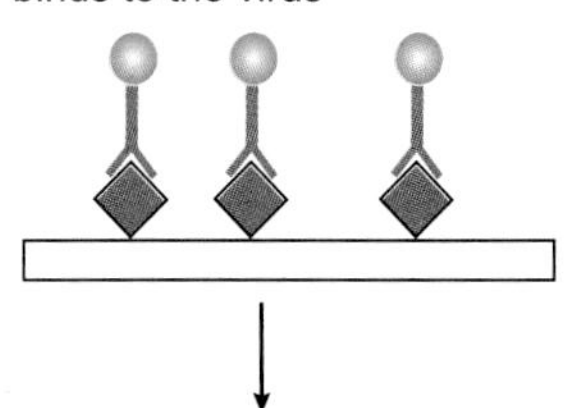

The substrate for the enzyme is added and is changed to a product that has a different colour

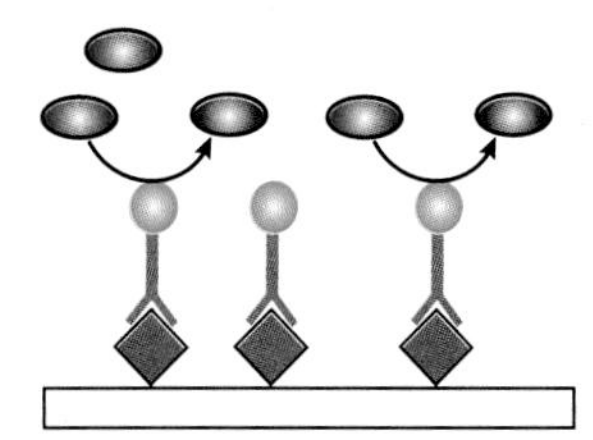

Key

 Virus (antigen)

 Enzyme

 Antibody

 Substrate for enzyme

 Coloured product

Figure 17 *The ELISA test*

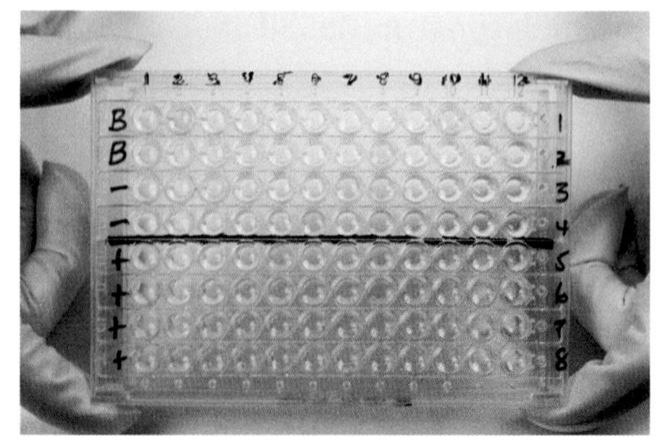

Figure 18 *The result of an ELISA test for HIV. The wells that are colourless did not contain the virus, but those that are yellow did.*

QUESTIONS

11. Explain the meaning of the term 'monoclonal'.

12. Monoclonal antibodies attached to drugs are sometimes called 'magic bullets'. Suggest why they are given this name.

Stretch and challenge

13. Suggest why, in an ELISA test, some wells are filled with a liquid known not to contain the antigen being tested for, and some are filled with a liquid known to contain it.

14. Not all antibodies used in research and assays are monoclonal. Some antibodies are known as polyclonal. Suggest what you think is the difference between monoclonal and polyclonal antibodies. Why might you not want to use polyclonal antibodies for ELISA tests?

15. Why might you choose to use polyclonal antibodies instead?

KEY IDEAS

- Monoclonal antibodies are identical antibodies made by a clone of plasma cells.
- Monoclonal antibodies can be used to target medication to specific cell types.
- Monoclonal antibodies can also be used to diagnose pregnancy or diseases, by detecting the presence of particular antigens in a sample, for example using an ELISA test.

ASSIGNMENT 4: REDUCING THE USE OF ANTIBIOTICS

(MS 3.1, PS 1.2, PS 2.1, PS 2.2, PS 2.3, PS 3.1, PS 3.3)

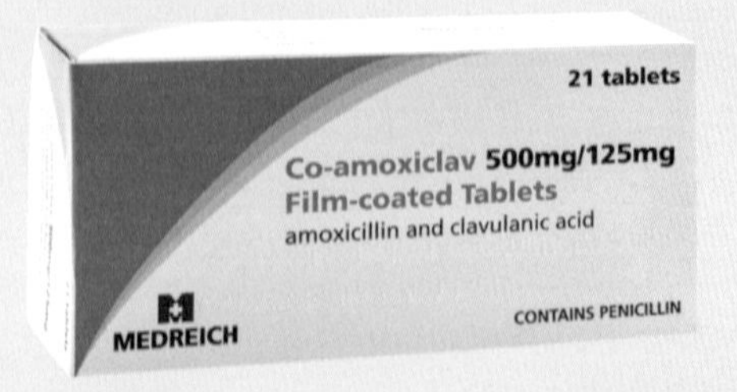

Figure A1 *Can antibiotics cure the common cold?*

Most children will have from three to eight colds per year.

Many scientists have investigated how useful antibiotics are in treating children who have the common cold.

The results of some of these studies are shown in Table A2. Symptomatic treatment includes 'self-treatment' using medicines, such as paracetamol and decongestants.

Study	Number of children in study	Comparison groups	Outcome
A	2177	Penicillin or symptomatic treatment	Required return outpatient visit(s): penicillin group 26%; symptomatic-treatment group 20%
B	217	Antibiotic or placebo	Rate of all infectious complications: antibiotic group 15%; placebo group 15%
C	845	Antibiotics or symptomatic treatment	Rate of all infectious complications: antibiotic group 14%; symptomatic-treatment group 9%
D	781	Antibiotic or symptomatic treatment	Rate of complications (for example, acute ear infection): antibiotic group 3.5%; symptomatic-treatment group 2.6%
E	261	Penicillin or placebo	Not improved or complications: antibiotic group 5%; placebo group 5%
F	212	Doxycycline or placebo	Runny nose at day 5: doxycycline group 14%; placebo group 30%
G	197	Amoxicillin, co-trimoxazole or placebo	Runny nose at day 8: amoxicillin 6%, co-trimoxazole 4%, placebo 15%. Normal activity at day 8: amoxicillin 89%, co-trimoxazole 95%, placebo 97%

Note: A placebo is a tablet or medicine that looks like the active drug but does not contain any active ingredient. It is used in comparisons to determine how many patients would get better anyway without active treatment.

Table A1

Questions

A1. Explain the need for campaigns to reduce the number of prescriptions for antibiotics.

A2. Plot these data on a suitable graph.

A3. Suggest, with reasons:

a. which study gave the most reproducible data

b. which studies gave the least valid data.

A4. What conclusions can be drawn from these studies?

A5. Evaluate the ethics involved in these studies.

Worked maths example: Disease and probability

(MS 0.3, MS 1.1, MS 1.4)

- The probability of a woman in the UK getting breast cancer is 0.1. What is the percentage risk?

Probability values are expressed on a linear scale: 0 is impossible and 1 is definite. To calculate a value on this scale as a percentage, just multiply by 100.

So, a probability of 0.1 is the same as $0.1 \times 100 =$ **10%**

- Having mutations in genes can increase the risk of getting cancer. Between 50% and 85% of women with a faulty breast cancer gene will go on to develop breast cancer. What are these percentages as a probability range? Remember the probability scale goes from 0 (impossible) to 1 (definite).

To convert a percentage to a probability, you need to divide by 100.

So, a range of 50–85% is the same as **0.5–0.85.**

- Look back at the data in Table A1, Assignment 2, page 132. What percentage of total cancer sufferers were female?

Cancers of the vagina, vulva and cervix all affect women. So, the number of cases of female cancer as a percentage of the overall number of cases is:

$$\frac{(11976 + 3136 + 729)}{16878} \times 100$$

= 93.8 (to three significant figures)

Cancer is called a *polygenetic* disease. This means that it involves lots of different genes. Estimating the probability of inheriting faulty genes can be complicated. Some diseases, such as cystic fibrosis (CF), are *monogenetic* – they only involve one gene.

- A couple are going for genetic testing before trying for a baby. The woman has CF. She has the genotype **aa**, where **a** represents a mutated CF allele. The man does not have CF, but he does have a mutated CF allele. His genotype is **Aa**, where **A** represents the healthy dominant CF allele. CF is a recessive disease. What is the probability of their children having the disease?

We can use the information to create a table (called a Punnett square):

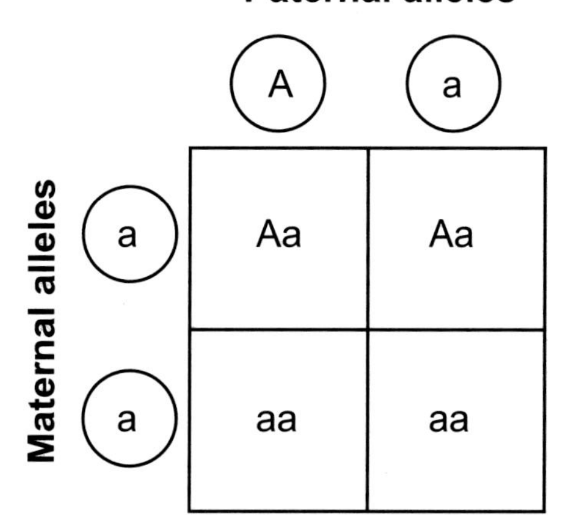

From the table, we can see that there are only two possible allele combinations for the children:

Aa is healthy because the presence of a dominant healthy allele means that this combination will not result in disease. It could, however, carry the recessive allele on to the next generation.

aa means the offspring has CF because both alleles of the CF gene are mutated.

So, there is a **50:50 chance** that their children may get cystic fibrosis, which is a **probability of 0.5.**

PRACTICE QUESTIONS

1. a. Describe the roles of macrophages and lymphocytes in:
   i. the humoral response
   ii. the cellular response.
   b. Figure Q1 shows an antibody molecule.
   i. Explain how an antigen–antibody complex is formed.
   ii. Explain how the structure of an antibody molecule is related to its function.

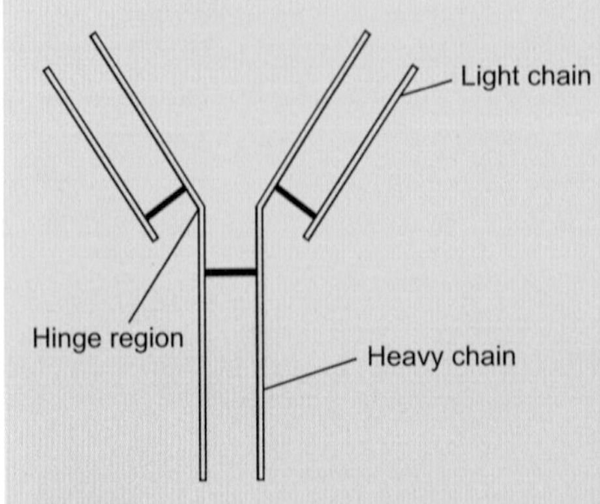

Figure Q1

   c. i. Explain what is meant by the term 'monoclonal antibody'.
   ii. Describe and explain one way in which monoclonal antibodies can be used to treat disease.

2. Wells **A**, **B** and **C** were cut into an agar plate and the following substances were put into them:
   - Well **A** – antigens from one strain of the influenza virus
   - Well **B** – a 'cocktail' vaccine containing antigens from several strains of the influenza virus
   - Well **C** – blood plasma from a human volunteer.

   After ten hours, precipitation lines were observed in the agar. Precipitation lines appear where there is agglutination. Figure Q2 shows where these lines occurred.

   a. Describe how the precipitation line between wells **A** and **C** was formed.
   b. Explain what the precipitation lines suggest about the volunteer's plasma.

   AQA January 2004 Unit 7 Question 4

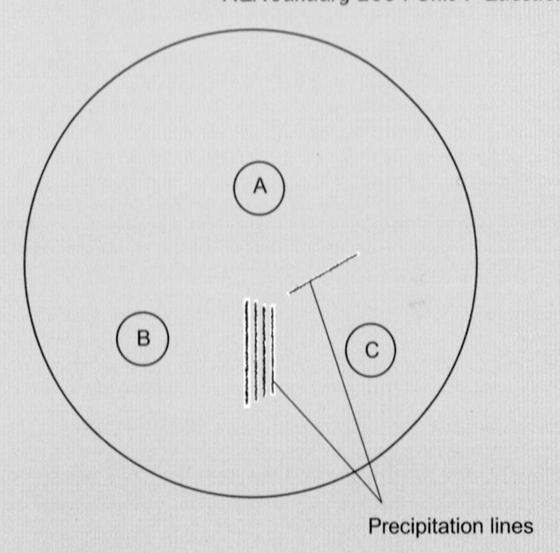

Figure Q2

3. Measles is an infectious disease that can cause serious complications in children. In countries where measles is uncommon, a combined measles, mumps and rubella (MMR) vaccine is given at 15 months. In a country where measles is common, a single measles vaccine (MV) may be given at nine months, followed by MMR at 15 months.

   In an investigation, the efficiency of the two vaccination programmes was compared in a country where measles is common. The amounts of measles antibody in the blood of children before vaccination and after completing vaccination were measured. The graph in Figure Q3 shows the results. All differences are statistically significant.

   a. What was the effect of vaccination in the MMR only group? Express your answer as the percentage increase in the amount of measles antibody in the MMR group after vaccination. Show your working.

(Continued)

b. The MV + MMR group had more measles antibodies in their blood before vaccination than the MMR only group. Suggest an explanation for this.

AQA January 2005 Unit 7 Question 4

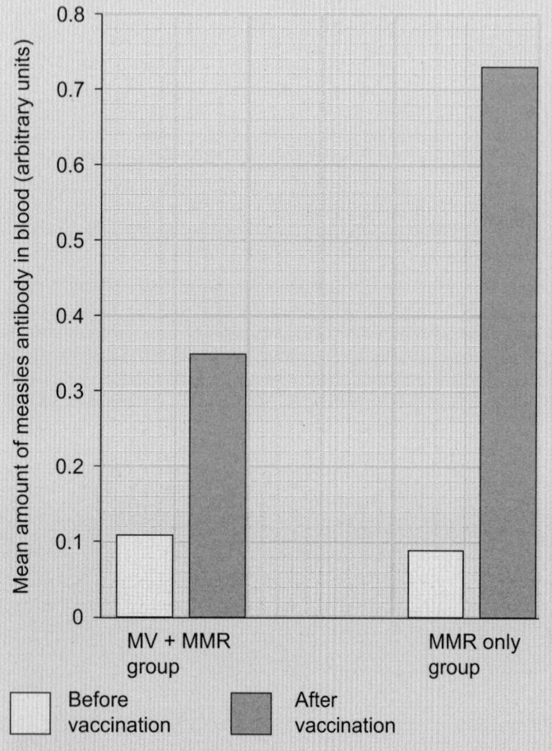

Figure Q3

4. The human immunodeficiency virus (HIV) leads to the development of acquired immunodeficiency syndrome (AIDS). Eventually, people with AIDS die because they are unable to produce an immune response to pathogens.

Scientists are trying to develop an effective vaccine to protect people against HIV. There are three main problems. HIV rapidly enters host cells. HIV causes the death of T cells that activate B cells. HIV shows a lot of antigenic variability.

Scientists have experimented with different types of vaccine for HIV. One type contains HIV in an inactivated form. A second type contains attenuated HIV, which replicates in the body but does not kill host cells.

A third type uses a different, non-pathogenic virus to carry genetic information from HIV into the person's cells. This makes the person's cells produce HIV proteins. So far, these types of vaccine have not been considered safe to use in a mass vaccination programme.

Use the information in the passage and your own knowledge to answer the following questions.

a. People with AIDS die because they are unable to produce an immune response to pathogens. Explain why this leads to death.

b. Explain why each of the following means that a vaccine may not be effective against HIV.

   i. HIV rapidly enters host cells.

   ii. HIV shows a lot of antigenic variability.

c. So far, these types of vaccine have not been considered safe to use in a mass vaccination programme. Suggest why they have not been considered safe.

AQA January 2013 Unit 1 Question 8

8 EXCHANGE WITH THE ENVIRONMENT

PRIOR KNOWLEDGE

You may remember that the rate at which diffusion takes place is affected by the area of the surface across which it takes place and the length of the diffusion pathway. You may also recall that monosaccharides and amino acids are water-soluble, polar substances, which cannot move freely through the phospholipid bilayer but instead must pass through carrier proteins. They can move by facilitated diffusion, or by active transport.

LEARNING OBJECTIVES

In this chapter, we look at how different kinds of organisms are adapted for exchanges of substances between their bodies and the environment, including the uptake and loss of gases for respiration and photosynthesis, and the uptake of nutrients.

(Specification 3.3.1, 3.3.2, 3.3.3)

Lake Titicaca is a huge, high-altitude freshwater lake, on the borders of Peru and Bolivia in the Andes mountains. Its surface is 3812 metres above sea level.

In this lake lives a giant frog, *Telmatobius culeus,* which is found nowhere else in the world. The frog looks like no other. Firstly, it is huge – adults can weigh up to 1 kg and measure 50 cm long. Secondly, it is covered with enormous folds of skin.

The frog uses its skin for gas exchange. Lake Titicaca's cold waters are always saturated with oxygen – gases dissolve more easily in cold water than warm, and the winds that constantly blow across the lake ruffle its surface and make it easy for oxygen to enter from the air. The frog's folds of skin increase the surface area across which oxygen and carbon dioxide can diffuse. These folds contain large networks of blood capillaries, which contain particularly large numbers of very small red blood cells. The small red cells again provide a relatively high surface area : volume ratio, increasing the rate at which oxygen can enter and leave them by diffusion. The frog does have lungs, but these are very tiny compared with those of other frogs, and it seems they are not used for gas exchange. The skin alone is sufficient, and the frog does not need to come to the surface for air. If, in a laboratory, it is put into water where there is significantly less dissolved oxygen than in its native habitat, it stands on its hind legs and bounces up and down, which makes its skin extensions flap around and helps to bring them into contact with more oxygen.

Not long ago, this frog was common in Lake Titicaca, but it is now critically endangered. The main threat appears to be human consumption. The frogs are collected and used to make a 'frog shake', by blending them with water to make a drink that is supposed to cure various human ailments.

8.1 SURFACE AREA : VOLUME RATIO

All organisms must constantly exchange substances between their bodies and their environment. You, for example, are taking in oxygen from the air in your lungs, and losing carbon dioxide into that air, which you then breathe out. A plant exchanges those gases at night, but during the daylight hours, when it is photosynthesising, it takes up carbon dioxide and gives out oxygen.

QUESTIONS

1. Copy and complete this table.

Length of side of cube/cm	Surface area of cube/ cm^2	Volume of cube/ cm^3	Surface area : volume ratio
1	6	1	6 :1
2			
3			
4			

2. Describe how surface area : volume ratio changes, as the size of the cube increases.

The ease with which an organism can exchange gases with its environment depends very much on its size. Imagine an animal whose body is a perfect cube. All of the cells inside the cube are using up oxygen and producing carbon dioxide, in respiration. These gases are being exchanged through the surface of the cube. If the size of the animal increases, its volume increases much more than its surface area.

Now let's put this into the context of real animals. An elephant has a much larger surface area and volume than a shrew (Figure 1). If we calculate the ratio of surface area : volume in each animal, we find that the surface area : volume ratio for the elephant is much smaller than the value for the shrew. This is a general rule: the larger the organism, the smaller its surface area : volume ratio.

This relationship between size and surface area : volume ratio affects the way in which organisms can exchange substances with their environment. Think about exchanging gases for respiration. The quantity of oxygen that is needed, and of carbon dioxide that is produced, depends on the volume of the animal. The rate at which these gases can move into and out of its body depends on its surface area. So, the smaller the surface area : volume ratio, the greater the challenge for the animal to supply all of its cells with the oxygen that they need, and to remove all of the carbon dioxide that they produce.

Figure 1 *A baby elephant has a much larger body than a shrew, and its surface area : volume ratio is smaller.*

ASSIGNMENT 1: INVESTIGATING HEAT LOSS

(MS 2.2, MS 2.3, MS 2.4, MS 4.1, PS 3.2)

The manatee and the walrus are both aquatic mammals and have many features in common but they live in completely different environments. The manatee lives in warm seas near the West Indies, while the walrus lives on the Arctic ice shelf off the coast of Alaska. Both maintain a constant body temperature. Table A1 shows the relative body sizes of the manatee and the walrus, and a large land mammal, the elephant, for comparison.

Figure A1 A manatee (left) and a walrus (right)

	Manatee	Walrus	Elephant
Approximate body length/metres	4	3.5	3
Approximate body width/metres	0.75	1	2
Approximate body mass/kilograms	1000	1700	5000

Table A1 Relative body sizes of the manatee, walrus and elephant

Questions

Stretch and challenge

A1. a. Assume the body of each animal has a cylindrical shape. Using the dimensions in Table A1, calculate the surface area of each animal, using:

$$\text{surface area} = 2\pi rh + 2\pi r^2$$

where:

r is the radius of the cylinder, and h is its length.

b. Animals have approximately the same density as water (1000 kg m^{-3}). Estimate the volume of each animal, using:

$$\text{density} = \text{mass} / \text{volume}$$

c. Calculate the surface area : volume ratio of each animal.

d. Which aquatic mammal is adapted to live in the Arctic? Explain your reasons.

A2. Do mammals lose heat more quickly in air or in water at the same temperature? Suggest a reason for your answer.

Large animals compensate for this by having body features that help them to exchange substances rapidly, and to distribute them around their bodies. Even animals as small as a shrew have special exchange features – they are still large animals compared with a single-celled organism such as an amoeba. These features usually include:

- having a body shape that increases surface area without increasing volume – for example, having lungs or gills that are specialised for gas exchange
- having a transport system that moves gases and other substances from exchange surfaces to other parts of the body.

The metabolic rate of an organism affects the rate at which it exchanges substances with its environment. An organism with a slow metabolic rate survives with a relatively small surface area : volume ratio, whereas a very active animal will have a higher one.

In this chapter, we will look at some of the specialised exchange surfaces in mammals. In the next chapter, we will consider transport systems.

KEY IDEAS

- The larger an organism, the smaller its surface area : volume ratio.

- The rate of use or production of substances that are exchanged with the environment is related to the volume of the organism, but the rate of exchange of these substances is related to the surface area.
- Large organisms have evolved special adaptations that increase the surface area across which exchange takes place.

Worked maths example: Surface area to volume ratio

(MS 0.3, MS 2.1)

Imagine you are investigating respiration of different animals. You have found records of the volume and surface area of some animals in the archives of a museum.

Which of the following algebraic statements are correct? (Here, the term s : v represents the surface area : volume ratio.)

- s : v (mouse) > s : v (lion)
- s : v (cat) = s : v (elephant)
- s : v (zebra) > s : v (shrew)

We know that, as the size of the animal increases, so surface area : volume ratio decreases.

Since a mouse is much smaller than a lion, it will have a larger surface area for its volume, so the **first statement is correct**. An elephant is much larger than a cat, so its surface area : volume ratio is unlikely to be the same. And a zebra is much larger than a shrew, so its surface area : volume ratio is likely to be smaller than a shrew's.

We can sketch a rough graph to show the relationship between surface area and volume in animals:

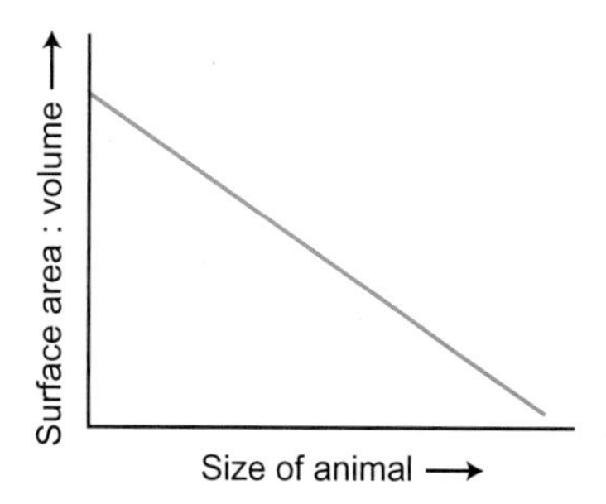

This is a rough sketch. We don't have the data to determine whether the relationship is linear, but we know that, as a general rule, the larger the animal, the smaller its surface area compared to its volume.

8.2 GAS EXCHANGE

Gas exchange in a unicellular organism

Figure 2 shows a micrograph of a single-celled (unicellular) organism called *Amoeba*. These are relatively large compared with most single-celled organisms, and you can just see them with the naked eye as tiny white spots.

Amoebas are constantly respiring, using up oxygen and producing carbon dioxide. This means that the concentration of oxygen inside the cell is always low (because it is being used up) compared with the concentration of oxygen outside the cell. Oxygen therefore diffuses into the cell through its cell-surface membrane. Similarly, carbon dioxide diffuses out, down its concentration gradient.

As the amoeba is so small, its surface area : volume ratio is relatively high. The rate at which oxygen and carbon dioxide are exchanged across its cell-surface membrane is easily enough to supply the cell's need. It does not need to have any special structures for gas exchange. There is also a very short diffusion pathway from its exchange surface to all parts of the cell.

QUESTIONS

3. Explain why oxygen and carbon dioxide are able to diffuse easily through a cell-surface membrane.

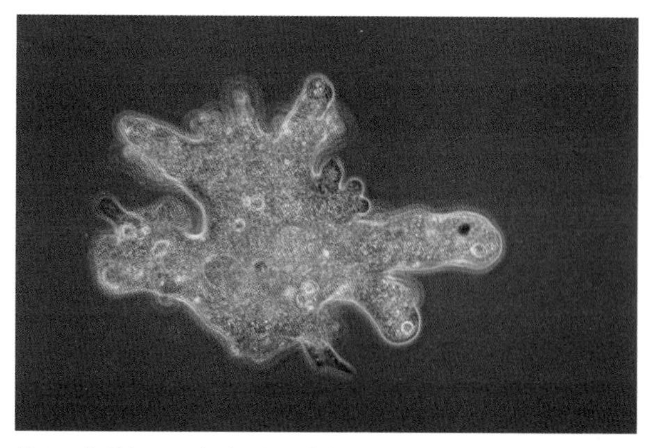

Figure 2 *This amoeba is about 0.1 mm across. Its tiny cell has a relatively large surface area : volume ratio for gas exchange with the water in which it lives.*

Gas exchange in insects

Insects were one of the first groups of animals to adapt to terrestrial life. Living on land poses two big problems compared with living in water: firstly, how to support your body when you do not have water to help with this support; and secondly, how to avoid losing too much water vapour from your body, especially across your gas exchange surface.

Larger organisms have specialised gas exchange surfaces, that effectively increase the surface area across which gas exchange can take place, resulting in an increase in surface area : volume ratio. But, for a terrestrial organism, increasing the area of the gas exchange surface also increases the area from which water can evaporate. Terrestrial animals have adaptations to avoid excessive water loss in this way, or they would dehydrate.

One adaptation that contributed to the success of insects on land was the evolution of a tough, supporting exoskeleton made of chitin, covered by a **cuticle**. The cuticle is impermeable to water so it helps the insect to conserve water. However, the cuticle is also almost totally impermeable to oxygen and carbon dioxide. It cannot be used for gas exchange.

Insects did not evolve lungs, nor did they evolve a blood system to transport respiratory gases. Instead, they have small openings in the cuticle called **spiracles**. These are connected to the inner organs by a system of highly branched, gas-filled tubes called **tracheae** (Figure 3). The smallest tracheae have a diameter of less than 1 μm. This system has a very large surface area, but it is protected inside the body, and so avoids too much water loss. The spiracles can be closed entirely to cut down water loss even more if need be. The air inside the tracheae also contains quite large quantities of water vapour, and so has a high water potential, which reduces the loss of water by evaporation from the gas exchange surface.

Gas exchange by diffusion between the air in the tracheal system and the body cells mainly occurs at the tips of the smallest branches, which are called **tracheoles**. Because no cell in the body is far from a tracheole, and there are numerous spiracles, diffusion pathways are usually short. Furthermore, diffusion is in the gas phase, which is faster than when dissolved in water.

Very active insects, especially those that fly, have muscles that pump air into and out of their tracheal systems. This active movement of air to the gas exchange surface is called **ventilation**. Ventilation leads to mass transport of oxygen and carbon dioxide – which maintains diffusion gradients for exchange.

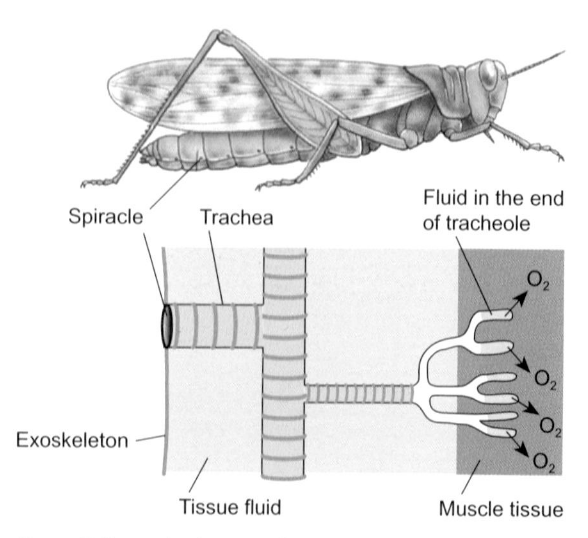

Figure 3 *The tracheal system of an insect*

QUESTIONS

4. Where is the gas exchange surface of an insect?
5. Explain how insects have increased the surface area for gas exchange, and yet manage to control the quantity of water that is lost from their bodies.
6. In many insects, the actions of their flight muscles help to move air through the tracheal system. Explain the benefits of this arrangement.

Stretch and challenge

7. One of the largest insects living today is the titan beetle (*Titanus giganteus*). Suggest why insects are generally not as big as mammals.
8. In prehistoric times, some species of insects used to grow a great deal larger. Suggest why.

ASSIGNMENT 2: INVESTIGATING INSECT GAS EXCHANGE

(PS 1.2, PS 3.1)

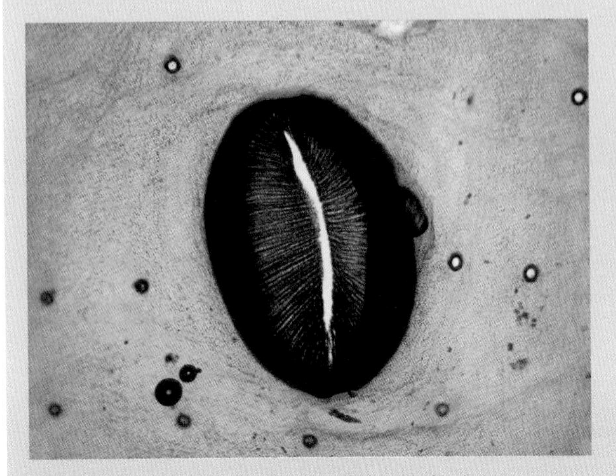

Figure A1 A microscopic view of a silkworm spiracle

The spiracles in an insect cuticle behave like valves, opening and closing to allow or restrict the insect's gas exchange. Scientists have observed a peculiar rhythmic respiratory behaviour referred to as the **discontinuous gas-exchange cycle (DGC)**. In insects exhibiting DGC, the spiracles close for long periods (up to several hours or even days) and open occasionally for only a few minutes. This unusual respiratory pattern has been observed in many adult insects, as well as in butterfly and moth pupae. Two main hypotheses have been proposed to explain why some insects display DGC:

- to reduce water loss through the spiracles
- as an adaptation to living underground.

Experiments carried out by scientists Hetz and Bradley provide evidence that the DGC prevents an insect taking up too much oxygen. Using the pupae of the moth *Attacus atlas*, these researchers varied the environmental oxygen concentrations from partial pressures of 5 kPa to 50 kPa (the normal atmospheric oxygen partial pressure at sea level is about 21 kPa). They found that the concentration of oxygen levels in the resting pupae remained low, close to 4 kPa, across the whole range of partial pressures. It seems that the moth pupa limits the amount of oxygen taken in by keeping the spiracles closed and only opening them to get rid of the accumulated carbon dioxide.

In 2004, scientists Gibbs and Johnson found that metabolic rate varied with the type of gas exchange – insects using the DGC had the lowest metabolic rate, those using continuous gas exchange had the highest metabolic rate, and insects using cyclic gas exchange had an intermediate metabolic rate. In their investigation, they used the burrowing ant *Pogonomyrmex barbatus*, which uses all three types of gas exchange.

Questions

A1. How might the DGC help an insect that spends much of its time living underground to get rid of carbon dioxide?

Stretch and challenge

A2. The graphs show three different patterns of gas exchange: graphs Ai and Aii show the DGC; graphs Bi and Bii show cyclic gas exchange; graphs Ci and Cii show continuous gas exchange.

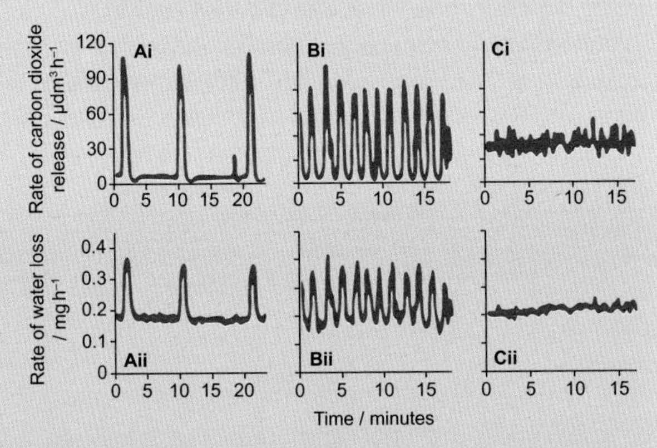

Figure A2

a. What do the data show about the relationship between water-loss rate and carbon dioxide release in each of the three patterns of gas exchange?

b. Do these data support the hypothesis that the DGC reduces the rate of water loss by insects? Explain the reason for your answer.

c. What evidence is shown in the graphs to support the findings of Gibbs and Johnson?

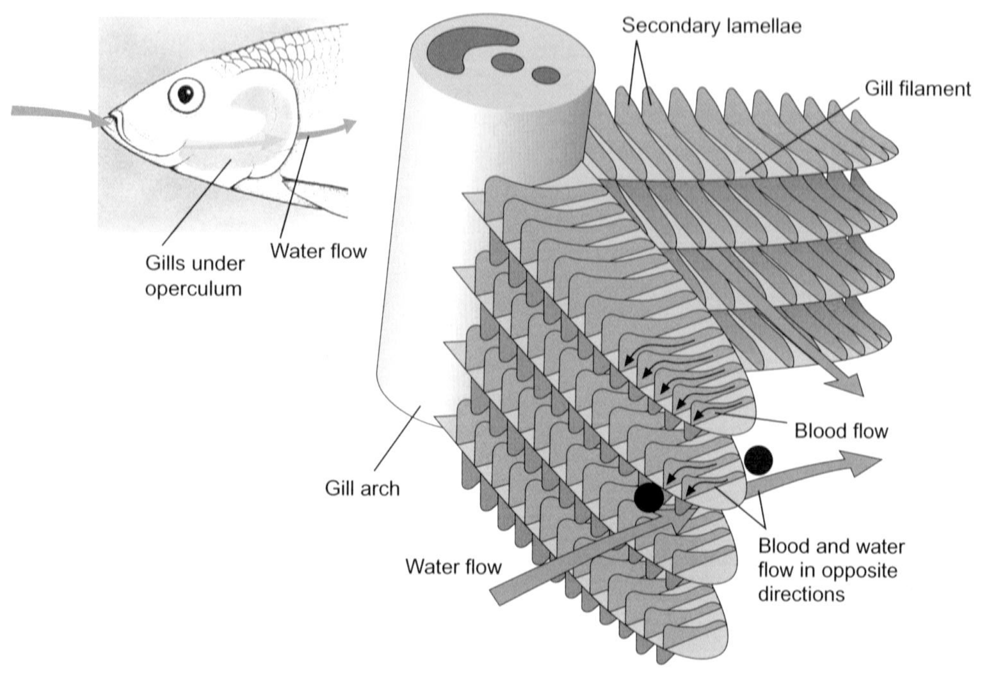

Figure 4 *The structure of a fish gill*

Gas exchange in fish

Fish are vertebrates and – like all vertebrates and many other groups of animals – they use blood to transport substances around their bodies. This includes oxygen and carbon dioxide. At their gas exchange surface, oxygen diffuses into the blood and carbon dioxide diffuses out of it.

Fish obtain their oxygen from dissolved oxygen in the water. Water is passed over the **gills**, which are their gas exchange surface. Inside the gills, projections called **secondary lamellae** (Figure 4) provide a large surface area. The diffusion path for oxygen is short, because the blood that flows within the lamellae is separated from the water outside them by a very thin layer of cells.

However, obtaining oxygen from water rather than air presents particular problems for aquatic animals. Oxygen is not very soluble in water and so water contains only one thirtieth as much oxygen per unit volume as air. Moreover, oxygen diffuses more slowly through water than through air. Water also has a higher density than air, and is therefore harder to move over a gas exchange surface during ventilation. Fish have evolved **a countercurrent system** (Figure 5), which maximises the efficiency of oxygen uptake from water.

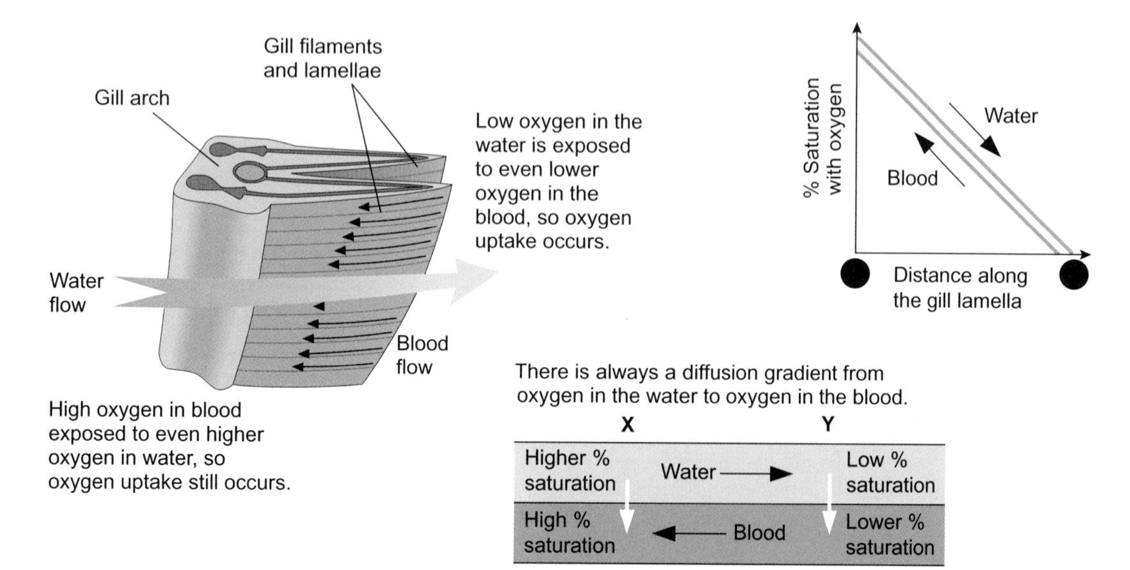

Figure 5 *Countercurrent flow in fish gills*

How does this work? The water that flows over the secondary lamellae, and the blood that flows through them, travel in opposite directions. This is the countercurrent flow. It increases the efficiency with which oxygen can diffuse into the blood because:

- at point X in Figure 5, the water that has just entered the fish's gills is saturated with dissolved oxygen, so there is a large diffusion gradient for oxygen between the water and the blood
- at point Y in Figure 5, although much of the dissolved oxygen has already diffused from the water into the blood, the blood here has just entered the gills. It therefore has a very low concentration of oxygen (because it has come from the fish's body, where the cells have been using oxygen in respiration) and so there is still a large diffusion gradient that favours the movement of oxygen from the water to the blood.

The countercurrent system is very effective: up to 80% of the oxygen dissolved in water is extracted by the gills. In comparison, our lungs can extract only a maximum of 25% of the oxygen from the air we breathe. The diffusion of carbon dioxide from the blood into the water is also helped by the countercurrent flow.

Note that the countercurrent mechanism works only when the ventilation current moves in one direction only. We cannot have such a mechanism in our lungs because, as you will see, air moves in and out of our lungs by going back and forth along the same pathway, rather than as a continuous flow across the gas exchange surface.

QUESTIONS

9. When a fish is out of water, surface tension sticks the gill lamellae together. Explain why this can quickly be fatal.

Gas exchange in plants

Plants do not move from place to place and so do not require as much energy (per unit of mass, per unit of time) as active animals. This means that they do not need anywhere near as much oxygen as an animal of equivalent size requires. But plants do need surfaces for gas exchange because the processes of cell respiration and photosynthesis both involve an exchange of gases with the atmosphere.

Most plants have a much more branching shape than most animals, which increases their surface area : volume ratio. The main gas exchange surface in the aerial parts of plants is inside the leaves (Figure 6). Cells inside the stem obtain their oxygen by diffusion from the air around them; roots obtain theirs by diffusion from air spaces in the soil.

A leaf has a large internal surface area in relation to its volume. Oxygen and carbon dioxide enter and leave a leaf mainly via microscopic pores called stomata (singular = stoma). There are huge numbers of stomata – several thousand per cm^2. After diffusing into the leaf through the stomata, gases diffuse through the intercellular spaces in the mesophyll. The leaves are usually very thin and the

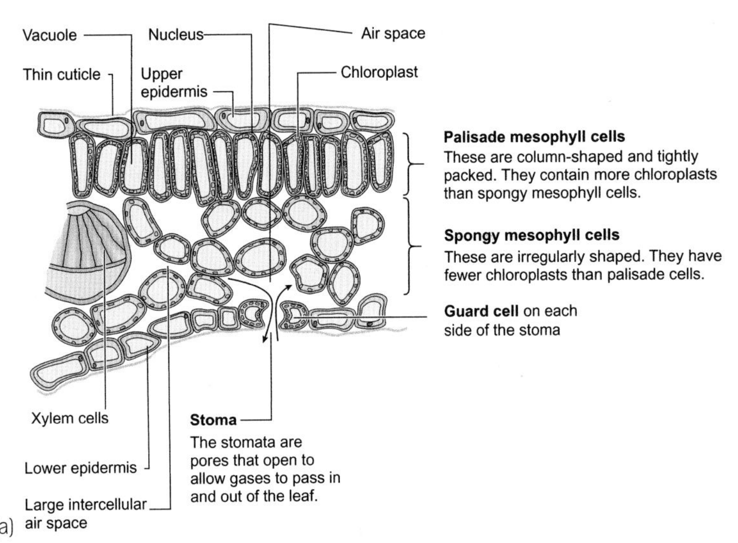

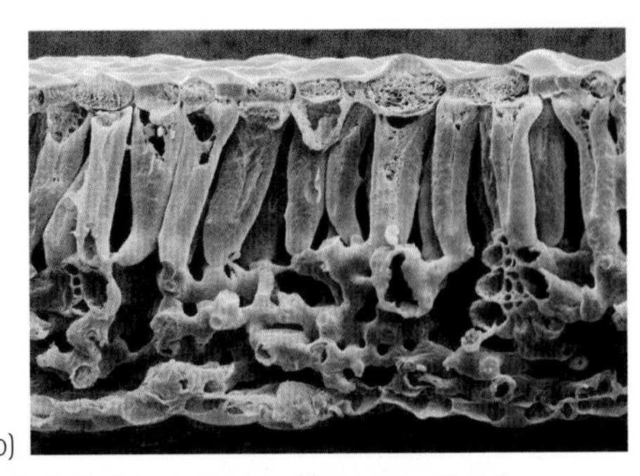

Figure 6 *(a) Gas exchange in a leaf; (b) Scanning electron micrograph of a transverse section of a leaf of a potato plant. You can see that the mesophyll cells have a large surface area in contact with the air spaces around them. This is the gas exchange surface.*

large number of stomata means that no cell is far from the external air, so the diffusion pathway is short. The surface of the mesophyll cells acts as the gas exchange surface. The cell walls are moist and permeable to gases. Because diffusion is in the gas phase, it happens quite rapidly.

Like all terrestrial organisms, plants have the problem of losing water to the air around them. The cell walls of their mesophyll cells are moist, as water moves down a water potential gradient out of the cells and evaporates into the air spaces around them. Water vapour can then diffuse out of the leaf through the stomata. However, excessive loss of water into the atmosphere is avoided because the exchange surface is inside the leaf, and the stomata can be closed. Each stoma is surrounded by two guard cells that can vary the width of the stoma, allowing more or less gas to enter or leave. Air in the air spaces is saturated with water vapour; this creates a high water potential, so less water evaporates from the cell walls. Some plants, called xerophytes, have evolved adaptations that greatly reduce water loss from their gas exchange surfaces, allowing them to live in very dry environments.

QUESTIONS

10. During daylight, plants respire and photosynthesise. The rate of photosynthesis is greater than the rate of respiration. During darkness, photosynthesis stops but respiration continues.

    Which gases will move across the gas exchange surface, in which direction, during the day and at night?

KEY IDEAS

- Single-celled organisms have a large surface area : volume ratio, and gas exchange across their cell-surface membrane is sufficient.
- Insects have a system of tubes, the tracheal system, that delivers air deep inside their bodies. Gas exchange occurs by diffusion at the ends of the tracheoles.
- Insects do not use their blood system to transport oxygen.
- Insects do not lose too much water vapour from the gas exchange surface because their gas exchange surface is deep inside the body, and their spiracles can close.
- Fish have gills, with lamellae that provide a large surface area for gas exchange.
- Gills have a good blood supply, which maintains a steep diffusion gradient across the exchange surface.
- A countercurrent mechanism, in which the directions of blood flow and water flow are opposite to one another, ensures that a diffusion gradient is maintained along the whole length of the exchange surface.
- Plants have branching shapes, which gives them a relatively large surface area : volume ratio.
- The gas exchange surface in a plant leaf is the surface of the mesophyll cells. These are surrounded by air spaces that link to the outside air via stomata.
- Stomata can close, which limits the amount of water vapour lost from the gas exchange surface.

8.3 THE HUMAN GAS EXCHANGE SYSTEM

Humans are relatively large organisms, and our external surface area : volume ratio is small. However, our lungs, tucked away deep inside the body where they cannot easily dry out, provide a huge surface across which gas exchange can take place. It is estimated that the surface area of one lung of an adult is more than 50 m^2.

Figure 7 overleaf shows the gross (large scale) structure of the human gas exchange system. A tube with walls reinforced with cartilage, the **trachea**, runs from the mouth and nose down into the thorax. Here, it branches into two **bronchi** (singular: bronchus) which in turn divide into numerous smaller **bronchioles**. These branch again, eventually terminating in millions of tiny air-filled sacs called **alveoli**.

Gas exchange in the lungs

The gas exchange surface is the surface of the alveoli. These are tiny sacs with walls made up of

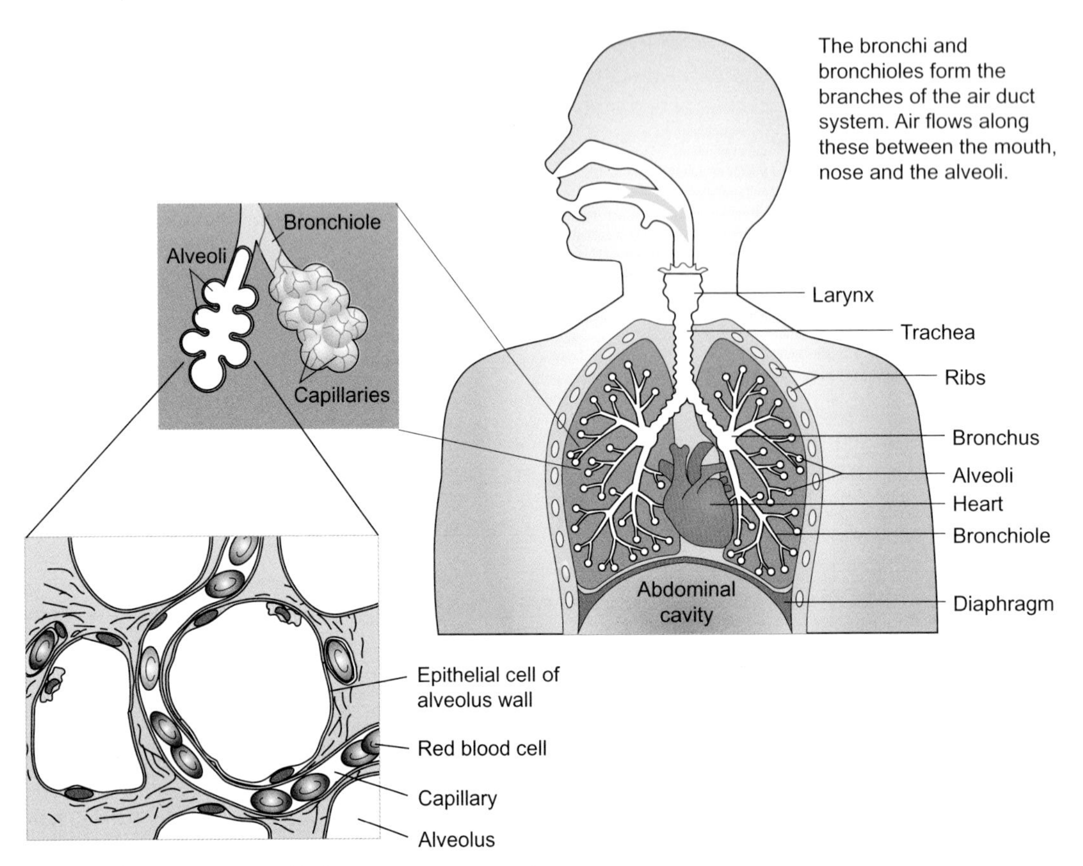

Figure 7 *The gross structure of the human gas exchange system*

a single layer of cells, the **alveolar epithelium**. (An epithelium is a tissue that covers a surface.)

When we breathe in, air enters the alveoli and the oxygen in that air diffuses across the thin walls and into the blood. At the same time, carbon dioxide from the blood diffuses into the alveoli to be breathed out.

Alveoli are well adapted to their function as an exchange surface for gases.

- The epithelial cells that make up the walls of the alveoli support a rich network of blood capillaries (Figure 8). This blood supply is essential to carry oxygen away from the air and bring carbon dioxide to the alveoli. As the oxygen-rich blood flows away, it is replaced by oxygen-poor blood, which maintains a steep diffusion gradient so that oxygen will continue to diffuse in. Similarly, the constant arrival of blood containing high concentrations of carbon dioxide maintains a diffusion gradient that allows carbon dioxide to diffuse out of the blood and into the air inside the alveoli.
- The alveolar walls are very thin and the cells of the wall are flattened. This ensures a short diffusion pathway and so a rapid rate of diffusion.
- The walls of the alveoli are permeable. This allows oxygen to pass into the cells but also lets water out, so the exchange surface is always moist. Oxygen dissolves in the layer of moisture and then diffuses through the wall into the blood.
- The air in alveoli are saturated with water vapour. This reduces the rate of water loss from cells by evaporation.

The inner surface of the alveolus has to be moist, so that the cells that make up the epithelium do not dry out and die.

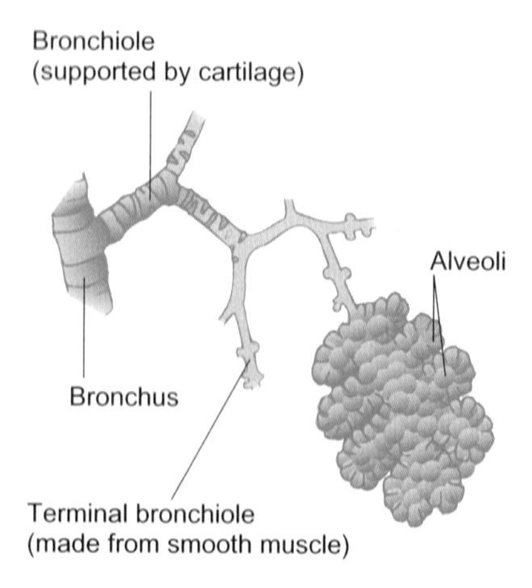

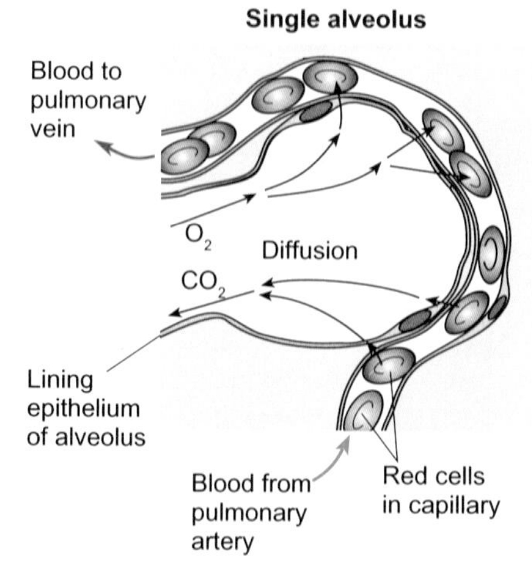

Each alveolus is a tiny air sac that has thin, flat walls. Oxygen from the air dissolves in the liquid that lines the alveolus and then diffuses across into the blood capillary. Carbon dioxide leaves the blood by the reverse route.

Figure 8 *Gas exchange across the alveolar epithelium*

QUESTIONS

11. Will the layer of moisture on the inner surface of the alveoli speed up or slow down the diffusion of oxygen across the alveolar wall? Explain your answer.

Ventilation of the lungs

Because the lungs are so deep inside the body, it would take too long for gases to diffuse from air to the gas exchange surface. Air is therefore moved in and out of them by breathing movements. This is called ventilation. It is an example of mass transport of gases.

Breathing draws fresh air into the lungs and forces stale air out. Ventilation ensures there is always a good supply of 'fresh' air inside the lungs. This maintains steep diffusion gradients for oxygen and carbon dioxide between air and blood. Steep diffusion gradients mean efficient gas exchange.

Notice the difference between ventilation, gas exchange and respiration. Respiration is a series of oxidation reactions that occur in cells. Gas exchange is the diffusion of gases across the gas exchange surface. Ventilation, brought about by breathing, is a muscular action that moves air into and out of the lungs.

Because the flow of air through the lungs occurs in both directions – in and out – we say that the ventilation is **tidal**. Air passes into and out of the alveoli via the trachea, bronchi and bronchioles. Except for the narrowest bronchioles, the walls of these three air tubes contain rings of cartilage. Cartilage is a strong, slightly flexible tissue. It helps to keep the tubes open during the pressure changes that take place as we breathe in and out.

Breathing in (inspiration) is always an active process (Figure 9, overleaf). It is brought about by two sets of muscles: the muscles of the **diaphragm**, and the set of muscles between the ribs called the **external intercostal muscles**. Both of these muscle groups contract to bring about inspiration. The contraction of the diaphragm muscles pulls it downwards, which increases the volume inside the thorax. The contraction of the external intercostal muscles pulls the rib cage upwards and outwards, which also increases the thoracic volume. The increase in volume reduces the pressure inside the thorax, so that the air inside the lungs is at a lower pressure than the air outside the body. Air therefore flows into the lungs, down its pressure gradient.

When the body is at rest, breathing out (expiration) is a passive process. During gentle expiration, the diaphragm muscles relax and the elastic recoil of the lungs and chest wall returns the thorax to its original shape and volume. This decreases the volume of the thorax, so that the pressure of the air inside the lungs is increased. Air therefore flows out of the lungs and into the air around us, down its pressure gradient.

When we are exercising, and the muscles use more oxygen to release energy through faster respiration, the rate and depth of breathing movements increases. Increasing the volume of the thorax decreases air pressure in the alveoli to below atmospheric pressure. Extra air therefore flows into the lungs with each breath. In addition, expiration becomes an active process. The internal intercostal

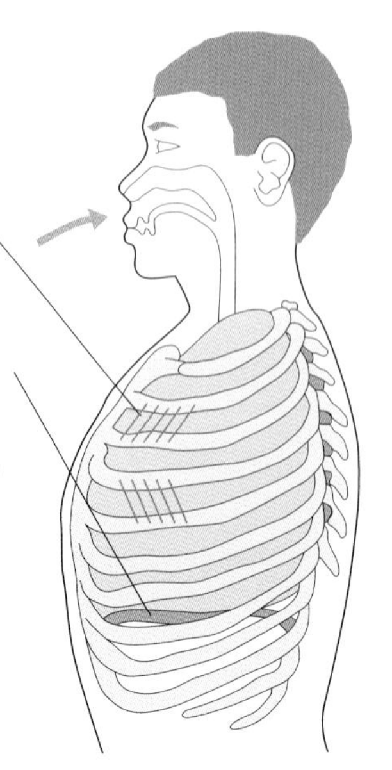

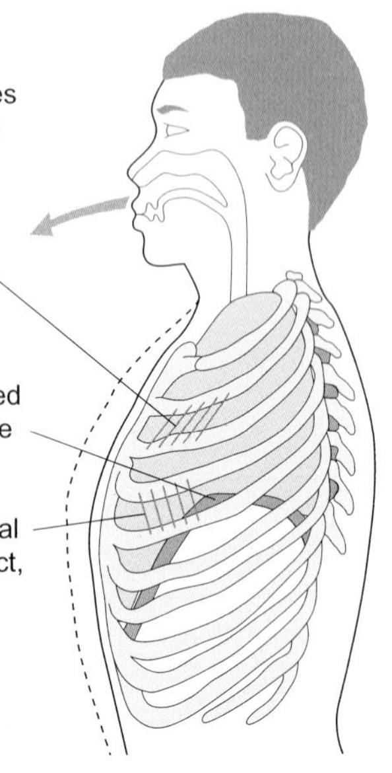

Figure 9 *Ventilating the lungs*

muscles contract, pulling the ribs downwards, making the volume inside the thorax even smaller than usual. At the same time, the muscles in the abdomen wall contract, pushing the diaphragm upwards even further. This increases the volume of air exhaled with each breath.

The external and internal intercostal muscles are an example of antagonistic muscles. The contraction of one set of muscles pulls in one direction, while the contraction of the other pulls in the opposite direction. When one set is contracting, the other set relaxes.

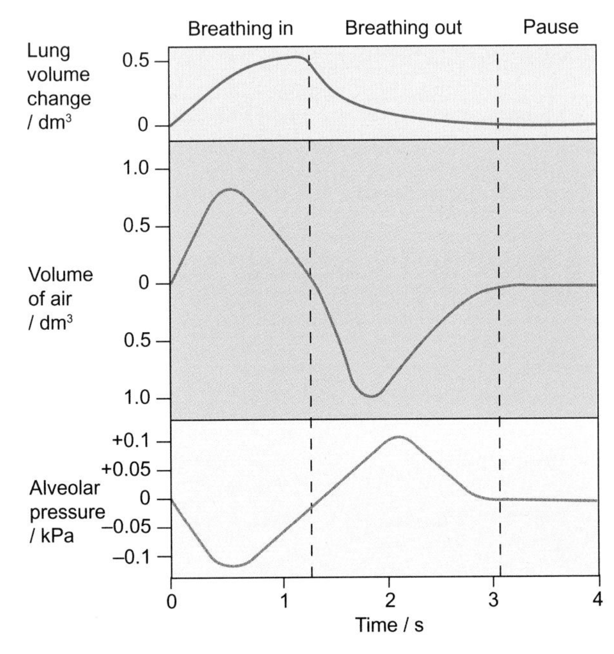

Figure 10 *Pressure changes and air movements during breathing*

QUESTIONS

12. List three differences and three similarities between the methods used by humans and fish to obtain oxygen.

ASSIGNMENT 3: EXPLORING ALTITUDE SICKNESS

(PS 1.2)

Climbing a mountain is hard work. As you climb, the percentage of oxygen in the air remains the same (about 21%) but the number of oxygen molecules that you can take in with each breath decreases as the pressure of air decreases. At sea level, air pressure is about 100 kPa, but at 3500 m (about 12 000 feet) this has fallen to about 63 kPa. Your body compensates by increasing your breathing rate, but if you ascend too quickly, you can start to develop altitude sickness. Climbing above 3000 m

Figure A1 *An inflated Gamow bag*

causes the symptoms of mild altitude sickness in many people. These symptoms include headache, dizziness, shortage of breath, nausea and disturbed sleep. They arise because when the air pressure falls to a level below the pressure in the lung capillaries, fluid leaks from the capillaries into the alveoli. Within a couple of days the body begins to compensate, and the symptoms start to disappear after about three days. This process is known as acclimatisation.

If you ascend very quickly to a very high altitude (4000–5000 m), the symptoms are much worse – shortness of breath even when resting, mental confusion and the inability to walk. This is acute altitude sickness and is caused by fluid leaking from the capillaries into the cavities in the brain. Someone with these symptoms should be returned to a lower altitude immediately. If this is not possible then they should be placed in a Gamow bag. The bag is inflated with a foot pump. Conditions inside the bag simulate lower altitudes and the patient quickly recovers. Most high-altitude climbing expeditions now carry at least one of these bags.

Altitude sickness can be prevented by acclimatising to the change in altitude in stages. This involves spending two to three days at a particular altitude before trying anything strenuous like rock climbing or skiing and before moving to higher levels. Acclimatisation occurs because when the body is at high altitudes it produces a hormone called erythropoietin. This hormone stimulates the body to produce more red blood cells to compensate for the reduced capacity to take in oxygen from the air. Many athletes train at high altitude. The increased number of red blood cells lasts for about two weeks and enhances performance, particularly in middle- and long-distance running events.

It is possible to inject erythropoietin to obtain the same effect, but athletics authorities forbid this. However, athletes who want to simulate the effects of going to high altitudes without having to climb a mountain can use a 'high altitude' sleeping chamber quite legitimately. An athlete training at sea level can gain the advantage of training at high altitude by sleeping in a Gamow bag set up to simulate high altitude.

Questions

A1. Explain why the body obtains less oxygen per breath at high altitude than at sea level.

A2. Explain how fluid leaking from capillaries into the alveoli affects gas exchange.

A3. Explain what is meant by acclimatisation.

A4. Explain how you would recognise if a companion has acute altitude sickness.

A5. Explain how the Gamow bag helps a person who has acute altitude sickness.

ASSIGNMENT 4: CORRELATION AND CAUSATION

Many biological investigations depend on a combination of observation and data analysis rather than on actual experiments. This is because it is often not practical to carry out proper controlled experiments with living organisms in the field, sometimes because of the complexity of interrelationships between organisms and the environment and sometimes for ethical reasons. You can't experiment on the effects of smoking by taking two groups of people and making one group smoke while keeping all other factors the same.

Investigators therefore have to look for causal relationships that occur in the normal course of events. However, care needs to be taken when drawing conclusions. The number of fish may decline in a lake affected by acid rain or some other pollutant, but this causal relationship does not necessarily prove that the pollution has caused the decline, or even that the two are connected. Further investigations could look for data on natural populations of particular fish species in water of different acidity. It would also be possible to carry out laboratory experiments to determine fish survival rates in water of different acidity. Results might well show that the lower the pH, the lower the survival rate. In this case there would be a **correlation** between pH and fish survival. This would still not prove that the decline in fish numbers in the lake was actually caused by the acidity.

If you counted, say, the number of nightclubs and pubs and the number of churches in several towns and cities and then plotted a graph of one against the other you would almost certainly find a correlation. But this would obviously not prove that churches cause nightclubs and pubs to be built or the other way round. The correlation is likely to be the result of completely separate factors, such as the size of the town or city.

Nevertheless, it is only by searching for correlations and investigating them further that biologists can increase their understanding. A correlation can be either positive or negative. When one factor increases as another increases it is a positive correlation; when one increases as another decreases it is a negative correlation.

Epidemiology

Causal links between human disease and factors such as environmental pollutants, diet, smoking and other aspects of lifestyle are equally hard to establish by experiment. Most causal relationships have been established by studies of the incidence of disease or disorder on large groups of people. Looking for patterns of disease in human populations is called epidemiology. Many of the suggested links have been matters of controversy and some have caused considerable confusion in the minds of the public. Consider issues such as the danger of using mobile phones; the possible link between taking contraceptive pills and various cancers; or fat consumption and heart disease. There are still some people who refuse to accept the causal links between smoking and cancer, despite the overwhelming experimental evidence. The stages in establishing the cause of a non-infectious disease are:

- searching for a correlation between a disease and a specific factor
- developing hypotheses as to how the factor might have its effect
- experimental testing to determine whether there is evidence to support the hypotheses.

To establish a correlation means collecting data from large numbers of people. Because of the huge variability between people and their lifestyles, it requires comparisons to be made as far as possible between matched groups. For example, suppose you were looking for a correlation between beer consumption and heart disease. It would not be sufficient just to compare the rates of heart disease between 1000 beer drinkers and 1000 non-drinkers. The ideal comparison would be between groups of people where the only difference in lifestyle was whether or not they drank beer. In practice this would be virtually impossible to achieve. However, at least a much more valid comparison could be made between groups matched for age, sex, amount of exercise taken and major features of diet. The difficulty is to eliminate the possibility that any correlation found is not due to some other linked factor, such as that people who are tempted to indulge in beer-drinking are also consumers of excessive quantities of fish and chips, or simply that they have some genetic predisposition to heart disease. The latter is a particularly difficult argument to refute, and has regularly been used as an excuse for shedding doubt on the smoking/lung cancer correlation.

Once a correlation has been found, the next stage is to try to develop an hypothesis to explain how the factor causes the effect – usually based on existing scientific knowledge. This is often much more difficult. Many diseases, such as cardiovascular disorders and cancers, develop as a result of several interacting factors. The correlation between smoking and incidence of lung cancer has been established for many years. The search to isolate a specific **carcinogen** in cigarette smoke has still not been successful. However, the tar inhaled into the lungs contains a massive cocktail of organic compounds, many of which may have carcinogenic properties. Some of the effects may be additive – a combination of substances and other factors may be the main cause. Also, individuals differ in their susceptibility, probably due to genetic factors. Research has developed experiments to test the hypothesis involving detailed chemical analyses of tar, experiments on animals and with tissue cultures, and comparisons between many genetically distinct groupings. Until a precise mechanism is discovered, the arguments will no doubt continue, and only after that will it be possible to produce effective countermeasures apart, of course, from not smoking! The chart in Figure A1 shows the risk of developing coronary heart disease during the next 10 years in relation to a number of risk factors.

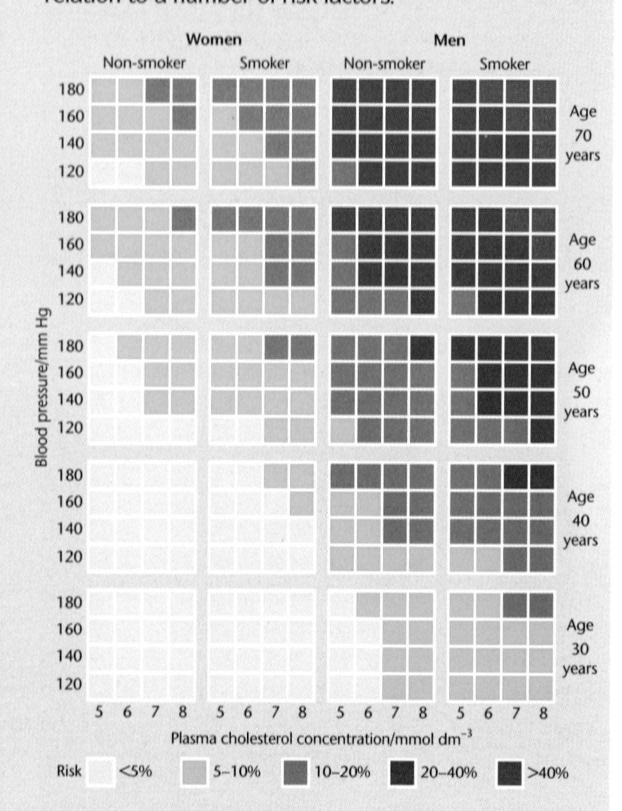

Figure A1 *Coronary heart disease and risk factors.*

Questions

A1. Which risk factors have been taken into account in this chart?

A2. According to the chart, what are the characteristics of the people with the highest risk of developing coronary heart disease?

A3. A group of 500 men aged between 40 and 50 years is studied. They all have high blood pressure of over 180 mm of mercury, and plasma cholesterol concentration between 7 and 8 mmol dm^{-3}. They all smoke. How many of these men would be expected to get coronary heart disease over the next 10 years? How many would get it if they were all non-smokers?

A4. Explain the limitations of using the chart as a way of predicting whether an individual will develop coronary heart disease.

A5. The data used to construct the chart have a number of limitations as a means of predicting risk. Suggest three weaknesses in the data.

A6. Suggest how experimental (and epidemiological) evidence linking smoking to a range of diseases has influenced the governments of numerous countries.

KEY IDEAS

- A correlation in data is where there appears to be a link between one factor and another, such as between the number of fish in a lake and its acidity.
- One factor may increase as the other increases (a positive correlation), or one may decrease while the other increases (a negative correlation).
- A correlation is not proof that a change in one factor is the direct cause of the change in the other. This can only be supported by experimental evidence.
- In complex situations, such as in ecosystems and human populations, it is difficult to establish cause and effect because many factors may interact.
- In human populations, epidemiologists look for causal relationships between factors and try to establish the likely causes of diseases by studying data from matched groups.
- The gas exchange surface in a human is the surface of the alveoli in the lungs.
- Alveoli are very small, and are present in the lungs in very large numbers. This provides a very large surface area for gas exchange.
- The walls of alveoli, and of the blood capillaries next to them, are each only one cell thick, minimising diffusion distance.
- The constant flow of blood, and the ventilation of the lungs with air, maintains a diffusion gradient between the blood and air in the alveoli.
- Ventilation is brought about by the action of the muscles in the diaphragm and the intercostal muscles.
- During breathing in (inspiration), the diaphragm muscles and the external intercostal muscles contract, increasing the volume of the thorax, decreasing the pressure within it and therefore causing air to flow in down a pressure gradient.
- During breathing out (expiration), the diaphragm muscles and the external intercostal muscles relax, decreasing the volume of the thorax, increasing the pressure within it and therefore causing air to flow out down a pressure gradient.
- During forced breathing out, the internal intercostal muscles and abdominal muscles contract, pulling the rib cage even further down and pushing the diaphragm even further up, increasing the pressure gradient.

8.4 DIGESTION AND ABSORPTION

So far in this chapter, we have concentrated on how organisms obtain gases from their environment, and return waste gases to it. But gases are not the only materials that organisms obtain from their environment. They also obtain nutrients.

We are mammals, and – like all mammals – we obtain nutrients by taking food into the alimentary canal (Figure 11). This canal is a long tube that runs from the mouth to the anus. Spatially, it is actually part of the outside world. An object such as an accidentally swallowed key can pass all the way through the alimentary canal from one end to the other, without ever crossing a cell-surface membrane and entering a body tissue.

In order to enter the body tissues, the nutrients move across the cells that make up the wall of the alimentary canal. This is called **absorption**. The molecules enter

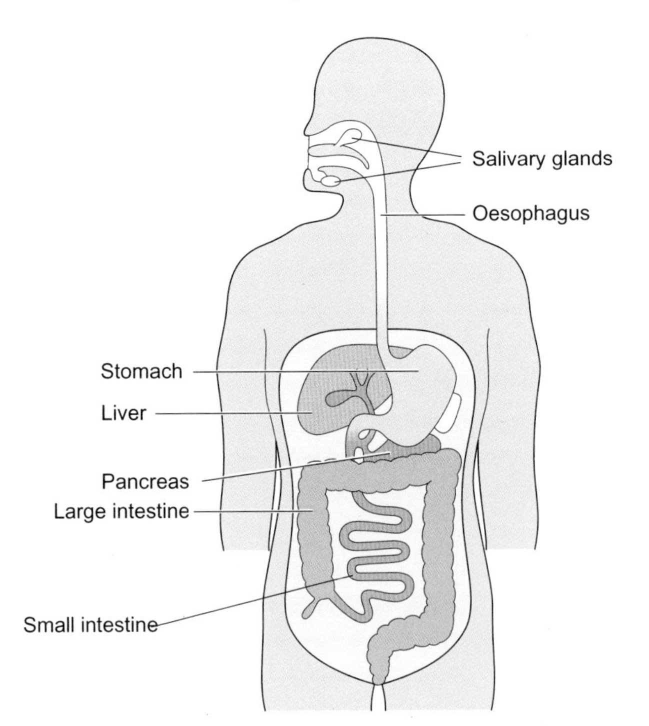

Figure 11 *The human digestive system. The digestive system is made up of the alimentary canal (a long, open-ended continuous tube) and other organs including the liver and pancreas.*

and leave these cells by crossing their cell-surface membranes by one or more methods – diffusion, facilitated diffusion, osmosis or active transport.

You will remember that it is difficult, if not impossible, for large molecules to move across cell-surface membranes. Yet many of the nutrients that we eat are made up of very large molecules. Starch, for example, has molecules that consist of hundreds of glucose monomers linked together. It is not possible for a starch molecule to cross a cell-surface membrane and enter a cell.

In order to allow nutrients to be absorbed, it is therefore necessary to break large molecules apart into smaller molecules. This is called **digestion**. You may remember that the break-up of large molecules such as starch into their smaller units involves hydrolysis reactions. These reactions are catalysed by hydrolase enzymes.

In humans, digestion happens in the mouth, stomach and small intestine. Absorption of organic molecules mostly takes place in the small intestine.

The alimentary canal

As food travels along the alimentary canal, it passes through several different regions. Each region is adapted to carry out a specific function:

- to break large lumps of food into smaller lumps, a process called mechanical digestion
- to hydrolyse large nutrient molecules, a process called chemical digestion
- to absorb the small molecules produced by digestion
- to get rid of undigested materials and other waste products.

Food enters the alimentary canal by the mouth. In the mouth cavity, chewing starts the process of mechanical digestion. Some chemical digestion also begins here since the saliva contains an enzyme that hydrolyses starch – although, most of the time, people do not chew their food for long enough for this to have much effect. The saliva softens and lubricates the food so that it can be swallowed easily.

From the mouth, the food passes down the oesophagus into the stomach. The rest of the alimentary canal is in effect a long tube with a variable diameter.

Digestion of carbohydrates

Starch is the main carbohydrate in our diet because it is such a common storage compound in plants. We eat small amounts of glycogen – animals use this polymer as their main carbohydrate storage compound, but they store only relatively small quantities in their liver and muscles. Another common carbohydrate polymer is cellulose, but humans do not have an enzyme to digest it. Cellulose therefore forms the bulk of the fibre in our diet, which is necessary to maintain a healthy colon and bowel.

The enzyme **amylase** hydrolyses starch into the disaccharide, maltose. Amylase is present in saliva, so starch digestion begins in the mouth. Some digestion of starch also occurs in the stomach as a result of the continuing action of the amylase that was added to food in the mouth. However, this tends to be short-lived as amylase is inactivated rapidly by the low pH caused by the presence of hydrochloric acid in the stomach.

No further digestion of starch takes place until the food reaches the small intestine. Here, digestive juices produced by the pancreas flow along the pancreatic duct and into the lumen (central space) of the small intestine. Pancreatic juice contains amylase, which continues to hydrolyse starch into maltose.

As the partly digested food continues to move through the small intestine, it encounters the highly folded surfaces of the intestine walls. The small intestine is narrow, so it is not easy for food to pass through without brushing against its walls. These walls are covered with **villi**, which look rather like the pile on the surface of a carpet (Figure 12). Each villus is covered with a layer of epithelial cells that have huge numbers of very thin, finger-like projections on their surfaces, called **microvilli**. The villi and microvilli greatly increase the surface area of the small intestine.

The cell-surface membranes of the microvilli contain the enzymes that break down disaccharides into monosaccharides. Maltase hydrolyses maltose to glucose. Lactase hydrolyses lactose into glucose and galactose. Sucrase hydrolyses sucrose to glucose and fructose.

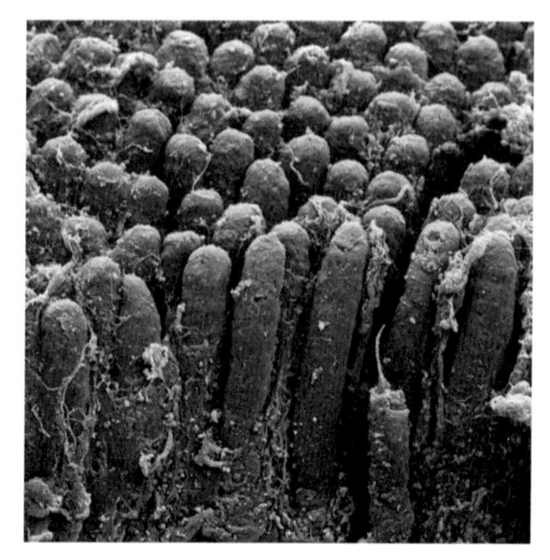

Figure 12 *Scanning electron micrograph of villi lining the small intestine. Each villus is about 1 mm long.*

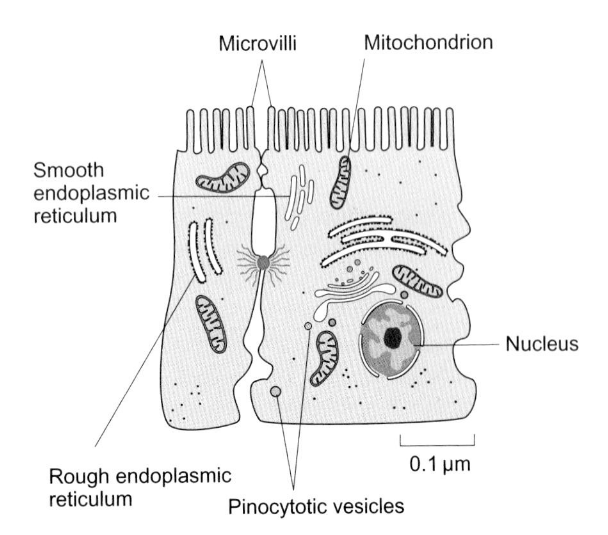

Figure 13 *Structure of a cell from the epithelium covering a villus*

QUESTIONS

13. Use your knowledge of enzymes to explain why sucrose can be digested by sucrase, but not by lactase.

14. Explain the difference between a villus and a microvillus.

Digestion of lipids

You may remember that lipids are insoluble in water. This makes them relatively difficult to digest. When they mix with the watery liquids in the alimentary canal, they tend to remain clumped together rather than spreading out as individual molecules. The individual lipid molecules are therefore not free to move around, so they do not make contact with the active sites of the lipase enzymes that can hydrolyse them.

However, when the food we have eaten reaches the small intestine, juices produced by the liver flow into it. These juices, called **bile**, are temporarily stored in the **gall bladder** before flowing along the **bile duct** into the small intestine. Bile does not contain any enzymes, but it does contain **bile salts**. Bile salts are molecules that have both hydrophilic and hydrophobic regions. Their hydrophilic parts associate with water molecules, while their hydrophobic parts associate with lipid molecules. The bile salts therefore interact both with lipids and with water, so that large lipid globules are broken up into much smaller droplets, which mix with water. The lipids still do not properly dissolve in the water – that is, their molecules do not completely separate from one another – but instead they form tiny droplets called **micelles**. This increases the surface area for lipase action. The mixture of micelles and water is called an **emulsion** (Figure 14).

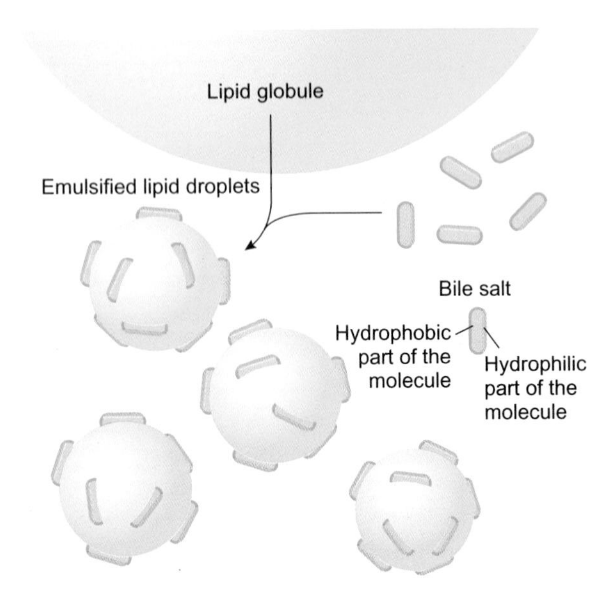

Figure 14 *How bile salts help lipids to disperse in water*

Once the lipid droplets have dispersed in this way, **lipase** enzymes are able to make contact with the individual lipid molecules in the micelles. Lipase hydrolyses lipids to fatty acids and monoglycerides – that is, a glycerol molecule with one fatty acid joined to it. Lipase is present in pancreatic juice, so lipid digestion takes place in the small intestine.

Digestion of proteins

Proteins are much easier to digest than lipids, because many proteins are water soluble, making it easy for protease enzymes to access them.

Protein digestion begins in the stomach, where the enzyme **pepsin** is present in the juices secreted by glands in the stomach walls, called **gastric juice**. This juice also contains hydrochloric acid, which produces a very low pH – often around pH 2 – in the stomach. This destroys many potentially harmful bacteria that may be present in food, and also provides the optimum pH for pepsin to work.

Pepsin is a type of protease called an **endopeptidase**. This means that it hydrolyses peptide bonds within polypeptide molecules, not near their ends (Figure 15). Pepsin therefore hydrolyses large polypeptides into smaller polypeptide chains, but does not release individual amino acids from them.

As the partly digested food leaves the stomach and moves into the small intestine, pancreatic juice is added to it. This contains another endopeptidase, called **trypsin**. Trypsin has a much higher optimum pH than pepsin, and so pancreatic juice also contains sodium hydrogencarbonate, which neutralises the hydrochloric acid from the stomach and increases the pH to just above 7.

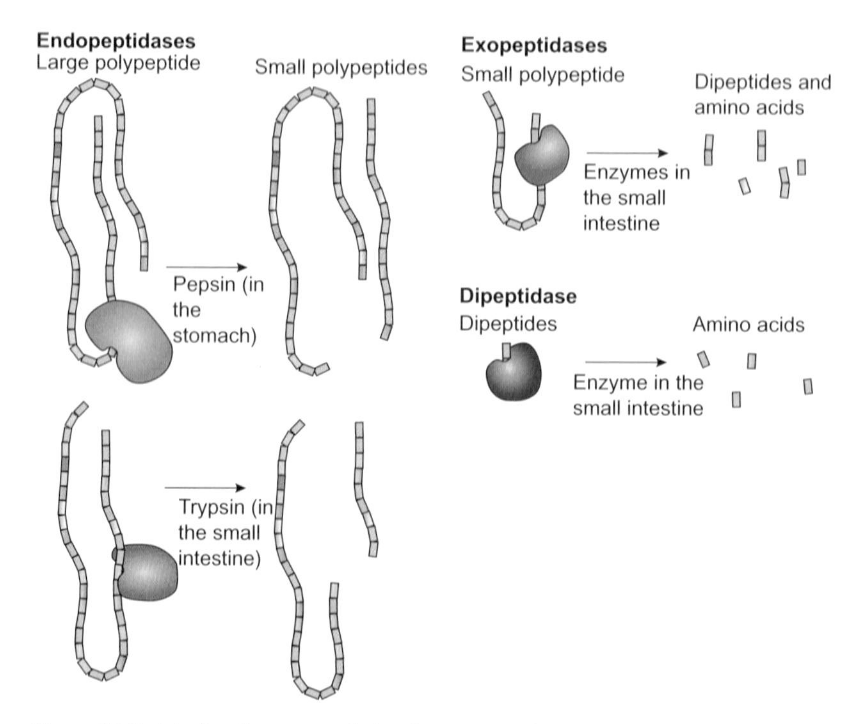

Figure 15 *Protein-digesting enzymes in the alimentary canal*

Protein digestion is completed by membrane-bound enzymes in the small intestine. Like the enzymes that hydrolyse disaccharides, these enzymes are part of the cell-surface membranes of the epithelial cells of the villi. They are **exopeptidases,** meaning that they break the last or penultimate (one before the last) peptide bonds at the ends of polypeptide molecules. This therefore breaks off either a single amino acid, or a dipeptide, from a polypeptide chain. Dipeptidases, also part of the membrane, then complete the hydrolysis of dipeptides to single amino acids.

Absorption from the small intestine

As we have seen, nutrients inside the alimentary canal cannot truly be said to have entered the body. In order to do this, they have to pass through the cells that make up the wall of the alimentary canal and enter the blood. This involves passing through a cell-surface membrane twice – once to enter the side of the cell that faces the lumen of the intestine, and then once to exit from the cell at its other side (Figure 16, overleaf).

Entering the epithelial cell

Carbohydrates are absorbed in the form of monosaccharides – glucose, fructose or galactose. Glucose is absorbed through the cell-surface membranes of the villi by facilitated diffusion, and also by active co-transport with sodium ions.

Proteins are absorbed in the form of amino acids. Most of these move through the cell-surface membranes of the villi by active co-transport with sodium ions, in a similar way to glucose.

The fatty acids and monoglyceride molecules, produced by digestion of lipids, tend to remain associated with each other and with bile salts as microscopic micelles (Figure 17, overleaf). They have their hydrophobic ends facing inwards, away from the watery liquids around them, and their hydrophilic parts facing outwards. Both fatty acids and monoglycerides are soluble in the phospholipid bilayer, so as they bump into the villi they break away from the micelles and pass into the cells by simple diffusion. Fatty acids can also enter the cell by facilitated diffusion, through specific carrier molecules in the cell-surface membrane.

Leaving the epithelial cell

All of these substances now pass out of the small intestine epithelial cells, and into the blood. Villi have thin-walled blood capillaries in close contact with their epithelial cells, so the distances to travel are small.

Monosaccharides and amino acids move into the blood by facilitated diffusion. Because they were transported into the cell by active transport, their concentrations inside the epithelial cell are relatively high. They can now move down their diffusion gradients into the blood, through carrier molecules in the cell-surface membrane.

Fatty acids and monoglycerides have a more complex method of absorption. Once they are inside the epithelial cell, they are resynthesised into triglycerides. These associate with proteins to form little droplets called **chylomicrons** (Figure 18, overleaf).

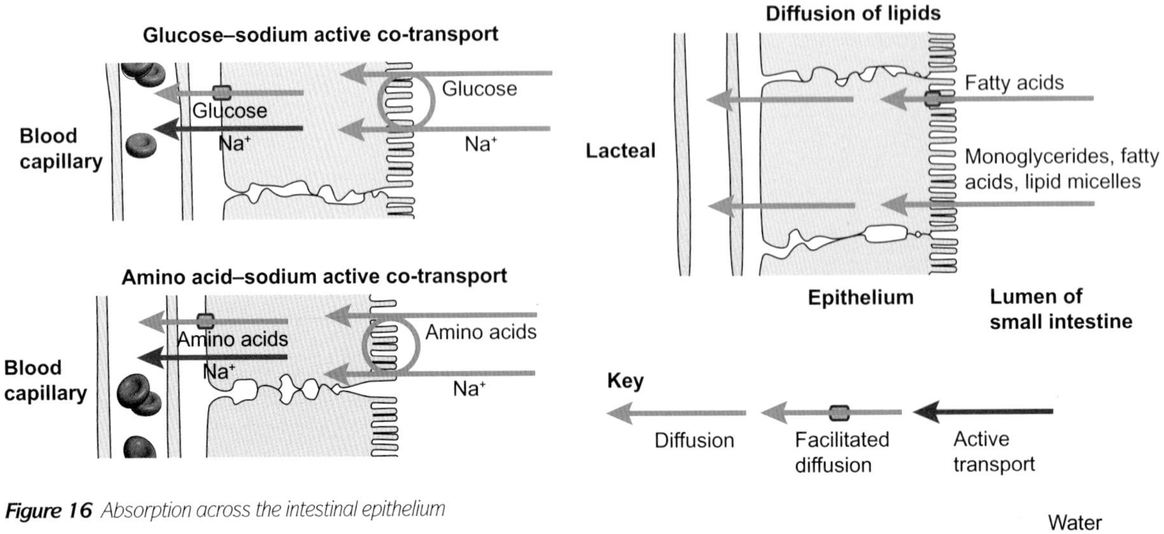

Figure 16 Absorption across the intestinal epithelium

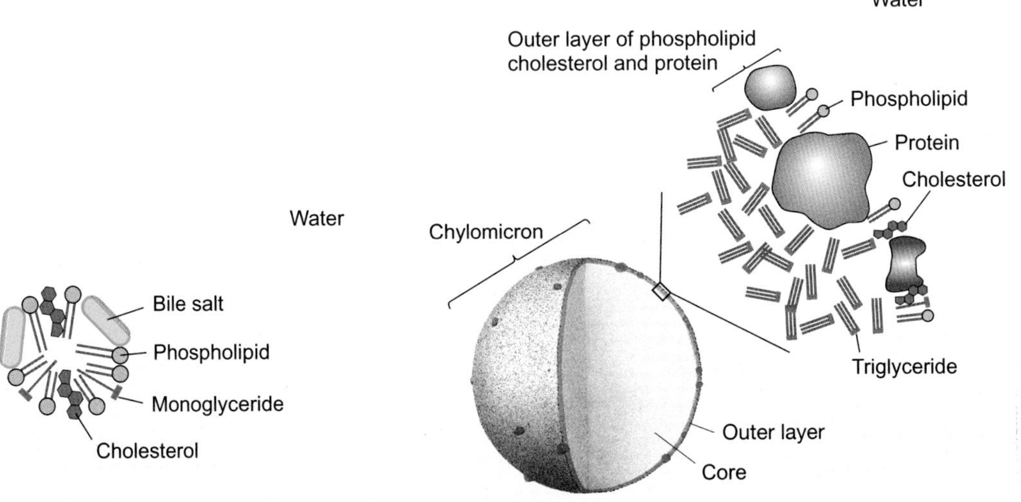

Figure 17 A micelle of fatty acids and monoglycerides

Figure 18 A chylomicron

QUESTIONS

15. Copy and complete this table to summarise the digestion and absorption of carbohydrates, proteins and lipids.

Nutrient	Enzymes that digest them	Substrate	Product(s)	Form in which absorbed	Method of transport into epithelial cell	Method of transport out of epithelial cell
Carbohydrate	amylase maltase sucrase lactase			monosaccharides (for example, glucose)		
Protein	pepsin trypsin exopeptidase dipeptidase					
Lipid	lipase					

KEY IDEAS

- For nutrients to be absorbed they have to be transported across cell-surface membranes of cells lining the small intestine (by diffusion, facilitated diffusion, active transport or co-transport).
- Only small molecules or ions can be transported. So, in digestion, large biological molecules are hydrolysed to smaller molecules that can be absorbed/transported.
- Starch is hydrolysed by amylase to maltose in the mouth and small intestine. Maltose, sucrose and lactose are hydrolysed to monosaccharides by enzymes that are part of the cell-surface membranes of the villi in the small intestine.
- Proteins are hydrolysed to polypeptides by the endopeptidases pepsin (stomach) and trypsin (small intestine). Polypeptides are hydrolysed to amino acids by membrane-bound exopeptidases and dipeptidases in the small intestine.
- Lipids are emulsified by bile salts, forming micelles. Lipase hydrolyses lipids to fatty acids and monoglycerides.
- Monosaccharides and amino acids are absorbed into the epithelial cells of the villi by active co-transport with sodium ions. They move out of the epithelial cells and into the blood by facilitated diffusion.
- Fatty acids and monoglycerides move into the epithelial cells by simple diffusion. They are reformed into triglycerides inside the cells, and move out of them into the lacteals in the form of chylomicrons, by exocytosis.

PRACTICE QUESTIONS

1. Figure Q1 shows part of the gut wall of an animal.

   a. Name the structure labelled **X**.

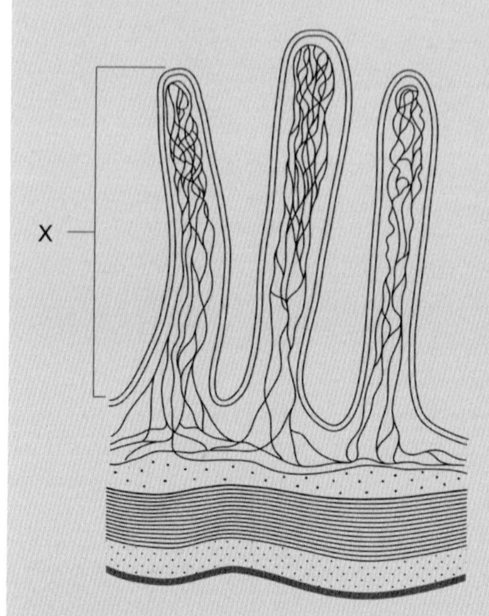

Figure Q1

   b. Describe and explain how **two** features shown in the diagram increase the rate of absorption of digestion food.

2. Figure Q2 shows a section through an alveolus and a blood capillary.

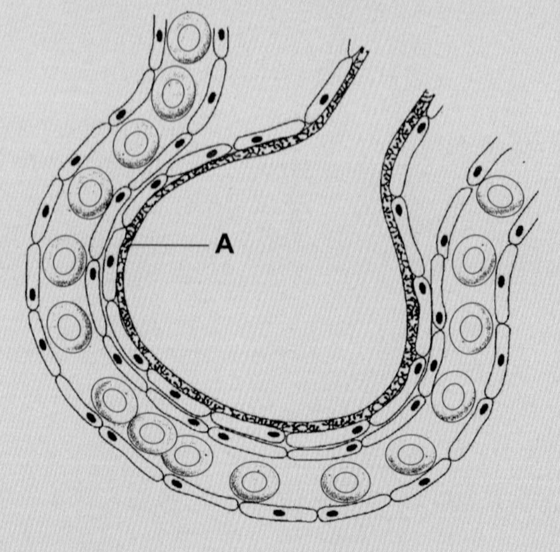

Figure Q2

(Continued)

a. What structural features, visible on the diagram, increase the rate of diffusion of oxygen into the blood?

b. Explain the function of the layer labelled **A**.

c. Explain how a diffusion gradient for oxygen is maintained between the air in the alveolus and the blood.

3. In fish, the flow of water over the gills and the flow of blood through the gills are in opposite directions (countercurrent flow) rather than in the same direction (parallel flow). The graph in Figure Q3 shows the effect of countercurrent flow on the oxygen saturation at different distances along a secondary gill lamella.

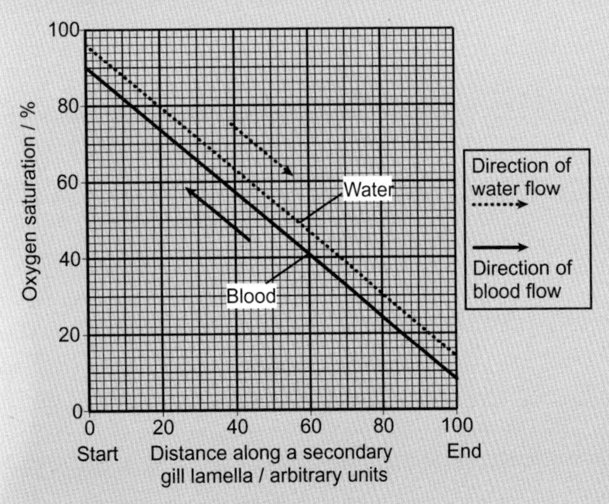

Figure Q3

a. What percentage of oxygen is removed from the water using this method of gas exchange?

b. Explain why countercurrent flow is more efficient for gas exchange than parallel flow.

c. The rate of diffusion across a membrane is related to a number of factors. These include surface area, membrane thickness and the difference in concentration on either side of the membrane. Use ticks to complete Table Q1 to show the effect of an increase in each factor on the rate of gas exchange.

4. Read the following passage.

Gluten is a protein found in wheat. When gluten is digested in the small intestine, the products include peptides. Peptides are short chains of amino acids. These peptides cannot be absorbed by facilitated diffusion and leave the gut in faeces.

Some people have coeliac disease. The epithelial cells of people with coeliac disease do not absorb the products of digestion very well. In these people, some of the peptides from gluten can pass between the epithelial cells lining the small intestine and enter the intestine wall. Here, the peptides cause an immune response that leads to the destruction of microvilli on the epithelial cells.

Scientists have identified a drug which might help people with coeliac disease. It reduces the movement of peptides between epithelial cells. They have carried out trials of the drug with patients with coeliac disease.

Use the information in the passage and your own knowledge to answer the following questions.

a. Name the type of chemical reaction which produces amino acids from proteins.

b. The peptides released when gluten is digested cannot be absorbed by facilitated diffusion. Suggest why.

Factor	Increase	No effect	Decrease
Surface area			
Membrane thickness			
Difference in concentration			

Table Q1

AQA January 2002 Unit 1 Question 4

(Continued)

c. The epithelial cells of people with coeliac disease do not absorb the products of digestion very well. Explain why.

d. Explain why the peptides cause an immune response.

e. Scientists have carried out trials of a drug to treat coeliac disease. Suggest **two** factors that should be considered before the drug can be used on patients with the disease.

AQA June 2012 Unit 1 Question 7

5. An investigation was carried out into the structure of the small intestine in three species of bats:

- *Miniopterus inflatus*, a bat that feeds on insects
- *Epomophorus wahlbergi*, which eats fruit
- *Lissonycteris angolensis*, which also eats fruit.

Table Q2 shows some of the measurements that were made on samples of bats from each species.

a. Calculate the ratio of intestinal length (in mm) to body mass (in g) for each species.

b. Calculate the ratio of intestinal surface area (in cm^2) to body mass (in g) for each species.

c. Explain how the structure of the microvilli results in the relatively high intestinal surface area in the two fruit bats, compared to the insectivorous bat.

d. Insects have a high protein content, while fruits have a high carbohydrate content, including a lot of cellulose which is difficult to digest. Suggest how the differences between the results for the insectivorous bat and the two species of fruit-eating bats could relate to their lifestyles.

Feature	*M. inflatus*	*E. wahlbergi*	*L. angolensis*
Mean body mass/g	8.9	76.0	76.9
Mean intestinal length/cm	20	73	72
Mean surface of intestine (including villi and microvilli)/m^2	0.13	2.7	1.5
Mean number of microvilli per intestine	4×10^{11}	33×10^{11}	15×10^{11}
Mean diameter of a microvillus/m	89	94	111
Mean length of a microvillus/μm	1.1	2.8	2.9
Mean surface area of a microvillus/μm^2	0.32	0.83	1.02

Table Q2

9 MASS TRANSPORT

PRIOR KNOWLEDGE

You may know about the general pattern of blood circulation in mammals. De-oxygenated blood from the body enters the right side of the heart, which pumps it out to the lungs for re-oxygenation and to dispose of waste products. The newly oxygenated blood returns to the heart through the left side and from here is pumped back around the body. The human circulatory system is an example of mass transport. You may remember that substances move into and out of cells by diffusion, osmosis, facilitated diffusion and active transport. You may also know that water is moved from the roots to leaves of a plant through xylem vessels, pulled by forces generated by evaporation of water from the leaves (transpiration pull). Gaseous exchange involved in the ventilation process is a further example of mass transport.

LEARNING OBJECTIVES

In this chapter, we look at how mammals and plants move bulk liquids around their bodies, in order to deliver requirements to all cells efficiently.

(Specification 3.3.4.1, 3.3.4.2)

If you look closely at a plant that has greenfly on it, you may also see ants scurrying back and forth, perhaps stroking the aphids with their legs or antennae. What is going on?

Plants, like most multicellular organisms, have transport systems that carry liquid substances from one part of their body to another. Unicellular organisms, such as bacteria, conduct their life processes within their one cell, so do not require a transport system. As it happens, the plant transport system that carries sugars (made by photosynthesis) around a plant lies close to the surface of the stem. So close, in fact, that a tiny aphid can push its syringe-like mouthparts (called stylets) through the outer cell layers and plug into the plant's sieve elements, which provide a steady source of sugary liquid. What's more, the liquid is under pressure, so all the aphid needs to do to get the sugary sap into its digestive system is just sit there; it does not even need to suck.

The plant has an automatic mechanism that plugs leaks from its sieve elements, rather as we use blood clotting to plug leaks from our blood vessels – the contents of both these mass transport systems are too valuable to be wasted. Feeding aphids seem to be able to block these protective responses by the plant, though just how they do this is not yet clear.

The sap that flows in the sieve elements is rich in sucrose, and therefore has a very low water potential. Taking in so much sucrose all at once could cause water to be drawn out of the aphid's cells by osmosis, so it quickly changes the sucrose molecules in its gut into short chains of sugars, called oligosaccharides, that have less effect on water potential. Aphids use a sucrase enzyme to do this, and if they are treated with a sucrase inhibitor they die – they actually shrivel up as they feed. Maybe this will one day be used as an aphid-specific insecticide.

So where do the ants come in? Such large volumes of sugar-rich liquid flow into the aphids' guts that they do not need all of it, and much simply emerges at the other end as honey dew. Ants love to feed on this, and they care for colonies of aphids rather like a farmer would look after a herd of cows. Recent research suggests that the ants secrete tranquillising chemicals from their feet that calm the aphids down, keeping them in place and ensuring that the ants have a constant supply of carbohydrate-rich food.

9.1 MASS FLOW

In Chapters 6 and 8, we have seen how individual molecules and ions can move across cell-surface membranes. During diffusion (including osmosis and facilitated diffusion), they move as a result of their own, individual, kinetic energy. In active transport, energy from ATP is used by the cell to move the particles in a particular direction.

These individual molecules and ions are unimaginably small, and – even when they have a lot of kinetic energy – it takes an appreciable amount of time for them to move even small distances. This is why gas exchange surfaces are very thin; if they were not, then it would take far too long for oxygen and carbon dioxide to diffuse across them.

Large organisms cannot rely on diffusion to move substances around their bodies. It would simply take too long. Instead, most large organisms use some kind of **mass transport** system to move substances around inside their bodies.

Mass transport involves the movement of huge numbers of molecules or ions of a substance in the same direction at the same time. We also use the term to mean the movement of people – a 'mass transport' system in a city could use trains to move people quickly from place to place. Instead of individual people walking to their destinations, many people get on a train that moves them, all together, from one point to another.

Another analogy that we can make is a river, in which uncountable billions of water molecules all flow together in the same direction (Figure 1). We call this type of movement **mass flow**. The mass of water molecules can move much, much faster in this way than if the individual water molecules simply moved around individually, by diffusion or osmosis.

Figure 1 *The water in a river moves by mass flow.*

Mass flow can only occur in fluids, not in solids. Gases and liquids – including solutions – can move by mass flow.

You will probably realise straight away that, in your body, blood is moved around by mass flow. Many animals have blood systems that transport substances rapidly and efficiently from one part of the body to another. Plants also use mass transport – they actually have two different systems, xylem and phloem, which transport different materials in different ways. In both animals and plants, mass flow transports substances to and from exchange surfaces, such as gas exchange surfaces or areas where nutrients are absorbed.

QUESTIONS

1. Which of these are examples of mass flow?
   - The random movement of molecules of scent through the air across a room.
   - The movement of air blown by an electric fan.
   - Water running out of a tap.
   - Pigment molecules from a tea bag spreading out into a cup of hot water.

KEY IDEAS

- Large organisms use mass transport systems to move substances to and from exchange surfaces.
- Mass flow involves the movement of large numbers of molecules or ions, in a fluid, all moving in the same direction together.

9.2 OXYGEN TRANSPORT IN MAMMALS

The main function of the mammalian blood system is to transport substances such as oxygen, nutrient molecules and mineral ions to the cells of the body, and to take wastes such as carbon dioxide and urea away. Blood consists of a liquid, called blood plasma, in which red blood cells, white blood cells and tiny cell fragments called platelets are suspended. Red blood cells transport oxygen, while white cells are involved in protecting the body from pathogens. Platelets are involved in blood clotting.

In this section we look at the role that haemoglobin plays in transporting dissolved gases in the blood, and then go on to examine how other materials are exchanged between the blood and the tissues.

Haemoglobin

Haemoglobin is a red pigment that transports oxygen around the body. It is this pigment molecule that gives red blood cells their colour.

A haemoglobin molecule is a globular, soluble protein with a quaternary structure. The four haem groups in one haemoglobin molecule can each combine temporarily with one oxygen molecule. An oxygen molecule, O_2, contains two oxygen atoms, so one haemoglobin molecule can carry eight oxygen atoms.

When oxygen concentration is high, such as in the capillaries of the lungs, haemoglobin and oxygen combine to form **oxyhaemoglobin**. When oxygen concentration is low, such as in the capillaries of exercising muscles, the oxyhaemoglobin dissociates: it splits up and releases the oxygen. A far greater mass of oxygen can be carried in the form of oxyhaemoglobin than can be carried in solution in blood plasma.

The reaction between oxygen and haemoglobin is summarised by the equation:

$$\text{oxygen} + \text{haemoglobin} \rightleftharpoons \text{oxyhaemoglobin}$$

The oxyhaemoglobin dissociation curve

To understand why haemoglobin is such an efficient molecule for transporting oxygen we have to consider the **oxyhaemoglobin dissociation curve**. This is shown in Figure 2. You will see that the *x*-axis of this graph is labelled *Partial pressure of oxygen*. This is a measure of the concentration of oxygen in the environment around the haemoglobin, and it is measured in kilopascals. The shorthand for 'partial pressure of oxygen' is ppO_2.

The *y*-axis is labelled *% saturation of haemoglobin with oxygen*. This represents the quantity of oxygen that the haemoglobin has combined with in a given volume of blood. If every molecule of haemoglobin is combined with the full four molecules of oxygen (O_2), then it is 100% saturated. If each molecule is combined with only two oxygen molecules, then it is 50% saturated. (This could also happen if 50% of the haemoglobin molecules are combined with four oxygen molecules, and the other 50% are not combined with any at all – but this would be very unlikely to happen.) So we can read the *y*-axis as showing us how much oxygen is combined with the haemoglobin.

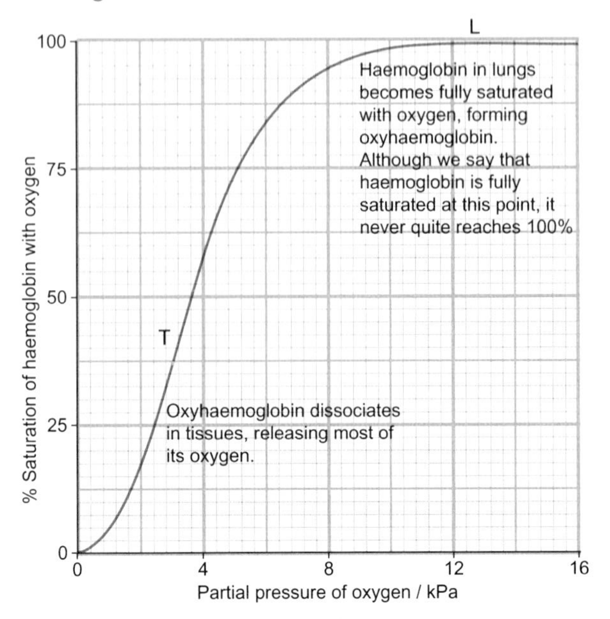

Figure 2 *The oxyhaemoglobin dissociation curve*

From Figure 2, you can see that the percentage saturation of haemoglobin increases as the external partial pressure of oxygen increases. At the oxygen concentration that occurs in the lungs, haemoglobin readily absorbs oxygen to form oxyhaemoglobin, becoming almost 100% saturated. Region L on the graph shows this. Even if the oxygen level in the lungs falls a little below normal, haemoglobin can still absorb oxygen – this means that the blood leaving the lungs is almost always completely saturated with oxygen.

In the body tissues, where respiration is using up oxygen, the partial pressure of oxygen is very low, as in region T on the graph. You can see that in this region the haemoglobin is much less saturated – that is, it is combined with much less oxygen.

Imagine what happens in the body as the blood flows from the lungs to an active muscle. As the blood passes through the lungs, it is in an environment where the partial pressure of oxygen is very high (right hand part of the graph, region L). Almost every haemoglobin molecule combines with its full quota of eight oxygen atoms. The blood then flows to the heart, from where it pumped out through the aorta and smaller arteries until it reaches the muscles. Here, suddenly, it is in an environment where the oxygen concentration is much lower (left hand part of the graph, region T). The haemoglobin rapidly gives up much of its oxygen, becoming much less saturated.

It's important to appreciate that the ability of haemoglobin to give up its oxygen in conditions where oxygen concentration is low is just as important as its ability to combine with oxygen where oxygen concentration is high.

The S-shaped curve

The curve in Figure 3 shows clearly that there is not a linear relationship between the partial pressure of oxygen and the percentage saturation of haemoglobin. Working from left to right, we can see that as the partial pressure of oxygen increases from 0 to 1 or 2 kPa, the percentage saturation of haemoglobin increases fairly gradually. But as the partial pressure of oxygen increases more, the percentage saturation of oxygen increases much more steeply, before finally flattening out at around 10 kPa.

That steep area in the middle of the curve is ideal for the needs of the body. It means that a slight decrease in oxygen concentration in the tissues produces a large increase in the rate of oxyhaemoglobin breakdown and release of oxygen. This delivers oxygen very effectively to the tissues that need it. Similarly, if we think in the other direction, a small increase in oxygen concentration produces a large increase in the quantity of oxygen that combines with the haemoglobin. This helps the haemoglobin to efficiently absorb oxygen in the lungs, even when the oxygen concentration is a little lower than normal.

A study of the properties of the haemoglobin molecule can explain why the dissociation curve is this shape. When one oxygen molecule combines with one of the haem groups in a haemoglobin molecule, it causes the haemoglobin to slightly change its shape. This makes it much easier for the next haem group to combine with an oxygen molecule, and this in turn helps the last two haem groups to combine with oxygen as well. The same thing happens in the reverse direction; the loss of one oxygen molecule makes it easier for the next one to be lost, and so on. In theory, losing the fourth oxygen molecule requires only a minimal pressure decrease. However, in reality it is hardly ever released under normal conditions.

The Bohr effect

Figure 3 shows the haemoglobin dissociation curve again (in red), just as in Figure 2. However, it also shows a slightly different dissociation curve (in blue). This blue curve is what we find when there is a high

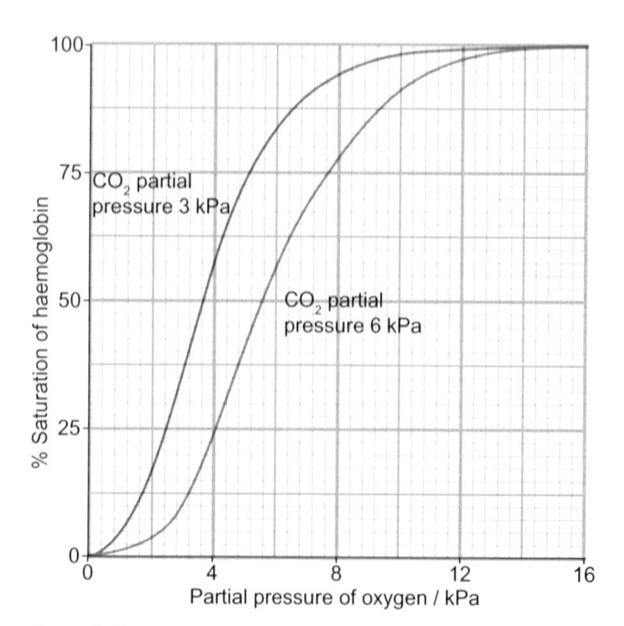

Figure 3 *Haemoglobin dissociation curves at different concentrations of carbon dioxide, showing the Bohr effect*

concentration of carbon dioxide in the environment around the haemoglobin.

In respiring body tissues, carbon dioxide is given off, and this carbon dioxide diffuses into the blood plasma. Some of it diffuses into the red blood cells.

When this happens, the carbon dioxide reacts with water to form carbonic acid, and the haemoglobin molecule changes shape – therefore, it rapidly lets go of the oxygen that was combined with it. This means that, when a lot of carbon dioxide is present, haemoglobin gives up its oxygen more readily than when there is only a little carbon dioxide.

This is exactly what is needed. When there is an increased level of respiration – during exercise, for example – there is also an increase in carbon dioxide levels. This increases the amount of oxygen released from the haemoglobin to maintain respiration in the tissues.

If you think carefully about the curves in Figure 3, you can see how they illustrate this effect of carbon dioxide on haemoglobin dissociation. Pick a particular value of partial pressure of oxygen on the *x*-axis, and then look at the percentage saturation of haemoglobin at this value, on the red curve and on the blue curve. You will find that the percentage saturation is less on the blue curve (high carbon dioxide) than on the red one (low carbon dioxide). The presence of carbon dioxide shifts the curve to the right. This is called the **Bohr effect**.

Different types of haemoglobin

Haemoglobin comes in many different forms. For example, in humans, we have a different type of haemoglobin when we are fetuses in the uterus, than when we are adults. Fetal haemoglobin has a higher affinity for oxygen than adult haemoglobin (that is, it combines more readily with oxygen), which makes it possible for a fetus to 'take' oxygen from its mother's blood. If we were to draw the dissociation curve for fetal haemoglobin, it would be to the left of the curve for adult haemoglobin.

Animals that live in places where oxygen concentrations are relatively low also tend to have haemoglobin with a higher affinity for oxygen than our haemoglobin. Figure 4 shows the dissociation curve for a worm (Figure 5) that lives in deoxygenated mud at the bottom of ponds or polluted streams, compared with the dissociation curve for an adult human. You can see that this curve lies to the left of the human haemoglobin curve, signifying that the worm's haemoglobin has a higher affinity for oxygen.

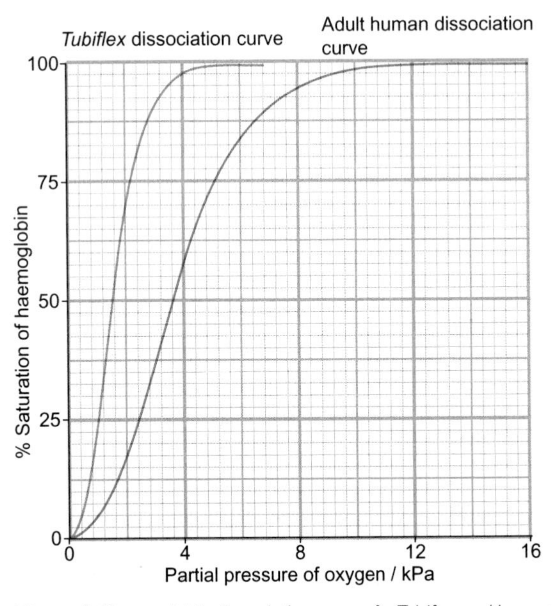

Figure 4 *Haemoglobin dissociation curves for* Tubifex *and humans*

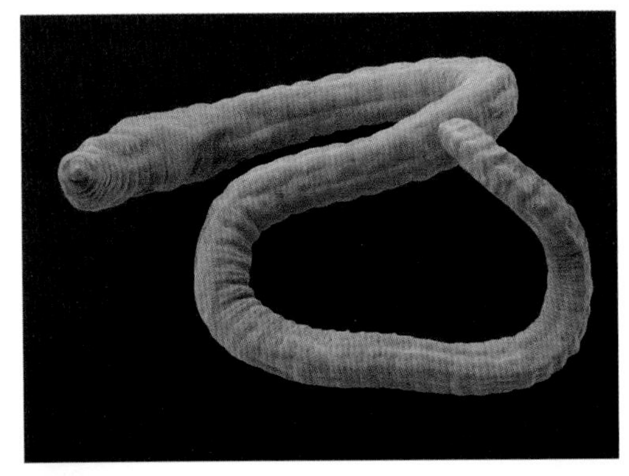

Figure 5 *A* Tubifex *worm*

QUESTIONS

2. What is the percentage saturation of adult human haemoglobin at partial pressures of oxygen of:
   a. 8 kPa
   b. 12 kPa
   c. 16 kPa?
3. What is the lowest partial pressure of oxygen at which all the haemoglobin molecules are fully combined with oxygen?
4. Blood moves from a partial pressure of oxygen of 16 kPa in the lungs to a value of 2.4 kPa in the muscles. What percentage of the oxygen is released?
5. What are the advantages and disadvantages to *Tubifex* in having haemoglobin with a dissociation curve of the shape shown in Figure 4?
6. Llamas have haemoglobin adapted for life at high altitude. Copy Figure 2, then draw on it the dissociation curve you would expect for a llama.

KEY IDEAS

- Haemoglobin is a protein with quaternary structure. Slightly different forms of haemoglobin are found in many different species of organisms.
- In mammals, haemoglobin is found inside red blood cells. Its function is the transport of oxygen from the gas exchange surface (alveoli in the lungs) to respiring tissues.
- Each haemoglobin molecule can reversibly combine with up to eight oxygen atoms.
- An oxygen dissociation curve shows the percentage saturation of haemoglobin with oxygen at different partial pressures of oxygen.
- The S-shaped curve for haemoglobin results from the cooperative binding of oxygen; once one oxygen molecule has bound to a haem group, it becomes easier for the next three oxygen molecules to bind to the other three haem groups.

- The steep part of the oxygen dissociation curve shows that a small change in partial pressure of oxygen produces a large change in the percentage saturation of haemoglobin.
- Carbon dioxide reacts with water inside red blood cells to form an acid (carbonic acid). This leads to a change in the shape of haemoglobin, causing it to release more oxygen. This is called the Bohr effect.
- Animals that live in environments in which oxygen is in short supply usually have a form of haemoglobin whose dissociation curve lies to the left of the curve for human haemoglobin.

9.3 THE HEART AND CIRCULATORY SYSTEM

In humans, as in all vertebrates, blood flows through a system of tubes, called blood vessels. The movement of the blood is produced by pressure changes caused by rhythmic contraction of muscles in the heart.

Mammals, including humans, have a **double circulatory system**. This means that the blood passes twice through the heart on one complete circuit of the body. One system, the **pulmonary system**, transports deoxygenated blood to the lungs, where it takes up oxygen, before returning to the heart. The other system, the **systemic system**, carries the oxygenated blood to the rest of the body, before returning once again to the heart. Try tracing these two pathways with your finger on Figure 6.

In Figure 6, deoxygenated blood is shown in blue, and oxygenated blood in red. This represents a change in colour of haemoglobin, which is bright red when it is combined with oxygen (as oxyhaemoglobin) and a blue-purple colour when it is not. You can see that in the systemic system, arteries contain oxygenated blood and veins contain deoxygenated blood. However, in the pulmonary system, the arteries carry deoxygenated blood and the veins carry oxygenated blood.

The structures and functions of the different kinds of blood vessel are described in Section 9.4.

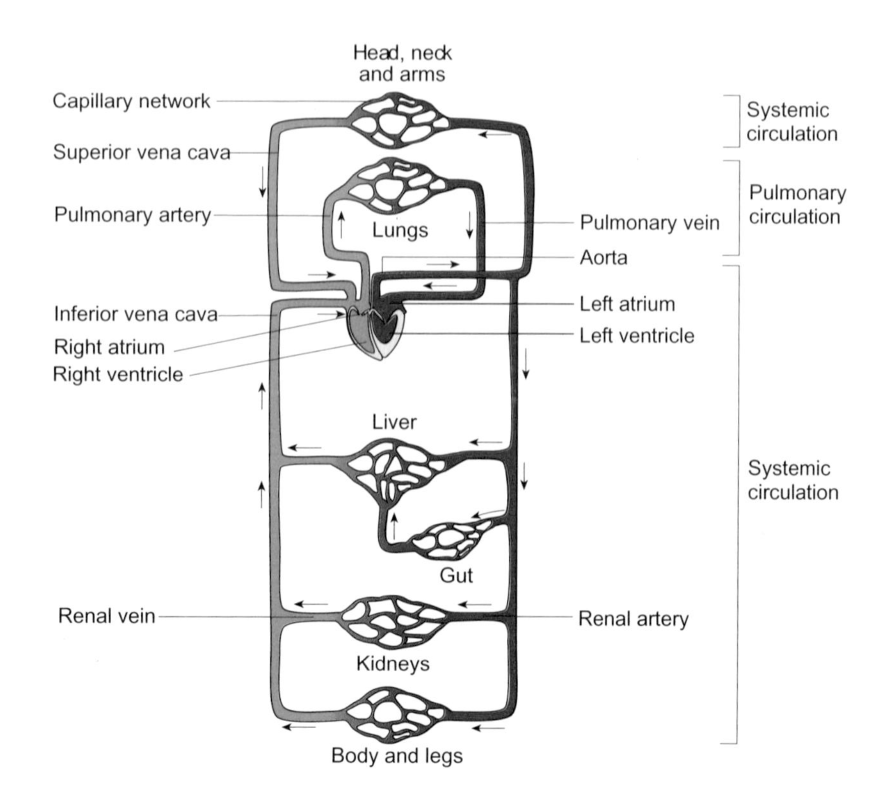

Figure 6 The double circulatory system

Structure of the heart

Figure 7 shows the structure of the human heart. It is made of a special type of muscle called cardiac muscle, which is able to contract and relax repeatedly.

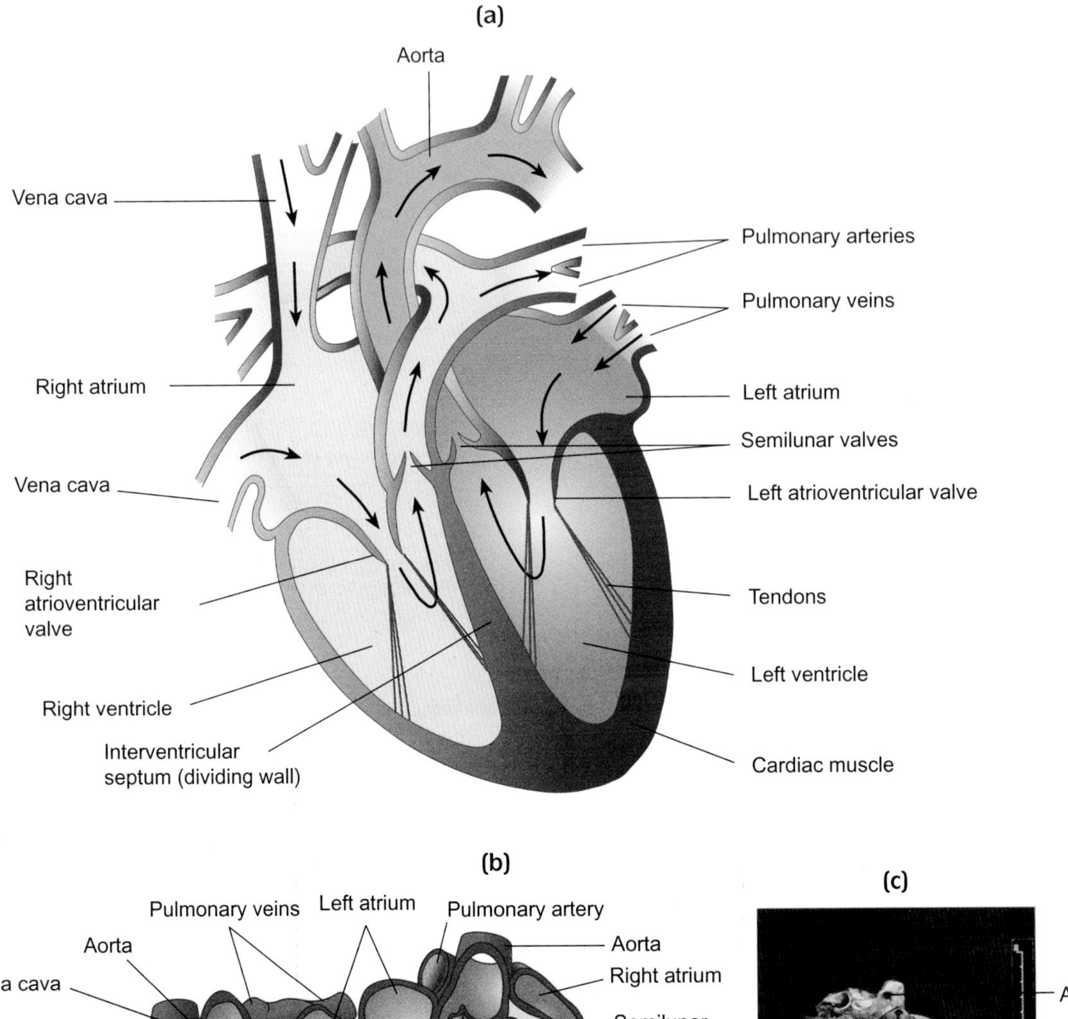

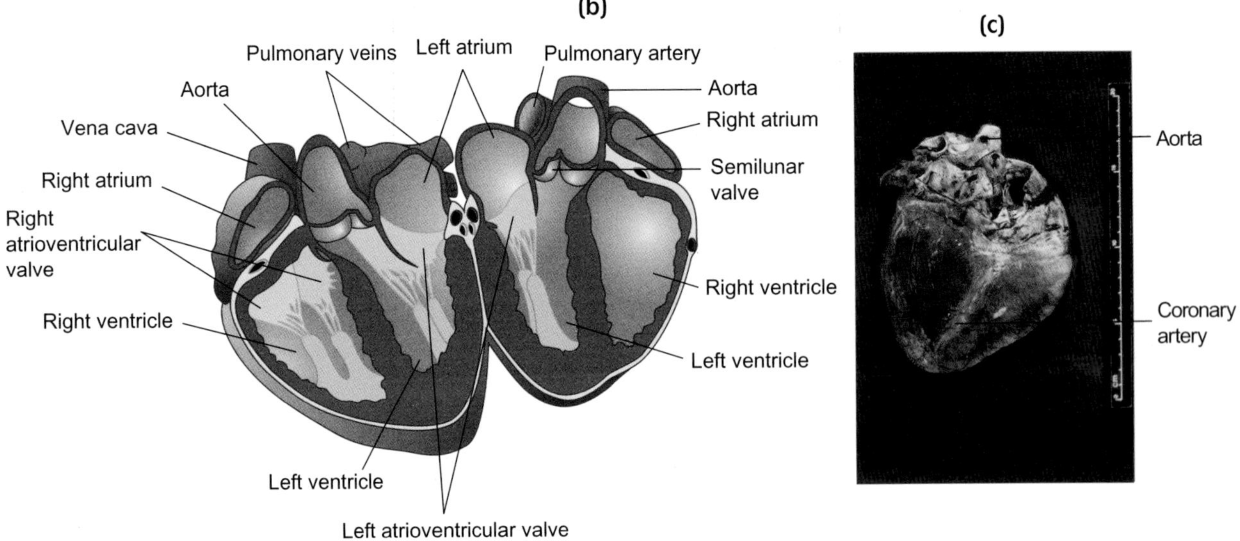

Figure 7 *(a) A vertical section through the heart; (b) The heart as it appears when cut open; (c) A post-mortem specimen of a whole human heart*

The heart is a pump – or, rather, two pumps side by side. Each pump has two chambers, an upper **atrium** and a lower **ventricle**. There is an **atrioventricular valve** between each atrium and ventricle. As you will see, these valves help to keep the blood flowing in the right direction. The valves are attached to the walls of the ventricles by tendons. Tendons are very strong cords that do not stretch. (You can feel a tendon at the back of your ankle, where it attaches the muscles of your calf to your heel bone.)

The right side of the heart receives deoxygenated blood from the body in the two **venae cavae** (singular: vena cava), and pumps it to the lungs along the **pulmonary artery**. The left side of the heart receives oxygenated blood from the lungs in the **pulmonary veins**, and pumps it to the rest of the body along the **aorta**. The pulmonary artery and the aorta contain half-moon shaped valves. The right and left sides of the heart are separated by the **septum**.

You may be able to see, in Figure 7a, that the muscle making up the walls of the atria is thinner than in the walls of the ventricles. It is the contraction of this muscle that produces the pressure to pump blood. The muscle in the walls of the atria needs only to produce enough force to push blood down into the ventricles, whereas the ventricles have to pump it out of the heart and around the rest of the body.

Similarly, the wall of the left ventricle is much thicker than the wall of the right ventricle. The right ventricle has to pump blood only as far as the lungs, whereas the left ventricle has to pump it all around the body. Thicker muscle is therefore required in the left ventricle wall so that it can generate a greater force when it contracts. If the right ventricle contracted with this great a force, it could damage the capillaries in the lungs.

The heart muscle is so thick that it needs its own blood supply. Arteries branch off from the aorta to form the **coronary arteries**, which can be seen snaking through the heart muscle, close to its outer surface.

The cardiac cycle

The cardiac cycle is the series of events that takes place during one heart beat. Both sides of the heart work in unison. This what happens on the left side of the heart (as shown in Figure 8):

Blood enters the left atrium from the pulmonary veins. The muscles of the atrium walls then contract (Figure 8a: Atrial contraction).

As the muscles in the atrium contract, they squeeze inwards on the blood in the atrium, increasing its pressure. When the pressure of the blood in the atrium becomes greater than the pressure in the left ventricle, the atrioventricular valve is forced open. Blood flows through into the ventricle.

Now the muscles in the walls of the ventricle contract (Figure 8b: Ventricular contraction). The strong muscular contraction greatly increases the pressure of

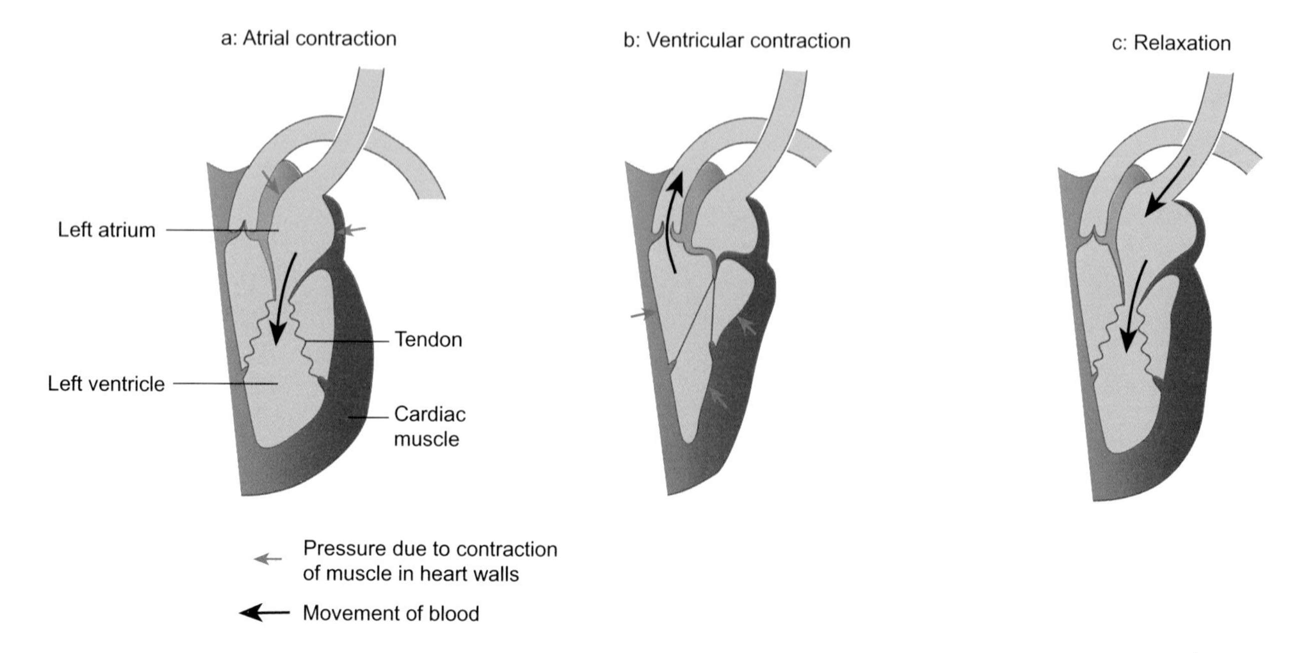

Figure 8 *The cardiac cycle*

the blood inside the ventricle, until it becomes greater than the pressure in the atrium. The force of the blood pushing up against the atrioventricular valve forces it shut. They are prevented from flipping right up into the atrium by the tendons that attach them to the ventricle wall. This action of the valves prevents blood from flowing back from the ventricle into the atrium.

When the pressure in the ventricle becomes higher than the pressure in the aorta, the (aortic) valve separating the ventricle from the aorta is forced open and blood flows into the aorta.

The heart muscle now relaxes (Figure 8c: Relaxation). The pressure of the blood in the atrium and ventricle decreases. The high-pressure blood in the arteries is now at a greater pressure than in the ventricle, and this causes the valves in the arteries to close, so that the blood cannot flow back into the ventricle.

Figure 9 shows the pressure changes in the left side of the heart during one cardiac cycle. This is a complex graph, and it is a good idea to begin by concentrating on just one line at a time. Start with the green line, which shows the pressure changes in the left ventricle.

QUESTIONS

7. Use Figure 9 to calculate how many complete cardiac cycles (heartbeats) there are in one minute.

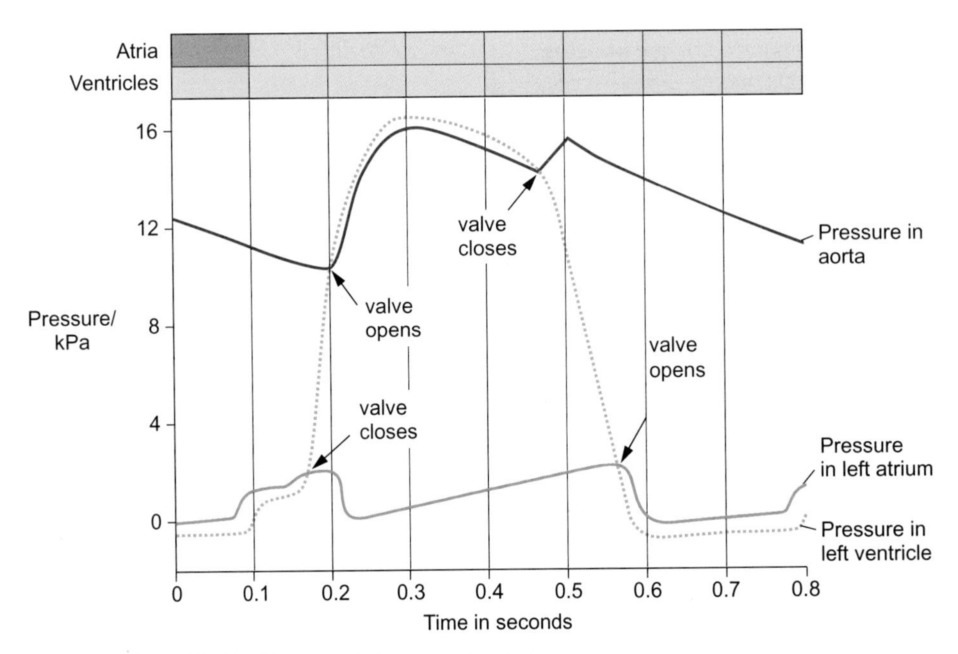

Figure 9 *Pressure changes in the left side of the heart during the cardiac cycle*

For the first 0.1 seconds, the pressure in the ventricle is low, as the muscle in its walls is relaxed. Between 0.1 and 0.2 seconds, the pressure increases a little, as blood is pushed into the ventricle from the atria, increasing the volume of blood contained within the ventricle. Then, just before 0.2 seconds, you can see that the pressure suddenly increases. This is caused by the contraction of the strong muscles in the ventricle wall, reducing the volume inside the ventricle, increasing pressure and, therefore, pushing the blood out into the aorta. The pressure stays high until about 0.3 seconds, when the muscles begin to relax. The recoil of the (thick) ventricle wall leads to a slight negative pressure in the ventricle, which adds to the pressure gradient for filling the ventricle. By 0.6 seconds, the pressure is back to its original level, and one complete heart beat has been completed.

Now look at the blue line. This shows the pressure changes in the left atrium. The first thing you will notice is that the pressure never gets anywhere near as high as in the ventricle. However, for the first 0.2 seconds, the pressure in the atrium is higher than in the ventricle, because this is when blood is flowing into the atrium, and when the atrial muscles are contracting. At this point, the atrioventricular valve will be open. Then, as the ventricle (green line) contracts, a point is reached when the pressure in the ventricle is greater than the pressure in the atrium, and this is the moment when the atrioventricular valve is forced shut. The valve only opens again at around 0.55 seconds, when the ventricles are fully relaxed and the pressure in the atrium is greater than in the ventricles.

Lastly, look at what is happening in the aorta (red line). The pressure begins to rise in the aorta at about 0.2 seconds, because the contraction of the ventricle is forcing blood into it. The semilunar valves in the aorta open at the moment when the pressure in the ventricle becomes greater than the pressure in the aorta. The pressure in the aorta continues to rise as the ventricle continues to contract, and then gradually falls as the ventricle relaxes. The semilunar valves close at the moment when the ventricular pressure falls below the pressure in the aorta.

QUESTIONS

8. The graph in Figure 9 shows the pressure in the left ventricle. How would you expect a line showing the pressure in the right ventricle to differ from this?

Controlling the rate of heart beat

You will be aware that your heart does not always beat at the same rate. When a person is relaxed, the heart often beats at around 60–70 times per minute. When you exercise, or when you are excited or nervous, the rate of heart beat increases. This happens so that oxygenated blood is delivered more quickly to your tissues, so that they can respire faster. Respiration releases energy that muscle cells can use for contraction.

Worked maths example: Stroke volume

(MS 0.1, MS 2.2, MS 2.3, MS 2.4)

Calculate the stroke volume of a patient with a heart rate of 70 beats per minute and a cardiac output of 4900 $cm^3\ min^{-1}$. Give your answer in dm^3.

Use this formula:

$$\text{cardiac output} = \text{stroke volume} \times \text{heart rate}$$

First, re-arrange the formula to:

$$\text{stroke volume} = \frac{\text{cardiac output}}{\text{heart rate}}$$

$$\text{stroke volume} = \frac{4900}{\text{heart rate}}$$

Remember: 1 dm^3 = 1 litre = 1000 cm^3

So, 70 $cm^3 = \frac{70}{1000} = 0.07\ dm^3$

This is about average for a healthy person.

(Note: You do not need to memorise the cardiac output equation for your examination).

ASSIGNMENT 1: ANALYSING THE BENEFITS OF EXERCISE

(MS 2.2, MS 2.3, MS 2.4, PS 1.2, PS 3.2)

People who exercise regularly are better off both physically and psychologically. Regular exercise increases muscle performance; it also affects the circulation and breathing systems of the body.

Regular exercise leads to the cardiac muscle becoming thicker and an increase in the size of the heart's chambers. This means that there is an increase in stroke volume.

If more blood is pumped with each beat, fewer beats per minute are needed to supply the body with the necessary oxygen and nutrients. So, the average heart rate of a very active person is lower than that of an inactive person. For most people, the resting heart rate is 60–70 beats per minute, whereas for athletes it can be as low as 40–50 beats per minute.

The volume of blood pumped by the heart per minute is called the cardiac output and it increases with activity.

The difference between the resting output and the maximum the heart can achieve is called the cardiac reserve.

Exercise also enlarges the blood vessels, so improving the blood supply to muscles and reduces blood pressure. It also improves the effectiveness of the breathing mechanism by strengthening the diaphragm and intercostal muscles that produce breathing movements. Gaseous exchange is more effective in both supplying oxygen and removing carbon dioxide.

Figure A1 *Regular exercise increases muscle performance.*

Questions

A1. a. An athlete has a resting heart rate of 45 beats per minute and a cardiac output of 5.25 $dm^3\ min^{-1}$. Calculate their stroke volume.

b. An average adult has a stroke volume of 0.06 dm^3. Explain the difference between this figure and your answer to part a.

A2. Study the data in this table.

	Cardiac output/$dm^3\ min^{-1}$	
Individual	**At rest**	**Vigorous activity**
Average adult	5.0	21.0
Athlete	5.25	30.0

a. Compare the difference in cardiac output of the athlete and average adult at rest and in vigorous activity.

b. Calculate the cardiac reserve of the average adult and the athlete.

c. Explain the effect of regular exercise on the cardiac reserve.

A3. Explain how each of the following helps to increase the rate of oxygen transport to tissues.

a. The heart

b. Blood vessels

c. The breathing system

KEY IDEAS

- Mammals have a double circulatory system, meaning that blood flows twice through the heart on one complete circuit of the body.
- The heart contains four chambers, two atria and two ventricles. The atria receive blood and pump it to the ventricles, which then contract strongly to push the blood out of the heart.
- The left side of the heart contains oxygenated blood, which is pumped out along the aorta. The right side contains deoxygenated blood, which is pumped out to the lungs through the pulmonary artery.
- During the cardiac cycle, the heart muscle contracts and relaxes.
- Valves between the atria and ventricles, and at the entrance to the aorta and pulmonary artery, prevent backflow of blood.

9.4 BLOOD VESSELS

Figure 10 shows the structures of the different types of blood vessels.

Arteries

Arteries are vessels that carry blood away from the heart. This blood, as we have seen, is at high pressure and travels in pulses, surging forward each time the muscles in the ventricular walls of the heart contract.

Like all blood vessels, the arteries are lined with a layer of flat, smooth cells making up the **endothelium**. The rest of the walls of the aorta and other large arteries is made up of a thick layer of elastic tissue, along with other connective tissues containing collagen fibres (fibrous tissue), and a small amount of smooth muscle. The elastic tissue is stretched by the pressure produced by the ventricles when the cardiac muscle contracts and blood is ejected. When the cardiac muscle relaxes and the heart re-fills with blood this elastic tissue recoils (goes back to its normal size), which squeezes on the blood, helping to keep its pressure high between heart beats. The expansion and recoil of the artery walls helps to smooth the surges in blood flow that are caused by contraction of the ventricles.

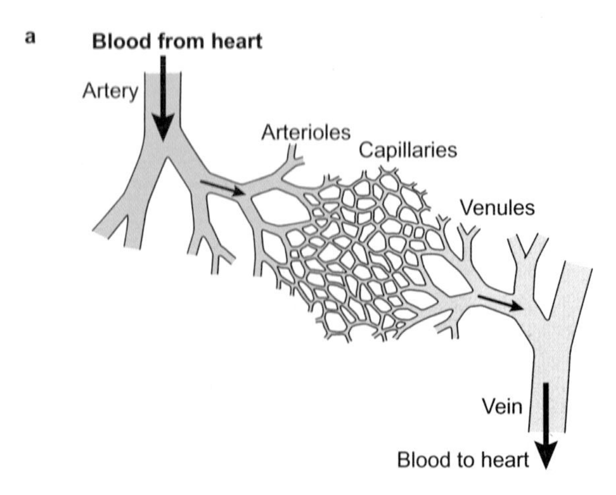

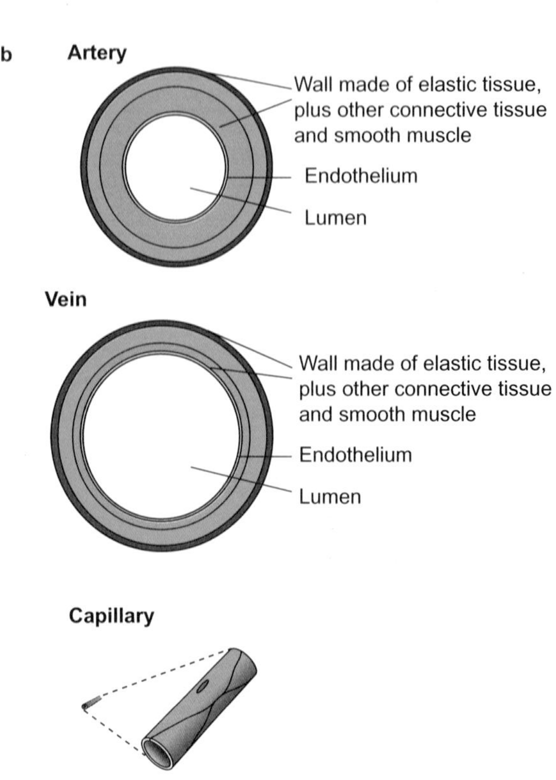

Figure 10 *The structure of blood vessels*

Nevertheless, we can still feel these surges as the pulse. This is particularly obvious at points where arteries pass over a bone near to the skin, such as at the wrist and the temple.

Arterioles

As the arteries get further from the heart, they split up into smaller vessels called arterioles. In the arterioles, the thickest layer of the wall consists of muscle fibres. This is a type of muscle called smooth muscle, which is able to contract steadily over long periods of time. This enables the arterioles to act as control points for the blood system. By contracting, the muscle narrows the lumen of the arteriole and can therefore restrict the flow of blood through a particular blood vessel.

As the arteries subdivide to form narrower arterioles, the pressure of blood in the vessels decreases. This is because the total cross-sectional area of all the smaller vessels is greater than that of all of the larger vessels (Figure 11).

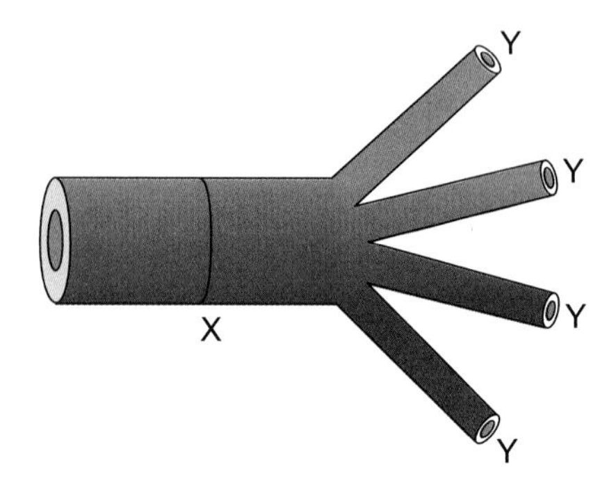

The total cross-sectional area at points Y is much greater than the cross-sectional area at point X.

Figure 11 *Cross-sectional areas of an artery and arterioles*

QUESTIONS

9. Suggest the effects of narrowing of the arterioles that supply the capillaries in the surface layer of the skin and that supply the capillaries in the villi of the small intestine.

Capillaries

The function of capillaries is to act as an exchange surface. Exchange of gases and other solutes between the blood inside the capillaries, and the other cells of the body, is made easy by the tiny size of the capillaries and their thin walls. They have a large surface area to maximise diffusion. They have a large surface area to maximise diffusion. Capillary walls are made up from just a single layer of endothelial cells. Capillaries have a diameter of 8 μm and a wall thickness of less than 1 μm.

Blood flow and speed are slowest through the capillaries. The speed of the blood describes how fast it is moving, while the flow describes how much blood moves past a particular point in unit time.

The flow depends on both the speed and the size of the blood vessel. You can see how a large slow-moving river (low speed) can shift much more water than a small fast-moving stream (high speed) in the same time. Blood flow follows the same rules (Figure 12).

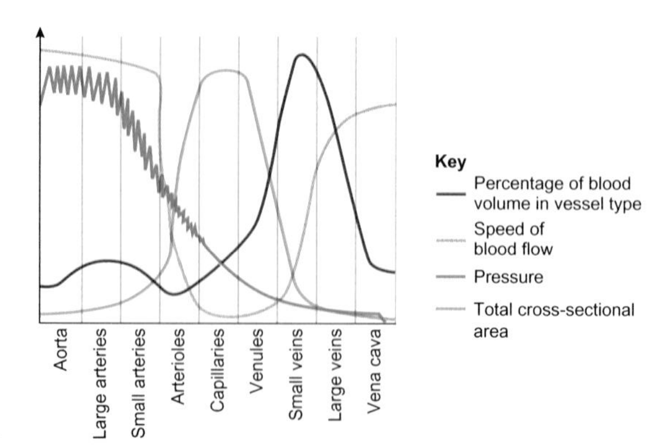

Figure 12 *Pressure, volume, cross-sectional area and speed of blood flow in different types of blood vessels*

So, although blood flows through the capillaries in an organ at a low speed, a large volume of blood flows through the many capillaries that serve that organ.

Exchange of materials such as oxygen, carbon dioxide and soluble nutrient molecules occurs between the blood and the tissues at the capillaries. Some capillaries have small gaps in between adjacent endothelial cells and also pores in the individual cells. These pores, called **fenestrations**, allow even faster rates of diffusion between the capillaries and the tissues. Fenestrations also allow some mass flow of liquid into the tissues. A thin fibrous connective tissue layer provides a physical filter, stopping cells and large proteins from leaving.

QUESTIONS

10. Use information from Figure 12 to explain why the speed of blood decreases as the blood flows from arterioles into capillaries.

11. Explain why the speed increases as the blood flows from the smaller veins into the larger veins.

12. How does the size of capillaries and the rate of blood flow through the capillaries suit them for their function?

Venules and veins

Venules and veins are vessels that collect blood that has flowed through capillaries, and return it to the heart. Generally, veins have thinner walls than arteries and contain much less elastic and muscle tissue. The pressure of blood in the veins is lower than in either arteries or capillaries, falling almost to zero in the largest vein, the vena cava, which carries deoxygenated blood back from the body into the heart. Unlike arteries, veins have valves to prevent backflow of blood.

Three main forces keep blood moving in the veins.

- The small residual pressure of blood coming from the capillaries.
- The action of the leg muscles and valves in the veins.
- The reduced pressure in the atria at atrial diastole.

When leg muscles contract to move the legs, the contractions squeeze the veins in the leg and this pushes blood upwards, rather like squeezing toothpaste along a tube (Figure 13). Valves in the veins prevent the blood from moving downwards. This explains why people can faint if they stand still for a long time – particularly on a hot day.

At atrial diastole, the muscles in the atrial walls relax and the atria increase in volume. This reduces the pressure and creates a pressure gradient between the vena cava and other main veins that pushes blood towards the heart.

Tissue fluid and capillary beds

We have seen that capillaries are tiny vessels that carry blood deep into the tissues of the body. Capillaries have so many branches that no cell is more than a short diffusion distance from a capillary Figure 15 on page 180 shows the composition of this blood. As the capillaries branch out among the cells in the tissues, they form a network called a **capillary bed**. The capillaries in the bed then gradually rejoin with each other, eventually forming a venule.

Capillary beds are exchange surfaces. Capillaries are leaky, allowing water and other substances with small molecules to seep out of them, into the spaces between the body cells. The fluid that collects between the cells is called **tissue fluid** (Figure 16, page 180).

Tissue fluid is, not surprisingly, very similar to blood plasma. It is mostly water, and contains many types of ions and small molecules in solution, which have

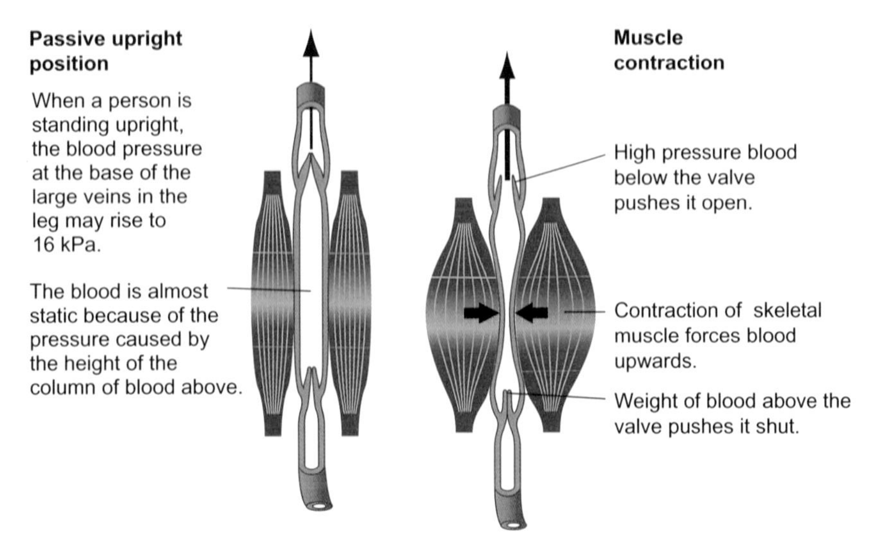

Figure 13 How leg muscles keep blood moving in the leg veins

	Aorta	Most arteries	Arteriole	Capillary	Venule	Most veins	Vena cava
Diameter	25 mm	4 mm	30 μm	8 μm	20 μm	5 mm	30 mm
Wall thickness	2 mm	1 mm	6 μm	<1 μm	1 μm	0.5 mm	1.5 mm
Endothelium Elastic tissue Smooth muscle Fibrous tissue							

Figure 14 Comparison of the structures of the different types of blood vessel

ASSIGNMENT 2: ANALYSING THE EFFECTS OF EXERCISE ON BLOOD FLOW

(MS 1.3, MS 3.2, PS 1.2, PS 3.1, PS 3.2)

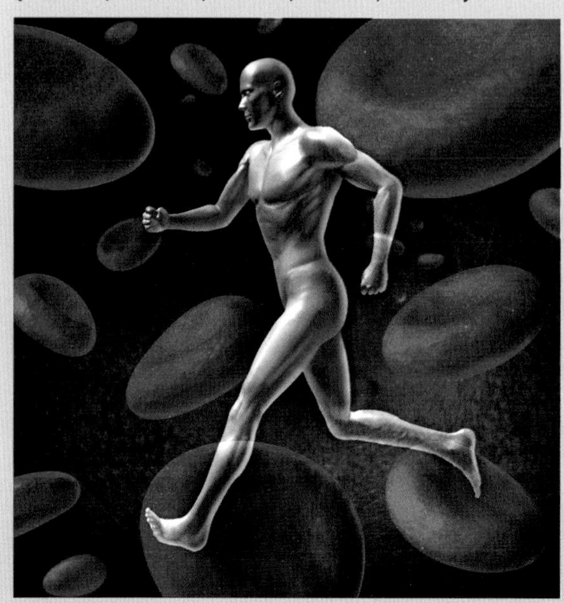

Figure A1 *How does exercise affect blood flow?*

Exercise has been shown to have many effects on the mind, including relief from stress, depression and anxiety, and improved self-esteem. The effects of regular exercise on the body are well documented and include improved cardiovascular performance and increased muscle mass and bone density.

In the short term, exercise has the greatest effect on blood flow. The table shows the volume of blood flowing through different parts of the body when resting and during exercise.

	Volume of blood flowing/cm^3 min^{-1}			
Part of body	When resting	Light exercise	Moderate exercise	Vigorous exercise
Brain	750	750	750	750
Digestive system	1400	1100	700	300
Heart muscle	250	350	600	1000
Kidneys	1100	900	600	250
Skeletal muscles	1200	4500	13 000	22 000
Skin	500	1500	1150	600
Rest of the body	600	400	400	600
TOTAL	**5800**	**9500**	**17 200**	**25 500**

Questions

A1. Which part of the body has the greatest blood flow at rest?

A2. In which part of the body does the blood flow decrease by 50% between rest and moderate exercise?

A3. By how much does the total blood flow increase as the exercise changes from moderate to vigorous exercise?

A4. Sketch graphs to show how the blood flow changes with exercise for:

a. the heart

b. the digestive system

c. the skin.

A5. Explain the changes in blood flow to:

a. the brain

b. the skeletal muscle.

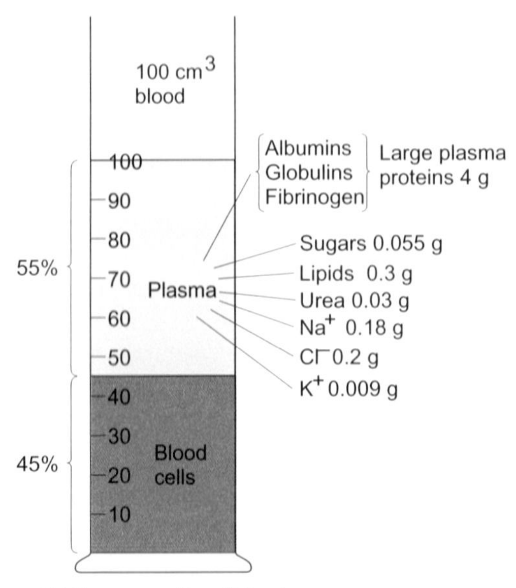

Figure 15 *The composition of blood*

either moved out of the capillaries, or diffused out of the body cells. It is through tissue fluid – the immediate environment of cells – that substances are exchanged between the body cells and the blood. So, for example, oxygen diffuses out of the red blood cells, through the blood plasma, through the tissue fluid and into a respiring cell. Carbon dioxide diffuses in the opposite direction. Water moves by osmosis in whichever direction its water potential gradient lies.

Blood cells are too large to escape from the blood capillaries, so these are not found in tissue fluid. The exception is macrophages, which are able to change their shape and actively crawl out through the gaps in the capillary walls. They patrol the tissues, engulfing and destroying pathogens by endocytosis.

Large protein molecules are also too big to escape from capillaries. Blood plasma contains a range of different proteins, called plasma proteins. These include, for example, the soluble, globular protein fibrinogen, as well as albumin and other globular proteins (sometimes called globulins). Tissue fluid is therefore rather like blood plasma, but without the plasma proteins. Like blood plasma, its composition varies depending on its position in the body. Tissue fluid in the small intestine, for example, contains a high concentration of monosaccharides in the couple of hours following a meal.

Most of the water in the tissue fluid (including substances it contains such as carbon dioxide) is returned to the capillaries. Any surplus water drains via the lymphatic system and back to the blood. The lymphatic system contains lymph vessels, which are part of the lymphatic system. These are blind-ending, and – like veins – they contain valves to help to keep fluid moving along them in one direction. Also like veins, they rely on contraction of nearby muscles to cause the lymph to flow. Lymph eventually empties back into the blood in the large veins that run close to the collarbone.

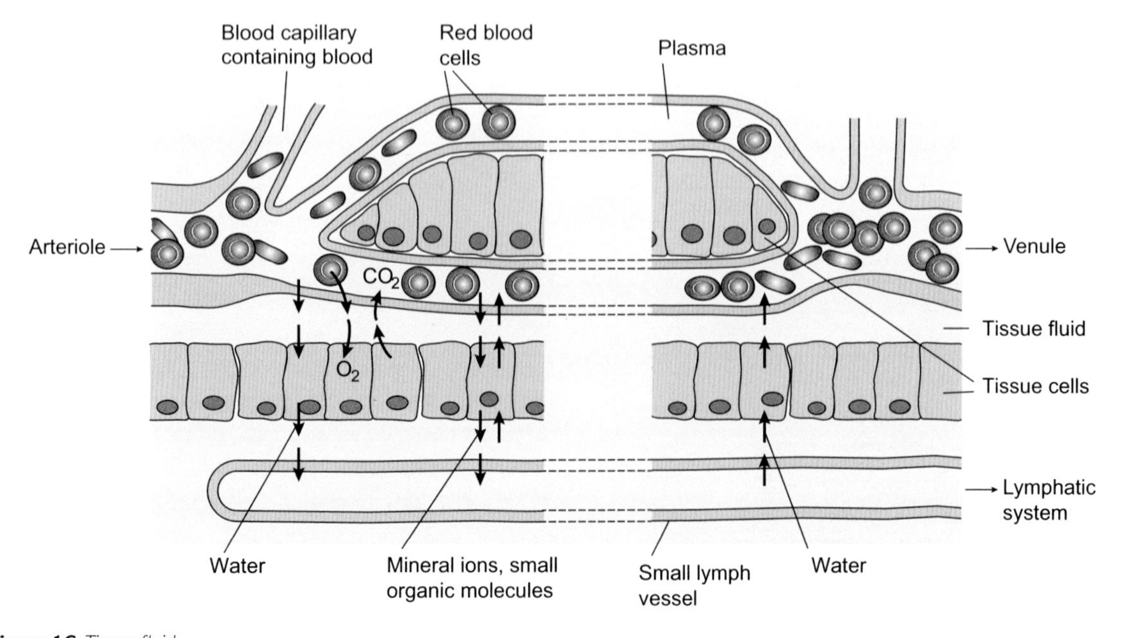

Figure 16 *Tissue fluid*

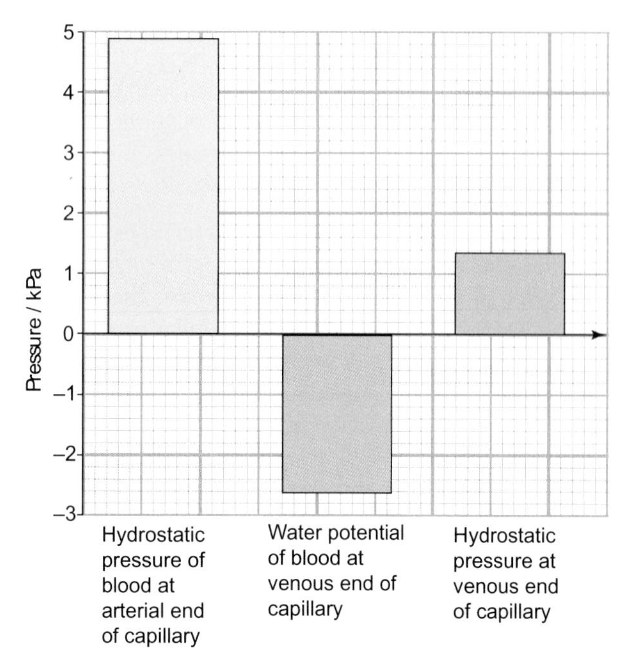

Figure 17 *Hydrostatic pressure and water potential in a capillary bed*

How tissue fluid forms

Two forces affect the exchange of water between the capillaries and the cells (Figure 17).

Hydrostatic pressure

This is the pressure that is generated by the pumping force of the heart. It tends to push water out of the capillaries when it first enters them. However, as the blood continues to flow through the capillaries, losing water all the time, its hydrostatic pressure decreases.

Water potential

As blood flows into the start of a capillary bed, the water potential of the plasma is higher than the water potential of the tissue fluid outside the capillary. Water therefore moves by osmosis out of the capillary, down its water potential gradient, into the tissue fluid. However, you will remember that many of the protein molecules dissolved in the plasma are not able to get out of the capillary. This means that, as the blood continues to flow along the capillary bed, its water potential decreases, because the steady loss of water increases the concentration of these dissolved proteins. By the time the blood reaches the far end of the capillary bed, its water potential is more negative than that in the tissue fluid, so water now moves back into the capillary, by osmosis.

So, by the time blood reaches the venous end of the capillary, the effect of the water potential drawing water in exceeds the hydrostatic pressure pushing water out. This means that there is a net re-entry of water into the capillaries.

QUESTIONS

13. Construct a table to compare the composition of blood, plasma and tissue fluid.

14. Oxygen, carbon dioxide, glucose, amino acids, urea and proteins are all carried in the blood. Which substances pass from capillaries into tissues and from tissues into capillaries at the following sites: intestinal villi; brain; leg muscles; liver?

15. Children suffering from protein deficiency often appear to be fat, but their abdomens are swollen because of the build-up of fluids in their tissues. Suggest why this fluid accumulation occurs.

ASSIGNMENT 3: USING HYPERBARIC OXYGEN

(PS 1.2)

Hyperbaric literally means 'increased pressure'. At sea level, a person experiences normal atmospheric pressure and breathes approximately 21% oxygen. In a hyperbaric chamber, this is increased to 100% oxygen and 1.5 × normal atmospheric pressure.

In the 1930s, the military started to test the use of hyperbaric oxygen to treat decompression sickness in deep sea divers. The link between decompression sickness and nitrogen bubbles in the blood had been discovered in 1878 by Paul Bert, who also discovered the pain could be relieved by recompression.

Decompression sickness occurs when nitrogen, which has dissolved in the blood after having been breathed in via the air tank, comes out of solution

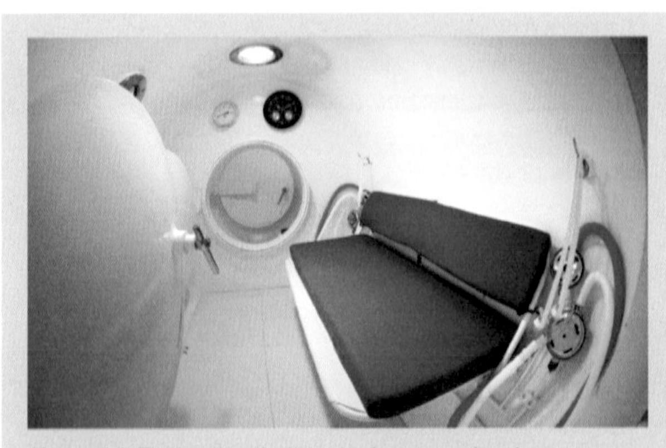

Figure A1 *The interior of a hyperbaric oxygen chamber*

and forms bubbles. These bubbles can cause pain and damage to tissues by blocking blood flow through small vessels.

In the 1950s it was discovered that the use of excess oxygen increased the effectiveness of radiation treatments for cancer, and there was much investigation into other medicinal uses. Increased oxygen enables the blood to carry more and deliver 15–25 times more to the tissues and organs of the body. This promotes faster and more efficient healing for a wide variety of diseases and ailments, including:

- decompression sickness ('the bends')
- severe carbon monoxide poisoning or smoke inhalation
- chronic wounds, diabetic ulcers and some infections
- radiation necrosis (body tissue dying after radiotherapy treatment)
- acute blood loss where a blood transfusion is not possible (for example, Jehovah's Witnesses believe that the Bible prohibits ingesting blood, so they cannot accept blood transfusions)
- sports injuries.

Under normal circumstances, oxygen is transported throughout the body by red blood cells. With hyperbaric oxygen treatment, oxygen is dissolved into all of the body's fluids, including the blood plasma and lymph. It can then be carried to areas where circulation is reduced, so extra oxygen reaches tissues.

Questions

A1. Explain how hyperbaric oxygen treatment can help in the case of acute blood loss.

A2. Describe how hyperbaric oxygen will help in cases of:

a. carbon monoxide poisoning

b. smoke inhalation.

A3. The increased pressure is important in treating decompression sickness. Suggest why.

A4. a. The use of hyperbaric oxygen shifts the oxygen dissociation curve to the left. Suggest why.

b. Name one other example of a situation in which the oxygen dissociation curve lies to the left of that for normal haemoglobin.

Cardiovascular disease

Cardiovascular disease (CVD) is a term that covers a range of different problems that can develop in the blood vessels and the heart. CVD generally affects arteries. You do not need to learn about any specific cardiovascular diseases. However, in an examination you may be given information and data which you have to analyse and/or evaluate, based on your knowledge of how the cardiovascular system works normally.

KEY IDEAS

- Arteries and arterioles carry blood away from the heart. They have relatively thick walls containing elastic tissue and smooth muscle.
- Contraction of the smooth muscle of arterioles reduces the size of the lumen, and this can be used to regulate the flow of blood to particular parts of the body.
- Capillaries are tiny vessels, with walls made of one layer of cells, whose role is to take blood close to every cell in the body.
- Veins and venules carry blood towards the heart. They have valves to prevent backflow of blood, which is now at low pressure. They have thinner, less muscular and less elastic walls than arteries.
- Tissue fluid is formed as blood plasma, minus many of the large plasma protein molecules, leaks from capillaries. Any tissue fluid not returned to the blood plasma drains back into the blood system via the lymph nodes.

REQUIRED PRACTICAL ACTIVITY 5: APPARATUS AND TECHNIQUES

(PS 4.1, AT j)

Dissection of animal or plant gas exchange system or mass transport system or of an organ within such a system

This practical activity gives you the opportunity to show that you can:

- Safely use instruments for dissection of an animal or plant organ.

Apparatus

Dissection is a key skill in biology. It requires you to be observant, patient and to have a steady hand. Safety, good hygiene and the ability to follow instructions are most important.

Apparatus	Use
Seeker and mounted needle	To probe tissues gently, lift flaps of tissue without tearing, or hold tissues out of the way.
Scissors	For cutting tissues; often more useful than a scalpel on tough tissue
Forceps	To hold or tease tissues; they may be blunt ended or sharp ended
Scalpel	To cut tissue; most useful for fine dissection. Note: great care must be taken as many scalpels are surgically sharp (but not surgically clean!)
Hand lens	To magnify, and enable you to see more clearly, what you are cutting
Wooden dissection board	Used for large specimens, such as the dissection of a mammalian gas exchange system. Also used when carrying out a whole organism dissection (a rat, for example) as pins can be placed in the wood to hold the organisms in place
Metal tray	Used for smaller specimens, especially those that may contain blood, such as a mammalian heart dissection. Some metal trays have a layer of wax in the bottom which can be used to pin tissues in place when dissecting an earthworm or insect, for example
White tile	Used for small specimens which are not going to produce significant amounts of fluid – when dissecting a plant stem, for example

***Figure P1** This girl is dissecting an eyeball on a metal tray. She is wearing protective gloves and is using a mounted needle.*

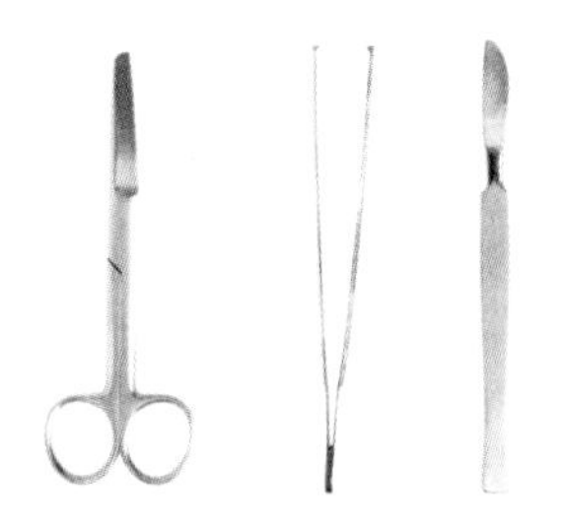

***Figure P2** Scissors, forceps and a scalpel*

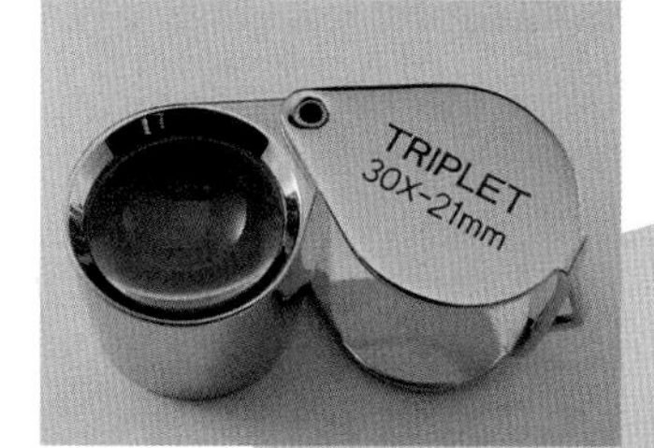

***Figure P3** A hand lens*

Techniques

Dissecting a heart

Dissection of the heart is relatively straightforward, provided you orientate the heart correctly before you start. Do not do any cutting until you have thoroughly examined the outside and determined which is the front. The front of the heart has a diagonal coronary artery running from top right to bottom left. This will enable you to distinguish the left and right sides of the heart.

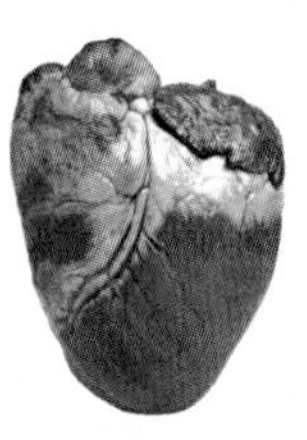

***Figure P4** The anterior of a pig's heart showing the coronary arteries*

If the heart you are dissecting has come from a butcher, you may not have complete atria and the major blood vessels will be difficult to distinguish. In this case, the vessels can be best identified after the heart has been opened and a seeker can be used to trace the aorta and pulmonary artery out of the ventricles.

To cut into the heart, a scalpel is usually best. The most common dissection is a simple cut from the left side of the heart to the right side or *vice versa*. The heart can then be opened up like a book. If you leave some tissue holding the two halves together you should be able to see the atrioventricular valve on at least one side.

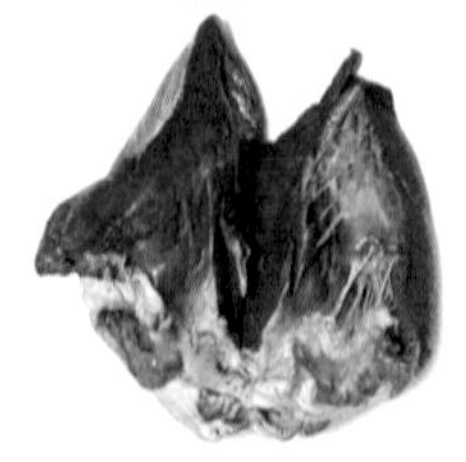

Figure P5 *A dissected pig's heart*

Dissecting lungs

As with the heart, it is important to get the lungs the right way round. Often the best way is to look for remains of the pericardium (the layer of tissue that surrounds the heart) which will show you where the heart was, and so which is the front. In addition you may be able to orientate the lungs by determining the position of the oesophagus at the back of the trachea.

Lung dissection usually involves cutting down the trachea and following the bronchial tree as far as possible. Scissors are best for this – a scalpel may just bend the cartilage if it's not very sharp. Other investigations may involve inflating the lungs. First, place the lungs in a plastic bag (this prevents the possible escape of aerosols contaminated with pathogens). Then, insert a rubber tube into the trachea and use a foot pump (not exhaled breath). You could also see how much air is in the lung by floating a piece of it in water.

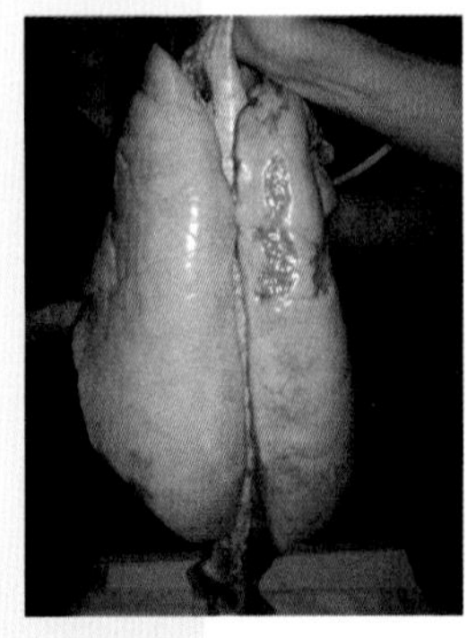

Figure P6 *These pig's lungs have been filled with air using a rubber tube.*

Dissecting a plant stem

The transport system of a plant can be investigated by dissecting a stem.

Plant stems (or roots) can be dissected length-wise to show the xylem and phloem tissues. The plant is usually put in water containing a harmless food dye for a period of time and this will stain the xylem tissue as the plant takes up the water. A scalpel is best for this dissection.

If you have access to a dissecting microscope or a magnifier of about × 15 you can investigate the detail of xylem tissue. Almost any plant stem can be used but fleshy stems are the easiest to both dissect and to see the transport vessels.

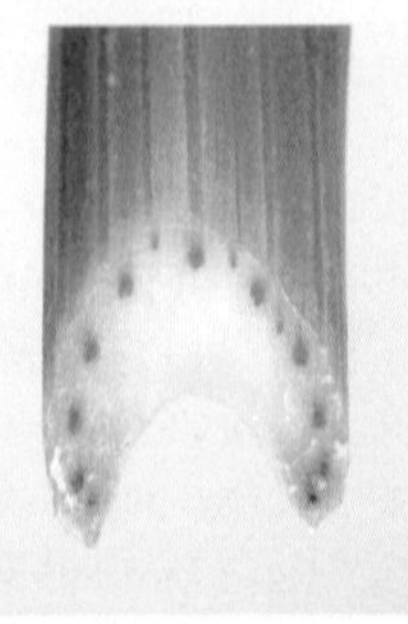

Figure P7 *A celery stalk that has been left in red food dye*

Dissecting a whole animal

If you are dissecting a whole animal to look at its transport or gas exchange system, you will need to use a wider range of dissecting instruments and be especially vigilant with health and safety. Although it is always important to wash your hands after handling biological material, with whole animals there is a greater opportunity for contamination with, for example, faecal matter.

Whole animal dissections are not done commonly, but the usual animal used for dissection is a rat. Under no circumstances should you attempt to dissect a dead animal which has been found. Only commercially bred animals should be used.

QUESTIONS

P1 When dissecting a heart, why should you probe the aorta from the left ventricle upwards?

P2 Why is a stem for dissection put into a dye?

P3 Why should you never dissect a dead animal you have found?

9.5 WATER TRANSPORT IN PLANTS

Plants have two mass transport systems. Both of them use mass flow to move substances from one part of the plant to another. One, made up of **xylem** tissue, carries water and dissolved mineral ions from the roots upwards, to all other parts of the plant. The other, made up of **phloem** tissue, transports organic substances that the plant has made – largely sucrose and amino acids – both upwards and downwards. Neither of these transport systems has a pump like the heart.

Water transport in xylem

Water transport in a plant requires absolutely no energy input from the plant itself. Yet water can be moved to enormous heights – all the way from the roots to the topmost leaves of the tallest tree. To understand how this happens, we need to start at the top of the water movement pathway, in the leaves.

In Chapter 8, we saw that plant leaves are adapted for efficient gas exchange. The large surface area of the mesophyll cells, which are surrounded by air spaces inside the leaf, makes it easy for oxygen and carbon dioxide to diffuse in and out of the mesophyll tissue. These air spaces connect to the air outside the leaf through stomata. This large surface area also means that water vapour, which evaporates from the wet cell walls of the mesophyll cells, can diffuse out of the leaf.

Figure 18 shows how this happens. The air spaces inside the leaf are normally saturated with water vapour, because liquid water in the cell walls constantly evaporates. This produces a high water potential within the air spaces, which reduces the rate of water loss from the mesophyll cells – confining water loss to stomata – which have a much lower surface area than the mesophyll. (The same principle is found in insect tracheal systems and lungs.) If the air outside the leaf is not saturated with water vapour, then there will be a water potential gradient between the leaf's air spaces and the external air. Water vapour will therefore diffuse out of the leaf, through the stomata, down its water potential gradient. This loss of water vapour from the leaf is called **transpiration**.

Factors affecting transpiration rate

Water molecules diffuse down a water potential gradient towards areas with a more negative water potential value. The greater the gradient, the faster the rate of movement. The air inside the leaf is always saturated with water vapour, so changes outside the leaf can alter the water potential gradient. Three factors increase this gradient and therefore the rate of transpiration:

- an increase in temperature
- a decrease in humidity
- an increase in wind speed.

An increase in temperature causes a higher rate of evaporation from the leaf surface by transferring more energy to the water molecules. The more energy a water molecule has, the faster it moves and the more likely it is to escape into the atmosphere outside the leaf and move away from the plant.

Humidity is a measure of the number of water molecules in the air. Any decrease in humidity in the atmosphere around the leaf decreases the number of water molecules in the air and so increases the water potential gradient from the plant to the air. Warm air can hold more water vapour molecules than cold air. On a hot day the air in a tropical jungle may become fully saturated with water vapour, reducing

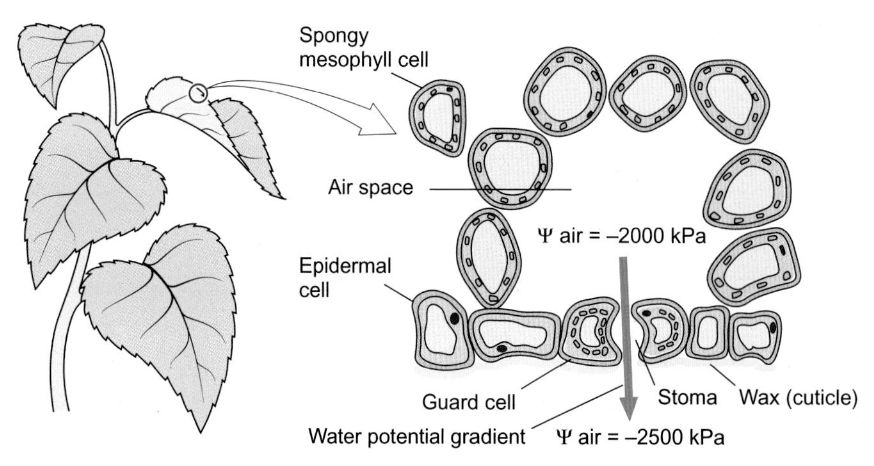

Figure 18 How transpiration happens

transpiration to zero, as there is no longer a water potential gradient between the plants and the air.

When wind moves the air and water vapour away from around a leaf, this increases the water potential gradient from the plant to the air.

Air always has a value of water potential more negative than that of the leaf cells, except when it is fully saturated with water vapour. So water molecules will always diffuse out through the open stomata. The only way to stop the loss of water is to close the stomata. Most plants close their stomata during the night which reduces water loss. If a plant has lost too much water during the day, it gets the chance to replenish its supplies during the night.

Xerophytes

Plants that live in dry places usually have adaptations to survive long periods of drought. Such plants are called **xerophytes**.

Xerophytes are highly adapted to deal with the conflict between gas exchange and water loss in an environment where water is in restricted supply, or potential water loss is high. They are therefore better able to conserve water than plants that live in conditions where water is not in short supply. For example, many desert trees, such as the quiver tree (Figure 19), drop their leaves to stop the whole tree from dying from desiccation. This certainly works, as it stops transpiration completely, but it is a drastic solution. A tree without leaves can no longer photosynthesise, so it must rely on the store of carbohydrates made during better times.

Cacti also have many adaptations that enable them to survive in deserts.

- The spines on the cactus, if closely packed, can trap a layer of air that is rich in water vapour next to the plant. This reduces the chances of wind moving the moist air layer away from around the plant. Many cacti have a dense covering of hairs that traps even more water vapour.
- The spines and stiff hairs on cacti also help to deter animals that try to eat the cactus to get at the stored water, and block a lot of direct sunlight (heat).
- The epidermal cells of cacti have a thick outer layer of wax that reduces water loss by evaporation.
- Many cacti store water in their stems.

Xylem

Cells that make up the **xylem vessels** (Figures 21 and 22 on page 188) have no living contents. Xylem tissue is made up of long, narrow cells with cellulose cell walls. As they grow, their walls are impregnated with a strong, impermeable substance called **lignin**. This causes them to die; their end walls break down so that, when the cells are aligned end to end, they form a continuous tube, all the way up from the roots to the leaves.

The tubes formed by the xylem cells function like the water pipes in your home – and, like your plumbing, they work only if there is nothing to block the flow of water.

In the leaves, water moves out through pores in the xylem vessels, into the mesophyll cells. It does this because the loss of water from the cell walls of the mesophyll cells, described previously, lowers the water potential in the cells. Water therefore moves down its water potential gradient from the xylem into the leaf cells. As water moves out of the xylem vessel in a leaf, the column of water behind it is pulled upwards.

Figure 19 *Quiver trees in Namibia can shed their leaves and store water in their branches to survive periods of extreme drought. They are called quiver trees because native people clean out the soft branches and used them as a quiver (a pouch) for their arrows.*

Figure 20 *Three different types of cacti growing in the Sonoran Desert, Arizona*

ASSIGNMENT 4: INVESTIGATING RESOURCEFUL XEROPHYTES

(PS 1.2)

Xerophytes are all adapted to dry conditions, but the strategies and adaptations they have evolved differ widely from plant to plant.

The sand dunes on the seashores in Britain have soil that often contains very little water, but some plants are very well adapted to living there. Perhaps the most successful is marram grass. When the soil is dry, hinge cells in the leaf make it roll into a cylinder. The air inside this cylinder quickly becomes saturated with water, resulting in a reduction in transpiration rate. When the soil is moist, the leaves uncurl.

A rather strange plant of the *Tortula* species, twisted moss, commonly called the 'all-screwed-up moss', also lives on sand dunes. It spends much of its life all screwed up, but when rain falls its leaves unwind.

Some xerophytic plants have different shapes of leaf in wet and dry seasons.

Collecting every drop of available water is vital to desert plants. Often the roots of these plants spread over a very wide area.

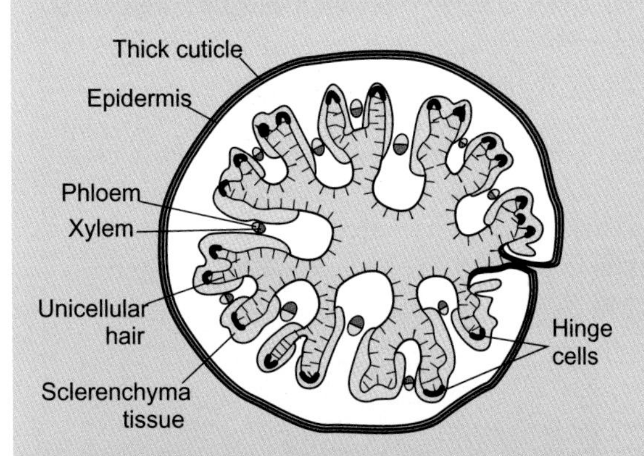

Figure A1 *Section across a rolled-up marram grass leaf*

Figure A2 *Twisted moss*

Questions

A1. Suggest two adaptations that reduce the rate of transpiration even when the marram grass leaves are uncurled.

A2. Explain how the behaviour of the all-screwed-up moss helps it to survive dry periods.

A3. Look at the structures of the grey sagebrush in Figure A3. Explain the advantage to the sagebrush of having smaller leaves during the dry season.

A4. Look at the root systems in Figure A4. Explain why they are xerophytic adaptations.

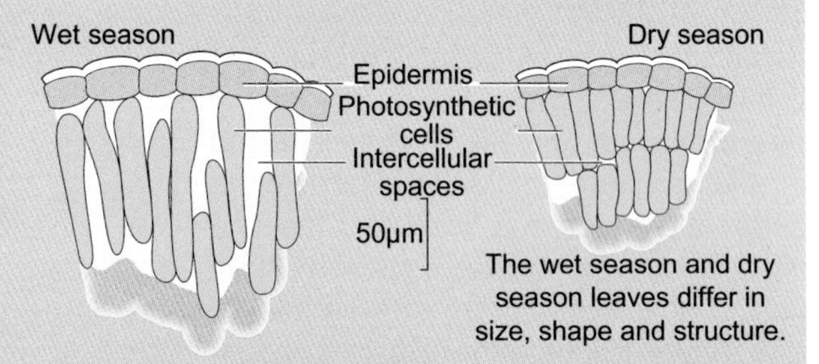

Figure A3 *Cross-sectional structure of the winter and summer leaf of the grey sagebrush*

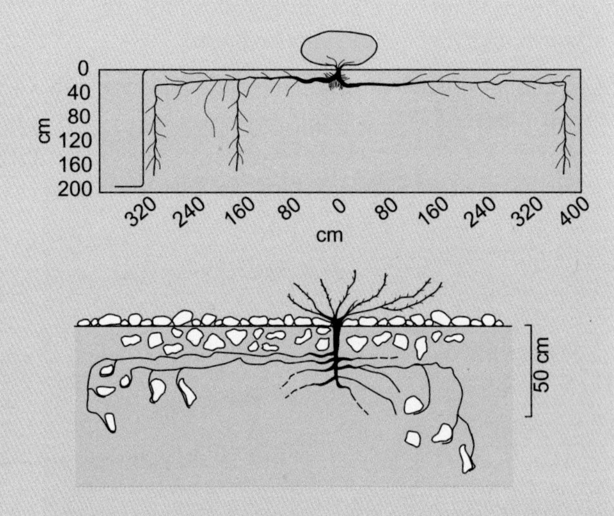

Figure A4 *A schematic drawing of the root system of the bean caper (top), and a schematic drawing of a hydrotropic root system, exploiting water pockets beneath stones (bottom)*

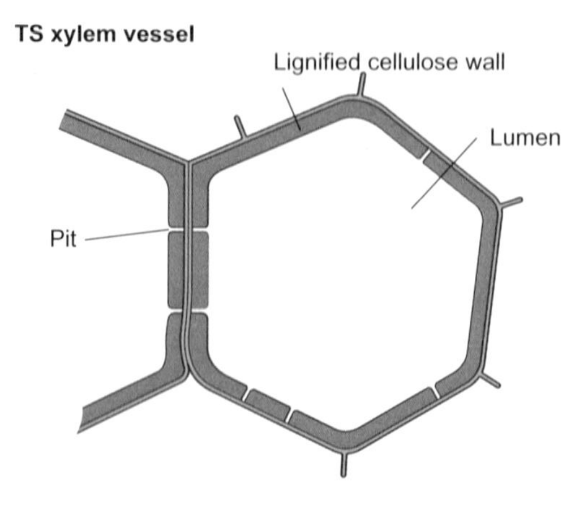

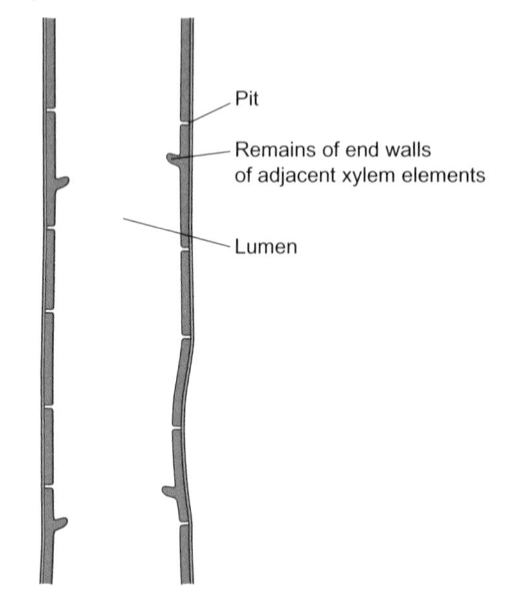

Figure 21 *The structure of xylem vessels*

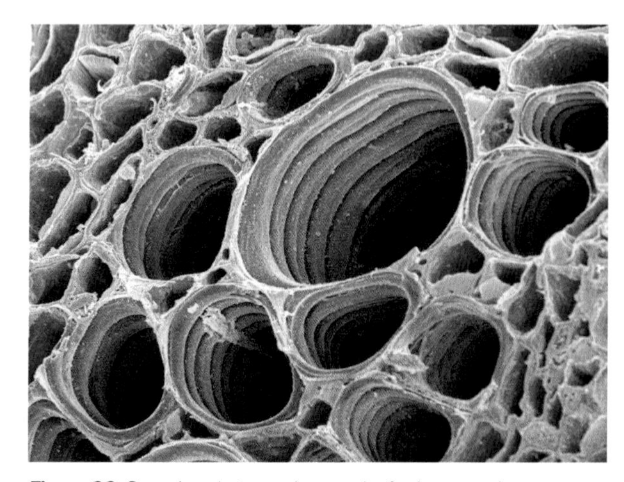

Figure 22 *Scanning electron micrograph of xylem vessels*

We have seen that there are forces, called hydrogen bonds, that attract one water molecule to another. When a water molecule moves out at the top of a xylem vessel, its neighbouring water molecules are pulled with it. The whole column of water tends to 'stick together' because of these hydrogen bonds. The force that holds the water molecules together is called **cohesion**.

There is another force acting on the water column – its own weight. Gravity pulls it downwards. With the effects of transpiration pulling the water column upwards and gravity pulling it downwards, the water column is under tension. If you pull a cylinder of modelling clay at both ends, the middle of the cylinder will get thinner and will eventually break. What stops this happening to the columns of water in the xylem vessels? Cohesion helps. Moreover, the water molecules are strongly attracted to the walls of the xylem vessels, a force called **adhesion**. This attraction between the xylem walls and the water is so strong that as the water columns are put under tension by evaporation of water from the leaves, they become thinner and actually pull the walls of the xylem vessels inward. One function of the thickening in the xylem walls is to allow them to resist this force, preventing the xylem vessels completely collapsing. However, the combined effect of the tension on all the xylem vessels in a tree trunk does produce a measurable narrowing of the trunk when the tree is transpiring (Figure 23).

This mechanism for the movement of water up the xylem is known as the **cohesion–tension** mechanism, since there is a cohesive force between

QUESTIONS

16. The graph in Figure 23 shows changes in the circumference of three different species of tree during a hot summer day. Suggest which of the trees is best adapted for reducing the rate of transpiration. Use evidence from Figure 23 to support your answer.

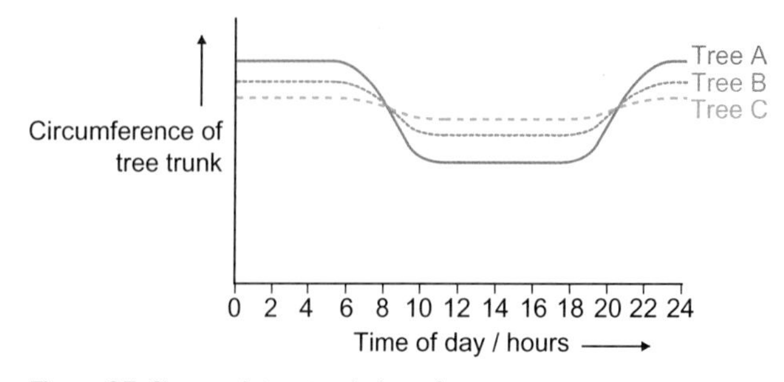

Figure 23 *Changes in tree trunk circumference*

the water molecules themselves, and the water columns are under tension because of the forces resulting from evaporation and gravity.

As water leaves the xylem at the top, the xylem cells at the bottom of the column pull water from the cells in the root cortex. The water potential of the root cortex now has a value that is more negative than that of the root epidermis and root hairs. This means that water moves into the cortex cells. The value of water potential for the epidermis is now more negative than that of the soil water and so water moves from the soil into the root.

ASSIGNMENT 5: GREENING THE DESERT

(MS 0.1, MS 1.3, MS 1.6, PS 1.1, PS 1.2, PS 3.2)

Marula is a large deciduous tree found in Southern Africa. The tree is highly prized by local people for its fruits. The fruits are plum-sized with a sweet-sour flesh that can be eaten fresh or used to prepare juices and alcoholic drinks. Researchers introduced marula trees into different sites in the Negev desert in Israel to evaluate its growth under different conditions, as shown in the table. At each of four sites, 30 one-year-old plants were planted.

The concentration of sodium chloride in the water supply was measured as its electrical conductivity, EC, in units of deci-siemens per metre, dS m^{-1}.

The plants were drip fertigated every one or two days in summer and every three to five days in winter. In the fourth year the volume of water supplied was determined by the evaporation rate and varied from 17 m^3 per tree per year at Besor and Ramat to 25 m^3 at Qetura and Neot. The table shows the mean dimensions of the trees after four years.

Figure A1 *Marula tree in Southern Africa*

The graph in Figure A2 shows the development of new leaves in the fourth year. The percentage of the tree canopy with new leaves was rated visually at the middle of each month on the following scale:

0 = no growth, 1 = less than 20%, 2 = 20–60%, 3 = 60–100%

Site	Temperature extremes	Water supply	Mean height of tree/cm (± standard deviation)	Mean trunk circumference/cm (± standard deviation)
Besor	Moderate	Fresh EC 1 dS m^{-1}	533 (± 60)	50 (± 2)
Ramat	Sub-freezing in winter	Brackish* EC 3.5 dS m^{-1}	290 (± 25)	54 (± 3)
Qetura	High summer temperatures, warm winters	Brackish EC 3.5–4.5 dS m^{-1} Ratio of Ca^{2+} to Na^+ 1.2	620 (± 30)	58 (± 2)
Neot	High summer temperatures, warm winters	Brackish EC 3.5–4.5 dS m^{-1} Ratio of Ca^{2+} to Na^+ 0.8	413 (± 39)	40 (± 4)

* Brackish water has a higher concentration of sodium chloride than fresh water, but not as high as sea water.

3.3 billion hectares of previously useful land are now classified as desert. If we can understand how plants manage their water economy we might be able to reclaim some of the land.

The above is an extract from research on introducing new crops into the desert (Adapted from A. Nerd and Y. Mizrahi 2003, pp. 496–9 in New Crops by Janick and Simon (eds), Wiley, New York).

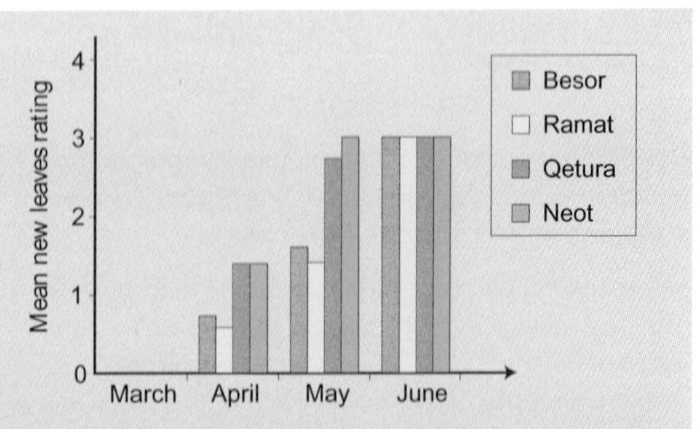

Figure A2

Questions

A1. The salinity (amount of mineral ions) of soil water is measured using electrical conductivity (EC).

a. What is the unit of conductivity?

b. Suggest why electrical conductivity is used to measure salinity rather than chemical methods.

c. Suggest what is meant by 'fertigated'.

A2. Suggest why the marula trees at Qetura and Neot were given much more water than those at Besor and Ramat.

A3. a. What was the height after four years of the tallest tree at Qetura?

b. What is the general relationship between the dimensions of the trees after four years and the conditions under which they were grown?

c. Use data from the table to suggest why there was such a big difference in dimensions of trees grown at Qetura and Neot.

A4. The appearance of new leaves showed different patterns at Besor and Neot.

a. Describe the patterns for the two sites.

b. Suggest an explanation for the different patterns.

KEY IDEAS

- Water moves passively from the roots to the leaves due to water potential gradients.
- The water potential in the leaves is lowered because water evaporates from mesophyll cells and diffuses out into the air. This is called transpiration.
- Xerophytes are plants that are adapted to live in dry conditions, and are able to reduce the rate of transpiration and therefore water loss.
- Xylem facilitates the upwards mass transport of water and mineral ions.
- The water column is pulled upwards as a result of transpiration, which produces a tension force. The water column holds together because of cohesion between water molecules.
- Xylem elements are elongated, empty, dead cells with lignin in their walls. They have no end walls and are arranged end to end to form continuous vessels.

9.6 TRANSPORT OF ORGANIC SUBSTANCES IN PLANTS

Plants move organic substances such as sucrose and amino acids around their bodies in **phloem tissue**. This movement is called **translocation.**

Unlike xylem, phloem tissue is made up of living cells. The individual cells that make up the 'tubes' through which substances are moved are called **phloem sieve elements** or sieve cells. Each of these has a **companion cell** closely associated with it (Figure 24).

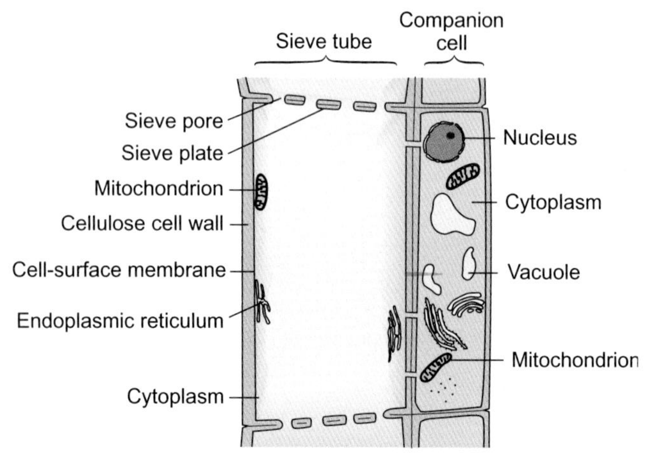

Figure 24 *The structure of phloem tissue*

Sieve elements have walls made of cellulose, without any lignin. They do contain cytoplasm, but this is only a thin layer close to their walls. They do not have a nucleus. Their end walls are perforated to form **sieve plates**. Long columns of these cells form many continuous tubes, called **sieve tubes**, that run through root, stem and leaf.

Companion cells, on the other hand, are full of cytoplasm, and this contains many mitochondria. They are connected to their neighbouring sieve element with many plasmodesmata.

QUESTIONS

17. Construct a table to compare the structure of xylem vessels and phloem sieve tubes.

The contents of phloem

If you could taste the contents of phloem sieve tubes, you would find that it tastes sweet. Indeed, many animals enjoy feeding on the sap that is contained in phloem and – because phloem sieve tubes generally run close to the surface of stems – it is not difficult for them to tap into it. Aphids, for example, sit motionless for many hours on a plant stem with their mouthparts plugged into the phloem.

The sugar in phloem is the disaccharide **sucrose**. During photosynthesis, leaves make glucose. Most of the glucose is used very rapidly for respiration, while some is changed into starch and stored. When other parts of the plant require carbohydrates, the starch is converted to sucrose and loaded into the phloem. During the summer months, when photosynthesis can happen at a great rate, many plants transport excess sucrose to their roots, where it is converted to starch and stored for use during the cold, dark days of winter. (Remember: roots do not carry out photosynthesis; the sugars moving down the phloem keep the root cells alive.)

So, in summer, the main movement of sucrose is downwards, from leaves to roots. There will also be some movement from the leaves to the flowers, where the sucrose is used to make nectar to attract insects for pollination, and also to newly growing leaves, and to fill the cells of developing fruits with sugars to attract animals to disperse their seeds. In winter, the main movement of sucrose is upwards, from stores in the root to other parts of the plant. An area *from* which sucrose is moving is called a **source**, and the area *to* which it is moving is called a **sink**.

Phloem also transports other substances that the plant has made, such as amino acids.

How translocation happens

We have seen that movement of water through xylem requires no energy input from the plant, relying purely on evaporation of water (driven by energy from the Sun) from the leaves to provide a pulling force. This is not true of phloem. The plant uses energy, obtained from the hydrolysis of ATP produced in respiration, to make phloem sap flow through the sieve tubes.

The energy is used in active transport at the source, to indirectly load sucrose into a sieve tube element. The mechanism by which this is thought to happen is shown in Figure 25. Proton pumps (carrier proteins that actively transport hydrogen ions) move hydrogen ions out of a companion cell. This produces a diffusion gradient for hydrogen ions, and they therefore diffuse back into the cell through co-transporter proteins, through which sucrose can also move. This increases the concentration of sucrose in the companion cell. The sucrose can readily move into the phloem sieve tube (through the plasmodesmata).

So now there is a high concentration of sucrose in the phloem sieve tube. This lowers the water potential,

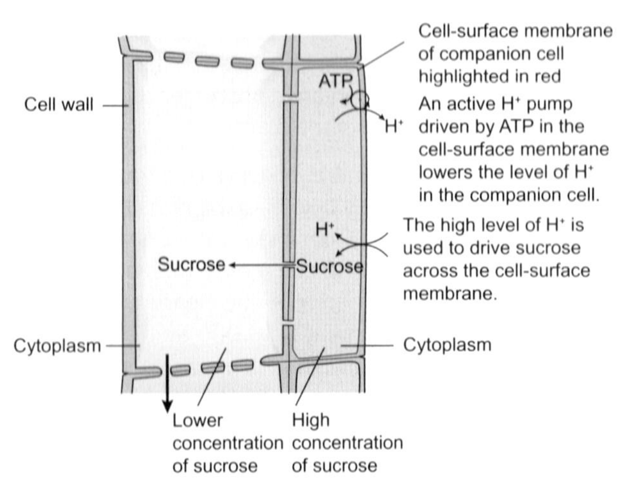

Figure 25 *Active transport of hydrogen ions out of a cell, followed by facilitated diffusion of sucrose into the cell through a co-transporter, is used to load sucrose into a phloem sieve tube.*

so water moves into the sieve tube, by osmosis. This extra water increases the hydrostatic pressure.

Liquids flow from areas of high pressure to low pressure. This lower-pressure area could be in a root. Here, root cells use sucrose – perhaps changing to starch for storage, or to glucose to use in respiration. Either way, they reduce the concentration of sucrose, and water consequently moves out of the phloem. This lowers the hydrostatic pressure. The phloem sap therefore moves by mass flow down through the sieve tube, from the region of high hydrostatic pressure where it is loaded with sucrose, to the region of lower hydrostatic pressure where sucrose is unloaded (Figure 26).

As it moves from source to sink, the sugary phloem sap passes through the pores in the sieve plates. It used to be thought that these might somehow provide energy to cause it to move, but this theory is now discounted. Instead, it is thought that the sieve plates are there to stop too much valuable sucrose being lost if the phloem vessel is damaged – for example, by a grazing animal. If the phloem is damaged, the pores in the sieve plates are quickly blocked by a substance called callose, rather as a broken blood vessel is blocked by the blood clotting.

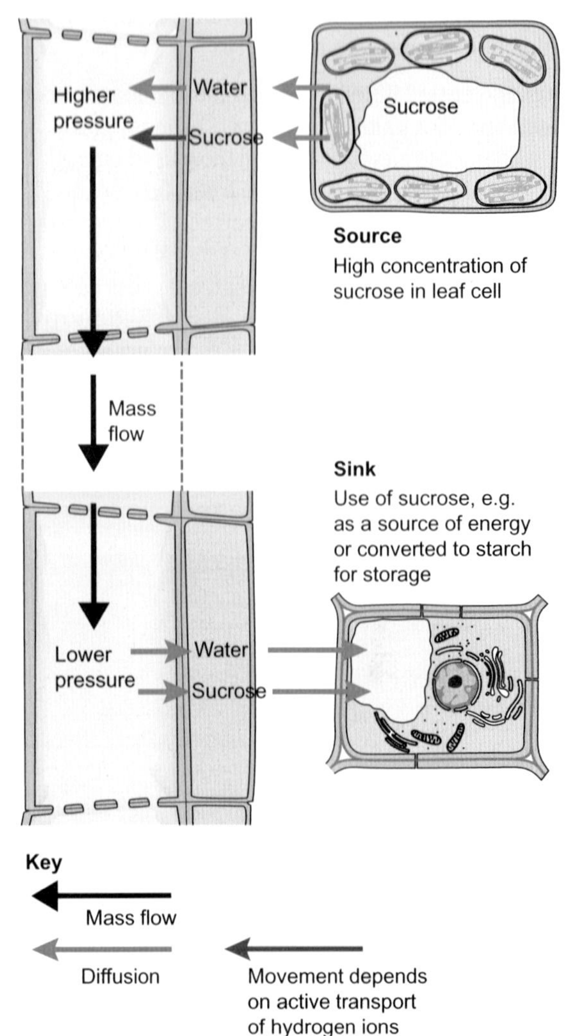

Figure 26 *Phloem sap flows from a high-pressure area at the source to a low-pressure area at the sink.*

Evidence for the mechanism of phloem transport

Some of the earliest evidence that sucrose was transported in phloem, rather than in xylem, came from ringing experiments. Phloem tissue lies close to the surface of plant stems and tree trunks, and if we remove a ring of bark from a young tree, we also remove the phloem just underneath it (Figure 27). This does not damage the xylem tissue, which lies deeper. After some time, in a tree that is photosynthesising, it is found that sucrose accumulates just above the

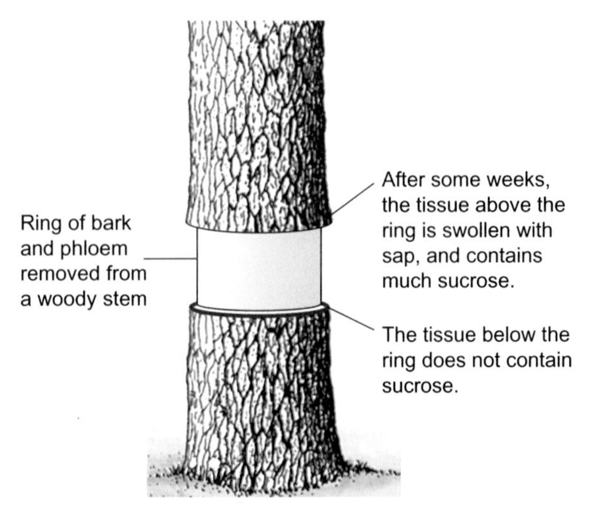

Figure 27 *Ringing experiments show that sucrose is transported in phloem.*

ringed area, while no sucrose can be detected in the phloem tissue beneath the ring.

Until the early 1980s, there was much debate about whether or not phloem sap moved by mass flow, or whether the sieve plates somehow used energy to make it move. One of the main reasons for this uncertainty was that many researchers felt that the sieve plates would not be there unless they were doing something useful – and that was most likely to be some kind of active transport that moved sucrose and other materials. Moreover, electron micrographs showed strands of protein running through these pores, from one cell to the next. Now, however, we know that these strands of protein are not present in living cells, but only appear when the tissue is prepared for viewing with the electron microscope. Their function appears to be to block the phloem sieve tubes if they are damaged, therefore reducing the loss of sucrose and other organic substances that would otherwise 'bleed' out of the plant.

There is now considerable evidence that translocation in phloem does indeed take place by mass flow. For example, radioactively labelled substances can be absorbed into phloem in one area, and their rate of movement to another area can be measured. We find that the rate of movement is 10 000 times faster than would be the case if anything other than mass flow (for example, diffusion) was involved. We can also measure pressure differences between sources and sinks, and work out how fast we would expect the liquid to move as a result. These calculations closely match the actual rates of flow that are measured.

ASSIGNMENT 6: USING CARBON-14 TO INVESTIGATE PHLOEM TRANSPORT

(MS 1.6, MS 3.1, MS 3.5, PS 1.2, PS 3.1, PS 3.2)

Most carbon in the atmosphere is in the form of carbon-12, ^{12}C. Carbon-14, or ^{14}C, is an isotope of carbon that emits beta radiation. This makes it possible to detect the presence of this isotope in a plant, by monitoring beta radiation.

In an investigation into the transport of sugars in phloem, researchers exposed the leaves of several tomato plants to carbon dioxide containing ^{14}C. There were several developing fruits on each plant. The researchers divided the tomato plants into three groups.

- Group **A** was kept in a steady temperature of 20 °C.
- Group **B** was kept in a steady temperature of 30 °C.
- Group **C** was kept in a steady temperature of 20 °C, but the tomato fruits were all removed.

The researchers cut sections across the stems of one plant from each group, a few hours after exposure to the labelled CO_2. These sections all showed radioactivity from ^{14}C in a ring just beneath the stem surface.

In the other plants in each group, the researchers measured how much of the radioactivity in the leaves remained in the 11 hours following exposure to the $^{14}CO_2$.

This graph shows the results:

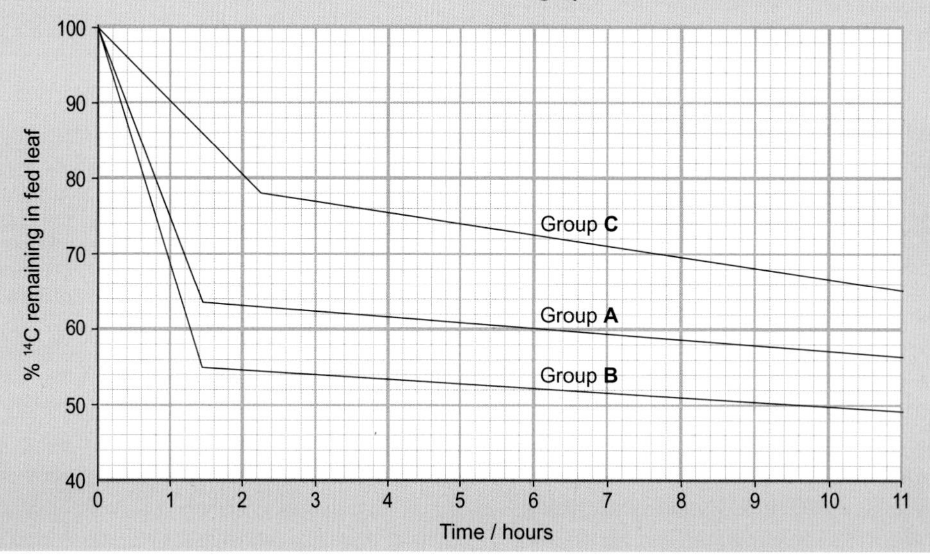

Questions

A1. Explain why radioactive ^{14}C atoms were present in a ring just beneath the stem surface, shortly after the leaves had been exposed to $^{14}CO_2$.

A2. a. Describe the results shown in the graph for the Group A plants.

b. Suggest reasons for the pattern of results that you have described.

A3. a. Calculate the mean rate of loss of ^{14}C from the leaf for Group A and for Group B.

b. Suggest an explanation for the differences between these two groups.

A4. a. Compare the results for the Group C plants with the Group A plants.

b. Suggest an explanation for the differences between these two groups.

A5. Suggest how the researchers could use this radioactive labelling technique to determine the speed of flow of assimilates through the phloem in the tomato stems.

KEY IDEAS

- Phloem is a tissue that transports organic substances from sources to sinks in plants.
- Phloem tissue contains phloem sieve elements and companion cells, closely associated with one another.
- Phloem sieve elements are living cells with small amounts of cytoplasm, no nucleus and cellulose cell walls. Their end walls form sieve plates.
- Companion cells have all the normal organelles, including many mitochondria. They are connected directly to phloem sieve elements through plasmodesmata.
- The main organic substances transported in phloem are sucrose and amino acids.
- Sucrose is actively loaded into phloem at the source. Water follows by osmosis, increasing hydrostatic pressure.
- Sucrose moves out of phloem at the sink. Water follows by osmosis, decreasing hydrostatic pressure.
- The sap in phloem therefore flows from the source to the sink, from high to low hydrostatic pressure.
- Ringing experiments provide evidence for transport of sucrose in phloem. Radioactive tracers provide evidence that this transport occurs by mass flow.

PRACTICE QUESTIONS

1. Figure Q1 shows changes in pressure in different parts of the heart during a period of one second.

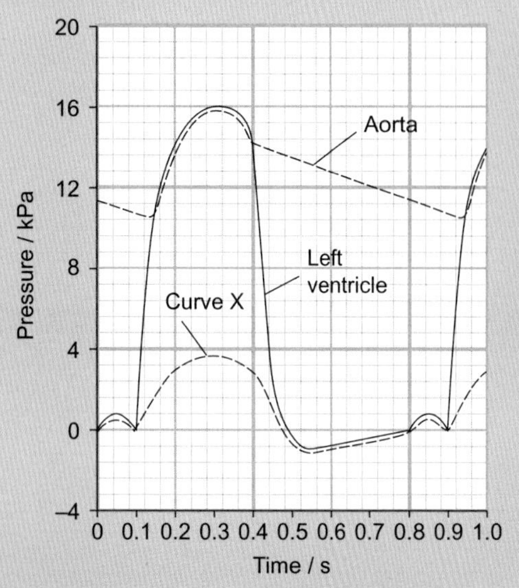

Figure Q1

a. i. At what time do the semilunar valves close?

ii. Use the graph to calculate the heart rate in beats per minute. Show your working.

iii. Use the graph to calculate the total time that blood flows out of the left side of the heart during one minute when beating at this rate. Show your working.

b. What does curve X represent? Explain your answer.

c. The volume of blood pumped out of the left ventricle during one cardiac cycle is called the stroke volume. The volume of blood pumped out of the left ventricle in one minute is called the cardiac output. It is calculated using the equation

cardiac output = stroke volume × heart rate

After several months of training, an athlete had the same cardiac output but a lower resting heart rate than before. Explain this change.

AQA June 2006 Unit 3 Question 2

2. Figure Q2 shows the oxyhaemoglobin dissociation curves for three species of fish.

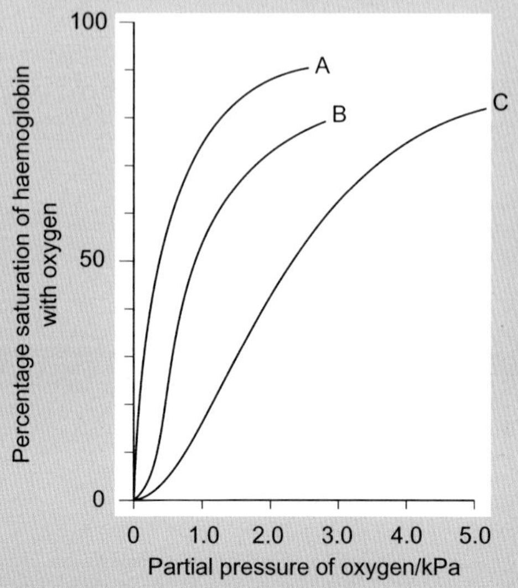

Figure Q2

a. Species **A** lives in water containing a low partial pressure of oxygen. Species **C** lives in water with a high partial pressure of oxygen. The oxyhaemoglobin dissociation curve for species **A** is to the left of the curve for species **C**. Explain the advantage to species **A** of having haemoglobin with a curve in this position.

b. Species **A** and **B** live in the same place but **B** is more active. Suggest an advantage to **B** of having an oxyhaemoglobin dissociation curve to the right of that for **A**.

AQA June 2006 Unit 3 Question 7

3. Figure Q3 shows the change in the speed of flow and pressure of blood from the start of the aorta into the capillaries.

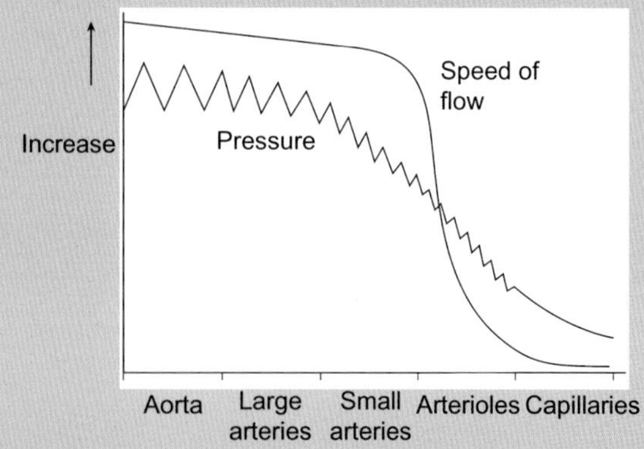

Figure Q3

a. Describe and explain the changes in the speed of flow of the blood shown in Figure Q3.

b. Explain how the structure of the arteries reduces fluctuations in pressure.

c. Explain how the structure of capillaries is related to their function.

d. In one cardiac cycle, the volume of blood flowing out of the heart along the pulmonary artery is the same as the volume of blood returning along the pulmonary vein.

Explain why the volumes are the same although the speed of flow in the artery is greater than in the vein.

AQA June 2006 Unit 3 Question 6

4. Some people have a form of heart failure where their heart is not pumping blood as well as it used to. Some people with heart failure are given an artificial heart to improve circulation of blood from the left ventricle. Figure Q4 shows where this type of artificial heart is connected.

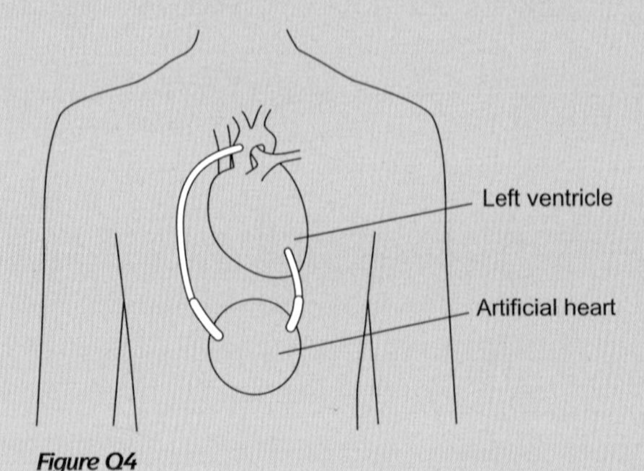

Figure Q4

a. Name the blood vessel to which the artificial heart is connected.

b. In these patients, the right ventricle still produces sufficient blood flow to keep the patient alive.

Suggest why the left ventricle requires the help of the artificial heart but the right ventricle does not.

Figure Q5 shows the internal structure of this type of artificial heart.

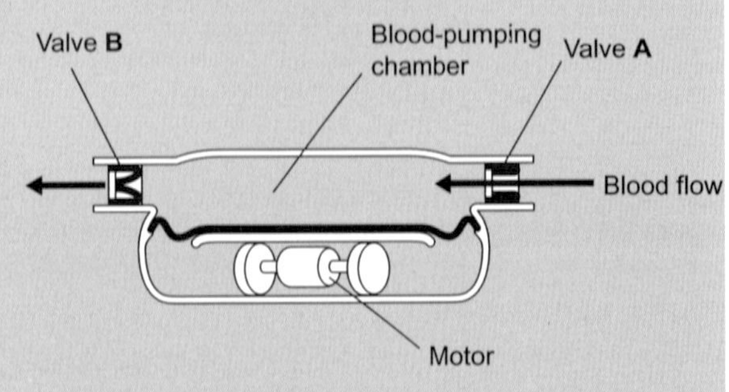

Figure Q5

c. Valves **A** and **B** have the same functions as heart valves involved in the cardiac cycle. Name the heart valve that has the same function as valves **A** and **B**.

d. There are different designs of artificial heart. Doctors compared results for patients who received two different types of artificial heart, **X** and **Y**. They recorded information two years after the artificial hearts were implanted. Their results are shown in Table Q1.

Which type of artificial heart was the more successful? Use calculations to support your answer.

AQA January 2013 Unit 1 Question 6

Type of artificial heart	Information recorded two years after artificial heart implanted		
	Number of patients surviving without replacement of artificial heart	Number of patients surviving but who required repair or replacement of artificial heart	Number of patients who died
X (119 patients)	62	13	44
Y (58 patients)	7	24	27

Table Q1

(Continued)

5. a. i. An arteriole is described as an organ. Explain why.
   ii. An arteriole contains muscle fibres. Explain how these muscle fibres reduce blood flow to capillaries.
   b. i. A capillary has a thin wall. This leads to rapid exchange of substances between the blood and tissue fluid. Explain why.
   ii. Blood flow in capillaries is slow. Give the advantage of this.
   c. Kwashiorkor is a disease caused by a lack of protein in the blood. This leads to a swollen abdomen due to a build up of tissue fluid. Explain why a lack of protein in the blood causes a build up of tissue fluid.

AQA January 2013 Unit 2 Question 2

6. A student measured the volumes of water absorbed by the roots of a plant and lost by transpiration over periods of four hours during one day. The bar chart in Figure Q6 shows the results.
   a. i. Describe the changes in the volumes of water absorbed and transpired between midnight and 1600.
   ii. Explain these changes in the volumes.

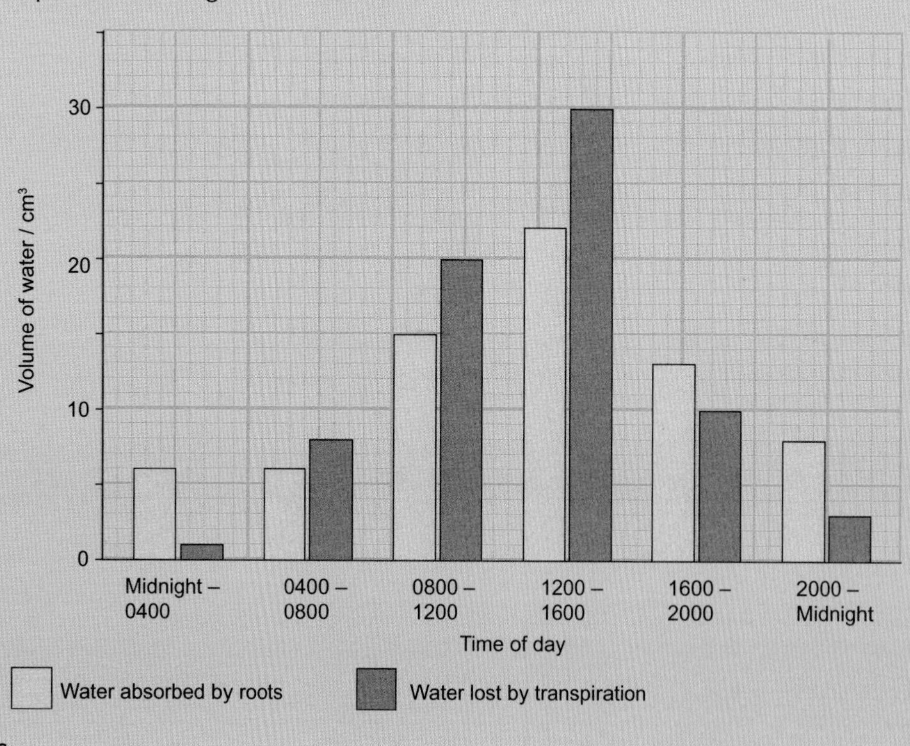

Figure Q6

   b. Use your knowledge of the cohesion – tension theory to explain how water in the xylem in the roots moves up the stem.

AQA June 2006 Unit 3 Question 3

7. Students investigated the effect of removing leaves from a plant shoot on the rate of water uptake. Each student set up a potometer with a shoot that had eight leaves. All the shoots came from the same plant. The potometer they used is shown in Figure Q7.
   a. Describe how the students would have returned the air bubble to the start of the capillary tube in this investigation.
   b. Give two precautions the students should have taken when setting up the potometer to obtain reliable measurements of water uptake by the plant shoot.
   c. A potometer measures the rate of water uptake rather than the rate of transpiration. Give two reasons why the potometer does not truly measure the rate of transpiration.

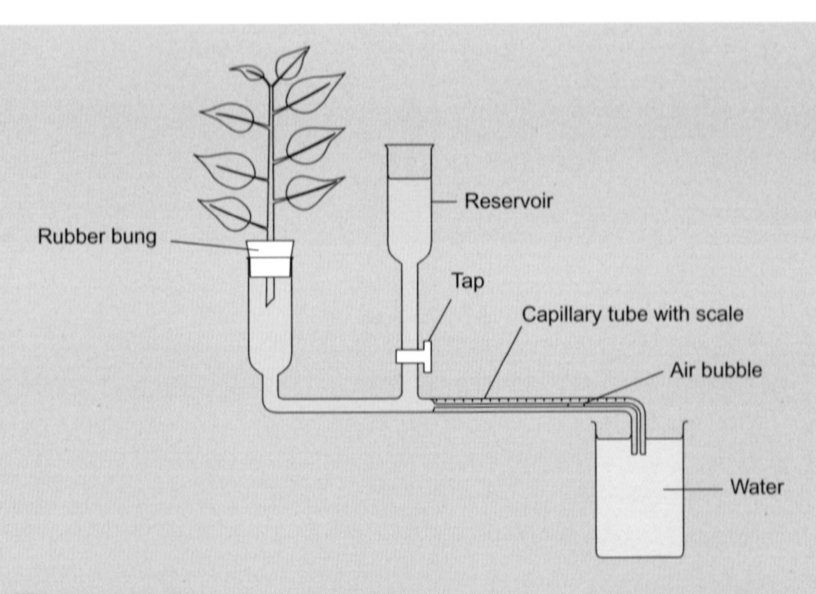

Figure Q7

d. The students' results are shown in Table Q2.

Number of leaves removed from the plant shoot	Mean rate of water uptake / cm^3 per minute
0	0.10
2	0.08
4	0.04
6	0.02
8	0.01

Table Q2

Explain the relationship between the number of leaves removed from the plant shoot and the mean rate of water uptake.

AQA January 2013 Unit 2 Question 5

10 DNA AND PROTEIN SYNTHESIS

PRIOR KNOWLEDGE

You may remember that DNA is a polynucleotide, made up of thousands of nucleotides linked together through phosphodiester bonds. Another polynucleotide found in cells is RNA, which differs from DNA in several ways. Proteins are also polymers, made up of 20 different amino acids arranged in different sequences, which affect their three-dimensional structure and therefore their functions.

LEARNING OBJECTIVES

In this chapter, you will find out how the sequences of bases in DNA molecules are used to guide the cell in making proteins, with the help of different types of RNA.

(Specification 3.4.1, 3.4.2)

The genetic code is an amazing method of controlling how organisms make proteins, and therefore controlling the structure and function of every living organism. It seems to have evolved right at the start of life on Earth, because every living thing has the same genetic code – that is, the same three-base sequences on the DNA and RNA molecules in every organism always code for the insertion of the same amino acid into a protein.

In 2013, researchers at Yale University in the USA managed to 'recode' a bacterium. One of the code-words (codons) in RNA is UAG, and this is actually more like a full stop rather than a word, as it means 'stop making a protein at this point'. The researchers removed all the UAG codons in the bacterium and replaced them with UAA codons, which also mean 'stop'. This freed up the UAG codon to code for something else. The researchers then reintroduced the UAG codons, together with a set of enzymes specifically designed to use this codon to introduce an amino acid that is not normally found in nature. In effect, they reassigned this codon to a new amino acid.

This bacterium can now make proteins that contain an amino acid that is never found in living organisms. The researchers speculate that perhaps one day their technique might be used as a weapon against viruses. Viruses make us ill by invading our cells and taking over our protein-manufacturing machinery so that we build new viruses instead of making our own proteins. Could we perhaps produce virus-resistant organisms by incorporating new genetic codes into them, that viruses would not be able to recognise and use? Indeed, these modified bacteria did prove to have increased resistance to a number of viruses, called bacteriophages, that normally infect them.

Many people worry about the introduction of genetically modified organisms into the wider environment. In fact the risk of these modified bacteria 'escaping' into the wild and causing problems is zero, because they will never be able to find the novel amino acids that they are given in the laboratory.

10.1 GENES AND CHROMOSOMES

All living organisms contain DNA. As we saw in Chapter 4, DNA codes for the production of proteins in a cell. The code works by using a sequence of bases in a DNA molecule to determine the sequence of amino acids that are strung together in the cell to build a protein molecule. The way in which a sequence of DNA bases 'stands for' a sequence of amino acids is called the **genetic code**. Amazingly, this code is almost identical in all living organisms on Earth, from bacteria to humans. The genetic code is said to be **universal.**

DNA is therefore of fundamental importance to every cell, and it must be carefully packaged and kept safe. The way in which this is done differs between eukaryotic and prokaryotic cells.

Chromosomes

In eukaryotic cells (such as our cells), DNA is found in linear **chromosomes**. A chromosome is an extremely long DNA molecule, associated with proteins called **histones**. Figure 1 shows a very small part of a chromosome. You can see that the DNA molecule coils around blocks of histones. In fact, this structure then coils again and again, so that a 'supercoiled' structure forms, in which enormous lengths of DNA become packed into a very small space. To give you an idea of how effective this supercoiling is, if you could stretch the DNA in just one of your cells out into a long, uncoiled string, it would measure about 1.8 metres.

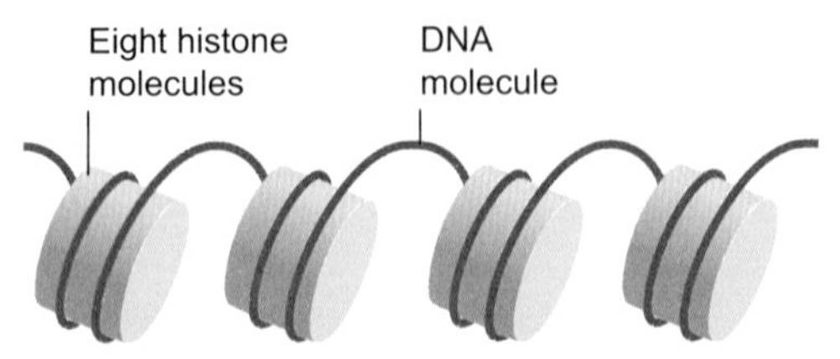

Figure 1 *DNA coils around histones to form chromosomes.*

Normally, the DNA in a cell is only partly coiled. The DNA and its associated histones are not wound into sufficiently thick coils for them to be visible as distinct chromosomes, even with an electron microscope. But, just before a cell divides, the chromosomes coil up much more tightly, becoming compact and thick enough for us to be able to see each individual one. You can even see the chromosomes in a dividing cell with a light microscope. At this stage, the DNA molecule has already undergone semi-conservative replication, and each chromosome is made up of two identical DNA molecules attached to each other at a centromere (Figure 2).

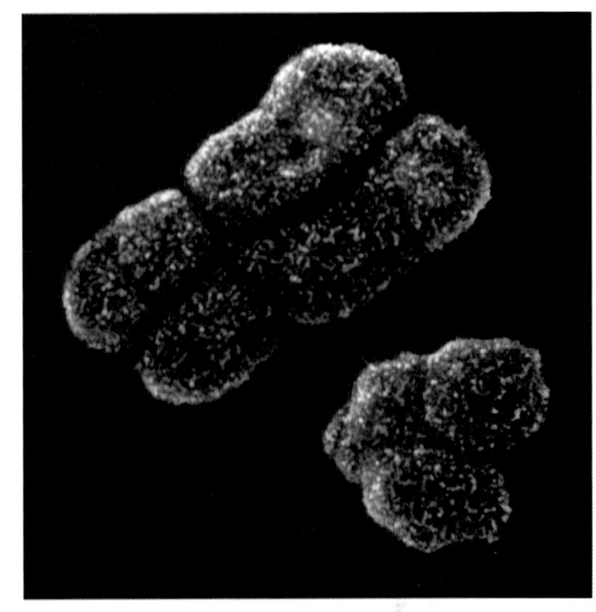

Figure 2 *This false-colour scanning electron micrograph shows the X and Y chromosomes from a cell of a man.*

DNA in prokaryotes

In prokaryotic cells, the DNA is organised differently. It still forms long molecules, but these are shorter than the DNA molecules in eukaryotic cells. Moreover, the DNA molecule is now joined into a complete circle, with no 'free ends' (Figure 3, overleaf). There are no proteins associated with it. For all of these reasons, it was originally considered incorrect to call the DNA in a bacterial cell a 'chromosome'. However, over time, scientific researchers have come to use the term '**bacterial chromosome**' to refer to these circular DNA molecules.

Many prokaryotic cells also contain one or more smaller DNA molecules, also in the form of circles. These small, 'extra' DNA molecules are called **plasmids**. Plasmids often contain genes that help a bacterium to survive in particular circumstances – for example, to give it resistance to an antibiotic. Plasmids can be transferred relatively easily between bacteria, even between species. Humans make use of this property of plasmids, employing them to transfer genes in gene technology.

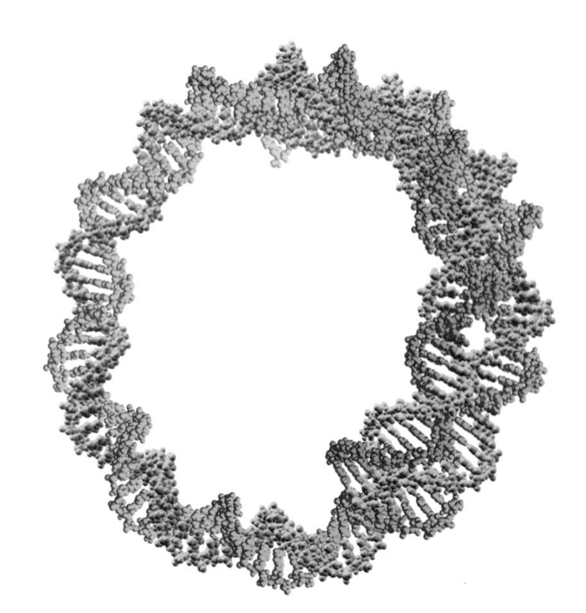

Figure 3 In prokaryotes, the DNA molecules are relatively short, circular and not associated with proteins.

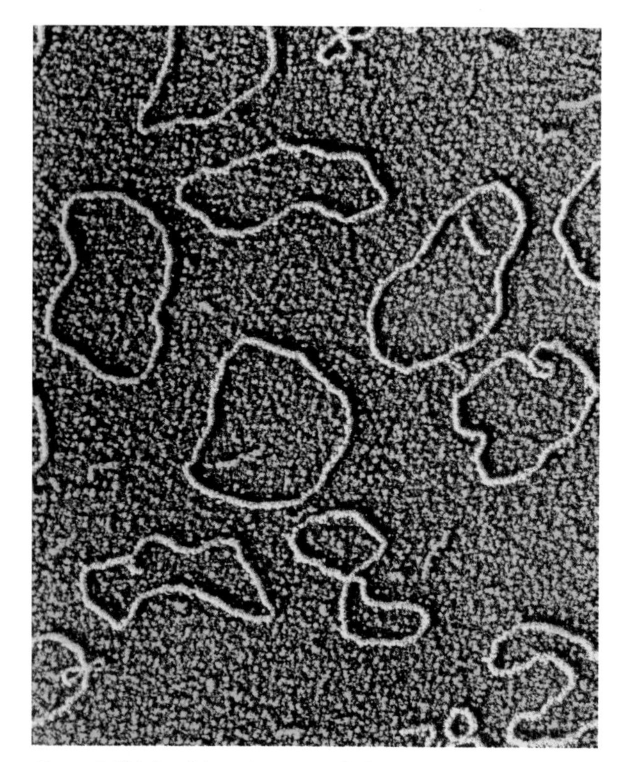

Figure 4 This is a false-colour transmission electron micrograph of plasmids from the bacterium Escherichia coli. These plasmids are often used in genetic engineering, to transfer genes from one organism to another.

If you read the information about the endosymbiont theory at the beginning of Chapter 5, you may remember that mitochondria and chloroplasts are now believed to have arisen from prokaryotes that invaded other cells, and that this is how eukaryotic cells began. Mitochondria and chloroplasts contain their own DNA, arranged in a circle and not associated with protein, just like that of prokaryotic cells. This DNA codes for some of the proteins that are used within the mitochondria and chloroplasts, but not all of them. These organelles also need proteins for which the genes are carried on the chromosomes in the nucleus of the cell.

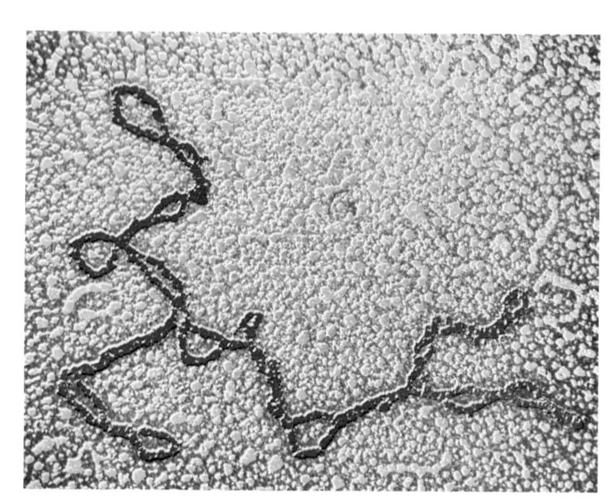

Figure 5 False-colour transmission electron micrograph of a circular DNA molecule from a mitochondrion. You can see that it is longer than the plasmids in Figure 4, but it still only codes for about 5000 amino acids, not enough to make all the proteins that the mitochondrion needs.

QUESTIONS

1. Construct a table to compare the DNA of a prokaryotic cell and a eukaryotic cell.

Genes

It is not easy to define exactly what is meant by a gene. A simple definition that worked well for a long time is that it is a length of DNA that codes for the production of one polypeptide or protein. However, we now know that, in order to produce a complete polypeptide or protein, other lengths of DNA may also be needed. All the same, it is still useful to think of a gene as a length of DNA with a base sequence that codes for the amino acid sequence of a polypeptide. As you will see later, on the way to making that polypeptide, an RNA molecule has to be made first.

We can therefore also think of a gene as a length of DNA with a base sequence that codes for a functional RNA (including ribosomal RNA and tRNAs).

In our cells, as in the cells of all other organisms, a gene for making a particular protein is always found in the same position on the same chromosome in every cell. We have 23 different chromosomes, with two copies of each one, making 46 in all. (In fact, that is not quite true, because in males the 23rd 'pair' is not a pair at all, but is made up of one X chromosome and one Y chromosome.) Our cells are diploid, meaning that they each contain two complete sets of chromosomes (Figure 6).

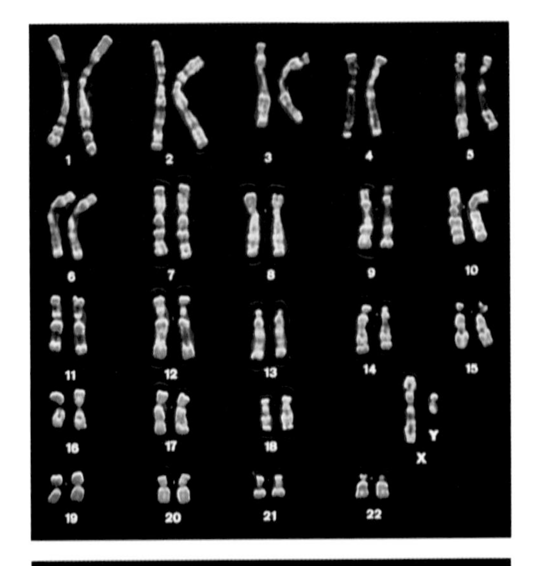

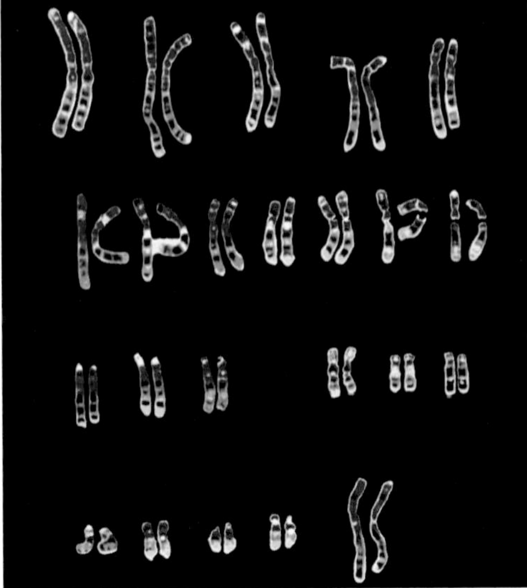

Figure 6 *These two images are micrographs of all the chromosomes in the cell of a human female and a male. First, photographs were made of the chromosomes in dividing cells. Then these photographs were digitally 'cut up' and the images of the chromosomes moved around to line them up in their matching pairs.*

QUESTIONS

2. In Figure 6, which image shows the chromosomes of a man, and which shows those of a woman? Explain your answer.

For example, the gene that determines the structure of the β chain polypeptide of the haemoglobin molecule is found on chromosome 11. The position of a gene on a chromosome is called its **locus**. There are many different loci on each chromosome – for example, the genes for the hormone insulin and the digestive enzyme pepsin are also on chromosome 11, at different loci. Genes for keratin (we have several different genes that contribute to making this important structural protein) are found on chromosomes 12 and 17.

KEY IDEAS

- In a eukaryotic cell, DNA is found as very long, linear molecules associated with proteins called histones.
- A single DNA molecule and the histones around which it is wound makes up a chromosome.
- In prokaryotic cells, there is a single, relatively short, circular DNA molecule. This is not associated with histones.
- Prokaryotic cells often have extra, even smaller, circular DNA molecules called plasmids.
- In eukaryotic cells, mitochondria and chloroplasts contain circular DNA molecules similar to those found in prokaryotic cells.
- A gene can be defined as a base sequence of DNA that codes for the amino acid sequence of a protein, or that codes for a functional RNA.
- The position of a gene on a chromosome is known as its locus. A particular gene is always found at the same locus.

Genes also occur in fixed positions on the circular DNA molecules in prokaryotic cells and in mitochondria and chloroplasts, including plasmids.

10.2 THE GENETIC CODE

Proteins are extremely important substances in living organisms. Many proteins are enzymes, controlling all of the metabolic reactions that take place in a cell. Others may be hormones, antibodies, oxygen-transporters (haemoglobin) or structural proteins (for example, collagen and keratin). The medley of proteins that a cell makes, and when it makes them, determines the structure and functions of that cell.

We have see that a protein is a polymer built up from units called amino acids. The order of amino acids in a protein (its primary structure) determines its three-dimensional structure (its tertiary structure) and therefore its function. The first protein to be sequenced was insulin, the hormone that regulates blood glucose concentration. In 1959, Frederick Sanger published the order of the 51 amino acids that make up the insulin molecule.

A triplet code

In Chapter 4, we saw that the sequence of nucleotides in a DNA molecule codes for the sequence in which a cell puts together amino acids when it builds protein molecules. There are only four different bases in a DNA molecule – A, C, T and G – but there are 20 naturally occurring amino acids. So how does the code work? A sequence of three bases, rather than a single base, codes for each different amino acid. A single-base code could code for only four amino acids. A two-base code could code for more amino acids, but a sequence of three bases is necessary to provide a code for all 20 amino acids. A sequence of three bases on a DNA molecule, coding for one amino acid, is called a **triplet**.

In fact, by using three bases (which gives 64 possible triplets), there are plenty of sequences to spare. Some triplets code for the same amino acid as other triplets, and this feature of the code is described as being **degenerate**. Some (three triplets) of the 'spare' triplet combinations of bases act as 'stop' triplets. If we think of the triplets that code for amino acids as being like words in a sentence, then the 'stop' triplets are like punctuation marks. Table 1 shows some examples of the triplets in DNA that correspond to specific amino acids (with their standard abbreviations in brackets).

The triplets in a DNA strand are read sequentially, and the code is said to be **non-overlapping** – in other words, three bases that make up one triplet code for one amino acid, and the next three code for the next one.

You will remember that a DNA molecule is made of two strands of bases wound together into a helix. Only the sequence on one of the strands is used as a code for making proteins. This strand is called the **sense strand**. We sometimes talk about 'base pairs' rather than just bases when describing the code carried by the DNA, meaning the bases on both of the strands rather than just on the sense strand – but it is important to remember that only the sequence of bases on one strand is actually used for making proteins.

Figure 7 shows a section of the gene that codes for the first four amino acids of insulin. The first triplet of bases, AAA, codes for phenylalanine, the next for valine and so on. So, this length of six bases in the DNA tells the cell: join the amino acid valine to the amino acid phenylalanine.

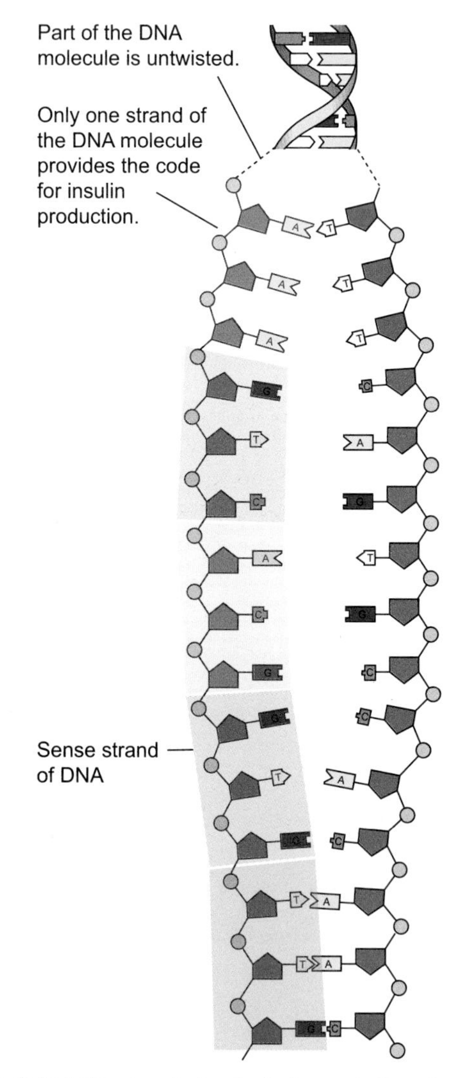

Figure 7 *Part of the gene that codes for the production of insulin*

DNA triplet	Amino acid
AAA	phenylalanine (Phe)
GTC	glutamine (Gln)
ACG	cysteine (Cys)
GTG	histidine (His)
TTG	asparagine (Asn)
GAG	leucine (Leu)
CAC	valine (Val)

Table 1 *DNA triplets and the amino acids for which they code*

QUESTIONS

3. Figure 7 shows the section of the insulin gene that codes for the first five amino acids of the insulin molecule. Using Table 1, decipher the code and list the first five amino acids.
4. The nucleotide sequence in a DNA strand is: A C G T T G G T G C A C G T G. What sequence of amino acids will this section of DNA add to a protein?

Multiple repeats and introns in eukarykotes

The complete set of genes in a cell is called its **genome**. When the complete human genome was first analysed and sequenced, at the start of the 21st century, scientists were surprised to find that only about 2% of the DNA actually codes for amino acids, which in turn make proteins. The remaining 98% of the DNA, originally known as 'junk DNA', has other functions. The roles of this 'non-coding' DNA are still being uncovered, and this research is revealing unexpected levels of complexity associated with the regulation of the activity of genes, and control over the amino acid sequences of the proteins that are synthesised. Some of this research is throwing up fresh ideas for treating diseases.

At some loci, there are multiple repeats of certain triplets of letters in the DNA sequence – for example, there may be a long sequence made up of GAT, GAT, GAT and so on. For example, this triplet of letters is repeated in the sequence of DNA at a locus on chromosome 3, where a marker called D3 is located. In different people, the number of repeats can be 12, 13, 14, 15, 16, 17, 18 or 19. Repeats like these may be part of a gene that codes for a polypeptide; if they occur in different numbers, then they make different forms, called **alleles**, of the gene. Often, however, these repeats occur in non-coding parts of the DNA, lying between gene loci.

Even within a gene, there are usually sequences of bases that do not code for amino acids that actually make up part of the polypeptide that is finally made. These sequences do code for RNA, but that RNA is not then used in the synthesis of proteins. The sections of a gene that code for amino acid sequences in the final protein are called **exons**, while those that do not are called **introns**. You will find out more about these in the next section.

KEY IDEAS

- A sequence of three DNA bases is called a triplet, and this codes for one amino acid.
- The genetic code is the same in all organisms, and is therefore said to be universal.
- More than one triplet codes for the same amino acid, so the code is said to be degenerate.
- The code is non-overlapping, meaning that the bases are read in non-overlapping sequences of three, all along a DNA molecule.
- In eukaryotes, much of the DNA does not code for amino acids. Some DNA is made up of repeated base sequences between genes. Even within genes, not all of the DNA codes for amino acids that will become part of proteins.

ASSIGNMENT 1: RECREATING DINOSAURS

(PS 1.2)

Figure A1 Amber – "the fossilised resin of prehistoric tree sap" – with an embedded insect

The novel *Jurassic Park*, by Michael Crichton, is based on the idea that DNA from dinosaurs could be used to recreate them.

In the book, the chief geneticist, Dr Wu, says they use the "Loy antibody extraction technique" to extract DNA from dinosaur bones. This, he says, recovers a 20% yield of protein – not enough to produce a dinosaur.

So, Dr Wu uses a different technique instead: he extracts DNA from red blood cells found in prehistoric insects that have been trapped and preserved in amber – "the fossilised resin of prehistoric tree sap" – for millions of years. And these insects may have once sucked the blood of dinosaurs!

Questions

A1. **a.** Explain why you would need a sample of its entire DNA in order to produce a dinosaur.

b. Wu suggests that only 20% of protein is recovered by Loy's procedure, and that this would not yield enough DNA. Explain his mistake.

c. The second method of obtaining dinosaur DNA was to extract it from the nuclei of red blood cells found in the insects preserved in amber. Explain why this would not work for mammals.

A2. In Jurassic Park, the dinosaur DNA that was extracted was placed into a crocodile egg from which the nucleus had been removed.

a. Explain why the egg nucleus would be removed.

b. Explain what the egg would provide that could make it possible for a young dinosaur to develop.

A3. The ideas in the book are not as far-fetched as they might seem. Similar techniques are being investigated to recreate more recently extinct animals – for example, the passenger pigeon. Suggest the practical difficulties that are likely to arise when trying to use these methods to recreate dinosaurs.

10.3 PROTEIN SYNTHESIS

An organism's complete set of genes is known as its genome. The complete set of proteins that are coded for by the organism's DNA is known as the **proteome**. In this section, we will see how the information on the DNA is used to make these proteins.

Ribonucleic acid (RNA)

DNA is immensely valuable – if it gets damaged, then the cell will not have correct instructions for making proteins, and is likely to die. The DNA is therefore kept remains in the nucleus. To transfer the information that it carries to the ribosomes, out in the cytoplasm, copies, or 'imprints', of the base sequence are produced.

This is done by building **messenger ribonucleic acid** molecules, almost always known as **mRNA**. They pass from the nucleus into the cytoplasm and are then used as guides to manufacture the protein encoded in their sequence of bases.

RNA molecules are well suited to their function. They use a similar four-base system to DNA, enabling the coded genetic information to be copied from DNA to mRNA. The bases are exposed on a single strand of mRNA, and this strand can be used to provide the information for assembling amino acids. The molecules of mRNA are small enough to pass through pores in the nuclear membrane. Unlike DNA, RNA molecules are quite short-lived; this enables the cell to change protein production to suit its needs.

As we saw in Chapter 4, the structure of RNA is similar to that of a single strand of DNA, except that the sugar ribose replaces deoxyribose, and the base thymine is replaced by another base called uracil. Uracil and thymine molecules are similar in size and shape, and uracil forms a complementary base pair with adenine.

Transcription

The process of using the coded information in DNA to form mRNA is called **transcription** (Figure 8). Transcription starts when an enzyme catalyses a reaction that makes the DNA of a gene untwist.

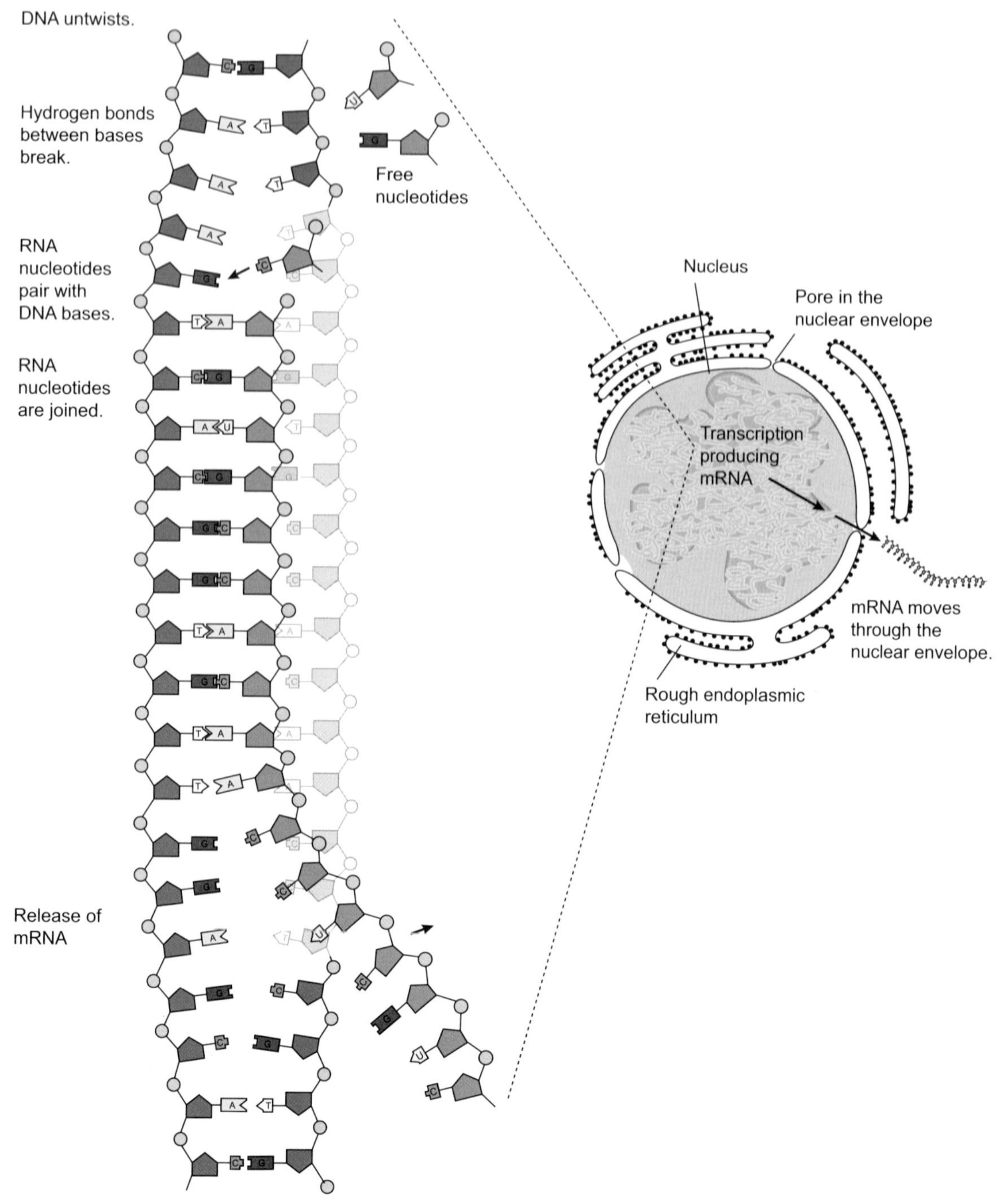

Figure 8 Transcription

The bases of RNA nucleotides then join, by complementary base pairing, with the exposed DNA bases on the sense strand. Another enzyme, **RNA polymerase**, then links the RNA nucleotides together by forming phosphodiester bonds between their ribose and phosphate groups. As the RNA polymerase moves along the sense strand, it produces a single stranded molecule of mRNA.

mRNA carries coded information in a similar way to DNA. The order of bases on an mRNA molecule is a 'mirror image' of those on the sense strand of DNA; the sequence of bases is therefore complementary to the sequence of bases in the DNA. Remember, though, that uracil bases are used in place of thymine. A sequence of three bases on a mRNA molecule, coding for one amino acid, is known as a **codon**.

The sections of the DNA sense strand and mRNA shown in Figure 8 have these bases:

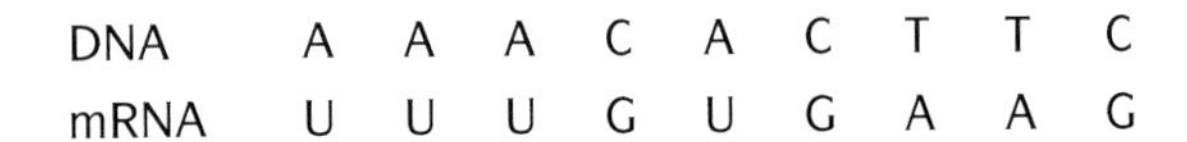

DNA	A	A	A	C	A	C	T	T	C
mRNA	U	U	U	G	U	G	A	A	G

The three mRNA codons are therefore UUU, GUG and AAG.

The mRNA detaches from the DNA and passes out of the nucleus through the nuclear pores and into the endoplasmic reticulum. The mRNA attaches to the ribosomes, which are also made of RNA. They have a specially shaped 'pocket'. The mRNA molecule fits into this pocket and the process of using the information on the mRNA to build a protein begins.

QUESTIONS

5. List four differences between molecules of DNA and mRNA.
6. What will be the order of nucleotides in the mRNA molecule produced from this section of a strand of DNA?
   A C G A T T G T G C A C G A G

Translation

Once the DNA code has been transcribed and the mRNA copies have passed out of the nucleus to the ribosomes, the code on the mRNA is used to assemble the amino acids of the protein in the correct order. This involves another type of RNA, called **transfer RNA** or **tRNA** (Figure 9).

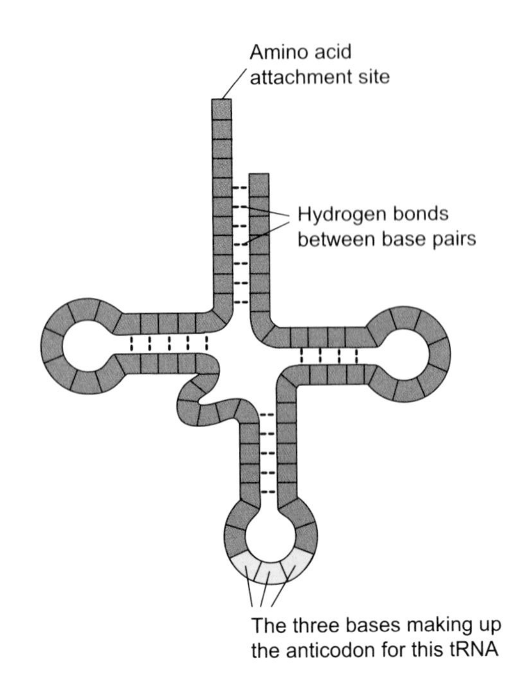

Figure 9 *A molecule of tRNA*

Every cell contains thousands of tRNA molecules. These are made of a single strand of RNA nucleotides, folded round to make three lobes, held together by hydrogen bonds between complementary base pairs. They are often described as being shaped like a clover leaf. Each tRNA has three unpaired bases, called an **anticodon**.

At the other end of the molecule is a site onto which an amino acid can be attached. There is a specific tRNA for each amino acid. Each attachment of a tRNA to an amino acid is carried out by a specific enzyme (called aminoacyl transferase); there are different versions of this enzyme for each amino acid. A tRNA with a particular anticodon will only bind with the amino acid that corresponds to that anticodon. For example, a tRNA molecule with the anticodon GAA can only attach to the amino acid phenylalanine. The loading of a tRNA molecule with its amino acid requires ATP.

The process of using the mRNA code to construct a polypeptide is called **translation** (Figure 10).

One end of the mRNA strand attaches to the ribosome. A tRNA molecule that has an anticodon with a complementary base sequence to the first

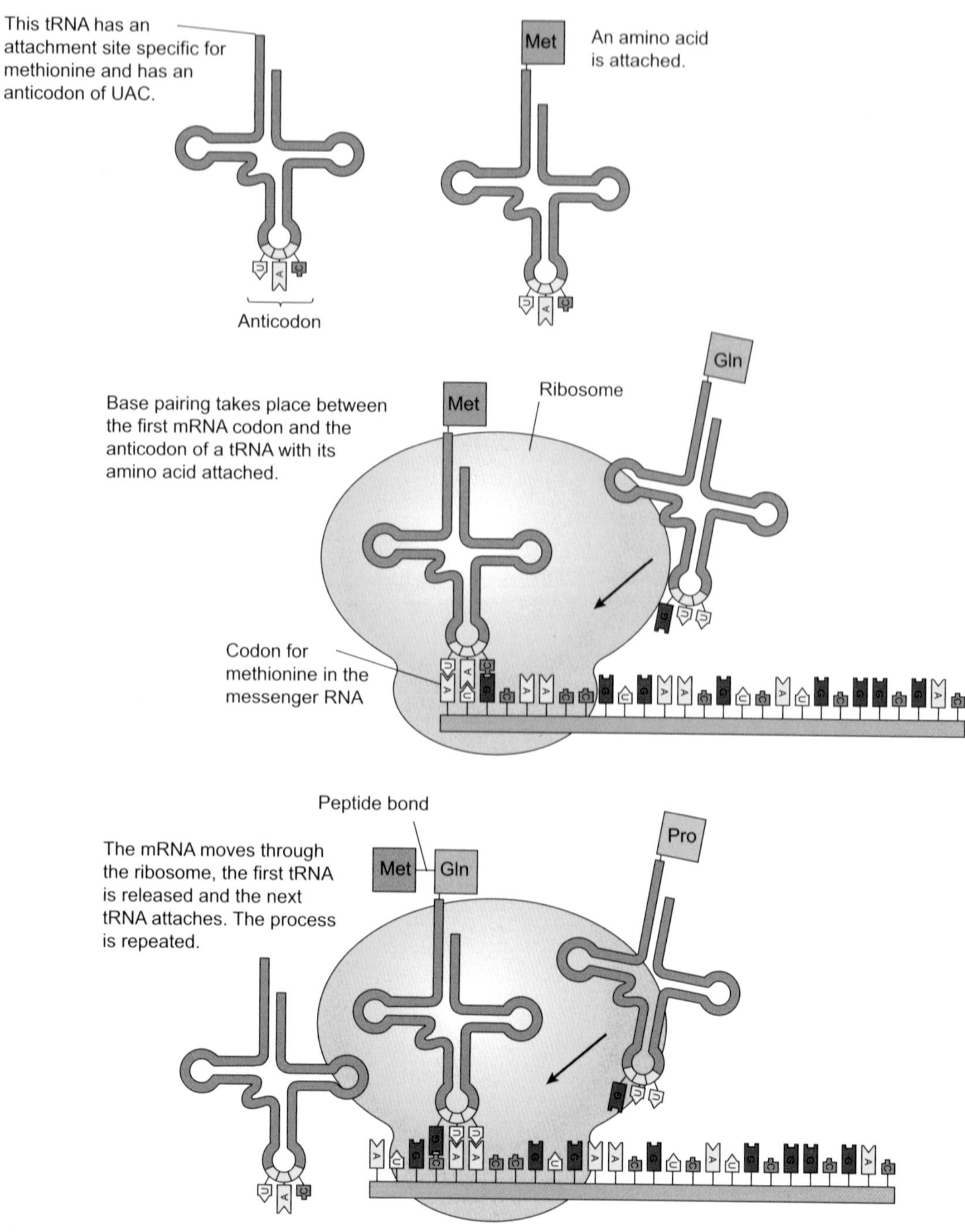

Figure 10 *Translation*

codon on the mRNA pairs with it, through hydrogen bonds between the bases. This tRNA will already have been loaded with its amino acid. A second tRNA molecule, also carrying its amino acid, then binds to the next codon on the mRNA. So, two tRNA molecules bind to the ribosome at once.

When they are both in place, the amino acids at the far end of the molecules are very close together. The amino acids are then joined together by a peptide bond. Energy from ATP is needed for this reaction to occur. The ribosome then moves along the mRNA. The first tRNA molecule, minus its amino acid, detaches and the second tRNA molecule takes its place. The next codon becomes available to bind a tRNA molecule with the next amino acid, which is then added to the growing polypeptide chain.

You can see that the order of codons on the mRNA molecule determines which tRNA molecules bind, and the tRNA molecules determine which amino acids are brought together. The whole system ensures that the amino acids are assembled in the correct sequence to make the polypeptide chain encoded by the original gene on the DNA molecule.

You can find out more about the structure and function of ribosomes in the assignment *Looking more closely at ribosomes* on page 211. Many ribosomes often work together on the same length of mRNA, each translating a particular part of it, so that multiple copies of a protein can be made simultaneously from just one strand of mRNA. A group of ribosomes working like this is called a polyribosome (Figure 11).

Figure 11 *This false colour transmission electron micrograph shows many ribosomes (blue) working on the same mRNA strand (red). The chains of amino acids they are making have been coloured green. The tRNA molecules are too small to be visible.*

QUESTIONS

7. Give two similarities and two differences between the structure of mRNA and that of tRNA.

8. Copy and complete Table 2, showing the codes at each stage of the process in assembling the first seven amino acids of an insulin molecule.

	Amino acid						
	Phe	Val	Asn	Gln	His	Leu	Cys
DNA triplet	AAA	CAC	TTG	GTC	GTG	GAG	ACG
mRNA codon	UUU	GUG					
tRNA anticodon	AAA	CAC					

Table 2

9. A polypeptide consists of 145 amino acids; 14 different amino acids are contained in its structure.

a. How many base pairs must there be in the gene that codes for this polypeptide?

b. How many nucleotides are there in the mRNA that is transcribed from this gene?

c. How many different types of tRNA are needed for the synthesis of this polypeptide?

10. Table 3 shows the complete genetic code, in terms of the amino acid coded for by each possible mRNA codon. (Note that you do not need to learn these.) To read the table, look up a codon using the first, second and third bases – for example, for the codon ACU, find the first base A (look down the first column), second base C (look along the top row) and third base U (look along the right hand column). This codon, ACU, codes for threonine.

First base	Second base				Third base
	G	A	C	U	
G	glycine	glutamic acid	alanine	valine	G
	glycine	glutamic acid	alanine	valine	A
	glycine	aspartic acid	alanine	valine	C
	glycine	aspartic acid	alanine	valine	U
A	arginine	lysine	threonine	isoleucine	G
	arginine	lysine	threonine	isoleucine	A
	serine	asparagine	threonine	isoleucine	C
	serine	asparagine	threonine	isoleucine	U
C	arginine	glutamine	proline	leucine	G
	arginine	glutamine	proline	leucine	A
	arginine	histidine	proline	leucine	C
	arginine	histidine	proline	leucine	U
U	tryptophan	stop	serine	leucine	G
	stop	stop	serine	leucine	A
	cysteine	tyrosine	serine	phenylalanine	C
	cysteine	tyrosine	serine	phenylalanine	U

Table 3

a. The genetic code is said to be degenerate. Use the information in Table 3 to explain what this means.

b. A length of mRNA has the base sequence: AAG CGC UCU GCA. What is the order of amino acids in the polypeptide it codes for?

c. Which anticodons on the tRNA molecules will attach to this section of mRNA?

d. The first stages in deciphering the genetic code involved making synthetic mRNA. The polypeptides they produced were then analysed. The researchers made mRNA in which all the bases were uracil. The polypeptide produced consisted entirely of phenylalanine. Explain why.

e. What amino acid would the polypeptide contain if the bases in the mRNA were all adenine?

f. The researchers then produced mRNA in which the bases uracil and cytosine alternated: UCUCUCUC. Which two amino acids would be present in the polypeptide synthesised, and in what proportions? Explain your answer. (Note that human genetic material is non-overlapping. This example refers to synthetic mRNA.)

RNA splicing

We have seen that, in eukaryotic cells, not all DNA codes for making proteins. Within the sequence of bases on the DNA that makes up a gene, only some of the triplets actually code for amino acids that will become part of the final protein molecule. These parts are known as exons, and the parts that do not code for amino acids are called introns (Figure 12, overleaf).

All of the gene, both exons and introns, is transcribed to form mRNA. So the mRNA contains both exons and introns. This form of mRNA is called **pre-mRNA**.

Before it leaves the nucleus, the pre-mRNA is treated to remove the introns and stick the exons together, a process known as **splicing**. The resulting mRNA is called sometimes called **mature mRNA**, or just mRNA.

Splicing of mRNA does not occur at all in prokaryotes because there are no introns. In prokaryotes, the mRNA that is produced by transcription goes straight to a ribosome and is translated in its entirety, to make a protein.

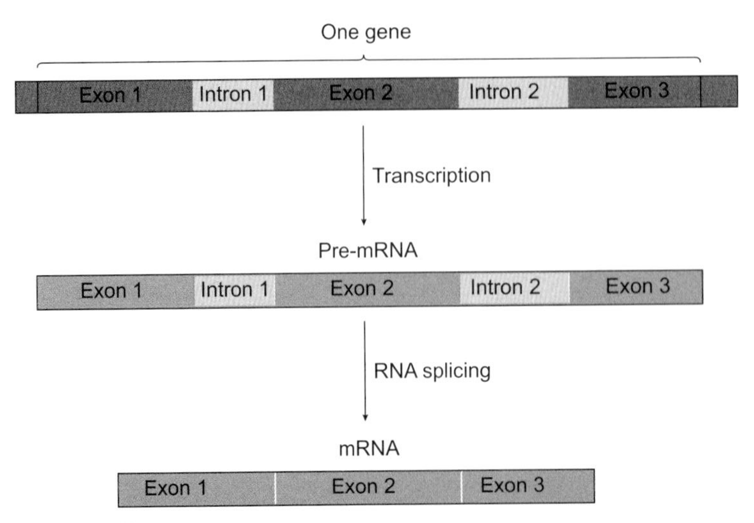

Figure 12 *Introns, exons and mRNA splicing*

ASSIGNMENT 2: LOOKING MORE CLOSELY AT RIBOSOMES

(PS 1.1, PS 1.2)

Ribosomes are a complex of about 80 protein molecules and ribosomal RNA molecules (rRNA). Eukaryote ribosomes have four rRNA strands and prokaryote ribosomes have three, making them smaller. In ribosome manufacture, the proteins enter the nucleolus and combine with the rRNA strands to create the two ribosomal subunits that will make up the completed ribosome. These subunits leave the nucleus through the nuclear pores and join together in the cytoplasm for the purpose of protein synthesis. When protein production is not being carried out, the two subunits of a ribosome are separated.

There are two types of ribosomes: those that are free in the cytoplasm and those that are bound to the endoplasmic reticulum. Ribosomes can also be found in the mitochondria and chloroplasts of eukaryotic cells. These ribosomes are much smaller and this is generally considered good evidence that mitochondria and chloroplasts evolved from symbiotic prokaryotes.

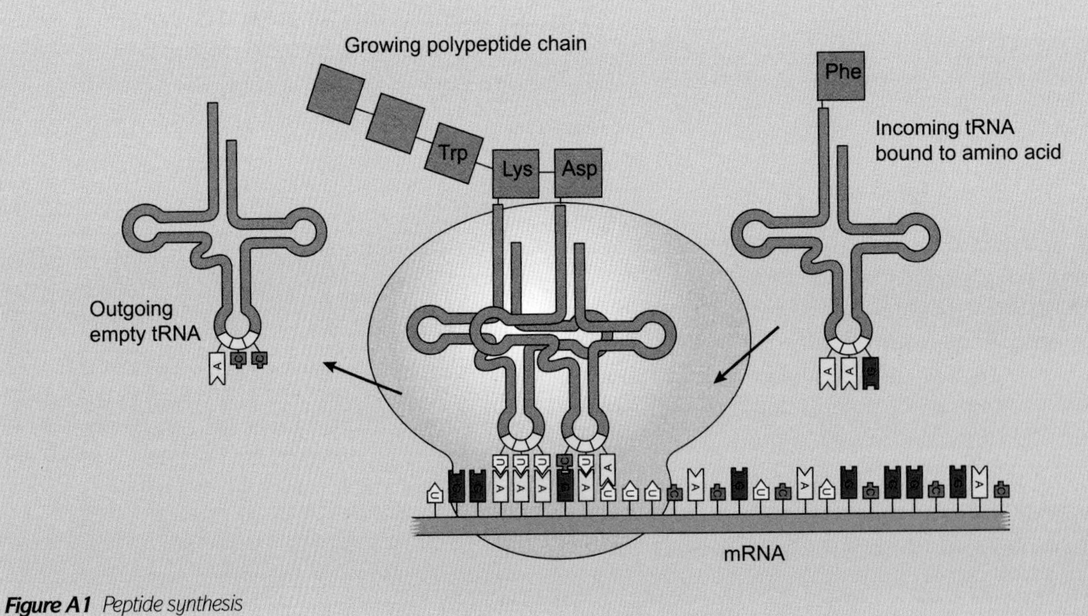

Figure A1 *Peptide synthesis*

There are three adjacent tRNA binding sites on a ribosome: the aminoacyl binding site for a tRNA molecule attached to the next amino acid in the protein, the peptidyl binding site for the central tRNA molecule containing the growing peptide chain, and an exit binding site to discharge used tRNA molecules from the ribosome.

The polypeptide chain is formed according to the codons on the mRNA, using the base pairing rule. Each tRNA molecule brings an amino acid to the ribosome and the amino acid joins the growing polypeptide chain based on the codon on the mRNA and the anticodon on the tRNA.

Questions

Stretch and challenge

A1. Suggest why more ribosomes are bound to the ER in cells which export proteins.

A2. What evidence suggests that mitochondria evolved from symbiotic bacteria?

A3. A short sequence of mRNA can be represented like this:

AUGGCCUCGAUAACGGCCACCAUG

a. What is the maximum number of amino acids which this piece of mRNA could code for?

b. How many different types of tRNA molecule would be used to produce a polypeptide from this piece of mRNA?

A4.

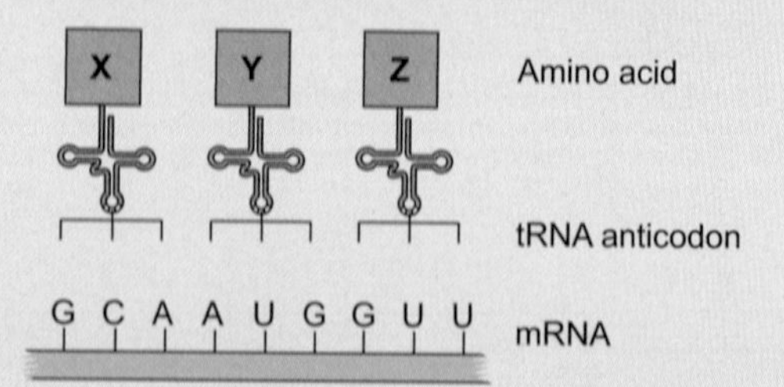

The diagram shows some molecules involved in translation.

Use this information to identify amino acids X, Y and Z:

tRNA anticodon	Amino acid
UAC	met
CAG, CAA	val
CGU, CGG, CGA	ala

KEY IDEAS

- The complete set of genes in a cell or organism is known as its genome. The complete set of proteins for which DNA in a particular cell codes is known as the proteome.
- Protein synthesis involves different types of single-stranded RNA. Messenger RNA is a linear molecule, while transfer RNA is folded to form a clover shape.
- A sequence of three bases on an mRNA molecule is known a codon. A sequence of three bases on a tRNA molecule is an anticodon.
- Protein synthesis begins with transcription of the DNA code of a complete gene to a complementary mRNA molecule, in the nucleus.
- The DNA making up a gene contains introns and exons, and these are copied onto the mRNA, which at this stage is known as pre-mRNA. The introns are then removed from the mRNA and the exons spliced together to make mature mRNA.
- The mRNA then moves out of the nucleus to a ribosome, where translation takes place.
- Two codons are exposed in the ribosome at one time.
- Each tRNA molecule is loaded with its appropriate amino acid, a process involving the use of ATP.
- The tRNA molecules with complementary anticodons to the first two mRNA codons bond with them in the ribosome. This brings their amino acids close together, and a peptide bond is formed between them. This requires energy from ATP.
- The ribosome moves one step along the mRNA, exposing the next codon ready to bond with the next tRNA.

PRACTICE QUESTIONS

1. Table Q1 shows the sequence of bases on part of the coding strand of DNA.

Base sequence on coding strand of DNA	C	G	T	T	A	C
Base sequence on mRNA						

Table Q1

a. Complete Table Q1 to show the base sequence of the mRNA transcribed from this DNA strand.

b. A piece of mRNA is 660 nucleotides long but the DNA coding strand from which it was transcribed is 870 nucleotides long.

i. Explain this difference in the number of nucleotides.

ii. What is the maximum number of amino acids in the protein translated from this piece of mRNA? Explain your answer.

c. Give **two** differences between the structure of mRNA and the structure of tRNA.

AQA January 2006 Unit 2 Question 6

2. Figure Q1 shows part of a DNA molecule.

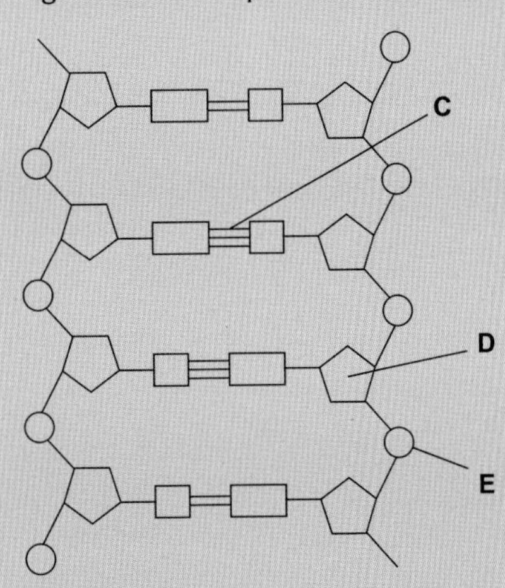

Figure Q1

a. i. DNA is a polymer. What is the evidence from the diagram that DNA is a polymer?

ii. Name the parts of the diagram labelled C, D and E.

iii. In a piece of DNA, 34% of the bases were thymine. Complete Table Q2 to show the names and percentages of the other bases.

Name of base	%
Thymine	34
	34

Table Q2

b. A polypeptide has 51 amino acids in its primary structure.

i. What is the minimum number of DNA bases required to code for the amino acids in this polypeptide?

ii. The gene for this polypeptide contains more than this number of bases. Explain why.

AQA Unit 2 June 2012 Question 5

3. a. Table Q3 shows some of the events which take place in protein synthesis.

A	tRNA molecules bring specific amino acids to the mRNA molecule
B	mRNA nucleotides join with the exposed DNA bases and form a molecule of mRNA
C	The two strands of a DNA molecule separate
D	Peptide bonds form between the amino acids
E	The mRNA molecule leaves the nucleus
F	A ribosome attaches to the mRNA molecule

Table Q3

i. Write the letters in the correct order to show the sequence of events during protein synthesis, starting with the earliest.

ii. In which part of a cell does C take place?

iii. Which of A–F are involved in translation?

b. Table Q4 shows some mRNA codons and the amino acids for which they code.

mRNA codon	Amino acid
GUU	Valine
CUU	Leucine
GCC	Alanine
AUU	Isoleucine
ACC	Threonine

Table Q4

i. A tRNA molecule has the anticodon UAA. Which amino acid does the tRNA molecule carry?

ii. Give the DNA base sequence that codes for threonine.

AQA A June 2004 Unit 2 Question 1

4. a. Figure Q2 shows the exposed bases (anticodons) of two tRNA molecules involved in the synthesis of a protein.

AGC	UUC

Figure Q2

Complete the boxes to show the sequence of bases found along the corresponding section of the coding DNA strand.

b. Describe the role of tRNA in the process of translation.

c. This is the sequence of bases in a section of DNA coding for a polypeptide of seven amino acids: **TACAAGGTCGTCTTTGTCAAG**

The polypeptide was hydrolysed. It contained four different amino acids. The number of each type obtained is shown in Table Q5.

Amino acid	Number present
Phe	2
Met	1
Lys	1
Gln	3

Table Q5

Use the base sequence above to work out the order of amino acids starting with met in the polypeptide.

AQA B January 2006 Unit 2 Question 5

11 GENETIC DIVERSITY

PRIOR KNOWLEDGE

You may remember that each organism carries two copies of each gene, which may have different forms called alleles. These alleles affect the phenotype (characteristics) of the organism. You may want to remind yourself about the structure of DNA, and how the sequence of triplets codes for the sequence of amino acids in a protein. You may also like to look back at how mitosis takes place.

LEARNING OBJECTIVES

In this chapter, we look at how genetic variation in a population of organisms provides the raw material on which natural selection can act, and how this can – over time – bring about changes in the characteristics of that population.

(Specification 3.4.3, 3.4.4)

You may remember that red blood cells contain the protein haemoglobin. This is a soluble, globular protein, made up of four interlinked polypeptide chains. There are two α chains and two β chains. Each one contains a haem group, which is capable of combining with oxygen when oxygen concentrations are high, and releasing it when oxygen concentrations are low. This allows red blood cells to pick up oxygen in the lungs, carry it in the blood to the tissues, and release it where oxygen concentrations are low, such as in respiring muscles.

There is an allele of the gene that codes for the β chain of haemoglobin that has an incorrect base in it. This results in a different amino acid being present in the chain. This amino acid makes the haemoglobin molecules less soluble, and also causes them to stick together. This happens especially when oxygen concentrations are low. The haemoglobin molecules inside the red blood cells clump together, pulling the cell into a kind of sickle shape. In this condition, the haemoglobin cannot carry oxygen and the red blood cell cannot get through capillaries. The person's tissues are starved of oxygen, and he or she also experiences pain caused by the blood cells stuck in their capillaries. The person is said to be having a sickle cell crisis. Without treatment this is often fatal.

The allele that causes this is known as the sickle cell allele, and we can use the symbol H^S to represent it. The normal allele is H^A. A person who inherits the H^S allele from both of their parents has sickle cell anaemia, and suffers crises as described previously. A person who inherits the normal allele from one parent, and the sickle cell allele from their other parent, has a mixture of two sorts of haemoglobin in the blood. Usually the person is fine, but problems may arise if he or she is doing something very strenuous, such as climbing at high altitude where oxygen levels are low. The person is said to have sickle cell trait (SCT).

Why has this harmful allele not been eliminated from the human population? The reason is that a person with one copy of the allele has a survival advantage in some parts of the world. In many tropical and subtropical countries, malaria is a common and frequently fatal disease. Many children die from this disease in sub-Saharan Africa and in tropical Asia. And it is here that we find the highest frequencies of the sickle cell allele. There is a direct connection between them. If these populations did not have a high incidence of the sickle cell allele, even more people would die. So, the sickle cell allele gives a survival advantage when malaria is present.

11.1 MUTATION

We saw in Chapter 4 that, before cell division, all of the DNA in a cell is replicated. During mitosis, chromosomes are shared out equally between the daughter cells. But, as in any complex process, errors do occur. These can result in changes to the base sequence in the DNA, or the number or structure of the chromosomes in cells. These changes are called **mutations**.

Gene mutation

The enzymes that control the replication of DNA normally only allow the 'correct' nucleotide to slot into place in the growing strand of nucleotides that builds up against the original strand. But, occasionally, there are mistakes.

For example, one nucleotide in a strand may be replaced by another. A nucleotide containing the base A may be added, where it should be one with the base T. Another possible error is that an extra nucleotide may be added, or one may be missed out completely. As a result, a change in the base sequence occurs. This is called a **gene mutation**; it affects the gene in that specific part of the DNA sequence.

There are three types of gene mutation:

- base addition – one or more extra nucleotides are inserted, so extra bases are added to the sequence
- base deletion – one or more nucleotides are removed
- base substitution – a nucleotide is replaced by a nucleotide with a different base.

What effect will each of these types of mutation have on the 'meaning' of the code carried in that gene? We can illustrate this using a sentence in which changes in some of the letters represent changes in bases.

Original – THE OLD MEN SAW THE LAD

Addition – THE COL DME NSA WTH ELA D

Deletion – THE LDM ENS AWT HEL AD

Substitution – THE OLD HEN SAW THE LAD

The effect of the mutation depends on how much the code is disrupted. Remember that the base sequence is read in sets of three – a triplet of bases. An addition of one nucleotide to that code, like the addition shown in the example sentence, changes all of the subsequent triplets. The entire sentence becomes meaningless. The same is true for a deletion. A substitution, however, changes only one triplet.

Addition and deletion mutations, therefore, tend to have serious effects on the protein for which that piece of DNA is coding, unless they happen to be very close to the end of the sequence. All the triplets after the mutation are changed, so a whole string of 'wrong' amino acids will be used to build the protein on a ribosome. The protein will have a completely different primary structure, and therefore its three-dimensional (3D) structure and its function will be totally lost or altered.

Substitution mutations, on the other hand, may have no effect at all. You may remember that several different triplets code for the same amino acid. A substitution of just one base can therefore mean that there is no change in the amino acid for which that triplet codes. There are two amino acids for which this doesn't apply – they will always be affected by a substitution mutation. Tryptophan and methionine only each have one triplet code, so any substitution of the bases in their triplets will change the amino acid. Whichever the original amino acid, if the new triplet does code for a different one, it is only that amino acid that is altered. All of the subsequent triplets remain unchanged, so all of the rest of the protein will be as it should be. There is a good chance that, even with one 'wrong' amino acid, the protein may still be able to carry out its normal function, depending on where the amino acid change occurs in the protein.

QUESTIONS

1. One strand of DNA has the following sequence of bases:

   CATCATAGATGAGAC

   a. Which type of mutation could have produced each of the following mistakes during replication of the original DNA sequence?

      CATCGTAGATGAGAC

      CATCATAAGATGAGACC

      CATCAGAGATGAGAC

      CATATAGATGAGAC

   b. Use the genetic code in the Table 3 in Chapter 10 to work out the amino acid sequence that the original code and each of the mutations would code for.

      Do not forget that the table shows the mRNA codons, not DNA.

   c. Describe the effect that each of these mutations would have on the polypeptide produced by the gene.

The consequences of gene mutation

A mutation can produce a different form of the gene. This is how all of the different alleles of a gene were originally produced.

We have seen that some mutations do not have a significant effect on the amino acid sequence and, therefore, RNA or DNA, despite changing the base sequence. However, changing the base sequence via a mutation can mean that a different polypeptide, and hence a different protein, is produced by a gene. This protein may not have the same three-dimensional shape as the original, and often does not work in the same way. It may still be functional (remember, mutations can also produce non-functional proteins) and may have slightly different characteristics to the original – such as an enzyme's optimum pH.

In animals such as ourselves, it is thought that most mutations are 'corrected' by enzymes that 'check' the new DNA molecules that are produced. Of those that escape this process, some affect cells so that they produce different antigens on their surfaces, and these are likely to be attacked and destroyed by the T cells in our immune system. So only a small proportion of the mutations that occur in our DNA actually have any effect on our cells.

However, mutated DNA does sometimes 'survive' in a cell. When the mutated DNA replicates before cell division, the new form is copied, so the mutation passes on to the new cells. But even then, this will usually have no effect at all. Imagine, for example, that a mutation occurs in the gene for hair colour or keratin production in a cell in your heart. Neither of those genes is expressed (switched on and used to make proteins) in that cell, so mutations in them have no effect whatsoever.

Even if a mutation occurs in a gene that is expressed in a cell, this still may not have much effect. For example, skin cells use a gene that codes for the production of an enzyme that allows the cells to make melanin, a substance that gives black colouring to skin, hair or feathers. If this gene mutates, then a functioning enzyme is not made and therefore no melanin. If this happens in just one cell, there will be no noticeable effect. If the cell divides, passing on its faulty gene to its daughter cells, then the animal may end up with a little patch of white on its skin, hair or feathers, like the blackbird in Figure 1. Moreover, because the mutated allele is only in its skin, not in its ovaries or testes, this blackbird cannot pass the allele on to its offspring.

If a mutation occurs in an ovary or testis as the gametes are being produced, then there is a completely different story. This new allele may become the only copy of that gene that is present in an egg cell or a sperm cell. When this egg or sperm is fertilised, the zygote that is formed contains this allele. As the zygote divides to form an embryo, every cell inherits the faulty allele. The adult organism therefore has the faulty allele in all of its cells, as in the completely albino blackbird in Figure 2. The allele will also be present in all of the cells in its ovaries and testes, and so can be passed into its gametes, and hence to its offspring.

Figure 1 *A mutation in the gene coding for melanin in a skin cell can produce small areas of white skin and feathers.*

Figure 2 *A mutation in a cell that goes on to form a gamete, and then a zygote at fertilisation, can affect all the cells in an organism.*

So, mutations that originate in ovaries or testes can be passed on to offspring, and may spread to many individuals. An allele will not usually spread into a population if its affects are too damaging. (It can on occasion; for example, it may do so if the damaging affects do not appear until after reproductive age, as in Huntingdon's disease (HD).) For example, if the mutation leads to a loss of function of a vital protein, and the mutation exists in alleles from both parents, the zygote will not develop. Even genetic mutations that do no harm may not necessarily spread into a population. The albino blackbird in Figure 2 may have a poorer chance of breeding than the other blackbirds because it looks different; also, its colour may make it more vulnerable to predators, so it may not survive to reproductive age. Its mutation isn't harmful, but it may affect the likelihood of the bird being able to pass it on to offspring.

Occasionally, depending on environmental circumstances, mutations can increase survival chances; such mutant alleles provide the genetic variation that permits natural selection and evolution. You can read more about this in section 11.3.

Chromosome mutation

Mutations can also occur in whole chromosomes, or large pieces of them. These are called **chromosome mutations**. One form of chromosome mutation is **non-disjunction**. This happens during cell division, most commonly during meiosis when homologous chromosomes fail to separate. Instead of the chromosomes being distributed evenly into the daughter cells, one cell gets an extra chromosome, leaving another with one chromosome less than it should have.

A cell missing a chromosome is unlikely to survive, but the one with the extra chromosome may. For example, sometimes non-disjunction during meiosis in a human testis or ovary produces a sperm or an egg that has two chromosome 21s, instead of one. When this gamete fuses with another, the resulting zygote has three chromosome 21s. The child that eventually develops from this zygote has three chromosome 21s in all of his or her cells. He or she will have Down's syndrome, a condition that affects mental development and also increases the risk of other health problems, for example affecting the heart.

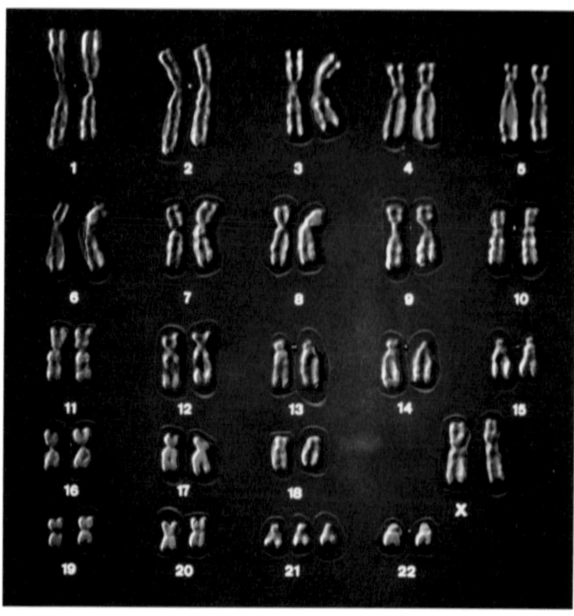

Figure 3 *The karyotype of a person with Down's syndrome, caused by non-disjunction during the formation of a gamete*

QUESTIONS

Stretch and challenge

2. In a metabolic pathway a series of reactions takes place. Each reaction is catalysed by a different enzyme. Look at the following pathway:

substance W —Enzyme A→ substance X —Enzyme B→ substance Y —Enzyme C→ substance Z

a. Substance Z acts as an inhibitor of Enzyme A. Explain how this could help to regulate the quantity of substance Z produced in the cell.

b. An addition mutation occurs in the gene that codes for enzyme B.

Explain why enzyme B produced using instructions from this mutated gene is unlikely to be able to catalyse the conversion of substance X to substance Y.

c. Predict and explain what is now likely to happen to the quantity of substance Z produced in the cell.

d. Predict and explain what is now likely to happen to the quantity of substance X produced in the cell.

e. Explain what would happen if Substance Y were then supplied to the cell.

Causes of mutation

Mutations occur spontaneously, at random. Gene mutations can happen at any time that a cell is replicating its DNA, which can be just before either mitosis or meiosis take place. Chromosome mutations can happen any time that a cell divides, especially when it divides by meiosis. It seems that both gene and chromosome mutation become a little more likely as we get older.

The risk of gene mutation is also increased with exposure to factors called **mutagenic agents**. These include ionising radiation, such as alpha, beta, gamma and X-rays, which damages DNA. Ultraviolet radiation, for example from the sun or sun lamps, damages DNA in the skin. Many different chemicals also act as mutagens. For example, cigarette smoke contains numerous mutagenic agents.

KEY IDEAS

- Gene mutations involve a change of the DNA in a chromosome.
- Gene mutations involving base substitution often have no effect, because of the degenerate nature of the genetic code.
- Base addition and base deletion are more likely to have an effect, because they change the whole set of triplets in DNA that follow them, and therefore greatly affect the sequence of amino acids in a protein that is built.
- Changes in the number of chromosomes in a cell, called chromosome mutation, can occur if the distribution of chromosomes into daughter cells during mitosis or meiosis goes wrong. This is called non-disjunction.
- Mutations that occur in body cells cannot be passed on to offspring, but mutations that occur in ovaries or testes can be passed to gametes, and therefore to offspring.
- Gene mutations and chromosome mutations can occur spontaneously. The risk is increased by mutagenic agents.

11.2 MEIOSIS

Gene mutation produces new alleles of genes, and therefore introduces new variations into cells and organisms. However, variation also arises through reshuffling gene alleles that already exist. This is done during meiosis. Meiosis maintains the chromosome number from one generation to the next in sexually reproducing organisms.

Cell division by meiosis

Humans, like other sexually reproducing organisms, inherit their genes from their parents' sex cells. Each of the cells in our body has two complete sets of chromosomes – 23 from our mother and 23 from our father, making 46 in all. Cells with two sets of chromosomes are **diploid** cells. Each chromosome has a matching partner. There are two chromosome 1s, two chromosome 2s and so on. You can see these in Figure 3. The 'matching' chromosomes are called **homologous chromosomes**.

Gametes (sex cells – such as egg cells and sperm cells) have only one set of chromosomes. They are said to be **haploid**. When the nuclei of two haploid gametes fuse together at fertilisation, they form a cell with two sets of chromosomes – a diploid zygote. It is important that one copy of each chromosome, and hence of each gene, is passed on from each parent; otherwise there would be an incomplete set of instructions in the zygote.

Meiosis is a type of nuclear division that produces haploid cells from diploid cells. In humans, it only takes place in the ovaries and testes, to produce eggs and sperm. It involves two cell divisions. Figure 4 shows how meiosis fits into the life cycle of a sexually reproducing animal, and the assignment *Investigating life cycles* on page 221 looks at its role in a more complex life cycle. Figure 5 shows a simple summary of how meiosis happens, using an imaginary diploid cell with just two chromosomes as an example.

We can use the letter n to show the number of chromosomes in a set. A haploid cell therefore has n chromosomes, and a diploid cell has $2n$ chromosomes. In humans, n is 23. Different species have different chromosome numbers; in fruit flies, for example, n is 4.

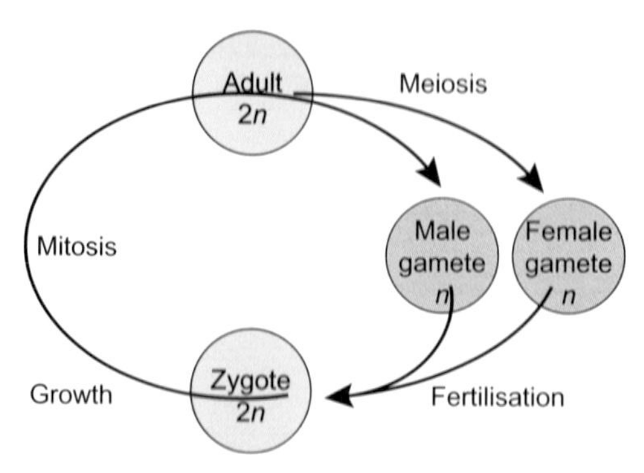

***Figure 4** The role of meiosis in the life cycle of a sexually reproducing animal*

In humans, a cell about to begin meiosis has 46 chromosomes. Just before meiosis begins, the DNA in each chromosome duplicates, forming two identical chromatids that remain attached at the centromere. So far, this is just like mitosis, but now something different happens. Each chromosome finds and pairs up with its homologous partner, so that homologous pairs of chromosomes are formed. The pair of homologous chromosomes is called a **bivalent**.

As they lie side by side, the chromatids of the homologous chromosomes make contact with one another and become wrapped around each other. As we shall see, this is very important for producing variation among the gametes that are eventually formed.

The pairs of homologous chromosomes are now separated from one other, so that one chromosome from each homologous pair goes to opposite ends of the dividing cell. The cell divides. This produces two new cells, each with a nucleus containing 23 chromosomes.

Each of these cells now divides again, exactly as happens during mitosis. The two chromatids of each chromosome are separated from one another. By the end of the second division, there are four nuclei, each with 23 chromosomes.

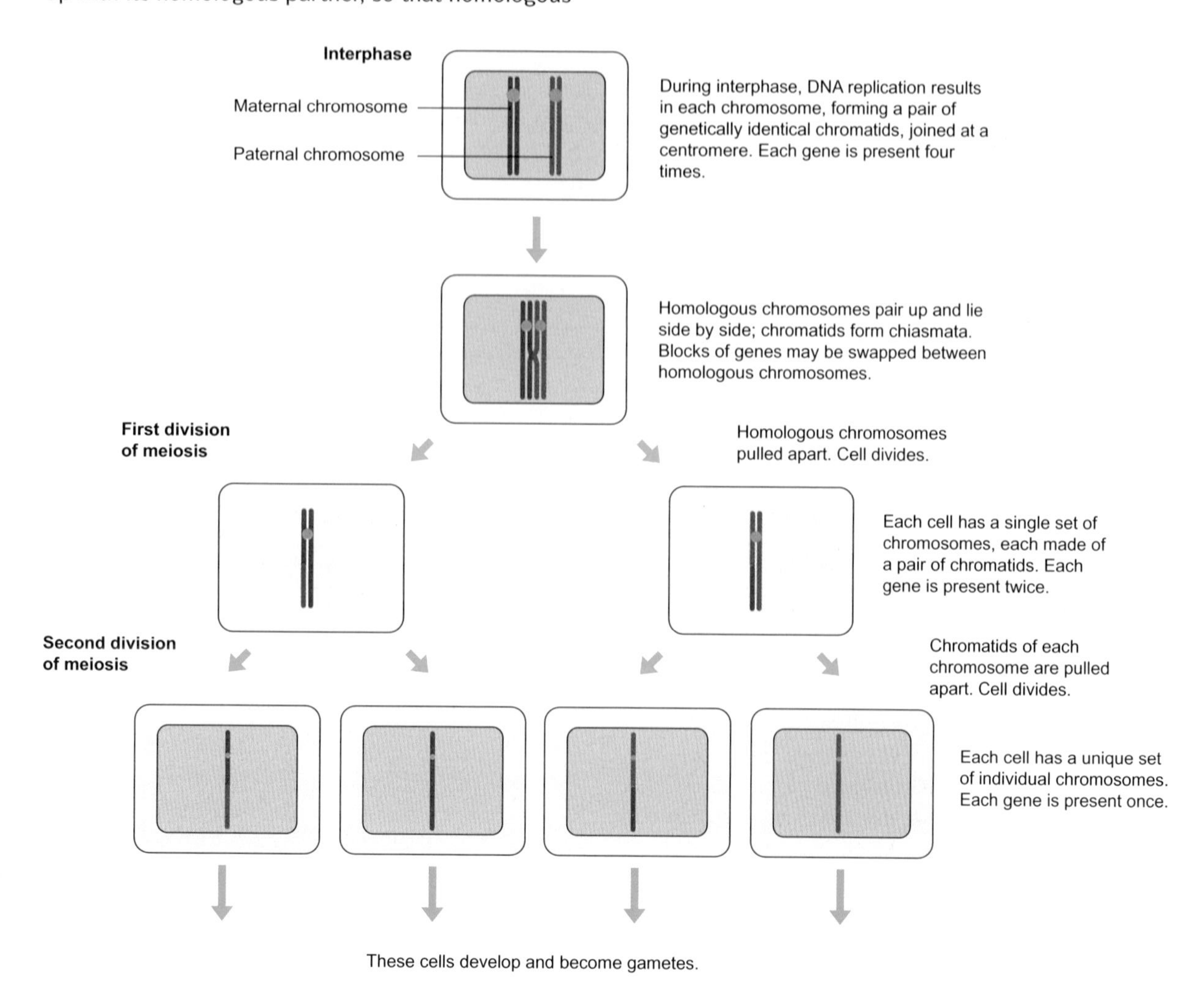

***Figure 5** An overview of meiosis*

QUESTIONS

3. a. Copy and complete Table 1 to show the differences between mitosis and meiosis.

Mitosis	Meiosis
one division	two divisions
two daughter cells produced	
daughter cells have the same number of chromosomes as the parent cell	
homologous chromosomes do not pair up	
no crossing over takes place	
daughter cells are genetically identical with the parent cell	

Table 1

b. List three similarities between mitosis and meiosis.

ASSIGNMENT 1: INVESTIGATING LIFE CYCLES

(PS 1.2)

Dinoflagellates are mainly marine protists with two flagellae. (A flagellum is a long, whip-like extension of a cell, like a very long cilium.) During summer, upwelling occurs in the ocean; this brings nutrients up from the bottom of the ocean that trigger a 'bloom' of photosynthetic dinoflagellates. There can be more than 20 million dinoflagellates per litre of water, which can make the ocean look red – this is called a 'red tide'.

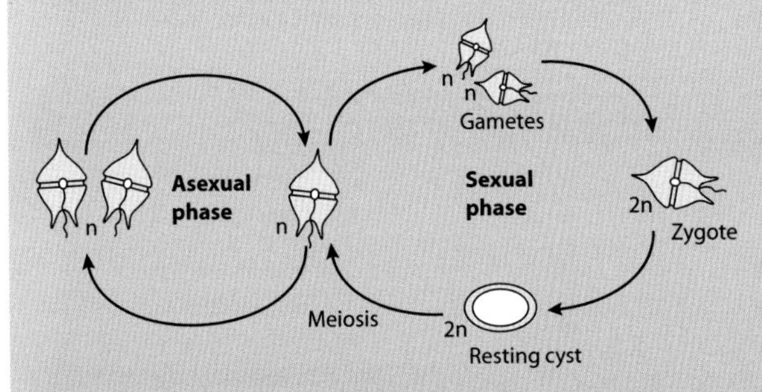

Figure A1 *A typical dinoflagellate's life cycle*

Dinoflagellates usually reproduce asexually. Mature dinoflagellates are haploid; when the sexual cycle begins, gametes are formed. When two gametes fuse, a planozygote – an actively swimming zygote – may be formed. During the zygote stage, a cyst may form under unfavourable conditions. This is a dormant capsule that protects the dinoflagellate until favourable conditions return.

Questions

A1. Is a dinoflagellate vegetative cell haploid or diploid?

A2. What process of cell division is involved in gamete formation?

A3. At what stage of the life cycle does meiosis occur?

A4. Suggest conditions that may lead to

a. gamete formation

b. encystment

Sources of variation

New combinations of alleles may be produced in three ways: independent assortment, crossing over or random fertilisation.

Independent assortment

You will remember that a zygote has two complete sets of chromosomes, one from its mother (maternal) and one from its father (paternal). This means that, in each homologous pair of chromosomes, one chromosome is derived from the mother and one from the father. During meiosis, these maternal and paternal chromosomes can be reshuffled in any combination – they are independently assorted. This process occurs at metaphase and anaphase of the first division of meiosis, because the maternal and paternal chromosomes of one homologous pair behave quite independently of all the other pairs (Figure 6).

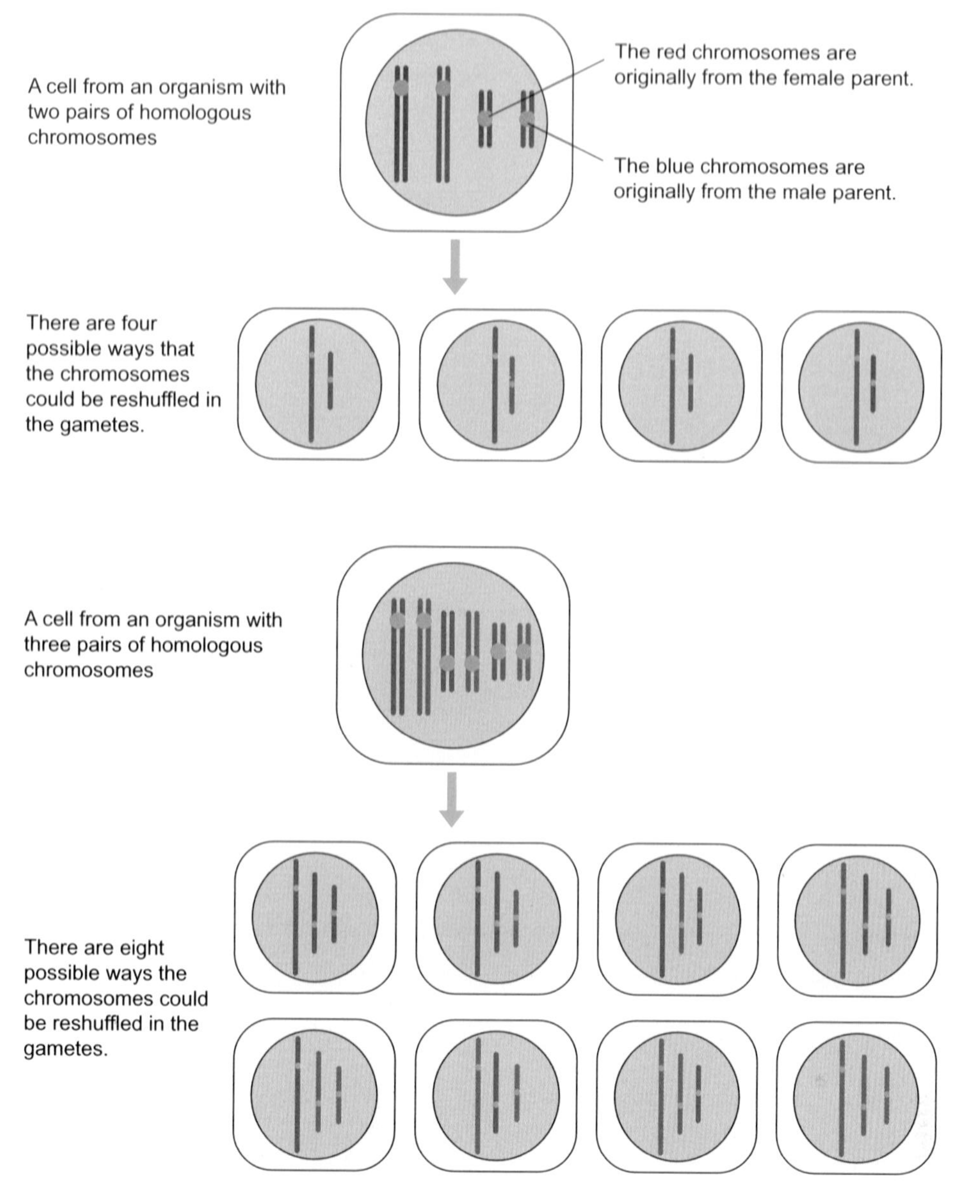

Figure 6 *Independent assortment of maternal and paternal chromosomes happens during the first division of meiosis.*

QUESTIONS

4. Use the pattern shown in Figure 7 (overleaf) to work out how many different combinations are possible if there are:
   a. four pairs
   b. 23 pairs
   of homologous chromosomes.

Crossing over

Even more variation is introduced during prophase I by a process called crossing over, as shown in Figure 7. Once a homologous pair of chromosomes has formed a bivalent, two of the chromatids from the two different chromosomes coil together like mating snakes. At the points where the chromatids cross over each other, they may break and rejoin. The positions at which non-sister chromatids appear joined to each other are called **chiasmata** (singular – chiasma). When the chromosomes separate again during anaphase, parts

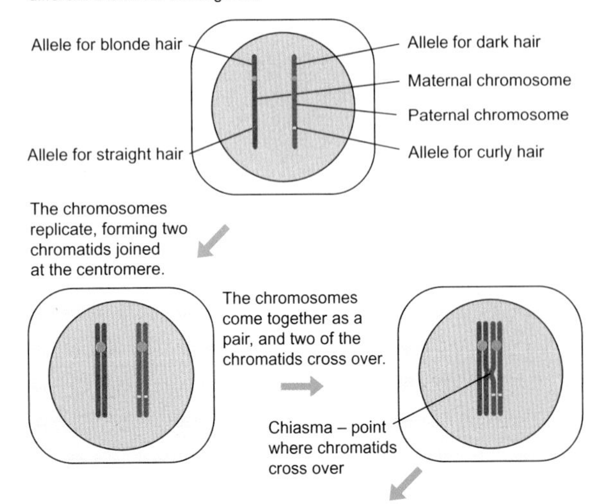

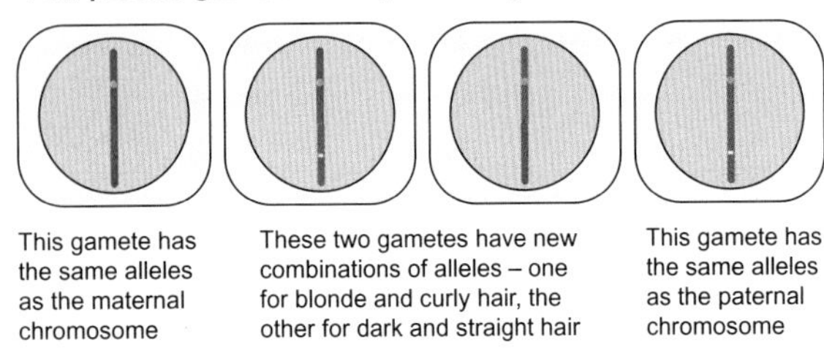

Figure 7 *Crossing over in meiosis*

of the chromatids are swapped from one chromatid to another. As a result, the new chromatids have some sections that have been copied from the maternal chromosome and some sections that have been copied from the paternal chromosome, and will therefore carry mixtures of maternal and paternal alleles. Figure 8 shows a close-up of this process.

Random fertilisation

Fertilisation is random – any female gamete can join with any male gamete. The gametes are from different individuals, each with chromosomes carrying different alleles of many of the genes. In humans there are about 30 000 genes, many of which have several alleles, so the number of possible combinations is astronomical.

QUESTIONS

5. A chromosome has three genes on it. Each gene has two alleles (A and a, B and b, C and c). Crossing over occurs at two chiasmata, as shown in Figure 8.

   Use coloured pens to draw diagrams to show the chromosomes in the gametes that would be produced.

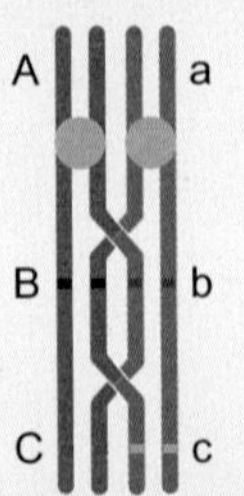

Figure 8 *How crossing over can affect the alleles on a chromosome*

KEY IDEAS

- Meiosis is a type of nuclear division that produces haploid cells from diploid cells. In some diploid organisms, such as animals, it is used in the formation of gametes.
- Meiosis has two divisions. In meiosis 1, homologous chromosomes pair up and are then separated from one another. In meiosis 2, the chromatids of each chromosome are separated.
- Meiosis produces four genetically different daughter cells and new combinations of alleles.
- During meiosis 1, pairs of homologous chromosomes behave independently of one another, so there is independent segregation of paternal and maternal chromosomes into the daughter cells.
- During meiosis 1, crossing over can occur between the paternal and maternal chromatids of homologous pairs of chromosomes.
- Random fertilisation of gametes introduces more chances of genetic variation.

11.3 NATURAL SELECTION

It is rare to find two living organisms that are identical. Even if they have exactly the same set of alleles – such as identical twins – we can usually pick out some differences between them (Figure 9). Some differences may be anatomical, and so are easily visible. Other differences may be physiological or behavioural.

Figure 9 People who know identical twins well can almost always tell them apart.

These differences between organisms of the same species are known as **variation**. We have seen that a great deal of variation is caused by organisms having different alleles of particular genes. Variation is also caused by effects of the environment. For example, a person who trains hard to be an athlete will develop much larger and stronger muscles than his or her identical twin who is not interested in sport at all.

In this section, we will concentrate on variation caused by the different combinations of alleles of genes. This different genetic make-up in the organisms in a population is called **genetic diversity**. We can define genetic diversity as the number of different alleles of genes in a population. In turn, a population can be defined as a group of organisms of the same species, that live in the same habitat at the same time and can interbreed with one another.

ASSIGNMENT 2: ANALYSING HUMAN MIGRATION

(MS 0.3, PS 1.1, PS 1.2)

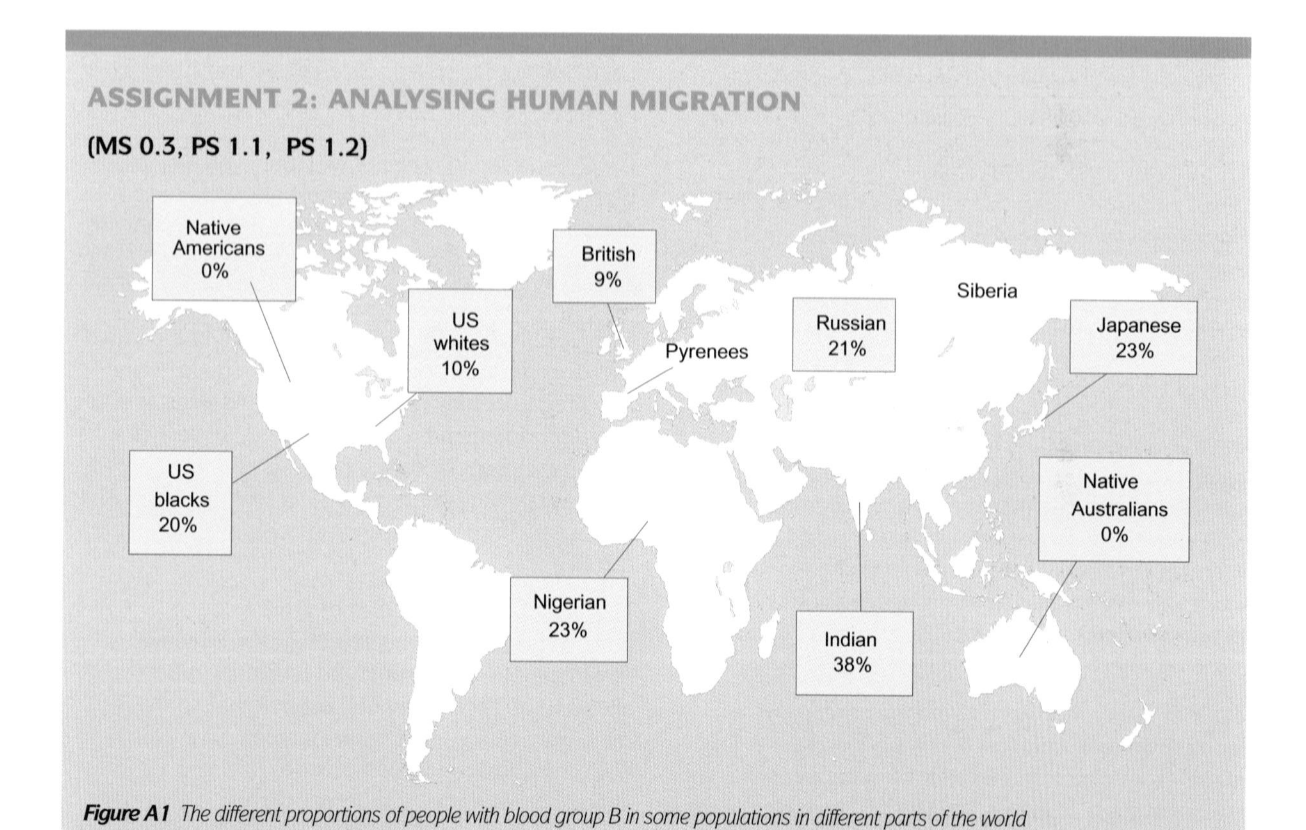

Figure A1 The different proportions of people with blood group B in some populations in different parts of the world

Human white blood cells (leucocytes) carry a set of antigens called HLA (human leucocyte antigens), which vary between individuals. Red blood cells also carry antigens, which determine blood group. All of these are determined by a person's genes.

By studying the distribution of blood groups and HLA types it is possible to build up a picture of the complex patterns of migration that have taken place in human history. It seems that the first humans originated in Africa and then spread around the world. Studies of the genes of the native (original) inhabitants of North and South America suggest that a small group of people, with a limited range of alleles of some genes, reached North America from Siberia and then gradually populated the continent. Analysis of mitochondrial DNA (mtDNA), which is inherited only from the mother, shows that all Polynesian people can trace their ancestry back to Southeast Asia. Mitochondrial DNA has also been used to prove that the ancestors of native Americans really did cross from Siberia about 13 000 years ago.

Questions

A1. It is suggested that none of the people that reached North America from Siberia carried the allele for blood group B. What evidence from the map supports this suggestion?

A2. a. Is the occurrence of blood group B correlated with skin colour? Use evidence from the map to explain your answer.

b. How might the different proportions of group B in US blacks and US whites be explained?

A3. Most West European populations have about 9 or 10% with blood group B. However, this percentage is much lower in the Basque people, who live in the Pyrenees.

a. What does this suggest about the origins of the Basque people?

b. What additional information could be used to help confirm relationships between peoples from different parts of the world?

A4. Why is mitochondrial DNA passed down through the maternal line?

The advantages of variation

Despite the wide range of variation within our species, *Homo sapiens*, people of all nationalities are quite clearly human beings. Human beings always give birth to human beings; cats always produce cats; earthworms produce earthworms; and so on.

Why, then, do so many species reproduce sexually and thereby boost the amount of variation? What are the advantages of individuals of the same species being different? Why do members of the same species not become more and more alike until they have produced the perfect specimen? Alternatively, why do members of a species not become more and more varied until they are no longer remotely similar to one another?

Figure 10 shows a theoretical model of how a group of rodents on an isolated island may change as a result of genetic variation over a period of time. This model is simple compared with what happens in a real ecosystem. However, you can see that even quite a small selective advantage or disadvantage can, over just a few generations, have a significant effect on the characteristics of the organisms in the population.

Despite the broad range of variation, you may predict that, in time, all the unfavourable alleles would disappear from the population and that the rodents would all have the favourable alleles. In practice, however, things change.

In the example in Figure 10, the climate of the island may become even colder or it may warm up. Some new predators may reach the island. As a result of the rodents feeding on the plants, there may be a change in the plant populations; for example, it may be that only very prickly shrubs or shrubs that produced fruits with a very hard shell, or poisonous flesh, were able to survive.

This may mean that alleles that were disadvantageous in one situation now become advantageous. If a population has many different alleles, then there is a good chance that some of them may help some organisms to survive even if environmental conditions change. Having a large gene pool, with lots of different alleles, is a kind of insurance policy for the future, no matter what it may bring.

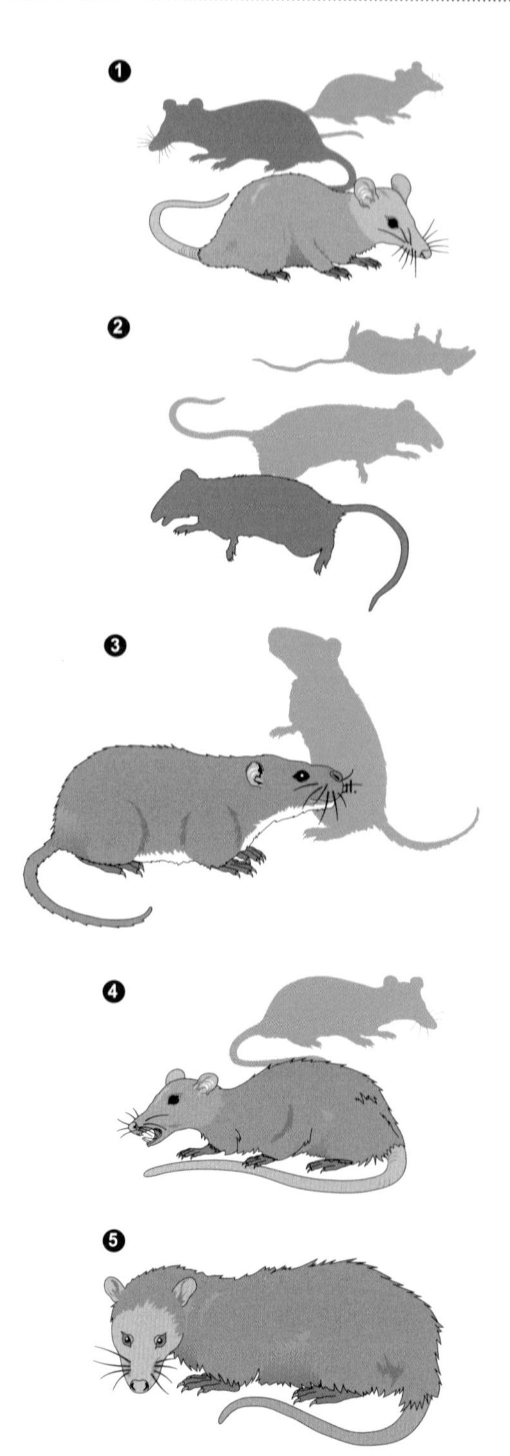

Imagine a small island on which just shrubs and grasses grow. A small population of rodents of the same species reaches the island, perhaps floating to it on tree trunks blown out to sea by a storm. The rodents feed on the seeds of the grasses and shrubs. The climate of the island is somewhat colder than the mainland from which the rodents came. There are no predators on the island.

What would you expect to happen to the size of the population of rodents during the first few years?

With no predators and plenty of food, the rodents do very well. There is, however, a limit to the number of rodents that can survive on the island. Many of the young rodents will die before they can breed.

What may cause the rodents to die?

Some may die from sheer bad luck, such as a rock falling on them. Some may inherit a pair of recessive alleles that cause a fatal disorder. As a result of variation caused by different combinations of alleles, some will be smaller than others, some will have thicker layers of fat under the skin, some will have longer fur, some may be better at finding seeds, some may be able to stretch higher to reach seeds on the shrubs, some may produce enzymes that help them to digest other parts of the plants, and so on.

Which of these features will help them to compete more successfully for food?
Which will help the rodents survive the colder climate?

The gene pool of the rodents includes two alleles for a hair length gene: long, H, and short, h. On the mainland, from where the rodents originally came, the long-haired and short-haired individuals happened to be present in exactly equal numbers. The longer hair length gives better insulation against the cold, so on the island the short phenotype becomes less favourable.

What will happen to the rodents with genes for features that favour survival?
What will happen in the population to the frequency of the alleles that favour survival?

It seems a reasonable prediction that the rodents with alleles that help them to survive are more likely to be successful. They are therefore more likely to breed and pass on these alleles to their offspring.

Figure 10 *A theoretical model of how genetic diversity may affect the survival of a population*

QUESTIONS

6. a. In the theoretical model in Figure 10, what factors could limit the size of the population of rodents that can live on the island?
   b. Suppose that each pair of rodents produces about 20 young in a year. What would happen to the population if all of these young survived and bred?
   c. If the adults live on average for only a year, how many of these young must survive if the size of the population is to stay roughly constant?

Stretch and challenge

7. The allele that gives the rodents on the island long hair is dominant, and the short hair allele is recessive. Rodents with the allele combination HH or Hh have long hair, while those with hh have short hair. The visible effect of these alleles on the rodents is called their phenotype.

Table 2 shows the percentage of short–haired rodents in the population on the island when different percentages fail to breed in each generation.

Generation	Percentage of organisms with alleles hh in the population when:		
	2 % of the hh organisms fail to breed	10 % of the hh organisms fail to breed	50 % of the hh organisms fail to breed
0	50	50	50
5	48	40	11
10	46	31	4
15	44	24	2
20	42	18	1

Table 2

a. Draw a graph to show the data in Table 2.
b. Assuming that the population stayed constant at 10 000, how many individuals would show the phenotype of the recessive allele after 20 generations at each of the three selection pressures?
c. What would happen to the frequency of the h allele in the population in each case?

How natural selection works

The situation described in the example in Figure 11 shows the main steps in the theory of natural selection.

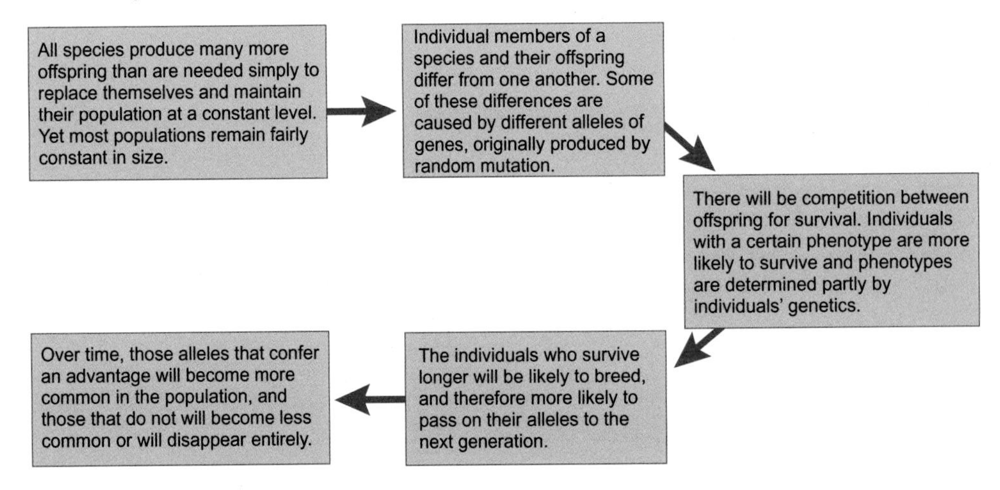

Figure 11 *The process of natural selection*

The ideas in Figure 11 form the basis of Charles Darwin's theory of natural selection. Alfred Russel Wallace independently came up with much the same hypothesis, and it was his pressure that persuaded Darwin to publish his ideas in 1858. The idea of evolution had been around for many years; the ancient Greeks had proposed it well over 2000 years before Darwin. Darwin's theory was put forward as an explanation of how evolution could have occurred. Other ideas had been suggested before, but Darwin was the first scientist to supply detailed supporting evidence for his theory.

Darwin pointed out that selection by humans of features when breeding animals can produce startling changes within a few generations. Today, the breeds of dogs that have been produced in the last 100 years or so are a good example of this (Figure 12). At the time, however, Darwin could not provide direct evidence that natural selection had produced change in a particular species, and many people challenged his explanations. Until Darwin could provide this evidence, his explanation remained a hypothesis.

Figure 12 *The selection of desired features over many generations has produced a wide diversity of different breeds of dogs.*

Evidence for natural selection

Since Darwin's time, huge amounts of evidence have been accumulated in support of his theory. One well-documented example of change involves the peppered moth, *Biston betularia*. The typical form of peppered moth has white wings speckled with black scales in an irregular pattern. Collections of moths from about 1850 show that at this time almost all peppered moths in Britain were this speckled form.

A major cause of mortality in these moths is predation by birds. The moths rest on tree trunks during the day, and rely on camouflage for protection. Speckled moths are superbly camouflaged on tree bark covered by lichens, whereas black moths are very visible. They are better camouflaged on trees with few lichens, especially if the tree trunks have soot deposits on them.

The melanic form of peppered moth has wings that are almost entirely black. A small number of specimens first appear in moth collections made after 1850, just after the industrial revolution had begun. By the end of the century nearly all the peppered moths in some areas, such as around Manchester, were melanic. Over the same period of time, sulfur dioxide killed most of the lichens (compound organisms consisting of an association of fungi and algae) living on tree trunks in industrial areas, and deposits of soot particles blackened the bark.

It is important to realise that the melanic form of the moth developed as a result of a chance mutation. There were probably always a few melanic moths around, but in times before pollution darkened the trees they would mostly have been eaten before they bred. It was because the environment became black and sooty that black moths could not be seen by predators, and so survived more successfully than speckled moths. The mutation was entirely random; the moths did not purposely change their colour (any more than you would be able to do), nor did they become black because they were covered by soot. The change in the environment simply shifted the selection

Figure 13 *The normal speckled form, and the melanic (dark) form of the peppered moth,* Biston betularia

pressure – it was now an advantage to be black rather than speckled.

We now know that wing colour in the peppered moth is controlled by a single gene, the melanic allele being dominant to the speckled allele. Studies of the distribution of the typical and melanic forms showed that by the 1950s the melanic form was by far the more common in the heavily polluted parts of Britain, whereas in western areas with very little pollution almost 100% of the peppered moths were still the typical speckled form.

QUESTIONS

8. Suggest how the theory of natural selection could explain the change from speckled to melanic forms of peppered moth in industrial areas of Britain.

Although the increase in the proportion of melanic moths correlated with the increase in pollution, this did not prove that natural selection had occurred. It was just possible that there was some other explanation, such as that a pollutant was directly causing black pigment to be produced in the moths. The first step was to check that the melanic moths really did have a selective advantage in polluted areas. A biologist called Henry Kettlewell released equal numbers of marked moths of each type – melanic and speckled – in an unpolluted wood in Dorset and in a polluted wood near Birmingham. He then compared the proportions of each type recaptured after a few days. The assumption was that the more moths had survived, the more he would recapture. The results are listed in Table 3.

Site where moths were released	Percentage of moths that were recaptured	
	Speckled	Melanic
Dorset (unpolluted)	12.5	6.3
Birmingham (polluted)	15.9	34.1

Table 3

QUESTIONS

9. Describe how the results listed in Table 3 support the hypothesis that natural selection occurred.

Next, Kettlewell attempted to establish whether the difference in survival rates really was due to differences in predation. The fact that people can see the speckled form more easily on polluted tree trunks does not prove that birds find the moths more easily. So Kettlewell placed moths of each type on tree trunks and then filmed them. He found that birds did indeed catch more of the speckled form on blackened trunks, and vice versa. The evidence strongly suggests that the melanic form has been selected in industrial areas mainly because it survives predation more successfully than the speckled form.

QUESTIONS

10. As sulfur dioxide and soot pollution are reduced, in many areas the lichens are returning and tree trunks are becoming less black. What do you predict will be the effect of natural selection on the peppered moth populations in these areas?

Worked maths example: Analysing fruit flies and understanding chi-squared

(MS 1.9, PS 3.2)

The biological species concept defines a species as members of populations that actually or potentially interbreed in nature.

In 1989, scientist Diane Dodds conducted experiments with fruit flies. She took many individuals of a single species of fruit fly (*Drosophila pseudoobscura*) and split them into two groups.

The two groups were kept isolated and fed on different foods – one group was fed starch-based food (Diet 1), while the other group was fed maltose-based food (Diet 2).

After the fruit flies had bred for four generations, 30 females from the Diet 1 group, and 29 females from the Diet 2 group, were put together with males from both groups.

Dodds's results are shown in this table, which gives the numbers of flies that mated together.

		Females	
		Diet 1	Diet 2
Males	Diet 1	22	9
	Diet 2	8	20

Of the 30 females fed on Diet 1, 22 preferred to mate with males fed on Diet 1 and only eight preferred to mate with males fed on Diet 2.

Of the 29 females fed on Diet 2, 20 preferred males fed on Diet 2 and only nine preferred males fed on Diet 1. Dodds suggested that this could be the beginning of speciation.

Data similar to that of Dodds's can be analysed using a statistical test called the chi-squared test:

The chi-squared test

Chi-squared is a statistical test that can be used to determine whether or not the results we have obtained are significantly different from those we would expect if our hypothesis is correct. Note: You do need to remember the equation for chi-squared, but you may be tested on your ability to select and use statistical tests.

In order to be able to use the chi-squared test, the data must be in categories. We then must generate a null hypothesis, which states that there is no difference in the proportions in the different categories. We do not have to believe in this hypothesis – it is just something that we use in order to carry out the test.

Here is a worked example, using **observed data** on eye colour in boys and girls.

	Blue eyes	Brown eyes	Total
Girls	67	41	108
Boys	34	73	107
Total	101	114	215

First, we construct the null hypothesis. This would be:

There is no difference in the proportion of boys and girls with blue or brown eyes.

Next, we use this hypothesis to calculate the expected numbers for each cell, using

$$\frac{\textit{column total} \times \textit{row total}}{\textit{grand total}}$$

So, for blue-eyed girls, the expected number is

$$\frac{101 \times 108}{215}$$

$$= 50.73$$

where 101 is the total of the 'Blue eyes' column, 108 is the total of the 'Girls' column, and 215 is the grand total (all columns and rows).

We then do the same calculation for blue-eyed boys, brown-eyed girls and brown-eyed boys. This can be displayed in a new table:

	Observed blue eyes	Expected blue eyes	Observed brown eyes	Expected brown eyes	Total
Girls	67	50.73	41	57.27	108
Boys	34	50.27	73	56.73	107
Total	101	101	114	114	215

Chi-squared can now be calculated as:

$$\Sigma \frac{(O - E)^2}{E}$$

where:

O is the observed result and E is the expected result.

So, in our example:

$$\textit{chi-squared} = \frac{(67 - 50.73)^2}{50.73} + \frac{(41 - 57.27)^2}{57.57}$$

$$\frac{(34 - 50.27)^2}{50.27} + \frac{(73 - 56.73)^2}{56.73}$$

$$= 5.22 + 4.62 + 5.27 + 4.67$$

$$= 19.78$$

(Note that squaring the difference between 'O' and 'E' gets rid of any negative numbers.) The next step is to work out the degrees of freedom. This is equal to the number of columns in the original table minus one, multiplied by the number of rows in the table minus one.

In this table, there are two rows of raw data (one of boys' and one of girls' eye colour) and two columns into which the colours are separated (one for blue eyes and one for brown eyes). Therefore, the number of degrees of freedom is:

$$Df = (2 - 1) \times (2 - 1) = 1$$

This can now be compared with the critical values on a chi-squared table:

	Probability			
Degrees of freedom	0.95	0.10	0.05	0.01
1	0.004	2.706	3.841	6.635
2	0.103	4.605	5.991	9.210
3	0.352	6.251	7.815	11.345
4	0.711	7.779	9.488	13.277

If the chi-squared value is greater than or equal to the critical value: There is a significant difference between our observed results and our expected results. That is, the difference between actual data and the expected data is probably too great to be attributed to chance. So, we conclude that our sample supports the hypothesis of a difference, and reject the null hypothesis. We usually look at the value at a probability of $p < 0.05$.

If the chi-squared value is less than the critical value: There is no significant difference between the observed and expected data, and the difference is likely to be due to chance. So, we conclude that our sample does not support the hypothesis of a difference and we accept the null hypothesis.

Comparing the value of 19.78 with the critical values on the chi-squared table for 1 degree of freedom we can see that it is greater than 6.635, so the probability of this being due to chance is less than 0.01.

Directional selection

The increase in the frequency of the melanic allele in the peppered moth population, during the industrial revolution, is an example of **directional selection**. A change in the environment changed the selection pressures on the population. Instead of the speckled form having a better chance of survival, the melanic form was the better adapted. Directional selection shifts the selection pressure towards one end of a variation range for a characteristic by favouring more extreme phenotypes over others.

Another example of directional selection is the evolution of antibiotic resistance in bacteria. Many diseases that were once thought to be under control are now making a reappearance. This is because the pathogenic bacteria have evolved genetic changes that allow them to survive antibiotic attack. The widespread and sometimes indiscriminate use of antibiotics for 40–50 years has provided a selection pressure that has favoured the survival of resistant bacteria.

Not all bacteria are susceptible to all antibiotics. For example, it is pointless to use penicillin to treat an infection caused by bacteria that have a capsule, as the capsule prevents penicillin from entering the cell. There are a number of different mechanisms by which bacteria are able to resist the action of antibiotics (Figure 14; note that you do not need to memorise these mechanisms for your examination).

The most common mutations change the binding site for an antibiotic (for example, the ribosomes) so that

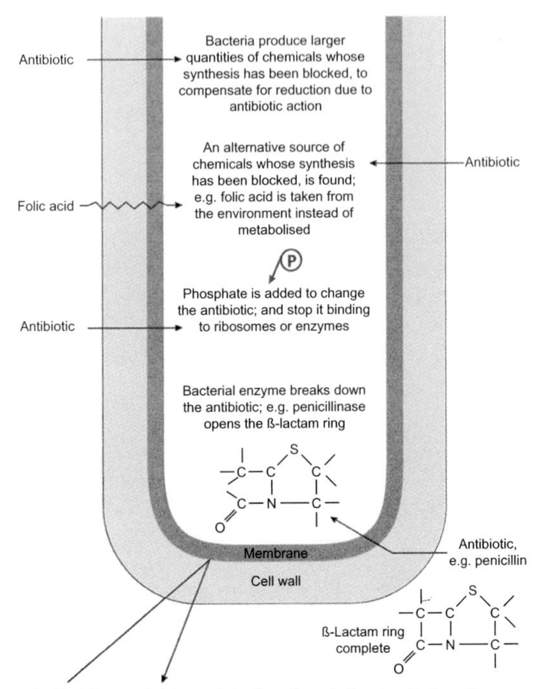

Figure 14 *Mechanisms of bacterial resistance to antibiotics*

the antibiotic cannot bind. In a host that is being treated with an antibiotic, these surviving mutants are at a selective advantage because non-mutated bacteria are inhibited or killed by the antibiotic. The mutants can then grow and reproduce with much less competition for nutrients.

It is important to remember that antibiotics do not *cause* mutations for resistance to happen in bacteria. Spontaneous, random mutation occurs in bacteria just as in other organisms, and this produces new alleles of genes. Some of these may produce an enzyme or other protein that gives an individual bacterium resistance to an antibiotic. If that bacterium happens to be in an environment where that antibiotic is being used, then it will survive while the other members of the population are killed.

Hospitals are a focus for bacteria. People going in for treatment often have infectious diseases, and take the bacteria into hospital with them. Sometimes these bacteria include resistant strains that are at a selective advantage in hospitals because the generally high usage of antibiotics allows the resistant strains to multiply as the rest fall victim to the effects of the drugs. Today, a major reservoir of resistant strains is found in hospitals. This reservoir of resistant bacteria carries a pool of resistance genes, many of which are in their plasmids. (You may remember that plasmids are small circles of DNA, separate from the main DNA molecule in a bacterium.) A single plasmid can contain genes for resistance to more than one antibiotic. Bacteria often swap plasmids with one another, even with members of different species, so this means that resistance to several antibiotics can spread rapidly through the bacterial populations. Pathogens with multiple resistance have caused difficult-to-treat infections called super-infections.

Where resistance seems likely, doctors now use combinations of two or more antibiotics to reduce the possibility of resistant bacteria surviving. Before antibiotics, tuberculosis (TB) was a serious disease for which there was no cure. Then antibiotics brought it under control and patients recovered. But the bacteria that cause TB have now evolved resistance and TB has re-emerged as a serious infection.

QUESTIONS

11. In intensive farming of cattle, sheep and hens, antibiotics have sometimes been routinely given to all of the animals in order to prevent infectious diseases, even when these diseases are not present. Explain how this use of antibiotics can increase the chance of the appearance of new resistant strains of bacteria that can infect humans.

Stabilising selection

Most of the time, natural selection does not cause change. If the environment does not change, and if the population is already well adapted to it, then selection tends to keep things as they are. This is called **stabilising selection**. In Devon and Cornwall, for example, where there was no air pollution during the industrial revolution, the peppered moth population remained almost entirely of the speckled type, with hardly any melanic moths.

Human birth weights are a good example of stabilising selection. Birth weight is affected by many different genes, and is therefore said to be a **polygenic** characteristic. (Birth weight is also affected by the environment, but for now we will consider only the genetic component.) Moreover, each of these genes has many alleles. This means that there is a wide range of possible birth weights, with a baby having any possible weight between the two extremes of the range (Figure 15).

Figure 16 also shows the mortality rates of babies with different birth weights. (Note that, in the UK, the actual numbers of babies that die is very small.) You can see that the highest mortalities are at the extreme lower end of the range. Mortality is lowest in the middle of the range, and increases again as birth weight gets greater. This means that there is selection against babies with birth weights that are lower or greater than the middle of the range. This means that babies who have inherited alleles that give them a 'normal' birth weight are more likely to survive, and they will then have a chance of passing these alleles on to their babies, in turn. Over time, there is no change in the pattern of birth weights in a population.

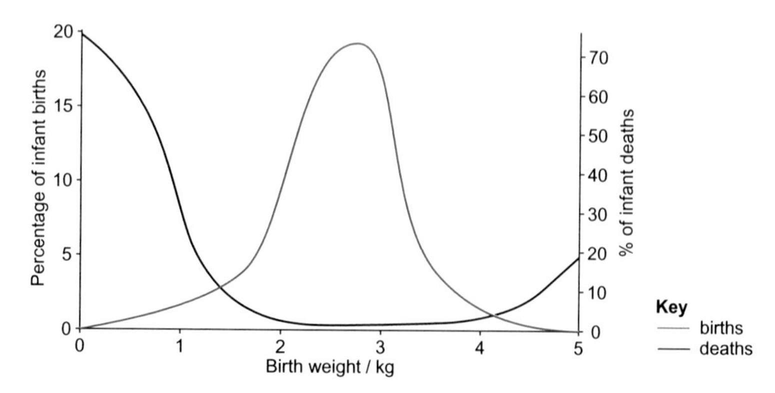

Figure 15 The range of human birth weights

REQUIRED PRACTICAL ACTIVITY 6: APPARATUS AND TECHNIQUES

(MS 2.5, PS 4.1, AT c, AT i)

Use of aseptic techniques to investigate the effect of antimicrobial substances on microbial growth

This practical activity gives you the opportunity to show that you can:

- Use laboratory glassware apparatus for a variety of experimental techniques including serial dilutions
- Use microbial aseptic techniques including the use of agar plates and broth

The use of correct aseptic technique is essential for all microbiology investigations to prevent contamination and to maintain health and safety.

Health and safety

- Always ensure you wash your hands thoroughly after any microbiology practical.
- Your work area should be swabbed with a commercial disinfectant, such as Virkon®, or with ethanol (highly flammable) before and after any microbiology work.
- You should only use commercially supplied cultures of microbes for investigations. If you collect microbes from the environment, there is always the risk that you will culture something that is pathogenic.
- You should not open agar plates once they have been incubated even if the culture has been commercially supplied as there may be contamination. All inoculated plates should be labelled and dated immediately before or after inoculation, taped with three or four pieces of tape, and autoclaved after use.
- All equipment that has been used should either be sterilised or disposed of safely.

Apparatus

Some of the equipment used in microbiology will be familiar to you, but other equipment is quite unique. For example:

- Inoculating loops are used to transfer microbes from a culture bottle and produce a streak plate.
- Mounted needles are used when culturing fungus as it is often easier to pick up a small piece of agar plus hyphae.
- Glass or metal spreaders are used to produce a lawn plate.
- Graduated pipettes or Pasteur pipettes are used to transfer cultures, sterile media and sterile solutions.

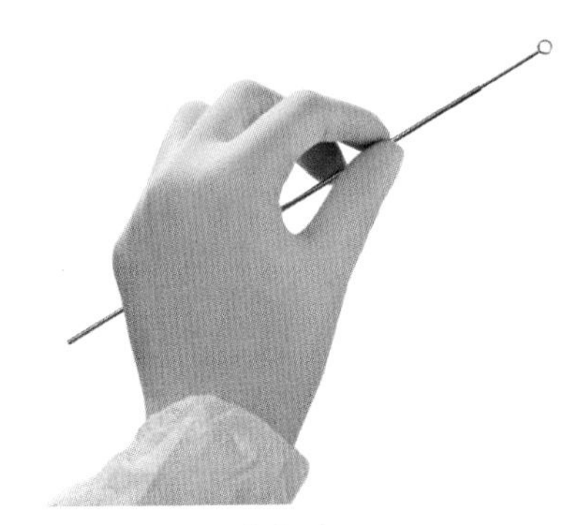

Figure P1 *An inoculating loop*

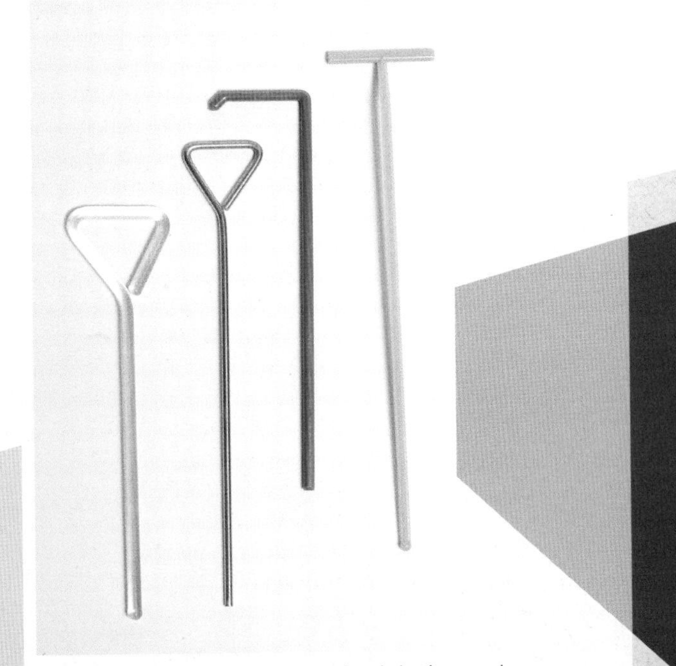

Figure P2 *A selection of glass, metal and plastic spreaders*

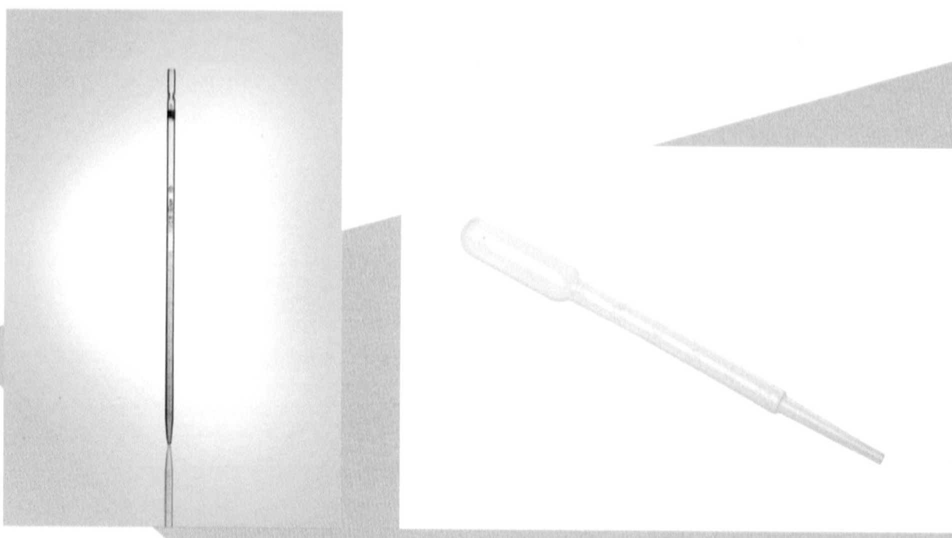

Figure P3 A graduated pipette (left) and a Pasteur pipette (right)

Techniques

Sterilising equipment

To flame a wire loop or mounted needle, you should put the metal into the light blue cone of a Bunsen burner flame and draw it slowly into the hottest region – immediately above the blue cone. Hold there until it is red hot and then allow it to cool for a few seconds and use immediately. Re-sterilise the loop immediately after use.

A glass rod is usually sterilised by putting it in ethanal (highly flammable), as a Bunsen burner flame could damage it.

Pastuer pipettes are usually disposable plastic and are provided already sterile in sealed packets. Glass pipettes can be sterilised by heating, or with alcohol.

The necks of glass bottles and test tubes can also be flamed.

This ensures that no microorganisms enter to contaminate the culture or the medium. It produces a convection current away from the opening by passing the neck forwards and back through a hot Bunsen burner flame. It is usual to use your free hand to remove the top of the tube or bottle, by turning it with your little finger.

Transferring cultures

This should be done as quickly as possible, with tubes and plates open to the air for the minimum length of time to reduce the chance of contamination.

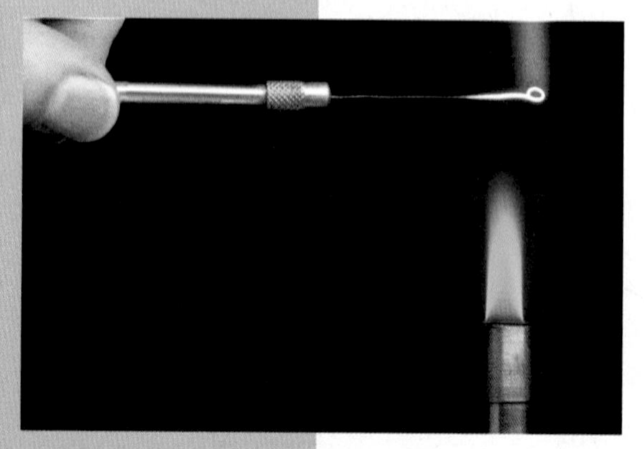

Figure P4 Flaming an inoculating loop

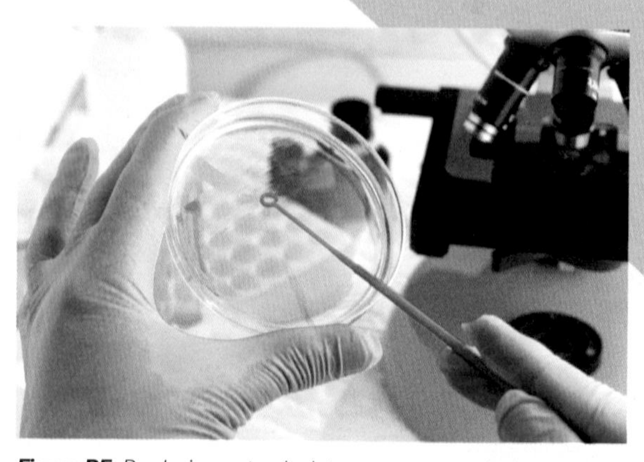

Figure P5 Producing a streak plate

The neck of a tube or bottle must be immediately warmed by flaming in a Bunsen burner flame to kill any microbes around the opening.

A sterile inoculating loop can be used to transfer microbes from a culture bottle and produce a streak plate. This is generally used when you wish to identify microbes or to produce a colony for sub-culturing.

A lawn plate is commonly used when investigating the effect of substances on the growth of bacteria. A small amount of culture broth is put onto the agar plate either directly from the bottle or using a pipette. This broth is then spread evenly over the top of the agar plate to produce a uniform covering of bacteria.

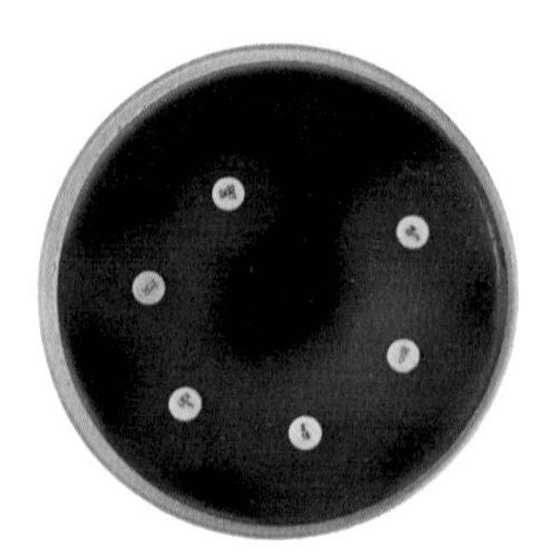

Figure P6 *An antibiogram showing the effectiveness of antimicrobial substances on* Streptococcus pneumoniae *bacteria. The larger the clear areas around the disc, the more effective the substance.*

Using antimicrobial substances

You could investigate a wide range of antimicrobial substances including mouthwashes, toothpastes, hand washes, disinfectants or antibiotics.

Antibiotics are normally investigated using commercially produced discs called multidiscs. These are a series of linked discs impregnated with different antibiotics. They can be used to compare the effectiveness of different antibiotics on a given microbe.

Other antimicrobial substances can be investigated by soaking small filter paper discs in the substance. A common investigation is to look at the minimum concentration needed to kill bacteria.

Making a serial dilution

In microbiology, a serial dilution is often used to either dilute the concentrations of microbes in a broth, or to dilute the concentration of an antimicrobial substance.

To make a serial dilution:

- Add 1 cm^3 of solution A (the stock solution) to 9 cm^3 distilled water. This produces solution B, a dilute solution with a 1 in 10 dilution (in other words, there is 1 cm^3 of stock solution in 10 cm^3 of solution B).
- You can then further dilute this by adding 1 cm^3 of solution B to 9 cm^3 of distilled water. This produces solution C, a dilute solution with a 1 in 100 dilution.
- You can continue to dilute in this way to produce 1 in 1000, 1 in 10 000, 1 in 100 000 and so on.

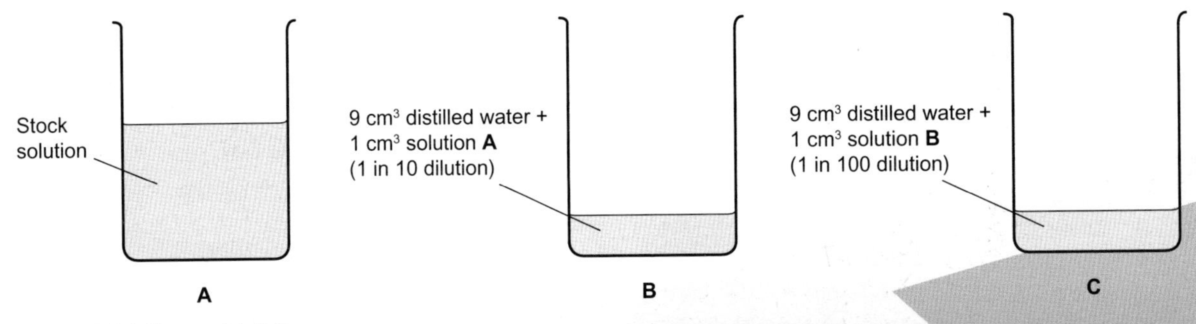

Figure P7 *Making a serial dilution*

Incubating your culture

All Petri dishes should be closed with three or four strips of tape and labelled on the underside with the date, the microbe used, your initials or name and other relevant information.

Cultures should be incubated with plates inverted at a temperature of 25 °C and not opened after incubation.

QUESTIONS

P1. Would you produce a lawn plate or a streak plate to investigate the effect of different antibiotics on a bacterium?

P2. Why must all equipment be sterilised before use?

P3. How would you make a dilution of 1:10 000 using solution C?

P4. Why is a culture incubated at 25 °C and not at 37 °C?

KEY IDEAS

- Genetic diversity is the number of different alleles of genes in a population.
- Genetic diversity provides variation on which natural selection can act.
- New alleles are produced by random mutation of genes. Most mutations are harmful, but some may be beneficial.
- Organisms with phenotypes (features) that give them a better chance of survival are more likely to reproduce and pass on their alleles to their offspring.
- Over many generations, the advantageous alleles will increase in the population.
- Where new mutations occur that provide a new advantage, or where environmental conditions change, natural selection may produce a directional change in a population; this is directional selection and it results in a change in the frequency of alleles in the population.
- Where environmental conditions remain constant, natural selection tends to keep things much the same; this is stabilising selection.

PRACTICE QUESTIONS

1. Figure Q1 shows the life cycle of an alga.

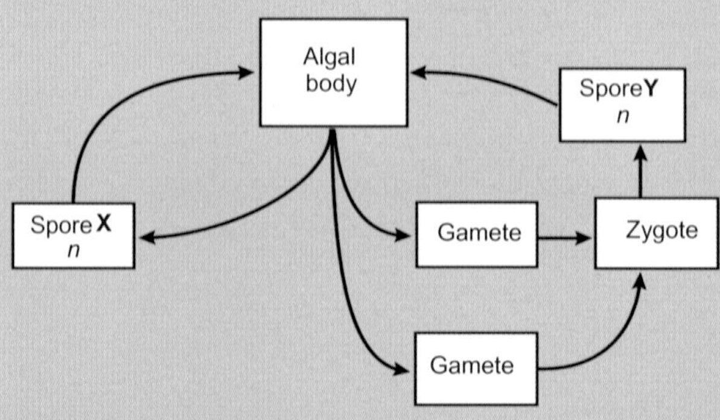

Figure Q1

   a. Copy the diagram, and then on the diagram
      i. mark with an H a haploid cell
      ii. mark with a D a diploid cell
      iii. mark with an M where meiosis occurs.
   b. i. The algal body grown from spore X will be genetically identical to its parent. Explain why.
      ii. The algal body grown from spore Y may be different from its parent. Explain why.

2. Two pairs of alleles – A and a, and B and b – are found on one pair of homologous chromosomes. A person has the genotype AaBb. Figure Q2 shows the chromosomes at an early stage of meiosis.

   The position of two of the alleles is shown.

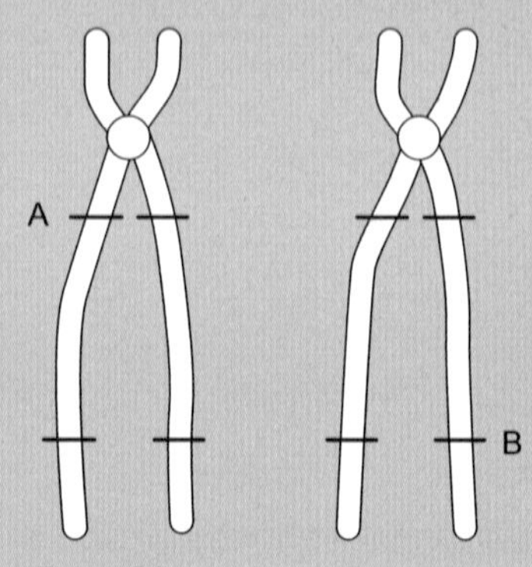

Figure Q2

   a. Copy and complete Figure Q2 to show the alleles present at the other marked positions.

b. Crossing over occurs as shown in Figure Q3. From Figure Q3, give the genotypes of the gametes produced containing the chromatids

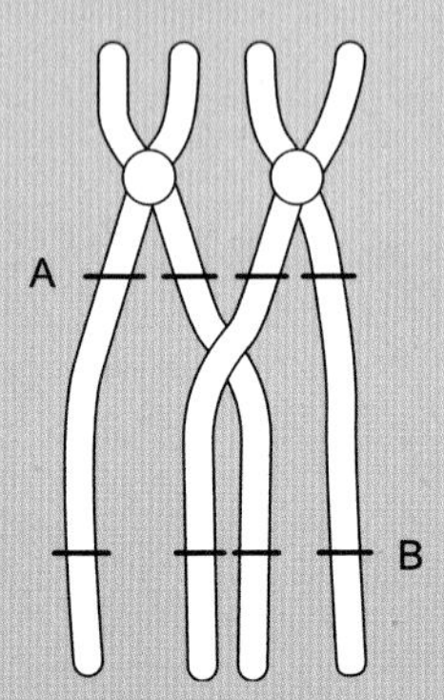

Figure Q3

i. that have not crossed over;

ii. that have crossed over.

c. Give two processes, other than crossing over, which result in genetic variation. Explain how each process contributes to genetic variation.

AQA January 2005 Unit 4 Question 8; modified

3. a. Explain the difference between a gene mutation and a chromosome mutation.

b. The base sequence on part of a DNA molecule is:

AAT GGC GGC TAT TGA

Explain why a substitution of C for the base G at the start of the second triplet might have less effect on the phenotype of the organism than the deletion of this base.

c. Down's syndrome results from the presence of three copies of chromosome 21 in a person's cells. Explain how this condition arises.

4. The graph shows the proportion of bacteria resistant to the antibiotic vancomycin in samples taken from patients in an intensive care ward.

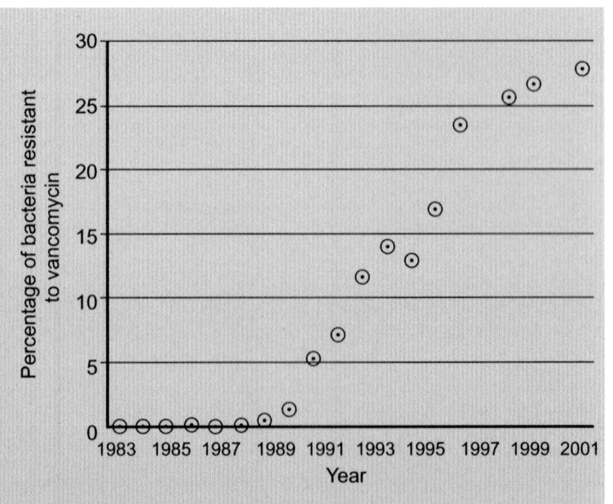

Figure Q4

a. Describe the trend shown by the data.

b. Suggest an explanation for this trend.

c. Due to the rise in the proportion of resistant bacteria, it was decided to trial the use of a different antibiotic in the intensive care ward. Discuss the ethical issues involved in a trial such as this.

5. Great tits are small birds. The graph in Figure Q5 shows the relationship between the number of breeding pairs in the population and the mean number of eggs per nest in different years in a wood.

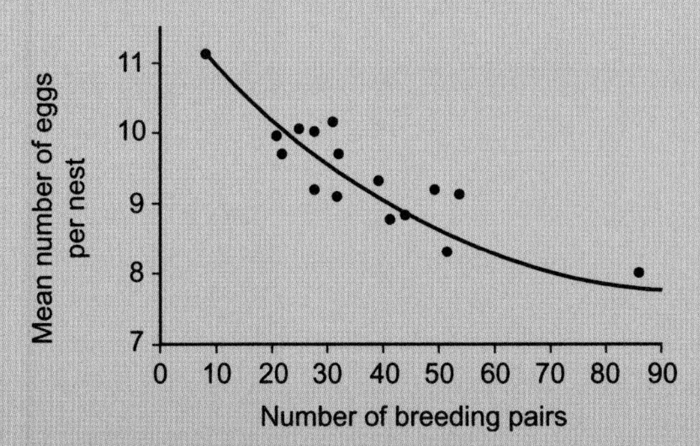

Figure Q5

a. Explain the relationship shown by the graph in Figure Q5.

b. Female great tits usually lay between three and 14 eggs in a nest.

i. In the same year, the birds do not all lay the same number of eggs. Explain how one factor, other than the number of breeding pairs, could influence the number of eggs laid by a great tit.

ii. Natural selection influences the number of eggs laid. Explain why great tits that lay fewer than three eggs per nest or more than 14 eggs per nest are at a selective disadvantage.

AQA June 2006 Unit 5 Question 5

6. Mycolic acids are substances that form part of the cell wall of the bacterium that causes tuberculosis. Mycolic acids are made from fatty acids. Isoniazid is an antibiotic that is used to treat tuberculosis. Figure Q6 shows how this antibiotic inhibits the production of mycolic acids in this bacterium.

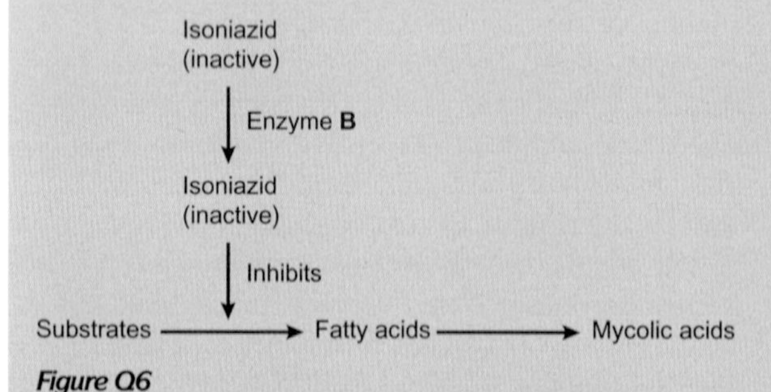

Figure Q6

a. Treatment with isoniazid leads to the osmotic lysis of this bacterium. Use information in Figure Q6 to suggest how.

b. Human cells also produce fatty acids. Isoniazid does not affect the production of these fatty acids. Use information in the diagram to suggest one reason why isoniazid does not affect the production of fatty acids in human cells.

c. A mutation in the gene coding for enzyme B could lead to the production of a non-functional enzyme. Explain how.

d. Using isoniazid to treat diseases caused by other species of bacteria could increase the chance of the bacterium that causes tuberculosis becoming resistant to isoniazid. Use your knowledge of gene transmission to explain how.

AQA Jan 2013 Unit 2 Question 6

12 TAXONOMY AND BIODIVERSITY

PRIOR KNOWLEDGE

You may recall that variation between the members of a species is caused by both genes and environment, and that this variation can be discontinuous (human blood groups, for example) or continuous (human height, for example). You may want to remind yourself about the way that the base sequence of DNA affects the proteins and therefore the characteristics of living organisms, and how natural selection favours the survival of organisms with alleles that best adapt them to their environment.

LEARNING OBJECTIVES

In this chapter, we consider how biologists classify living organisms, including the difficulties in determining whether or not two organisms belong to different species, and also consider how we can measure and analyse diversity within populations.

(Specification 3.4.5, 3.4.6, 3.47)

Scientists have so far given a name to about two million species of living organisms. More than half of these are insects. In Great Britain alone there are more than 3500 different species of beetle. In the rainforests of the world there may well be millions of still unnamed species; one estimate says that there could be 28 million species still to find.

You may wonder how it is possible to estimate the number of unknown and unnamed species. One way is to collect, say, a hundred different species of insect from a sample area of rainforest, or even from one species of rainforest tree. The insects are then checked against the databases of organisations such as the Natural History Museum in London to see whether they have been described and named previously. From the proportion of unnamed species in the sample, it is possible to calculate the likely number of unknown species in rainforests.

Most of the new discoveries are small and apparently insignificant organisms. (If you want to go down in history by giving your name to a newly discovered species, your best chance is to become an entomologist specialising in beetles.) But there are still occasional discoveries of much larger organisms. In 2013 alone, approximately 18 000 new species were named, including a new species of racoon-like mammal, the olinguito, which was discovered in the cloud forests of the Andes.

Studies made in other habitats and of other invertebrates, plants and microorganisms, such as bacteria, reveal our staggering ignorance about the vast majority of inhabitants of the planet. For example, we know little about the deep oceans, which may well conceal many species of fish and large invertebrates, such as types of squid, previously unknown to science. Huge numbers of species are destined to disappear without humans having any knowledge of them.

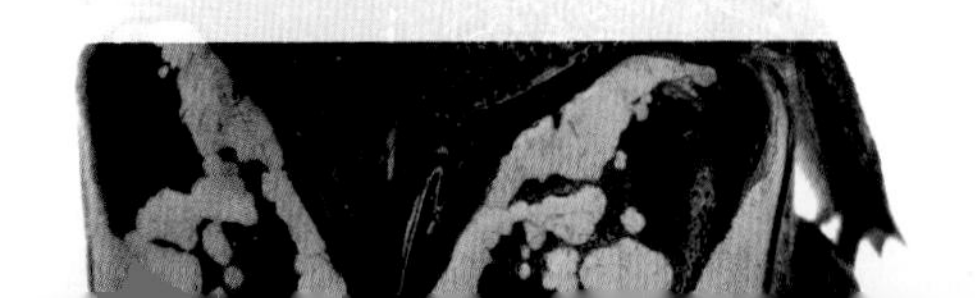

12.1 THE SPECIES CONCEPT

We can recognise a dog, a sheep and a human as being different species because they look completely different. However, difference in appearance is not enough to distinguish one species from another. The two dogs in the photograph are far from being lookalikes, yet they belong to the same species. We describe them as different breeds. Their distinctive features are the result of artificial selection, which involves processes similar to those seen in natural selection. Dogs with specific features have been chosen as mating partners so that those features are retained in the resulting litter of puppies. For example, the Great Dane was bred for hunting large animals such as wild boar and deer stags, and mating pairs were chosen for their size and strength. The terrier was selected for its ability to follow small prey into underground burrows.

Figure 1 *A Great Dane and a Yorkshire terrier*

Despite the great difference in size, the two dogs would potentially be able to breed together. The puppies that would result from such a mating would show a mixture of the features of the two breeds.

Dogs are a prime example of how members of the same species can look very different from one another. The opposite is also true – members of different species can look very similar. Chiffchaffs and willow warblers (Figure 2) are difficult to tell apart, even for experienced birdwatchers. They do, however, have quite different songs, so the birds themselves are not confused. Although they may live together in the same habitat, they do not interbreed.

These two examples lead us to one of the most important features of a species: members of the same species can interbreed to produce fertile offspring. Members of different species do not normally interbreed.

Figure 2 *A chiffchaff,* Phylloscopus collybita, *and a willow warbler,* Phylloscopus trochilus

There are many different kinds of barriers to interbreeding. For animals, successful mating is often the result of ritualised courtship behaviour, in which the two potential partners carry out a series of actions – often including both movement and sound – that stimulate one partner to accept the other as a mate. The 'correct' performance of the courtship ritual is unique to one particular species, so even if another species looks similar, mating between two organisms with different courtship behaviours will not take place. If you are able to visit a freshwater lake, you may be lucky enough to see pairs of great crested grebes performing an elaborate 'dance' together, in which the male presents the female with pieces of weed that he has dredged up from the water, and the two of them follow a series of movements together (Figure 3). Without this behaviour, the pair would not bond and mating would not occur.

Figure 3 *In this phase of their courtship dance, great crested grebes raise their neck ruffs, face each other, and move their heads from side to side.*

Figure 4 *Horses and donkeys (left and middle) belong to two different species, but are able to breed together to produce a mule (right).*

There are examples, however, where members of different species can and do mate and produce offspring. A mule (Figure 4) is the result of a mating between a male donkey and a female horse. Donkeys and horses would not normally mate in the wild, but will often do so when they are kept together. Genetically, donkeys and horses are quite distinct. Wild horses have 66 chromosomes but donkeys have only 62. While the mule offspring are generally healthy, they have one important deficiency; they are infertile, and so are unable to breed.

QUESTIONS

Stretch and challenge

1. Use the information about the numbers of chromosomes in the cells of a horse and a donkey, and your understanding of meiosis, to explain why mules are sterile.

Defining a species

From the examples we have just seen, we can say that a species is a group of organisms that interbreed and produce fertile offspring. This makes a good working definition that gives us a precise way of deciding whether two organisms belong to the same species or two different species.

But there are many problems with this seemingly simple definition. In particular:

- We may not know anything about the mating habits of a particular organism. For example, nothing is known about the life cycles of many species of insects.
- Some organisms have never been observed to reproduce sexually. For example, some species of plant always reproduce asexually, so we are not able to say whether or not they could breed with other species.
- Some organisms may live far apart from one another, and never get the chance to interbreed. We cannot know whether or not they would interbreed if they had the opportunity to do so.
- Some species are only known from museum specimens or fossils. We will never know anything about their breeding abilities.

In these cases, biologists often fall back on looking for similarities and differences between two organisms to decide whether they belong to the same species or different ones. Visible features are not always reliable – many species of insects and roundworms, for example, can only be distinguished by careful microscopic examination, and some microorganisms can only be distinguished by biochemical tests. We have also seen how chiffchaffs and willow warblers are very difficult to tell apart from their appearance, but have different songs. In practice, we have to take as many aspects as possible into account.

Taxonomy

As well as classifying each organism as belonging to a particular species, biologists classify species into different groups. Classifying organisms involves more than giving them a name and then producing a catalogue. It helps biologists to understand the relationships between organisms. But the basic system of classification, which is called **taxonomy**, was devised before ideas of natural selection, evolution and speciation were developed.

Taxon	Description	Example
domain	the largest group, containing three types of organisms with very different biochemistry	Eukaryota
kingdom	major groups of organisms within each domain	animals
phylum	a group of organisms that share a common body plan, for example having a segmented body, an external skeleton made of chitin, and joined limbs	arthropods
class	a major group within a phylum, for example all the arthropods with three pairs of jointed legs	insects
order	a subset of a class, for example all the beetle-like insects	beetles (Coleoptera)
family	a group containing organisms with very similar features	ladybirds (Coccinellidae)
genus	a closely related group within a family, for example red or orange ladybirds with black spots	*Coccinella*
species	a group of individuals that can breed with one another to produce fertile offspring, but not with other members of the same genus	*Coccinella septempunctata* (the seven-spot ladybird)

Table 1 *How organisms are classified.*

Carl Linnaeus, a Swedish biologist, devised the taxonomic system that we use today in the 18th century. He classified living things into groups based on obvious similarities. The large groups, such as 'plants' and 'animals', were subdivided into smaller groups that showed even closer similarities. For example, birds are clearly animals and not plants, and are similar in that they all have wings and feathers, features that distinguish them from other animals. Such a system, in which large groups are split into smaller and smaller groups, is called a **hierarchy**.

These are the basics of a taxonomic hierarchy:

- Organisms are classified into groups, which are further subdivided into yet more groups.
- Organisms are classified on the basis of similar or shared features. These features may be obvious, or they may be features determined by biochemical or genetic tests.
- There is no overlap between the groups; an organism must be either a bird or a reptile, for example – it cannot fall halfway between the two.

Naming species

In order to name a species precisely, biologists use both the **genus** and **species** names. The resulting two-word name is called a **binomial**. These names are international, understood by scientists in all countries in the world, and are often based on Latin or Greek. The binomial for the seven-spot ladybird, for example, is *Coccinella septempunctata* (the Latin meaning 'bright red with seven spots').

Notice that a binomial is always printed in italics; because you cannot do this when you write by hand, you should underline the words instead. The genus name starts with a capital letter but the species name (second word) has a small letter.

Phylogenetic classification

If we look back into the fossil record, we can find many examples of species that existed long ago but are now extinct. On the other hand, we can find species living today that clearly did not exist long ago. This is evidence that species can change; some species are permanently lost, while new ones come into existence.

The production of a new species is called **speciation** (you will study speciation in detail in the second year of your A-level studies). New species are produced when something happens to prevent two groups of the same species being able to breed together. For example, imagine a species of lizard living in rainforest in South America. Two lizards accidentally get carried down a river on a piece of floating wood, and are eventually brought to land on a distant island. They reproduce and produce a small population. But the conditions on the island are very different from the conditions in the rainforest, and so natural selection results in a different collection of characteristics in the island population. Eventually, they become so different from the rainforest lizards that they can no longer interbreed with them, and are classified as a new species.

This is only one of the ways in which new species can form. Anything that prevents two populations of the same species breeding together can result in speciation. Every new species forms from a pre-existing species. Every species therefore has a 'family history', and is related to other species.

Figure 5 shows an imaginary 'tree' representing the relationships between several different species. At

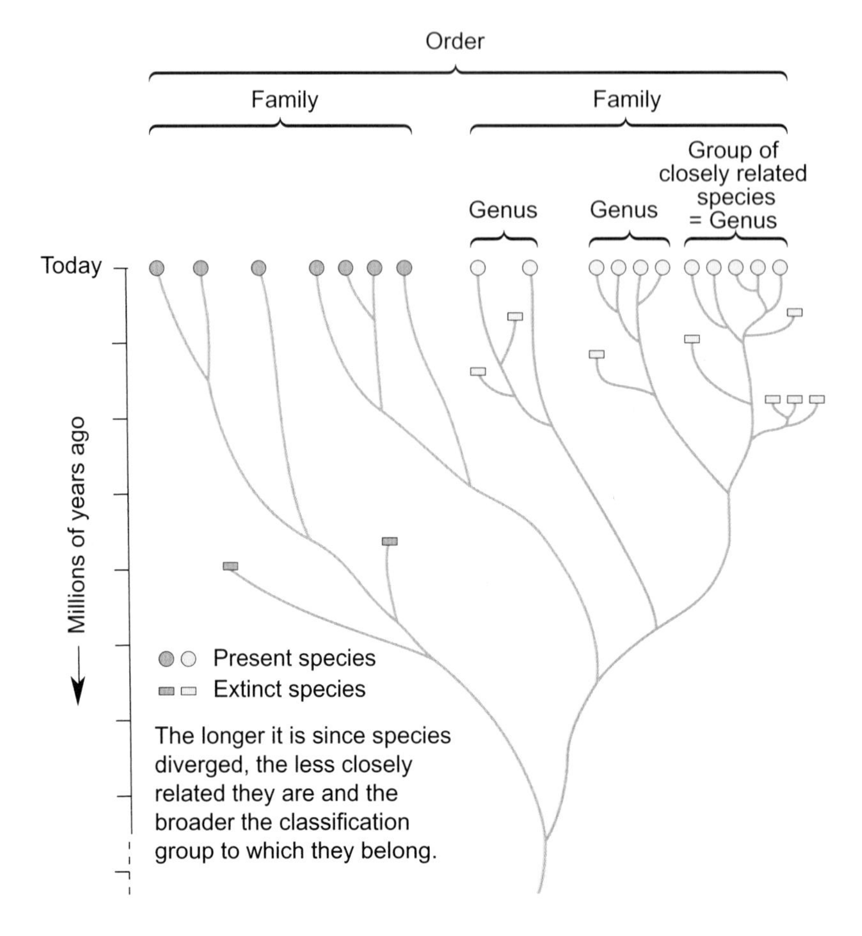

Figure 5 *An imaginary family tree for a group of species*

the very bottom of the tree is a single species from which they all arose – their **common ancestor**. This ancestor gave rise to two species, and then new species evolved from each of these branches. All of the species in the tree are therefore 'related' to one another – the more recently they branched off, the more closely they are related.

Modern biologists aim to reflect the ancestry of species when they classify organisms. This system of classification is said to be **phylogenetic** – it takes account of evolutionary history, as far as we can deduce it. To obtain evidence about evolutionary relationships, taxonomists study not only the anatomy of organisms and the fossil record, but also the structure of their proteins and DNA. You can read more about this in the section *Investigating diversity* on page 248.

QUESTIONS

2. Table 2 shows the classification of a hedgehog.

Taxon	Name
domain	Eukaryota
kingdom	Animalia
	Chordata
	Mammalia
	Erinaceomorpha
family	Erinaceidae
	Erinaceus
	europaeus

Table 2

a. Copy and complete Table 2 to show the missing names of the taxa.

b. What is the binomial for a hedgehog?

3. The binomials of three types of insect are:

Bombus pratorum

Bombus terrestris

Meromyza pratorum

Which two of these insects are thought to share the most recent common ancestor?

KEY IDEAS

- A species can be defined as a group of organisms that are able to interbreed to produce fertile offspring.
- In practice, we often cannot obtain enough evidence to apply this criterion, and have to use similarities in physical, physiological and behavioural characteristics to help to decide on whether two organisms belong to the same species.
- New species arise from previous species, generally by a process of evolution by natural selection.
- Biologists classify living organisms according to a phylogenetic system, which is intended to reflect their evolutionary relationships.
- The biological classification system is a hierarchical system, in which smaller groups are placed with larger groups, with no overlap between groups.
- Each group in the hierarchy is called a taxon. In the most commonly used system, the taxa (largest first) are domain, kingdom, order, class, family, genus and species.
- The name of the genus and species to which an organism belongs are used to form its binomial (two-word name). The name of the genus is given a capital letter, and the name of the species a small letter. The binomial is written in italics, or underlined.

12.2 BIODIVERSITY

Biodiversity literally means the diversity of life. The biodiversity in an area includes the number of different ecosystems and communities of organisms that there are, the number of different species, and also the degree of genetic diversity within the members of those species. (A community is all the different species that live together in a habitat.)

Biologists studying an ecosystem often want to know the range of different species that live there. A very simple statistic is the number of different species that live in a particular habitat. This is called the **species richness**. It is relatively easy to determine species richness, as you simply have to count the number of different species that you can find. It is not even necessary to identify them, as long as you are sure that they are different species.

However, species richness does not tell you anything about how common each of the different species is in the habitat. To take this into consideration, we need to find the **species diversity**. This takes into account not only the number of different species in a habitat, but also the number of individuals in each species. A meadow with 20 plant species, in which nearly all the plants belong to six species and the other 14 are very rare and are clumped in one corner, has the same species richness as another meadow in which all 20 species are fairly common and are found all over the meadow. However, the first meadow has a smaller species diversity than the second.

Although it is easy to get a general idea of what species diversity means, there are no universally accepted definitions of it. There are many formulae for calculating species diversity, including an index of diversity. One formula for calculating this is:

$$d = \frac{N(N-1)}{\Sigma n(n-1)}$$

where:

d = index of diversity
N = the total number of individuals recorded in the sample
Σ = the sum of
n = the number of individuals of each species.

The larger the value for the index of diversity, the greater the species diversity in the habitat.

Worked maths example: index of diversity

(MS 1.5, MS 2.3, MS 3.1, PS 3.2)

There is a growing concern that global warming will alter the natural diversity of animal and plant species around the world. Being able to quantify diversity is important for many biologists, including conservationists working to maintain biodiversity in the wild.

Species	Number of individuals recorded in the sample (n)	$n - 1$	$n(n - 1)$
grass species 1	80	79	6320
grass species 2	45	44	1980
clover	9	8	72
black medick	22	21	462
daisy	13	12	156
dandelion	3	2	6
speedwell	10	9	90
self-heal	14	13	182
moss	36	35	1260
TOTALS	**232**		**10 528**

Table 3 *Data taken from sampling a regularly mowed but rather weedy lawn*

To calculate the index of diversity, d, we apply the information in the table to the equation:

$$d = \frac{N(N-1)}{\Sigma n(n-1)}$$

The data in Table 3 show that N is 232, and $\Sigma n(n - 1)$ is 10 528

$$d = \frac{232 \times (232 - 1)}{10\,528}$$

= 5.09 to three significant figures

This gives us an idea of the spread and size of populations throughout the lawn. Biologists can use this formula to measure changes in diversity over time, or as a result of variables such as increased temperature.

QUESTIONS

4. A student collected the following data from a grass field grazed by sheep.

   Grass 185

   Thistles 28

   Stinging nettles 35

   Moss 2

   a. Arrange these data in a table that will allow you to calculate the index of diversity.

   b. Calculate the index, showing your working.

   c. How does the species diversity of the field compare with that of the lawn in the worked example? Suggest reasons for the differences between them.

The effects of farming on biodiversity

Any form of farming reduces biodiversity. Farmers clear the natural vegetation from the land in order to grow crops, removing the habitats of many species, thus decreasing the diversity index.

A farmer's first priority is to make the best possible use of the land, either for growing crops or raising animals. Traditionally, farmers produced food for themselves and for the local community by growing a variety of different crops and keeping a range of different animals. In many parts of the world, subsistence farmers are still doing this.

Today, there is increasing pressure on farmers to specialise and grow only a single crop, or maybe two or three, and to concentrate on one form of animal husbandry, such as dairy cattle or pig breeding. This is more economical because better use can be made of machinery, and organisations such as supermarkets have fewer farmers to negotiate purchases with. Also, transport is now much easier, so produce can easily be moved to more distant markets. Consequently, in many parts of the country, large areas are used for single crops, such as wheat or oil-seed rape. This is called **monoculture**.

Figure 6 *Only one species of plant grows in this wheat field.*

It is easy to see how growing crops in this way reduces biodiversity. Farmers usually spray weedkillers on to fields of wheat and other crops, to prevent other species of plants growing and competing with the crop plants for light, water or minerals from the soil. Because there is only one species of plant, there are very few species of animals that can live in the habitat, and even those that can may be killed by insecticides to prevent them from damaging the crop and reducing yields.

Clearly, we cannot stop farming, because everyone needs to eat. Modern farming techniques produce relatively cheap food. Many people are prepared to pay a little extra for food that is produced in ways that are more sympathetic to the environment, such as so-called 'organic' farming, but not everyone can afford this.

We have to try to find a balance between producing high-quality, relatively inexpensive food, and conserving biodiversity. In the UK, government grants provide some financial help to farmers in return for using land management regimes that help to conserve biodiversity. Even without such financial compensation, many farmers are prepared to forgo some profit in order to keep wildlife and natural vegetation on parts of their land. For example:

- Margins can be left around arable fields in which 'weeds' are allowed to grow (Figure 7).
- Stubble (the remains of the stems after a crop has been harvested) can be left on fields over the winter, providing a food source for birds (Figure 8).

Figure 7 *A margin has been left around this wheat field in which wild plants are allowed to grow. These provide food and habitats for many different species of insects, birds and other animals, greatly increasing biodiversity.*

Figure 8 *Cirl buntings,* Emberiza cirlus, *are rare in the UK. This one is feeding on insects and seeds in weeds growing in stubble left in a field over winter.*

- Hedges can be planted with many different species, and allowed to grow thick to provide food and habitats for insects and birds.
- Care can be taken to avoid contaminating water courses with fertilisers and herbicides, which can kill aquatic organisms.
- Stocking densities on pasture can be kept within sensible limits, so that there is less likelihood of the ground becoming compacted and overgrazed. Overgrazing can lead to soil erosion, which greatly reduces biodiversity.

- Meadows with high biodiversity can be mowed relatively late in the year so that plants have time to produce seeds; avoiding use of fertiliser on such meadows is also very beneficial to biodiversity because fertilisers allow a few species to grow so large that they outcompete all the rest for light (Figure 9).

Figure 9 *Avoiding the use of fertilisers, cutting late in the year and not overgrazing all help to conserve biodiversity in meadows.*

ASSIGNMENT 1: INVESTIGATING THE EFFECT OF HEDGES ON CROP YIELD

(MS 3.1, PS 1.2, PS 3.2)

During the second half of the 20th century, many farmers were encouraged to remove hedges, to provide more space for crops and in the belief that this would increase yields.

Figure A1 shows the effect of a two metre high hedge on crop yield and on some abiotic factors at various distances from the hedge. (Abiotic factors are those that are caused by non-living components of the ecosystem.)

Questions

A1. Describe the effects of the hedge on soil moisture and air temperature. Suggest explanations for these effects.

A2. How far from the hedge is the greatest increase in crop yield?

A3. Suggest why the yield close to the hedge is decreased.

A4. What is the evidence from the graph that the net effect is a gain in yield in the area within 40 metres of the hedge?

A5. Suggest why this gain in yield may not be found in practice.

A6. Suggest how the presence of hedges around fields affects biodiversity.

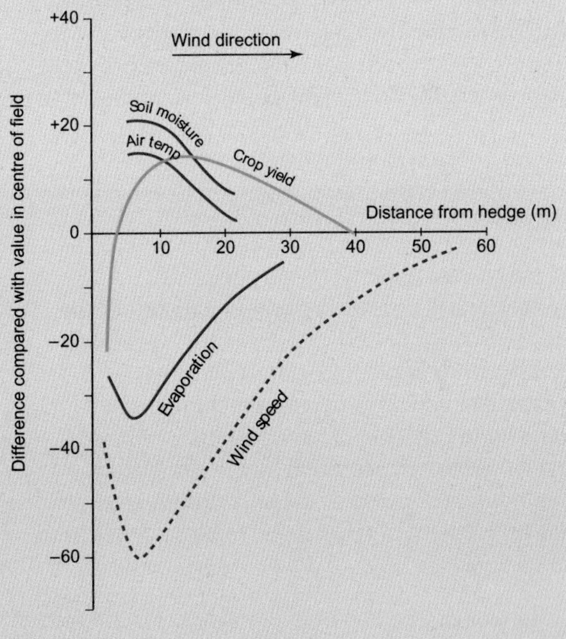

Figure A1

KEY IDEAS

- Biodiversity includes the variety of ecosystems, habitats and species in an area, as well as the genetic diversity among the members of each species.
- Species richness is the number of different species in a community.
- An index of diversity, takes into account the numbers of individuals within each species as well as the number of different species.
- Some modern farming techniques reduce biodiversity.
- Monocultures of crops allow large quantities of food to be produced relatively cheaply, but prevent almost any species other than the crop living in the habitat.
- Conservation means attempting to maintain or increase biodiversity.
- A balance can be struck between the need to produce affordable, high-quality food and conservation. Farmers can be given financial incentives to reduce food production on their land and implement management regimes that increase biodiversity.

12.3 INVESTIGATING DIVERSITY

Genetic diversity

We have seen that it is not always easy to determine whether or not two organisms belong to the same or different species. Breeding habits may not be known, and deciding just how similar they are in their appearance, physiology or behaviour is difficult and can be subjective. Increasingly, taxonomists are turning to analysis of an organism's DNA or proteins to make decisions about how closely related one organism or species is to another. Advances in gene technology have made such analyses much simpler and cheaper to carry out than they were even a few years ago.

Analysing DNA can also give us information about genetic diversity within a population. This is called **genome sequencing**. All of the organisms of a species have the same set of genes on their chromosomes, but those genes can have different alleles – that is, they can have slightly different base sequences. We can also analyse the base sequences of mRNA produced from the DNA, or even the amino acid sequences of the proteins that are made. Knowing the genetic triplet code is universal; if you know the protein primary structure, or the mRNA base sequence, you can determine the DNA base sequence that produced them.

Such analyses can tell us about the diversity within a species. They can also help us to decide whether two organisms belong to the same species or to different ones. Moreover, they can give us clues about how closely two different species are related to each other.

QUESTIONS

Stretch and challenge

5. A taxonomist wants to compare the base sequence of the insulin gene in two species. Suggest why it may be easier to do this using mRNA extracted from beta cells in the pancreas, rather than using DNA.

Working out relationships between species

We know that the evolutionary history of our species, *Homo sapiens*, shares its roots with the family tree of the apes, including gorillas, orangutans and chimpanzees. We can use analysis of the base sequences in DNA, and the amino acid sequences in proteins, to try to quantify these relationships.

Comparison of the DNA, mRNA and proteins of humans and chimpanzees shows that our chemistry is very similar to theirs. The DNA base sequence that can be directly compared between chimpanzees and humans is almost 99% identical. At the protein level, 29% of genes code for the same amino acid sequences in chimpanzees and humans. In fact, on average the proteins that are made in our cells each differ by only one amino acid from those made in the cells of chimpanzees. This evidence has been used to estimate that humans and chimpanzees diverged from a common ancestor about six million years ago.

To put this into perspective, the number of genetic differences between humans and chimpanzees is approximately 60 times less than that seen between humans and mice, and about 10 times less than between mice and rats. But the number of genetic differences between a human and a chimp is about 10 times more than between any two humans.

If we want to make comparisons between many different organisms, then we need to choose a

Figure 10 DNA and protein analyses show bonobo chimpanzees to be closely related to humans.

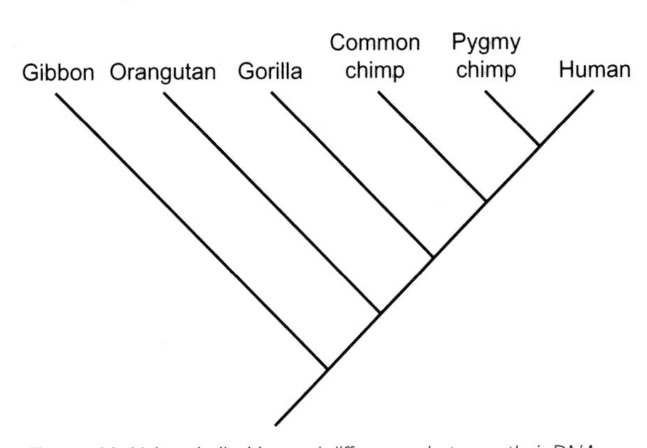

Figure 11 Using similarities and differences between their DNA, we can deduce the relationships between the different species of ape.

gene or a protein that is found in all of them. Most vertebrates, for example, produce the protein haemoglobin. We can therefore use the amino acid sequences of the β chain of haemoglobin to work out how closely related two species of vertebrates are. The more dissimilar the amino acid sequences, the longer ago their evolutionary lines diverged from a common ancestor. Table 4 shows the number of differences in the amino acid sequences in the β chain of the haemoglobin of each listed animal and a human.

Animal	Number of differences
gorilla	1
gibbon	2
rhesus monkey	8
dog	15
horse	25
mouse	27
grey kangaroo	28
chicken	45
frog	67

Table 4 Number of differences from human haemoglobin in the amino acid sequences of nine animals

QUESTIONS

6. What is the relationship between the number of differences in amino acid sequence of the β chain of haemoglobin shown in Table 4, and the closeness of kinship to humans?

While haemoglobin is useful for comparing different species of vertebrates, it does not occur in many other groups of organisms. We need to find a more universal protein to compare a wider range of organisms. Cytochrome c is a protein molecule found in the mitochondria of every eukaryotic cell, including all animals, plants and fungi. The amino acid sequences of cytochrome c in many species have been determined, and comparing them shows how closely the species are related.

Human cytochrome c contains 104 amino acids; 37 of these have been found at equivalent positions in every cytochrome c that has been sequenced. Scientists deduce that these molecules have descended from a precursor cytochrome in a primitive microbe that existed more than two billion years ago.

During the 21st century, there have been rapid advances in the speed and accuracy with which biologists can analyse the DNA and proteins of organisms. But using this evidence to work out the order (in time) in which the mutations that caused the differences may have occurred, and the relationship between different species, is still difficult. It's elaborate detective work, and the evidence is never complete and can often be interpreted in more than one way, so these are ongoing investigations. Conclusions about species' relationships often change when new evidence comes to light.

Results from different methods of comparison, such as genome sequencing, amino acid sequencing and immunology, do not always paint the same picture. This, and ever-increasing speed and sophistication of methods of analysis, mean that detailed classification of some species is being revised as new information about how those species have evolved comes to light. However, it is never possible to be certain about the actual pathway of evolution, and scientists often disagree about how a particular species should be classified.

ASSIGNMENT 2: ANALYSING AMINO ACIDS SEQUENCES

(PS 1.2)

Table A1 shows the 22 amino acid residues at the N-terminal end (the end of the amino acid chain that finishes with an amine group) of human cytochrome c. The corresponding sequences from five other organisms are aligned beneath. A blank cell in the table indicates that the amino acid is the same as the one found at that position in the human molecule.

Species	Sequence of the last 22 amino acids																					
	1	2	3	4	5	6	7	8	9	10	11	12	13	14	15	16	17	18	19	20	21	22
human	Gly	Asp	Val	Glu	Lys	Gly	Lys	Lys	Ile	Phe	Ile	Met	Lys	Cys	Ser	Gln	Cys	His	Thr	Val	Glu	Lys
pig											Val	Gln			Ala							
chicken			Ile								Val	Gln										
fruit fly									Leu		Val	Gln	Arg		Ala							Ala
wheat		Asn	Pro	Asp	Ala		Ala				Lys	Thr			Ala						Asp	Ala
yeast		Ser	Ala	Lys			Ala	Thr	Leu		Lys	Thr	Arg		Glu	Leu						

Table A1 *Differences in the amino acid sequences of cytochrome c between humans and other species*

1	2	3	4	5	6	7	8	9	10	11	12	13	14	15	16	17	18	19	20	21	22
Gly	Asp	Val	Glu	Lys	Gly	Lys	Lys	Val	Phe	Val	Gln	Lys	Cys	Ala	Gln	Cys	His	Thr	Val	Glu	Asn

Table A2 *The sequence of amino acids in the cytochrome of dogfish.*

Questions

A1. At which positions in the cytochrome c molecule are the amino acids always the same, in every species tested?

A2. Suggest what this may indicate about the importance of these amino acids in determining the function of the cytochrome c protein.

A3. Where would you place the dogfish in Table A2? Explain your decision.

Stretch and challenge

A4. Devise a way of using the data in the table to make a quantitative estimate of the relative closeness of relationship between each of the species. To what extent does your estimate support the way in which these species are classified?

Variation within species

So far, we have concentrated on how we can use differences between organisms to classify them into different species. But not every member of a species is identical to every other member. Individuals within a species can vary quite considerably from one another. We are most aware of these differences in our own species, but careful observation also shows that every species has some degree of variation in its characteristics.

Variation in a characteristic may be clear-cut, such as having different blood groups. This is **discontinuous variation**. All human beings have one of the four ABO blood groups: A, B, AB or O. Each of us must be one of these four possible phenotypes. You cannot be halfway between groups A and O, or just slightly blood group B. Discontinuous variation is caused by different alleles of genes. ABO blood group is determined by a single gene, one that has three alleles, I^A, I^B and I^o. Alleles I^A and I^B code for the production of a cell-surface protein, while allele I^o does not. Different combinations of these alleles give the four possible blood groups. Another example of discontinuous variation is spot colour in Dalmatians – a recessive allele codes for brown spots, while the dominant allele codes for black spots (Figure 12).

Figure 12 *Spot colour in Dalmatians is an example of discontinuous variation. What characteristics can you see that show continuous variation?*

More often, there is a range of variation (from tall to short, for example) and this is called **continuous variation**. There are no definite groups. A person can be any height between the two extremes of the height range. It is not only physical characteristics, such as mass, length of big toe, hair colour and nose shape that have a range of continuous variation, but also metabolic characteristics such as heart rate, speed of reaction, muscle efficiency and the ability of the brain to process information.

Characteristics that vary continuously are usually the result of several different genes acting in such a complex way that it is not possible to distinguish separate phenotypes. A person's height depends on many different factors, including the growth rate of several different bones, hormone production and metabolic rate. Each factor may be controlled directly, or indirectly, by several genes, each of which may have several different alleles. However, these characteristics are often also influenced by an organism's environment. You may have a collection of alleles that would allow you to grow tall, but if you have a poor diet while you are growing, you may never achieve your potential height.

Analysing continuous variation

Figure 13 is a graph showing the number of men in a population of different heights. This shape of curve is very commonly produced when we plot data for a characteristic that shows continuous variation. This symmetrical curve is called a **normal distribution curve**.

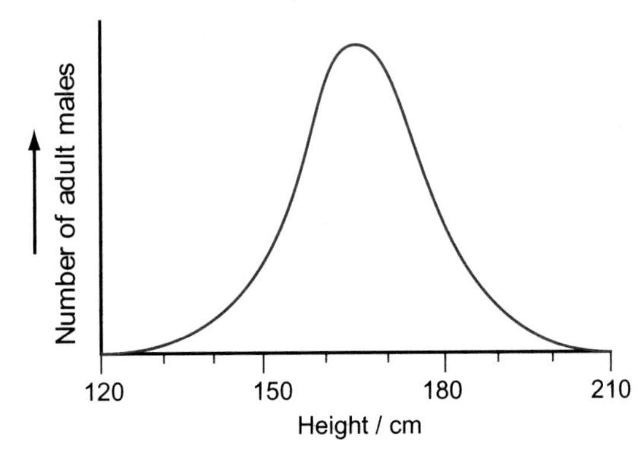

Figure 13 *A normal distribution curve for human height*

The exact shape of the normal distribution depends on the **mean** and the **standard deviation** (SD) of the distribution. To calculate the mean, we add up all of the individual values and divide them by the number in the sample. The standard deviation is a measure of how much the individual values are spread out from the mean. Calculating standard deviation manually is quite time-consuming, but many calculators, spreadsheets and websites will do it very quickly for you – you just have to type in the individual values in the sample. The greater the standard deviation, the greater the spread of the data on either side of the mean value.

In a normal distribution:

- measurements greater than the mean and measurements less than the mean are equally common
- small deviations from the mean are much more common than large ones
- 68% of all of the measurements fall within a range of ± 1 SD from the mean and 95% within ± 2 SDs.

Figure 14 shows the normal distribution of diastolic blood pressure for a sample of men. The mean blood pressure is 82 mmHg and the SD is 10 mmHg. (mmHg is unit of blood pressure commonly used by doctors. In an A-level examination you will see and use the unit kilopascal, kPa.) This means that 68% of the sample have blood pressure between 72 and 92 mmHg; 95% of the sample have blood pressure between 62 and 102 mmHg. It follows that if we select a man at random, there is only a 0.05 (1 in 20 chance) of him having a blood pressure less than 62 or greater than 102 mmHg.

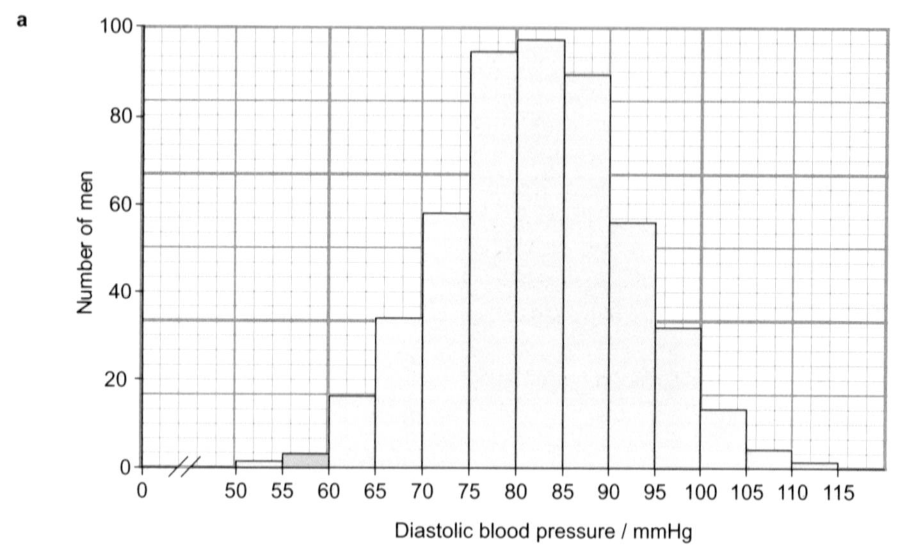

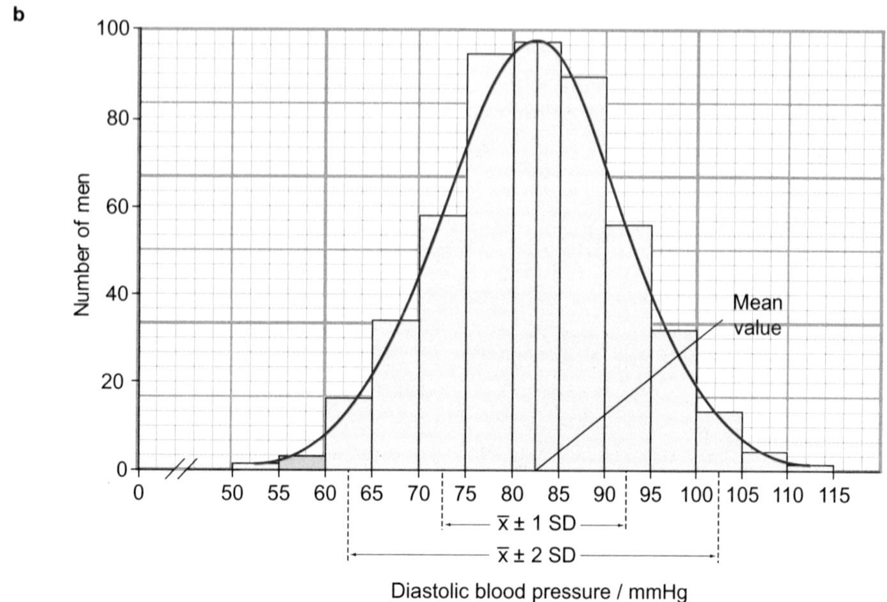

Figure 14 *Blood pressure in a sample of men: a data plotted as a frequency distribution (histogram); b the red line is a smooth curve drawn between the mid-points of the top of each bar. The mean value, $\bar{x}$, is 82 mmHg. The dotted lines show the range of values that lie within one or two times the standard deviation from the mean. (Note: In this chart, the unit for pressure is mmHg, which is still commonly used by doctors. In your A-level studies, however, you are more likely to come across the unit kPa. 1 kPa = 7.5 mmHg.)*

When collecting data in order to analyse variation, it is very important to take a **random sample** of a population. Simple random sampling is the basic sampling technique where we select a group of subjects (a sample) for study from a larger group (a population). Each individual is chosen entirely by chance and each member of the population has an equal chance of being included in the sample.

Worked maths example: Analysing variation in stem length in dandelions

(MS 1.2, MS 1.3, MS 1.6, MS 1.10, PS 3.2)

Imagine that we want to measure, record and analyse the variation in the length of flower stems of dandelions growing in a lawn.

Figure 15 *Dandelions growing in a lawn*

Frequency distribution and random sampling

The frequency distribution shows the variation in dandelion stem length in the lawn. Drawing the frequency distribution as a histogram is a useful way of getting an overarching impression of how many stems are a certain length. However, as can often occur in biology, we are dealing with huge numbers. It may not be feasible to measure every single plant. Biologists often get round this problem by taking a **sample**. This sample must be random – that is, every flower stem must have an equal chance of being measured. To do this, we can treat the edges of the lawn as though they were the axes of a graph. We can then use a calculator or computer to generate a pair of random numbers, which we use as coordinates to tell us where to take our sample.

Sorting measurements into groups

These are the measurements, in mm, of a randomly chosen sample of 50 flower stems.

63, 79, 71, 108, 82, 33, 40, 69, 95, 51, 58, 42, 56, 77, 67, 84, 60, 61, 87, 39, 65, 52, 42, 66, 78, 79, 87, 90, 118, 60, 65, 72, 78, 88, 18, 33, 64, 77, 91, 58, 40, 76, 68, 51, 41, 31, 67, 55, 59, 69

In order to plot a frequency distribution, we need to put these individual measurements into groups. We should have at least eight groups (so that we can draw at least eight bars on the graph) and the range of lengths in each group should be the same. Here, we could organise the groups like this:

Length of flower stem/mm	**10–19**	**20–29**	**30–39**	**40–49**	**50–59**	**60–69**	**70–79**	**80–89**	**90–99**	**100–109**	**110–119**
Tally	|		||||	𝍸	𝍸 |||	𝍸 𝍸 |||	𝍸 ||||	𝍸	|||	|	|
Number of stems	1	0	4	5	8	13	9	5	3	1	1

Drawing a histogram

To draw the frequency distribution or histogram, we label the x-axis with the length ranges and the y-axis with the number of stems. The bars are drawn touching one another, because we have a continuous scale (a continuous range of lengths) on the x-axis (Figure 16).

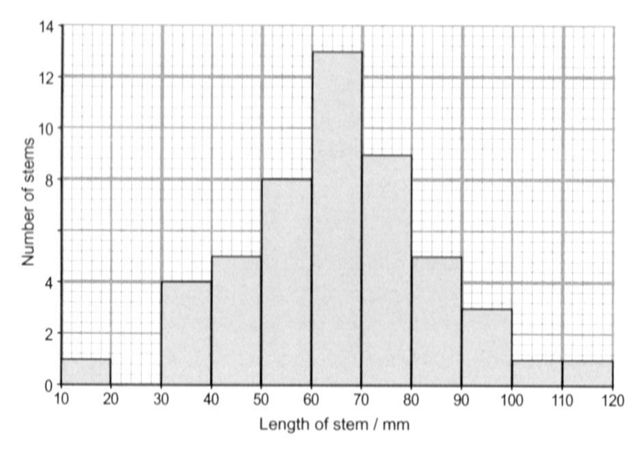

Figure 16 *Frequency distribution of flower stem lengths in a sample of dandelions*

The range

The range of stem length is the shortest value subtracted from the largest. The range tells us the variation between the minimum and maximum values. This is a basic measure of dispersion – how dispersed our data are. The range is the largest number (in this example, 118) minus the smallest (in this example, 18).

=118 – 18 = 100 mm

Calculating mean, median and mode

From the raw data, we can find different averages: mean, median and mode. The mean is the sum of the data, divided by the number of data points. It's the most commonly used average.

For our data:

$$\text{mean} = \frac{\Sigma n}{N} = \frac{3260}{50} = 65.2$$

The median is the mid-point in a set of data arranged in numerical order:

18, 31, 33, 33, 33, 40, 40, 41, 42, 42, 51, 51, 52, 55, 56, 58, 58, 59, 60, 60, 61, 63, 64, 65, **65**, 66, 67, 67, 68, 69, 71, 72, 76, 77, 77, 78, 78, 79, 79, 82, 84, 87, 87, 90, 91, 95, 108, 118

If there is an even number of data points, the median is the mid-point between the two middle values. For our data the median is 65.

The mode is the most commonly occurring data point. In this case, it is 33.

Calculating standard deviation

To calculate the standard deviation, we can use this formula:

$$s = \sqrt{\frac{\Sigma(x - \bar{x})^2}{n - 1}}$$

where:

s is the standard deviation

x is an individual reading

$\bar{x}$ is the mean

n is the number of readings

Note: you will not need to memorise this formula; it will be provided in the examination if it is needed.

This is quite time-consuming to calculate 'by hand', but you may have a function on your calculator that will work it out for you – all you have to do is to key in all of the individual readings.

In this case, the standard deviation is 20.3.

Standard deviation versus range

Standard deviation is more useful than range for measuring dispersion, because it measures the spread of *all* of the data around the mean. Range takes into account only the maximum and minimum values, so is vulnerable to skew if either of these are outliers.

ASSIGNMENT 3: ANALYSING VARIATION ON LUNDY ISLAND

(MS 1.2, MS 1.3, MS 1.6, MS 1.10, MS 3.2, PS 2.2, PS 3.1, PS 3.2)

Lundy Island (Figure A1) is off the north coast of Devon in the Bristol Channel. The waters around the island are of special importance for their marine wildlife containing the finest examples of rocky reefs in Britain, with an amazing diversity of sea life including some very rare species.

Figure A1 *In 2010, Lundy Island became England's first marine conservation zone.*

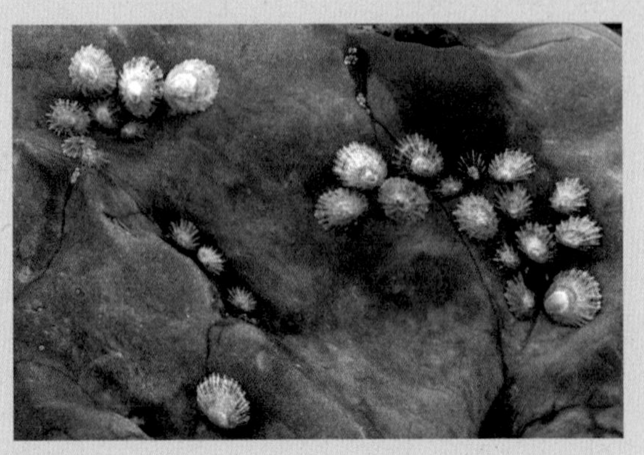

Figure A2 *Limpets on a rock*

Scientists measured the sizes of limpets, *Patella vulgate* (Figure A2), on Lundy Island to look at their variation. They measured the length of the shell and its height. Their data are given in this table:

length of limpet shell/mm	29.7	26.0	26.2	52.8	39.8	34.5	41.4	33.5	46.0	51.8	36.4	34.5
height of limpet shell/mm	12.4	10.0	8.0	18.6	19.7	16.2	23.7	14.0	21.5	26.5	9.5	14.0

Questions

Stretch and challenge

A1. a. Determine the mean, mode and median of both sets of results.

b. Determine the standard deviation for the length of limpet shells.

A2. Plot these data on a suitable graph. (Think carefully about the best type of graph to draw.)

A3. Describe what the data show.

ASSIGNMENT 4: ANALYSING DIVERSITY OF APRICOT TREES

(MS 1.2, MS 1.3, MS 1.5, MS 1.6, MS 1.10, MS 3.2, PS 1.2, PS 3.1, PS 3.2)

Figure A1 *Ripe fruit on an apricot tree,* Prunus armeniaca

A group of biologists plan to investigate the genetic diversity within two populations of apricot tree. They have to decide whether to measure variation in observable characteristics, or variation in the sequence of bases in DNA.

A second group of biologists plan to investigate fruit production from the trees. They measure the mass of 20 fruits from each population. Each measurement is made to the nearest gram. These are their results.

Population A: 18, 21, 20, 21, 19, 18, 17, 24, 20, 20, 21, 20, 19, 22, 20, 22, 20, 23, 21, 21

Population B: 22, 24, 22, 20, 23, 23, 21, 25, 22, 24, 23, 21, 23, 20, 24, 25, 22, 23, 21, 23

Questions

A1. Suggest some advantages and disadvantages of measuring

a. variation in observable characteristics

b. variation in the sequence of bases in DNA.

A2. **a.** Organise the second group's results into a table, so that you can plot them on two frequency distribution curves.

b. Plot frequency distribution curves (histograms) for each set of data.

c. Use an electronic calculator with a standard deviation function, or a website, to calculate the mean and standard deviation for each set of data.

d. Compare the data for the two populations.

Stretch and challenge

e. A student suggested that these data provide evidence that the two populations belong to different species. Do you agree? What could be done to test this suggestion?

KEY IDEAS

- We can measure genetic diversity within or between species by comparing the base sequences in DNA or mRNA, or the amino acid sequences in proteins, based on knowledge of the universal genetic code.
- Species with similar base sequences or amino acid sequences are likely to be more closely related to each other than those whose sequences differ more.
- The frequency of observable or measurable characteristics between organisms in a species, or between different species, can be used to infer genetic differences between them. However, many characteristics are also influenced by the environment, as well as genes.
- Variation within a species or population may be discontinuous or continuous. In discontinuous variation, there is a set number of clear-cut categories and no overlap between them. In continuous variation, the organism may lie anywhere between the two extremes of a range of values.
- To measure variation in a population, we should take a random sample, in which any individual has an equal chance of being measured.
- To calculate the mean value of the characteristic being measured, add up all of the values and divide by the number in the sample.
- Continuous variation often shows a normal distribution, in which a graph is a symmetrical curve peaking at the mean value.
- Standard deviation is a measure of how much the data spread on either side of the mean value.

PRACTICE QUESTIONS

1. Table Q1 shows the percentage similarity in the sequence of amino acids in the β haemoglobin chain of six different animals.

Use evidence from Table Q1 to explain whether bats are more closely related to gorillas or to birds.

Percentage similarity in β haemoglobin chains

	Indian short-nosed fruit bat	California big-eared bat	white stork	domestic pigeon	lowland gorilla	horse
Indian short-nosed fruit bat	100	89.7	68.5	69.9	85.6	84.9
California big-eared bat		100	66.4	67.1	84.2	82.9
white stork			100	86.3	68.5	65.8
domestic pigeon				100	68.5	69.2
lowland gorilla					100	82.9
horse						100

Table Q1

2. Read the following passage.

Higher animals seem never to have evolved cellulases. Ruminants, such as cattle, deer and camels, house certain types of bacteria in a four-chambered stomach. Cellulose is digested by the prokaryotes, which possess cellulases. The rabbit, *Oryctolagus cuniculus*, has, like the horse, an enlarged caecum in which cellulose breakdown occurs, also brought about by resident bacteria.

Cellulose is also digested in the intestine of the garden snail, *Helix pomatia*, although, in experiments, extracts of the digestive gland lack cellulase activity. A protoctist, *Trichonympha*, inhabits the intestine of wood-eating termites, and is responsible for their being able to digest cellulose in their diet of wood. Bacteria are also present here, but do not produce a cellulase. Instead they seem to be nitrogen-fixing, which may explain how termites are able to thrive on a diet so low in nitrogen-containing compounds.

a. Give the names of two genera mentioned in the passage.

b. A garden snail was mated with a snail from a different continent. Offspring were produced. What would you need to know about these offspring to be certain that both parents belonged to the same species?

AQA June 2002 Unit 4 Question 2

3. a. The cheetah, *Acinonyx jubatus*, and other cat species belong to the family Felidae. Complete Table Q2 to show the classification of the cheetah.

Kingdom	Animalia
	Chordata
	Mammalia
	Carnivora
family	Felidae
genus	

Table Q2

b. This system of classification is described as hierarchical. Explain what is meant by a hierarchical classification.

c. Despite differences in form, leopards, tigers and lions are classified as different species of the same genus. Cheetahs, although similar in form to leopards, are classified in a different genus.

i. Describe one way by which different species may be distinguished.

ii. Suggest two other sources of evidence that scientists may have used to classify cheetahs and leopards in different genera.

AQA June 2006 Unit 4 Question 3

4. Hummingbirds belong to the order Apodiformes. One genus in this order is *Topaza*.

a. i. Name one other taxonomic group to which all members of the Apodiformes belong.

ii. Name the taxonomic group between order and genus.

b. The crimson topaz and the fiery topaz are hummingbirds.

Biologists investigated whether the crimson topaz and the fiery topaz are different species of hummingbird, or different forms of the same species.

They caught large numbers of each type of hummingbird. For each bird they

- recorded its sex
- recorded its mass
- recorded the colour of its throat feathers
- took a sample of a blood protein.

Table Q3 shows some of their results.

Feature	Crimson topaz		Fiery topaz	
	Male	Female	Male	Female
Mean mass/g (± standard deviation)	13.6 (± 1.9)	10.8 (± 1.3)	14.2 (± 1.6)	11.6 (± 0.63)
Colour of throat feathers	Green	Grey edges	Yellowish green	No grey edges

Table Q3

i. Explain how the standard deviation helps in the interpretation of these data.

ii. In hummingbirds throat colour is important in courtship. Explain the evidence in the table that shows that the crimson topaz and the fiery topaz may be different species of hummingbird.

c. The biologists analysed the amino acid sequences of the blood protein samples from these hummingbirds.

Explain how these sequences could provide evidence as to whether the crimson topaz and the fiery topaz are different species.

AQA Unit 2 June 2012 Question 6

5. Scientists investigated the species of insects found in a wood and in a nearby wheat field. The scientists collected insects by placing traps at sites chosen at random both in the wood and in the wheat field.

Table Q4 shows the data collected in the wood and in the wheat field.

Species of insect	Number of organisms of each species	
	Wood	Wheat field
bird-cherry oat aphid	0	216
beech aphid	563	0
large white butterfly	20	0
lacewing	12	3
7-spot ladybird	36	0
2-spot ladybird	9	1
total number of organisms of all species	640	220

Table Q4

a. The scientists collected insects at sites chosen at random. Explain the importance of the sites being chosen at random.

b. i. Use the formula:

$$d = \sqrt{\frac{N(N - 1)^2}{\Sigma\, n(n - 1)}}$$

(Continued)

to calculate the index of diversity for the insects caught in the wood

where:

d = index of diversity

N = total number of organisms of all species

n = total number of organisms of each species

Show your working.

ii. Without carrying out any further calculations, estimate whether the index of diversity for the wheat field would be higher or lower than the index of diversity for the wood. Explain how you arrived at your answer.

c. A journalist concluded that this investigation showed that farming reduces species diversity. Evaluate this conclusion.

d. Farmers were offered grants by the government to plant hedges around their fields. Explain the effect planting hedges could have on the index of diversity for animals.

AQA Unit 2 June 2012 Question 7

6. Organisms can be classified using a hierarchy of phylogenetic groups.

a. Explain what is meant by:

i. a hierarchy

ii. a phylogenetic group.

b. Cytochrome c is a protein involved in respiration. Scientists determined the amino acid sequence of human cytochrome c. They then:

- determined the amino acid sequences in cytochrome c from five other animals
- compared these amino acid sequences with that of human cytochrome c
- recorded the number of differences in the amino acid sequence compared with human cytochrome c.

Table Q5 shows their results.

Animal	Number of differences in the amino acid sequence compared with human cytochrome c
A	1
B	12
C	12
D	15
E	21

Table Q5

i. Explain how these results suggest that animal **A** is the most closely related to humans.

ii. A student who looked at these results concluded that animals **B** and **C** are more closely related to each other than to any of the other animals. Suggest **one** reason why this may not be a valid conclusion.

iii. Cytochrome c is more useful than haemoglobin for studying how closely related different organisms are. Suggest **one** reason why.

AQA Unit 2, June 2013, Question 3

13 MATHS TECHNIQUES IN BIOLOGY

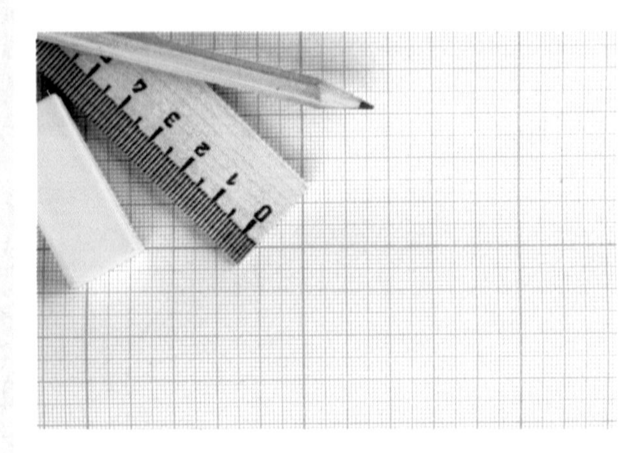

13.1 HANDLING NUMBERS

Ordinary and standard form

Numbers can be expressed in ordinary or standard form. For example:

Ordinary	Standard form
21 834	2.1834×10^4
0.0385	3.85×10^{-2}
0.0020 mol dm^{-3}	2.0×10^{-3} mol dm^{-3}

Notes:

- In standard form, there is one digit before the decimal point (d.p).
- To convert from ordinary to standard form, move the decimal point to left or right until you have one digit in front of it. Count how many moves you have to make to achieve this. This tells you the correct power of 10 to write in the standard form number.
- In the last example, note that we keep all of the significant figures during the conversion; therefore the zero that follows the two in the ordinary number is kept when this number is expressed in standard form.

Calculations using standard form

To **multiply** two numbers shown in standard form, multiply the two numbers together and then **add** the powers of 10.

Examples

$(6.93 \times 10^2) \times (2.18 \times 10^4) = 15.11 \times 10^6$
$= 1.511 \times 10^7$

$(3.85 \times 10^{-2}) \times (4.91 \times 10^{-3}) = 18.90 \times 10^{-5}$
$= 1.890 \times 10^{-4}$

$(8.01 \times 10^6) \times (9.52 \times 10^{-4}) = 76.26 \times 10^2$
$= 7.626 \times 10^3$

To **divide** two numbers shown in standard form, divide one number into the other and then **subtract** the powers of 10. (Remember that subtracting a negative number is the same as adding that number.)

Examples

$(2.63 \times 10^5) \div (1.94 \times 10^3) = 1.36 \times 10^2$

$(9.69 \times 10^8) \div (4.91 \times 10^{-3}) = 1.97 \times 10^{11}$

$(3.85 \times 10^{-2}) \div (4.91 \times 10^{-3}) = 0.78 \times 10^{-1}$
$= 7.8 \times 10^{-2}$

To **add or subtract** two numbers in standard form, convert the numbers to ordinary form and then add or subtract in the normal way.

Decimal places

When carrying out calculations, you should take care to give the correct number of decimal places in your answer. This is determined by the number of d.p. in the figures that you are using to do your calculation.

Example

Calculate the mean of these temperature readings: 18.4 °C, 21.6 °C, 17.2 °C

$$\text{Mean} = \frac{18.4 + 21.6 + 17.2}{3}$$

$$= \frac{57.2}{3}$$

$$= 19.1\ °C$$

Note:

- Your calculator will show you a figure of 19.06666. However, as the original values for temperature were read to just one decimal place, we should retain this number of decimal places in our value for the mean.
- In some cases, it is acceptable to use one more decimal place in your calculated value. For example, if you wished to calculate the mean of 1 and 2, it would be sensible to give your answer as 1.5, which has one more d.p. than the original numbers.
- If there are several steps in the calculation, do not round up or down until you get to the final answer. Keep all the d.p.s throughout all of the steps in the working.
- If the original measurements have different numbers of d.p.s, you should give your final answer to match the number of d.p.s of the **least** accurate measurement.

Significant figures

Significant figures are the digits in a number that carry genuine meaning.

Examples

The number 83.06 has four significant figures (sig figs or sf).

The number 22 584 has five sig figs.

Notes:

- A number such as 2000 is generally considered to have one sig fig, because only the 2 is telling us a particular value. The three zeros just tell us that the 2 represents thousands.
- However, now consider the number 2000 as part of a set of results in which you were counting or measuring something:

  2018, 3645, 1003, 2000, 3987

  Here, it is clear that the numbers are being counted or measured to the nearest whole number. We can now conclude that the number 2000 does mean exactly 2000, and therefore can consider it to have four sig figs.
- If you are doing calculations using data that have been collected to different numbers of sig figs, your final answer should be to the same number of sig figs as the **least** number of sig figs in the raw data used in the calculation.

Percentages

To calculate one number as a percentage of another, divide one number by the other and multiply by 100.

Examples

Out of a population of 235 animals, 43 have black fur.

Calculate the percentage that have black fur.

$$\text{Percentage with black fur} = \frac{43}{235} \times 100$$

$$= 18.3\%$$

A food package states that the food contains 15% fat. If the mass of the food is 200 g, how much fat does it contain?

$$\text{mass of fat} = 200 \times \frac{15}{100}$$

$$= 30\ \text{g}$$

Percentage change or percentage difference

To calculate a percentage change, first subtract one number from the other. Then calculate the difference as a percentage of the original number. State whether your answer represents an increase or a decrease.

Examples

A population of beetles increased from 23 to 96. What is the percentage increase in the population?

$$\text{Increase} = 96 - 23 = 73$$

$$\text{Percentage increase} = \frac{73}{23} \times 100 = 317\%$$

The mass of a piece of potato tissue immersed in a sucrose solution changed from 0.67 g to 0.52 g. What is the percentage change in the mass of the potato tissue?

$$\text{Change} = 0.67 - 0.52 = 0.15 \text{ g decrease}$$

$$\text{Percentage decrease} = \frac{0.15}{0.67} \times 100 = 22.4\% \text{ decrease}$$

Ratios

To represent two numbers as a ratio, divide the larger number by the smaller one.

Example

In a genetic cross between two flies, 24 offspring had red eyes and 9 had black eyes. What is the ratio of phenotypes in these offspring?

$24 \div 9 = 2.67$

The ratio is 2.67 red eyed : 1 black eyed

Probability

Probability tells us the likelihood of a particular event happening. It is normally expressed as a number between 0 (it cannot possibly happen) and 1 (the event will definitely happen). If a coin is tossed, we can state that the probability of it falling heads up is 0.5.

In genetics, we often use probability to predict the chance of an offspring having a particular genotype or phenotype. For example, if two parents with genotypes Tt and Tt are crossed, the probabilities of each genotype occurring in a particular offspring are:

TT 0.25 Tt 0.50 tt 0.25

The probabilities of all the possible outcomes add up to 1.

13.2 RECORDING AND DISPLAYING DATA

Results tables

This example highlights the features that you should aim for when constructing a table to record results obtained during an investigation. In this example, a catalase solution was added to hydrogen peroxide solution, and the volume of oxygen given off was measured at 30 second intervals for three minutes. The experiment was repeated three times.

The independent variable – the one you are changing – goes in the first row or column.

The dependent variable goes in the next rows or columns.

Time / s	Volume of oxygen released / cm^3			
	try 1	try 2	try 3	mean
0	0	0	0	0
30	2.5	3.1	2.8	2.8
60	5.8	6.0	6.2	6.0
90	7.7	7.9	7.7	7.8
120	9.0	8.8	8.8	8.9
150	9.8	8.6	9.6	9.7
180	10.2	10.0	10.0	10.1

Units are shown in the headings, not in the body of the table.

Each reading is taken to the same number of decimal places; this will normally be determined by the apparatus you are using.

Calculated means are shown to the same number of d.p.s as the original readings.

If a reading is very different from what was expected, it could be an anomaly. It may be helpful to make a note of it for future reference. Anomalies may indicate that further data needs to be gathered, or that the experiment method needs to be reviewed.

Line graphs

Where you have two sets of continuous variables, as in the previous results table, you can plot a line graph. When referring to lines on a graph, the correct terminology is actually 'curve', even if it is straight.

Axes should be drawn on a linear scale. The intervals should be chosen so that the scale uses most of the space available on the graph grid, and so that intermediate values can be easily read. For example, a scale that goes up in intervals of 7 would not be sensible.

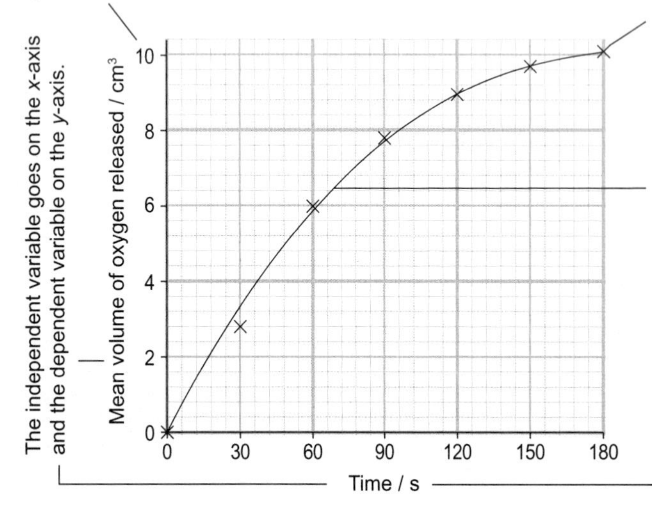

Plot points as small crosses, taking care to place them precisely. If you need to draw a second curve on the same axes, you can use a small dot with a circle drawn around it as an alternative way of plotting points.

Where you have good reason to believe that there is a smooth relationship between the quantity on the *x*-axis and on the *y*-axis, you can draw a best-fit curve. This does not necessarily go exactly through any of the individual points – not even the first and last ones – but you should take care that you have approximately the same number of points above and below the curve, and that you do not extrapolate the curve beyond the range of the data.

Both axes are fully labelled with the quantity and its unit. Note that these headings can be copied from the results table.

Notes:

- In biology, we are often unsure whether there is a smooth relationship between the quantities on the *x*-axis and *y*-axis. We may not be able to predict what the values would have been between our measured points. In this case, do not draw a best-fit curve. Instead, carefully join each point to the next using a ruler. This convention indicates that we have more 'trust' in our individual points that we do in any prediction of what may be happening between them.
- When you are planning an investigation that will generate data that you intend to plot on a line graph, try to have at least five different values of your independent variable, because fewer than five points do not allow you to plot a useful line graph. Your values should also have a well-chosen range, and equal intervals if possible.

Frequency diagrams or histograms

A frequency diagram or histogram shows the number of items that can be grouped into data intervals. In this example, a student measured the lengths of 20 leaves to the nearest mm. These were her raw results:

31, 33, 58, 9, 20, 39, 11, 51, 35, 24, 30, 16, 28, 44, 47, 36, 22, 38, 41, 36

She then placed the leaves into six classes:

Length / mm	0–9	10–19	20–29	30–39	40–49	50–59
Number of leaves	1	2	4	8	3	2

Notes:

- You should aim for a minimum of five classes.
- The range within each class should be the same.
- The **modal set of data** is the most common set in the results. Here, the modal set of data is 30–39 mm.
- The **median** is the middle value of all the values in all the sets of data. To find this value, you will need to arrange all the values in numerical order:

  9, 11, 16, 20, 22, 24, 28, 30, 31, (33) 35, 36, 36, 38, 39, 41, 44, 47, 51, 58

  Because there is an even number of values, the median lies between the two 'central' numbers in the list. The median is 33.5 mm.

This is the frequency diagram for these results:

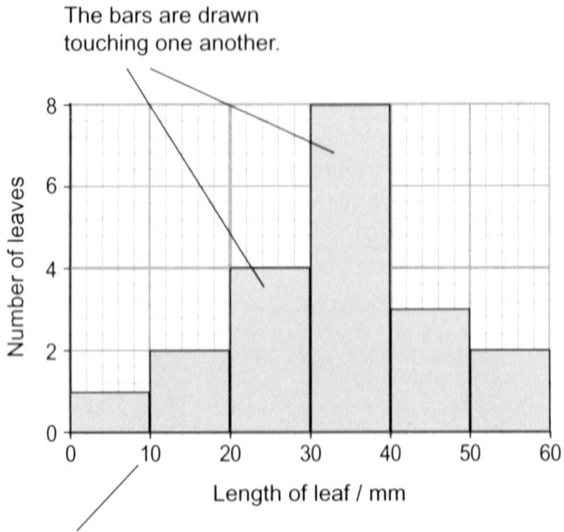

Bar charts

A bar chart is drawn when the x-axis scale is not continuous, but shows discrete categories. The bars do not touch.

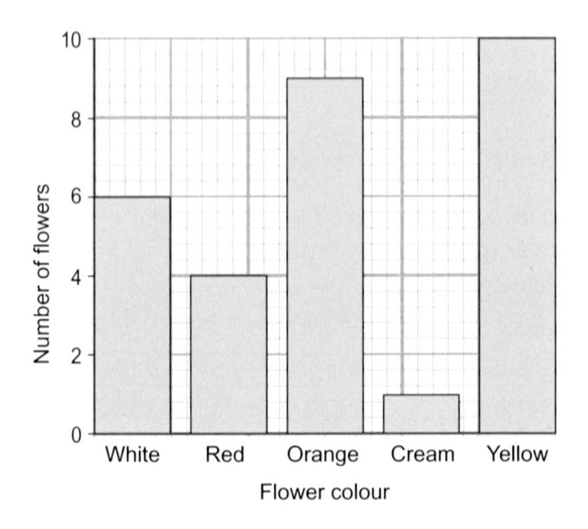

Scattergraph

A scattergraph is drawn when we do not have control over either of the variables we are plotting – that is, we cannot decide on our values for the x-axis variable. We simply take many paired measurements of the two variables. The scattergraph then gives us a visual representation of how one variable relates to the other, and can provide some indication of whether there may be any correlation between them.

This scattergraph shows data from measurements of hind leg length and maximum speed in a sample of 40 lizards.

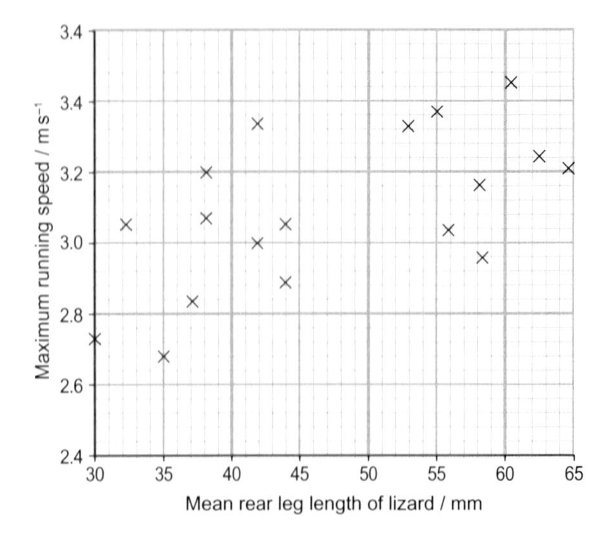

13.3 ANALYSING AND INTERPRETING DATA

Accuracy and precision

In everyday life, the terms 'accurate' and 'precise' have similar meanings. In science, they have different meanings, and you need to take care about when you use them. We can illustrate their meanings by thinking about a particular measuring instrument – a calibrated pipette.

Accuracy: A measurement result is considered accurate if it is judged to be close to the true value. For example, if you are measuring the volume of a liquid in the calibrated pipette, the accuracy of your measurement will depend on how perfectly the pipette has been calibrated. If the pipette has been calibrated correctly, then when the meniscus is at 15.6 cm^3, then the volume really *is* 15.6 cm^3.

Precision: Precise measurements are ones in which there is very little spread about the mean value. Precision depends only on the extent of random errors – it gives no indication of how close results are to the true value, hence the difference between precision and accuracy. Let's say that your pipette has not been calibrated properly, and when the scale reads 15.6 cm^3 the true volume is actually 15.7 cm^3. So long as the pipette *always* reads exactly 15.7 cm^3 when you put this volume of liquid into it, your measurements are precise. However, they are not accurate!

Calculating rates

You can use a line graph to calculate a rate or a rate of change.

If the curve is straight, the relationship between the two variables is said to be **linear**. The relationship between the two variables is represented by the equation:

$y = mx + c$

where:

y is the value of the variable plotted on the y-axis

m is the gradient

x is the value of the variable plotted on the x-axis

c is the y intercept – the point where the curve crosses the y-axis

The rate of change is calculated as:

$$\frac{\text{change in } y}{\text{change in } x}$$

This is shown by the gradient of the curve. Because the curve is straight and does not change its gradient at any point, the rate of growth is the same for the whole time shown on this graph. We can therefore use any part of the graph to calculate the rate. You can do this by drawing a right-angled triangle, using the graph curve as the hypotenuse, and then measuring the values of the vertical and horizontal sides of the triangle. Then divide the vertical value by the horizontal value.

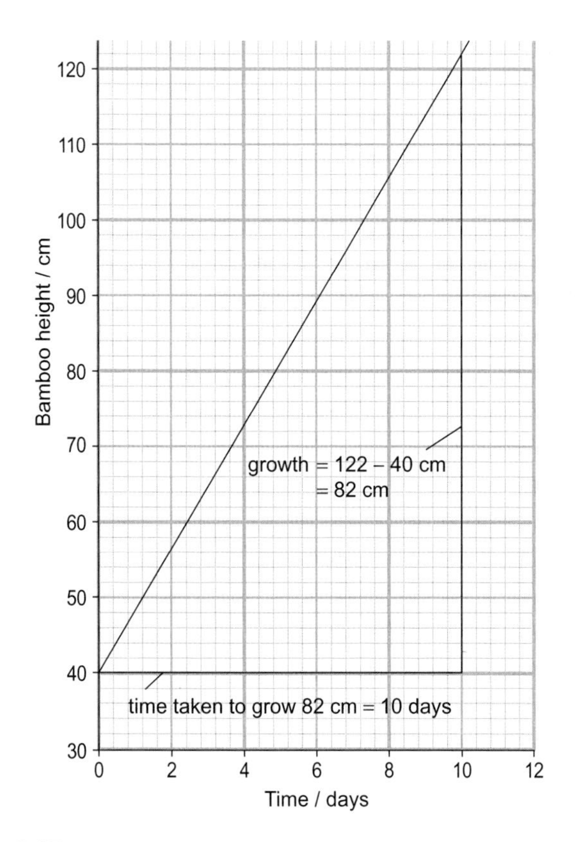

$$\text{growth rate} = \frac{\text{growth}}{\text{time}}$$

$$\text{growth rate} = \frac{82 \text{ cm}}{10 \text{ days}} = 8.2 \text{ cm day}^{-1}$$

If the curve is not straight, then this means that the rate of change is not constant. Here is the graph of oxygen given off against time for the catalase reaction again:

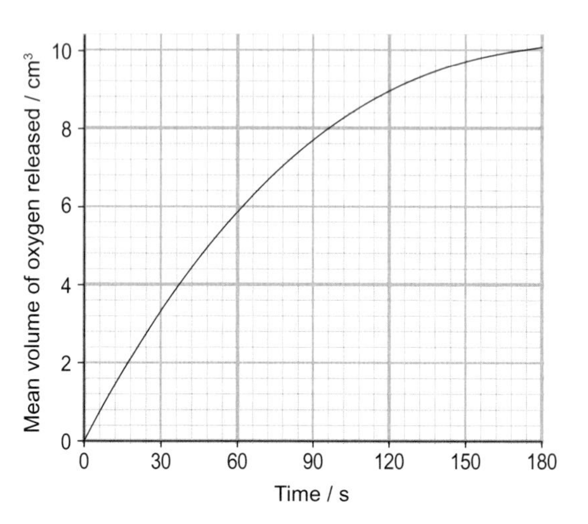

To calculate the rate at a given point in time, we draw a tangent to the curve at that point. We can then draw a right-angled triangle using the tangent as the hypotenuse, and calculate the rate in the same way.

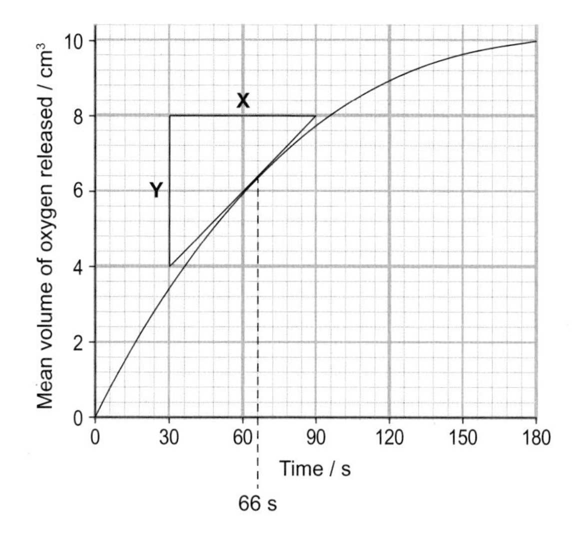

rate of oxygen release $= \frac{Y}{X}$

At 66 seconds:

Y oxygen released $= 8 - 4\ cm^3$
$= 4\ cm^3$

X time taken to release $4\ cm^3$ of oxygen
$= 90 - 30\ s$
$= 60\ s$

rate of oxygen release $= \frac{4\ cm^3}{60\ s} = 0.067\ cm^3\ oxygen\ s^{-1}$

Range and standard deviation

In biology, the data sets we work with often spread out quite widely. We say they are **dispersed**.

For example, you may have been measuring the masses of a set of fruits from a species of plant. Let's say that the smallest mass you measured was 1.7 g, and the largest mass was 4.6 g.

The simplest measure of dispersion is the **range** of the data – from the highest value, 4.6 g, to the lowest value, 1.7 g.

However, it is often more useful to calculate the **standard deviation.** Standard deviation tells us how much a set of data is spread out on either side of the mean. (Ideally, we only use standard deviation when the data show a normal distribution.) Standard deviation takes into account all of the values, not just the two most extreme ones. This can be a more useful measure if one of those extreme measurements was an 'outlier'. For example, it may be that the fruit with a mass of 4.6 g was an outlier, and all of the other fruits lay within the range 1.7 to 3.9 g. The value for the range is therefore rather misleading.

The formula for calculating standard deviation is:

$$s = \sqrt{\frac{\Sigma(x - \overline{x})^2}{n - 1}}$$

where:

s is the standard deviation

x is an individual reading

$\overline{x}$ is the mean

n is the number of readings

This formula may be different to the standard deviation formula you have seen before. Here we are calculating the standard deviation of a sample, in which case we divide by $n - 1$ (as shown above). If you are calculating the standard deviation of a population, you divide by n, not $n - 1$.

You may have a calculator with a standard deviation function, where you simply key in the list of individual readings and it does the whole calculation for you. There are also several free websites where you can do this.

Uncertainties in measurement

The **uncertainty** in a measurement is the interval within which the true value can be expected to lie, with a given level of confidence or probability.

For example, imagine that you have a thermometer with a scale whose smallest interval is 1 °C. You can probably read this scale to the nearest 0.5 °C. However, we cannot be absolutely sure that your reading is spot on. We have to assume that there is some margin of error, or uncertainty, in your reading. In general, we assume that this **uncertainty in measurement is half of the value of the smallest measurement on the scale**. In this case, if we read a temperature as 22.5 °C, we assume that we could be out by 0.5 °C in either direction. We show this by writing the reading as: 22.5°C ± 0.5 °C.

Now imagine that you are measuring a rise in temperature. You have to measure the temperature twice – once at the start and once at the end. You then subtract the initial temperature from the final temperature to find the temperature rise, which works out to be 14.0 °C. Here, we have *two* readings. It's possible that any errors in the readings would cancel each other out, but we have to imagine the worst and assume that they add up. So, we would write our results as 14.0 °C ± 1 °C.

Calculating percentage error

You can express uncertainties as a **percentage error**. To do this, divide the error by the actual measurement, and divide by 100.

For example, imagine you have used a graduated pipette to measure a volume of $23.0\ mm^3$. The scale was marked in $1\ mm^3$ intervals, so your uncertainty in measurement is $0.5\ mm^3$. The percentage error is therefore:

$$\frac{0.5}{23.0} \times 100 = 2.17\%$$

13.4 STATISTICS

Biological investigations often generate data sets that do not give a clear answer to the question we were investigating. We can use statistical tests to tell us whether or not our results support or do not support a particular hypothesis.

Null hypothesis

Statistical tests often tell you whether or not the results that you have obtained differ significantly from your hypothesis. Normally, we use a **null hypothesis**, which is a statement you are actually expecting to reject. You assume that the null hypothesis is correct unless your results prove otherwise. Some examples of null hypotheses could be:

- There is no difference between the effectiveness of the two drugs we are investigating.
- There is no difference between the biology test scores of students who went to bed at 2 am the night before the test, and students who went to bed before midnight.
- There is no difference between the results of this genetics cross and the results that I predicted.

Choosing a statistical test

There are three statistical tests you need to know about (although you do not need to memorise the equations for an examination):

- The chi-squared test is used to determine whether your observed results differ significantly from your expected results. It is especially useful for ecology investigations and in genetics.
- The student's t-test is used to determine whether the mean from two sets of results are significantly different from each other.
- The Spearman's rank correlation coefficient tells you how closely two variables correlate.

It is important that you think about which test to use before you start any experimental work, as part of the experiment design. Think about, for example, what you are measuring or recording, and how best data may be displayed. Here is a list of statistical tests and circumstances in which each one is useful:

Chi-squared

You have two sets of quantitative (numerical) results in which you have counted numbers of things in two or more categories – for example, the numbers of tomato plants with smooth leaves and with serrated leaves, or the numbers of people who smoke and who don't smoke. You want to see if your observed results differ significantly from the expected results that you predict from your null hypothesis.

Student's t-test

You have two sets of data, each with at least 25 items, that both show a normal distribution. You want to find out if the means of the two sets of data are significantly different from each other. Ideally, your two sets of data have similar standard deviations. For example, you want to know if the reaction times of students who have drunk coffee are shorter than the reaction times of students who have drunk water.

Spearman's rank correlation

You have two sets of paired data – for example, the surface area of a fruit and the time it takes to fall to the ground. A scattergraph suggests that there may be a relationship between them, not necessarily linear.

The chi-squared test

The formula for the chi-squared test is:

$$x^2 = \sum \frac{(O - E)^2}{E}$$

where:

$\sum$ means the sum of

O is a value for your observed results

E is a value for your expected results (assuming that the null hypothesis is correct)

Imagine that you want to find out if the number of honeybees visiting yellow flowers differs from the number visiting red flowers or white flowers. You sit patiently and count all the bee visits to the two types of flowers over one hour. Your results are:

number of bee visits to yellow flowers	81
number of bee visits to red flowers	65
number of bee visits to white flowers	64

This certainly looks as though there are more visits to the yellow flowers. But is this difference significant, or could it just be due to chance?

First, construct your null hypothesis. This is:

There is no difference between the numbers of bee visits to red flowers, yellow flowers or white flowers.

Now draw and complete a table like this. (If you have more than three categories, you can add extra rows so that you have one for each category.) You could use an Excel® spreadsheet to do this.

Type of flower	Observed number, O	Expected number, E	$O - E$	$(O - E)^2$	$\frac{(O - E)^2}{E}$
Yellow	81	70	11	121	1.73
Red	65	70	−5	25	0.36
White	64	70	−6	36	0.51
Total	**210**	**210**			**2.60**

To complete the **Expected number** column, assume that the null hypothesis is correct and that there is no difference between the number of visits to each type of flower. All you have to do is divide the total number of visits by three.

When you finally calculate the total of the last column, you have found chi-squared. The value of **2.60** is your value for chi-squared.

Now we have to find out what this means. To do this, you first need to work out many degrees of freedom you have in your results. This is one less than the number of categories of your results. Here, we have three categories (yellow, red or white) so we have two degrees of freedom.

Now look up your value of chi-squared in a chi-squared table. Here is part of one. The numbers in the table are chi-squared values.

Degrees of freedom	Probability of null hypothesis being correct			
	0.1	0.05	0.01	0.001
1	2.71	3.84	6.64	10.83
2	4.60	5.99	9.21	13.82
3	6.25	7.82	11.34	16.27
4	7.78	9.49	13.28	18.46

Find the row for two degrees of freedom, and go along the row to find a number close to your value for chi-squared. The nearest number is 4.60, and our number is much lower than this. So we can say that the probability of the null hypothesis being correct is more than 0.1.

In biology, we normally take a value of 0.05 as being the **critical value** of the test statistic (that, in our table, is 5.99 for two degrees of freedom). In other words, if the probability of the null hypothesis being correct is greater than 0.05 (and, in this case, we have found it to be between 0.1 and 0.2), we say that we have not disproved the null hypothesis – the difference between our observed and expected results could be due to chance. So in this case we have no firm evidence that there is any difference in the numbers of bees visiting the different coloured flowers. We would need to have a chi-squared value of 5.99 or more before we could say that the difference between our observed results and our expected results is significant.

The student's t-test

The formula for the t-test is:

$$t = \frac{\bar{x}_1 - \bar{x}_2}{\sqrt{\frac{s_1^2}{n_1} + \frac{s_2^2}{n_2}}}$$

where:

$\bar{x}_1$ is the mean of sample 1

$\bar{x}_2$ is the mean of sample 2

s_1 is the standard deviation of sample 1

s_2 is the standard deviation of sample 2

n_1 is the number of individual measurements in sample 1

n_2 is the number of individual measurements in sample 2

Note that you do not need to be able to recall this formula, nor any statistical formulae, in the examination.

You can see that you first need to calculate the standard deviation for your sample before you can do this calculation. The way to do this is explained in Section 13.3, *Analysing and interpreting data* on page 266.

Let's say that you are investigating whether the test scores of 12 students who went to bed at 2 am are significantly different from 12 students who went to bed at 11 pm.

The results were:

2 am students
12, 6, 14, 19, 5, 3, 13, 11, 12, 7, 16, 8

mean $\bar{x}_1 = 10.5$

standard deviation $s_1 = 4.777$

$s_1^2 = 22.82$

$\frac{s_1^2}{n_1} = \frac{22.82}{12} = 1.91$

11 pm students
12, 4, 19, 11, 9, 17, 2, 10, 8, 15, 12

mean $\bar{x}_2 = 10.8$

standard deviation $s_2 = 5.128$

$s_2^2 = 26.30$

$\frac{s_2^2}{n_2} = \frac{5.128}{12} = 0.43$

So,

$$t = \frac{10.5 - 10.8}{\sqrt{1.91 + 0.43}}$$

$= \mathbf{1.96}$

Now we need to look up this value for t in a table of probabilities. First, we need to decide how many degrees of freedom we have in our data. This is the total number of measurements minus 2, which in this case is 22.

The table below shows the values of *t* for 20, 22 and 24 degrees of freedom. The numbers in the table are values of *t*.

Degrees of freedom	Probability of null hypothesis being correct			
	0.10	0.05	0.01	0.001
20	1.73	2.09	2.85	3.85
22	1.72	2.07	2.82	3.79
24	1.71	2.06	2.80	3.75

Looking across the 22 degrees of freedom row, our value of *t* lies between 1.72 and 2.07. So the probability of the differences between the two sets of values being due to chance is somewhere between 0.10 and 0.05. Because this chance is greater than 0.05, we cannot say that the null hypothesis is not correct. Our results do not show a significant difference between the test results of the students who went to bed at 11 pm and those who went to bed at 2 am.

ANSWERS TO IN-TEXT QUESTIONS

1. WATER AND CARBOHYDRATES

1. carbon atom, amino acid molecule, protein molecule, animal cell
2. Hydrogen bonds attract water molecules to each other. When heat energy is added to water – for example, when infrared radiation from the Sun falls onto its surface – a lot of this energy is used to break the hydrogen bonds. Only then can the energy be used to increase the kinetic energy of the water molecules, which is what we measure when we measure temperature. Water, therefore, tends to heat up and cool down only very gradually, allowing fish to live in an environment with a relatively constant temperature, which also helps their own body temperature to remain constant.
3. **a.** The boiling point is affected by pressure. As altitude rises, atmospheric pressure decreases. At sea level water boils at 100 °C but on top of Everest it boils at about 71.7 °C. This is because the boiling point is the temperature at which the vapour pressure of the liquid is equal to atmospheric pressure.

   b. As the boiling point of water is lower, the maximum temperature the water can reach is lower. To ensure pathogens are killed the water should be boiled for longer.
4. **a** α-Glucose

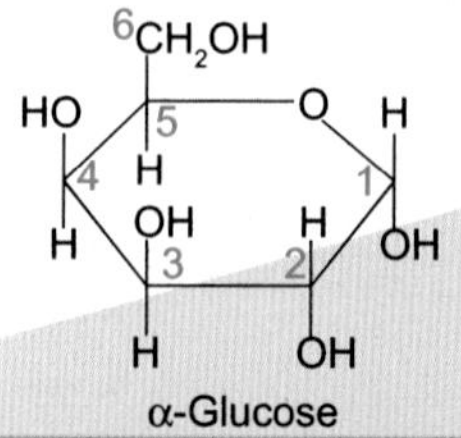

   b β-Glucose
5. Diagram should be the reverse of Figure 10.
6.

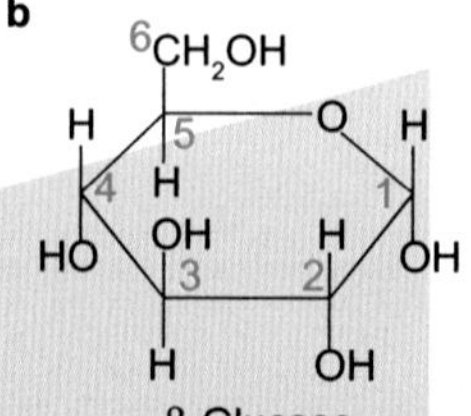

Feature	Starch	Cellulose
sugar units from which it is made	α-glucose	β-glucose
type of glycosidic bond	α1-4	β1-4
branching	much branching (at α1-6 links)	no branching
overall shape	spiral	straight
hydrogen bonding	only with in the molecule (holds the spiral in shape)	with other molecules (holds several parallel molecules together to form microfibrils)
solubility	insoluble	insoluble
ease of digestion	can be easily digested by amylase	more difficult to digest; relatively few organisms have cellulase
where found	inside plant cells	in plant cell walls
function	energy store	structural (gives cell walls their strength)

7. Maltose is a reducing sugar, but sucrose is non-reducing.
8. Boiling sucrose with acid hydrolyses it to glucose and fructose. These are reducing sugars, so Benedict's test now gives a positive result.
9. Prepare glucose solutions of known concentration (for example, 0.1%, 0.01%). This can be done by diluting a more concentrated solution. Carry out Benedict's test on each, using the same volumes of both the glucose and Benedict's solutions each time.

2. LIPIDS AND PROTEINS

1. Fatty acids and glycerol
2. An unsaturated fatty acid has at least one C = C double bond; a saturated fatty acid has all C–C single bonds.
3. In the phospholipid, a fatty acid molecule is replaced by a phosphate group.
4. Carbon, hydrogen, oxygen, nitrogen
5. Nitrogen
6. NH_2
7. COOH
8.

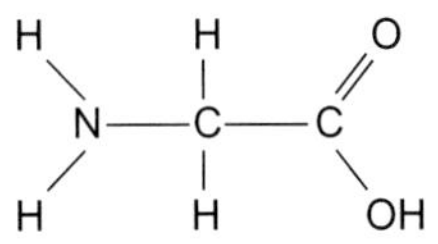

```
H       H     O
 \      |    //
  N-----C----C      Glycine
 /      |    \
H       H     OH

        H
        |
    H---C---H
H       |     O
 \      |    //
  N-----C----C      Alanine
 /      |    \
H       H     OH

        H
        |
        S
        |
H   H---C---H    O
 \      |       //
  N-----C------C      Cysteine
 /      |       \
H       H        OH
```

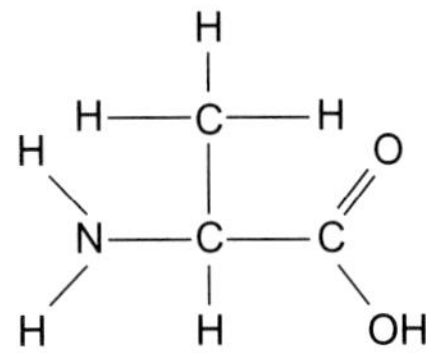

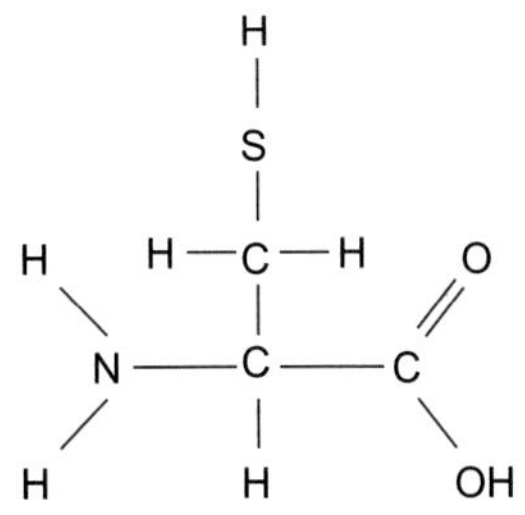

9. and 10.

$$NH_2CHCH_3CH_2COOH + NH_2CH_2COOH \rightarrow NH_2CHCH_3CONHCH_2COOH$$

Alanine + glysine → dipeptide

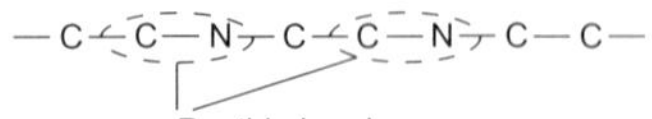

```
—C—C—N—C—C—N—C—C—
```

Peptide bonds

11. Cysteine
12. The three-dimensional shape is determined by the bonds that form between the R groups of different amino acids in the chain. The sequence of these amino acids (the primary structure) therefore affects where these bonds will form.
13. The tangle of collagen rods makes it less likely that the cartilage will tear and gives strength to the structure.
14. The cornea has to be transparent. The regular stacking of the collagen rods prevents light being scattered in all directions.

3. ENZYMES

1. The active site of an enzyme molecule has a very specific shape, determined by the tertiary structure of the molecule. This shape allows only one type of substrate molecule to fit and bind. The substrate molecule must have a complementary shape to the enzyme's active site.
2. The primary structure – that is, the sequence of amino acids in a protein molecule – determines its three-dimensional structure. This is because the primary structure affects where bonds can form between the different amino acids in the chain, and therefore the way that it folds. A small change in the primary structure of an enzyme can have a large effect on its overall three-dimensional structure, so its active site may be the wrong the shape to bind with its substrate.
3. Molecules have low kinetic energy, so they move slowly and are less likely to collide and react.
4. Approximately 2.5, 5 and 10 units
5. The rate of reaction approximately doubles.
6. Freezing would not denature the enzyme, so the rate of reaction would be unaffected.

7. a. 55–56 °C
   b. Clothes can be washed in warmer water without the enzyme being denatured. There is no need to presoak clothes in the washing powder.
8. Some proteins have large molecules that are not soluble in water. (Blood stains contain haemoglobin, which is usually soluble in water, but it often interacts with the molecules that make up fabrics and gets stuck to them.) The proteases in the washing powder break down the protein into amino acids, which are all soluble and are therefore washed away.
9. Some enzymes can operate optimally at very high temperatures so the peak of the curve would shift to the right. Some enzymes have a greater number of bonds between amino acids so the slope of the curve above the peak might be less steep as they are more resistant to denaturing.
10. As the concentration of enzyme increases, substrate molecules collide more frequently with active sites and the rate of reaction increases. The maximum rate is reached when all active sites are in use all the time. If the substrate concentration increases so much that there is excess substrate, the rate of reaction cannot increase further. This would normally only occur when enzyme concentration is low, as enzymes work so fast that only small amounts are needed.
11. a. X
    b. Y
    c. X
    d. Y
    e. X and Y
12. The structure of the molecules is very similar. The malonate is attracted to the active site but no reaction occurs. This reduces the number of sites available for the succinate to react.
13. The enzyme being inhibited catalyses a reaction that was unique to the pest; the inhibitor has no toxic effect on humans or other organisms; the inhibitor is stable, and does not break down into harmful substances; it can easily be administered to the pests in the right dose; it does not persist or build up to harmful levels in the environment.

4. NUCLEOTIDES

1.

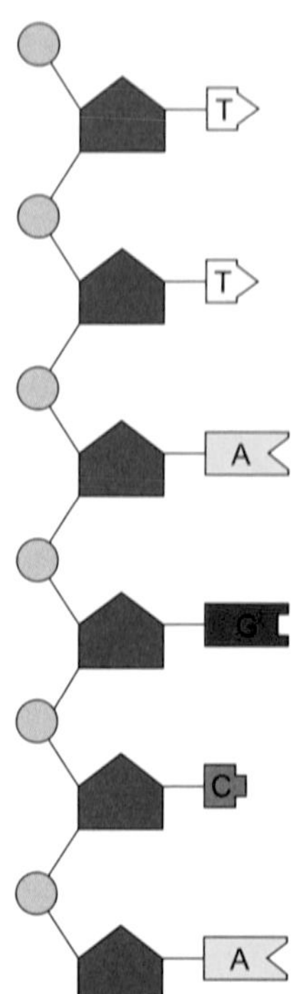

2.

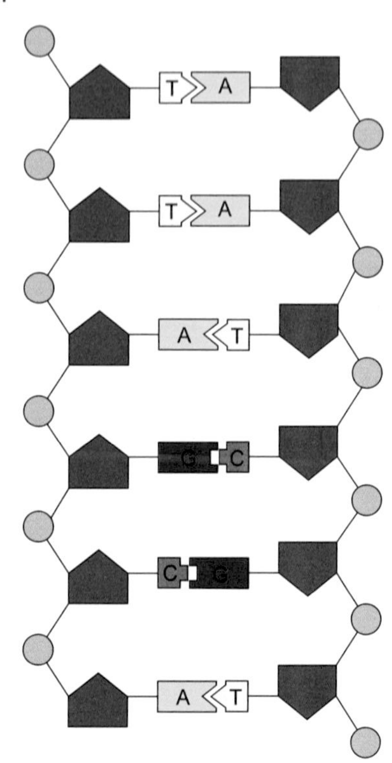

3. a. In each organism the percentages of adenine and thymine are the same, and the percentages of cytosine and guanine are the same. This is because they fit together as complementary pairs in DNA, so there must be the same number of adenine and thymine bases, and of cytosine and guanine.

   b. 24% (26% = adenine, so 26% = thymine. The remaining 48% are split equally between cytosine and guanine).

4.

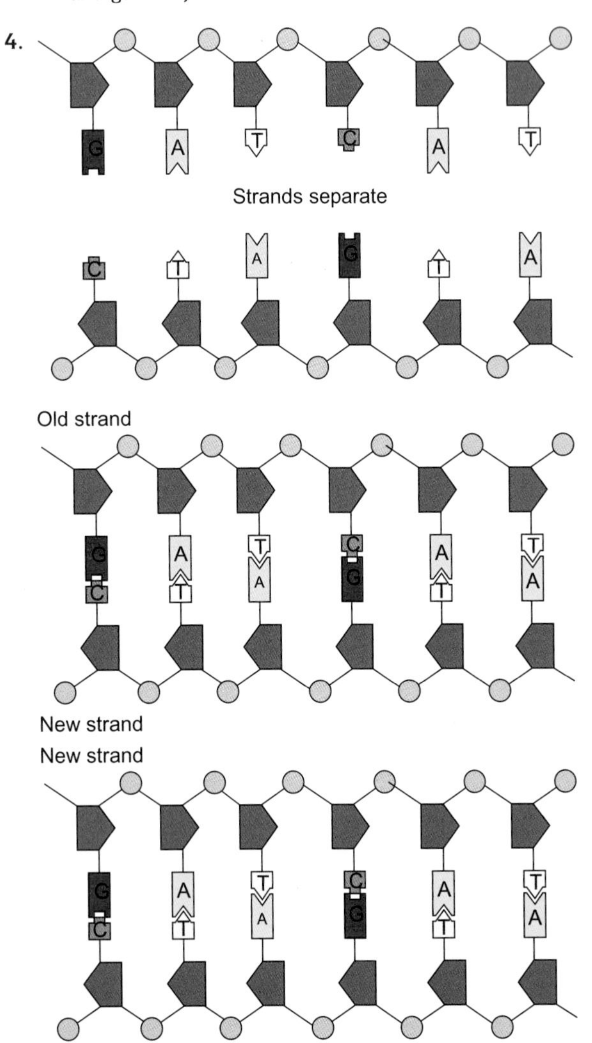

5.

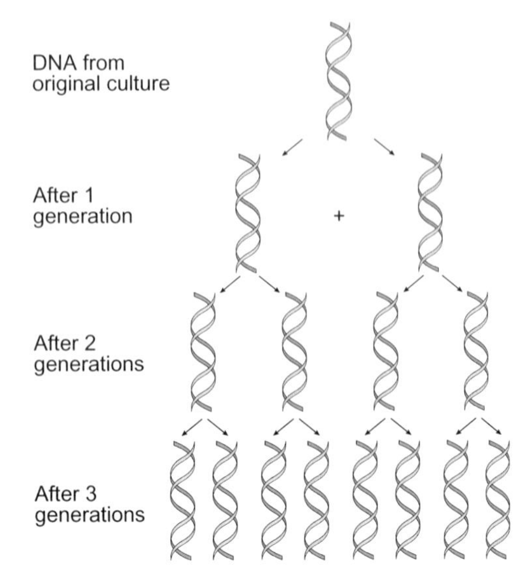

6. The graph should show DNA containing ^{15}N reducing to near zero, $^{14/15}$ N increasing then decreasing and ^{14}N increasing.

7. Two equal-sized bands, one at the top level (containing ^{14}N only) and one at the lowest level (containing ^{15}N only).

8. Condensation

5. CELLS

1. They would otherwise hydrolyse many substances within the cell, destroying the cell.

2. The active site of each enzyme has a complementary shape to part of its specific substrate molecule. Although there are many different molecules that can be hydrolysed, several of these molecules may have parts that are similar in shape to others, and so can bind with the same enzyme. For example, amino acids all have an amino group, so one enzyme with an active site that could bind and react with an amino group could process most different types of proteins regardless of their length or complexity of R groups.

3. Nucleus, mitochondria, chloroplasts

4. Both mitochondria and chloroplasts have two membranes (an envelope), which can be explained by the endosymbiont theory, which states that they have evolved from prokaryotic cells that came to live inside another cell. Both contain their own circular DNA, which can also be explained by the endosymbiont theory. This also explains the presence of ribosomes in both

organelles, which are of the smaller type characteristic of prokaryotic cells rather than eukaryotic cells.

In both of these organelles, membranes are folded to provide large surface areas. In mitochondria, this allows for the attachment of large numbers of molecules involved in the production of ATP in the final stages of respiration, while in chloroplasts it allows for the attachment of large numbers of chlorophyll molecules involved in the absorption of energy from sunlight.

Both organelles also have a 'background' material – stroma in chloroplasts, the matrix in mitochondria – that contain enzymes required to catalyse the reactions in certain stages of photosynthesis and respiration respectively.

5.

Prokaryotic cells	Eukaryotic cells
Have a cell-surface membrane	Have a cell-surface membrane
Have cytosol	Have cytosol
No nucleus; DNA is free in the cytosol	Have a nucleus bounded by a nuclear envelope
DNA is in the form of a circular molecule, not associated with proteins	DNA is in linear molecules, associated with proteins (histones) to form chromosomes
Have small ribosomes, always free in the cytosol	Have larger ribosomes, sometimes free in the cytosol and sometimes attached to endoplasmic reticulum
Have no membrane-bound organelles	Have membrane-bound organelles, including mitochondria, endoplasmic reticulum, Golgi apparatus, lysosomes and (in plant cells only) chloroplasts
Have a cell wall containing murein	Plant cells have a cell wall containing the polysaccharide cellulose, fungal cells have a cell wall containing a polymer called chitin, and animal cells do not have a cell wall at all
Much smaller than eukaryotic cells	Much larger than prokaryotic cells
May have small, circular DNA molecules called plasmids	Do not have plasmids
May have a capsule surrounding the cell wall	Do not have a capsule
May have a flagellum with a rotary motor	May have a flagellum, but not with a rotary motor

6. The combined magnification of the two lenses in an optical microscope produce a maximum magnification of only × 1500. More powerful lenses would not be able to resolve objects smaller than half the wavelength of light.

7. Magnification is the number of times larger the object appears; resolution is the ability to distinguish between two objects.

8. Multiply the eyepiece and objective together. So, the low power magnification is × 40, medium power is × 100 and high power is × 400.

9. It would allow them to examine real-time cellular changes and cell-cell interactions in 3D. Scientists have been able to see DNA replication and the cell cycle in action in real time using this technique.

10. a. To reduce the activity of enzymes to a minimum
    b. So that the organelles do not change size because of osmosis

11. a. Nucleus, mitochondria, ribosomes
    b. The nucleus in the bottom fraction, then mitochondria above, then ribosomes.

12. a. There will be twice as much at the end of interphase.
    b. 92
    c. So that they attach to the spindle together. The members of each pair are then pulled to opposite poles. This makes sure that each cell has one copy of each chromosome.

13. Metaphase

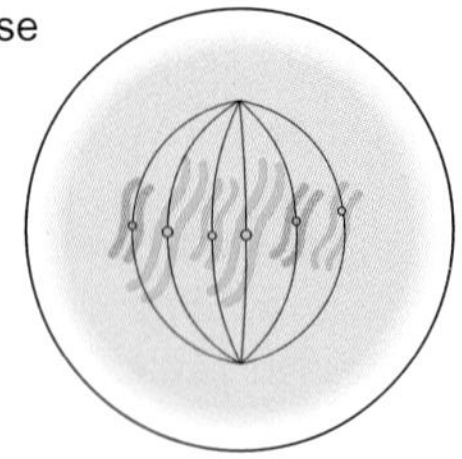

Anaphase

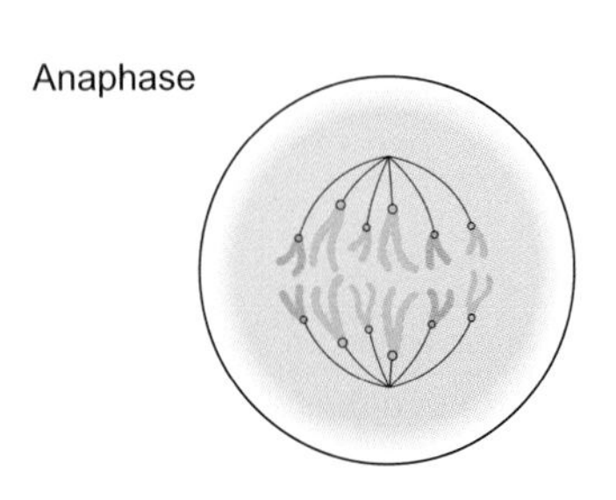

Telophase

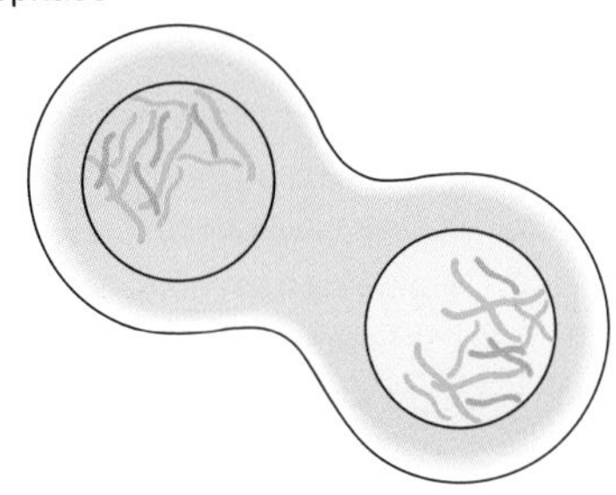

5. a.

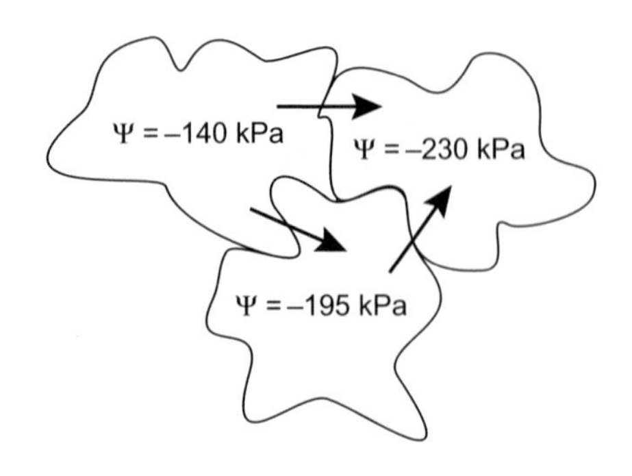

b. Most negative – 230 kPa; Least negative – 140 kPa

6. Molecules move in straight lines but in random directions. Relatively few molecules will be moving along paths that take them between the constantly moving phospholipid molecules. Also, phospholipids are hydrophobic in nature, which makes them more difficult for hydrophilic molecules (such as water) to interact with.

7. Because this has the same water potential as blood plasma.

8. Since blood takes glucose away from the cell, there will always be a concentration gradient down which glucose will diffuse. However, the sodium ion concentration in blood is greater than that in the cell, so active transport is needed to move the ions against their concentration gradient.

6. CELL MEMBRANES

1. a. Measured thickness is about 1.5 mm, which is equal to 1 500 000 nm. Magnification is × 370 000. Therefore, the actual thickness is 1 500 000 ÷ 370 000, which is about 4 nm.

   b. By making several measurements of the width, then calculating a mean

2. The length of a phospholipid molecule is half of the 4 nm calculated above, which is 2 nm.

3. The dark colour indicates a region of high concentration of solutes from the tea bag; lighter colours indicate regions of low concentrations of these solutes; there is a concentration gradient of these solutes from the dark colour to the lighter colours.

4. Osmosis is a special case of diffusion. It involves the movement of water through a partially permeable membrane. This membrane allows water molecules to pass through but not solute molecules. The water molecules move down their concentration gradient, but the membrane prevents the solute molecules moving down their concentration gradient.

7. THE IMMUNE SYSTEM

1. An enzyme that catalyses a hydrolysis reaction, that is a reaction in which water is added to a polymer (such as a protein) to break the bonds that hold the monomers together.

2. They are produced by mitosis, which always produced genetically identical cells – that is, cells that each carry identical DNA. The proteins in the cell-surface membrane, like all other proteins in the cell, are produced following the code carried in the DNA.

3. The cell-surface membrane controls what can enter and leave the cell. If holes are made in it, then substances inside the cell can escape, and substances outside can get in. This prevents metabolic reactions happening normally in the cell, so it dies.

4. It may be that the cancer cell does not display antigens that are recognised by any of the T cells as non-self.

5. The cellular response involves T cells, whereas the humoral response involves B cells. Cytotoxic T cells themselves destroy cells carrying their specific antigen, whereas plasma cells derived from B cells secrete antibodies that attach to the antigen and stimulate its destruction.

6. The secondary response is much faster than the primary response, generating antibodies in a much shorter time after entry of the antigen. In other words, the latent period is much shorter for the secondary response than for the primary response.

   The secondary response is greater than the primary response, producing much larger quantities of antibodies.

7. The primary response produces memory cells, each with receptors in their cell-surface membrane that can bind with the particular antigen that stimulated the response. If that same antigen enters the body a second time, these memory cells are already present in the blood in quite large numbers, and can respond quickly.

8. The first vaccination elicits a primary response and the second 'booster' dose elicits a secondary response. This produces more antibodies and more memory cells, providing better protection against the disease than could be achieved with a single injection.

9. Passive immunity results from being given antibodies, which do not last for very long in the body. Active immunity results from the body responding to the presence of an antigen by producing its own antibodies and memory cells. The memory cells stay in the blood for a long time.

10. a. As percentage vaccination increased between 1980 and 1982, the number of cases of diphtheria reduced considerably. Percentage vaccination has stayed fairly constant at about 85% since 1990, but the number of cases has fluctuated in that period.

    b. There was a slight decrease in the immunisation coverage between 1990 and 1991, from 84% to approximately 82%. 82% was not a high enough coverage to provide herd immunity and prevent outbreaks.

11. This means that the antibodies are produced by a single ('mono') set of genetically identical cells (a clone).

12. The monoclonal antibodies bind only to cells that have a specific antigen on their surfaces, and 'ignore' all the other body cells. They therefore deliver the drug only to this group of cells.

13. This helps to check that the test is working properly. If it is, then the cells containing the antigen will change colour, and those that do not, will not.

14. Monoclonals recognise a specific epitope and polyclonals can recognise multiple and similar epitopes on a molecule. (An epitope is a particular part of an antigen molecule that binds to an antibody.) If you are looking for a positive result using an Elisa test then you need to use monoclonals as they will only produce a positive result when bound to the target molecule, whereas polyclonal may attach to similar (but not the same) molecules producing a false positive.

15. Polyclonal antibodies can bind multiple epitopes of similar molecules so you may use them if you are not exactly sure what you are looking for and want to find out if a particular family of molecules (for example, if different proteins of the same family) are present or not. Also, if the molecule you are looking for is of very low abundance with hidden regions of sequences then a polyclonal can bind to other areas and amplify a signal.

8. EXCHANGE WITH THE ENVIRONMENT

1.

Length of side of cube/cm	Surface area of cube/cm^2	Volume of cube/cm^3	Surface area : volume ratio
1	6	1	6:1
2	24	8	3:1
3	54	27	2:1
4	96	64	1.5:1

2. As size increases, surface area : volume ratio decreases.

3. They are both quite small molecules, without charges. They are therefore able to move easily through the non-polar (hydrophobic) fatty acid tails of the phospholipids that make up the cell membrane.

4. At the tips of its tracheoles.

5. The network of tracheoles has a large surface area across which gases can diffuse to and from the body tissues. The amount of water loss is reduced because the endings of the tracheoles are deep inside the body, and also because the spiracles can be closed if necessary.

6. The rate of respiration of the insect's muscles is greatest during flight, when the muscles are very active. This is when they need most oxygen and produce most carbon dioxide. It is therefore advantageous to have greater ventilation of the gas exchange surface when the insect needs it most, produced by the same muscles that are involved in flight.

7. The external skeleton of insects is integral to its protection and the way its gas exchange system works. The larger the insect, the heavier its external skeleton and the greater the energy required to carry it around. Therefore, more oxygen would be required to perform respiration. This limits the size of insects. Mammals have an internal, and relatively light, skeleton.

8. In the prehistoric periods, such as the Carboniferous period 400 million years ago, larger insects existed due to the higher oxygen content of the atmosphere at the time.

9. The surface area of the lamellae is greatly reduced when they are stuck together, so not enough oxygen can be absorbed or carbon dioxide removed. (Also, the lamellae will quickly dry out when in air, killing the cells on their surfaces.)

10. During the day, oxygen diffuses out and carbon dioxide diffuses in. During the night, oxygen diffuses in and carbon dioxide diffuses out.

11. Oxygen diffuses more slowly through water than through air, so it will slow down the rate of diffusion.

12. **Differences**: Humans obtain oxygen from air, fish from water; humans have alveoli, fish have gill filaments; ventilation is tidal in humans, unidirectional in fish. **Similarities**: Both have a large gas exchange surface; both have a good blood supply to the gas exchange surface; both ventilate the gas exchange surface.

13. Sucrase has an active site of complimentary shape to sucrose. This is not the same shaped active site as on lactase. Lactase's active site is complimentary in shape to lactose, not sucrose.

14. A villus is a finger-like fold in the inner surface of the small intestine, made up of many cells. A microvillus is a tiny fold in the surface of a single cell, many times smaller than a villus.

15.

Nutrient	Enzymes that digest them	Substrate	Product(s)	Form in which absorbed	Method of transport into epithelial cell	Method of transport out of epithelial cell
Carbohydrate	Amylase	Starch	Maltose	Monosaccharides (such as glucose)	Active co-transport (with sodium ions)	Facilitated diffusion
	Maltase	Maltose	Glucose			
	Sucrase	Sucrose	Glucose and fructose			
	Lactase	Lactose	Glucose and galactose			
Protein	Pepsin	Protein	Polypeptides	Amino acids	Active co-transport (with sodium ions)	Facilitated diffusion and diffusion
	Trypsin	Protein	Polypeptides			
	Exopeptidase	Polypeptides	Amino acids and dipeptides			
	Dipeptidase	Dipeptides	Amino acids			
Lipid	Lipase	Lipids	Fatty acids and monoglycerides	Fatty acids and monoglycerides	Diffusion through phospholipid bilyaer	Exocytosis (in chlyomicrons)

9. MASS TRANSPORT

1. B and C
2. a. ~95%
   b. ~ 100%
   c. ~ 100%
3. 11 Approximately 12 kPa
4. 100 − 25 = 75%
5. The haemoglobin becomes fully saturated with oxygen even at the low concentrations of oxygen in the surroundings. The disadvantage is that it will not give up its oxygen until the oxygen concentration in the tissues is very low.
6. An S-shaped curve to the left of that for human haemoglobin; the partial pressure of oxygen will be low at high altitude and llama haemoglobin will become saturated with oxygen at these partial pressures.
7. 60 ÷ 0.8 = 75 beats per minute
8. The line would follow exactly the same pattern, changing gradient at exactly the same time. However, the highest pressure reached would be much less.
9. In the skin, reduced blood flow through the skin capillaries means that less heat is lost from the blood by radiation and the body conserves heat. In the villi, reduced blood flow reduces the rate of absorption of soluble nutrients from the blood.
10. The total cross-sectional area of capillaries is greater than that of arterioles.
11. The total cross-sectional area of large veins is less than that of small veins.
12. The slow flow rate allows more time for diffusion; they are thinner and also have very thin walls, reducing the length of the diffusion path.

13.

	Red blood cells	White blood cells	Fibrinogen	Other plasma proteins	Glucose, amino acids, ions, etc.
Blood	✓	✓	✓	✓	✓
Plasma	✗	✗	✓	✓	✓
Tissue fluid	✗	✗	✗	✗	✓

14.

	From capillaries to tissue	To capillaries from tissue
Intestinal villi	oxygen	carbon dioxide, glucose, amino acids
Brain	oxygen, glucose, amino acids	carbon dioxide
Leg muscles	oxygen, glucose, amino acids	carbon dioxide
Liver	oxygen, glucose, amino acids	carbon dioxide, urea (and glucose when in low concentration in the blood)

15. Lack of proteins in the blood plasma makes the water potential less negative. Less water returns to the capillaries at the venous end from the capillaries. Therefore, fluid accumulates in the tissues.
16. Tree C: its diameter decreases the least, showing that there is the least tension in the xylem vessels resulting from the transpiration pull.
17.

Xylem vessels	Phloem sieve tubes
Cells are dead, and have no contents	Cells are living, with small amounts of cytoplasm (but no nucleus)
Walls contain cellulose and lignin	Walls contain cellulose but not lignin
End walls have completely broken down	End walls are still present, perforated with pores to form sieve plates

10. DNA AND PROTEIN SYNTHESIS

1.

Prokaryotic DNA	Eukaryotic DNA
relatively short molecules	relatively long molecules
circular	linear
not associated with histones	wound around histones to form chromosomes

2. The top image is a woman's chromosomes, as you can see the X and Y at bottom right. The bottom image has two X chromosomes, so these are the chromosomes of a man.
3. Phenylalanine, glutamine, cysteine, histidine, asparagine
4. Cysteine, asparagine, histidine, valine, histidine
5. DNA molecules are longer; double stranded instead of single; have deoxyribose instead of ribose; have thymine instead of uracil.
6. U G C U A A C A C G U G C U C
7. Both are single stranded, both contain uracil instead of thymine. mRNA is a long strand, tRNA is a clover-leaf shape; tRNA has an anticodon, mRNA does not.
8. AAC CAG CAC CUC UGC

   UUG GUC GUG GAG ACG
9. **a.** 435

   b. 435

   c. 14
10. **a.** More than one codon codes for the same amino acid. For example, GGG, GGA, GGC and GGU all code for glycine.

    b. Lysine, arginine, serine, alanine

    c. UUC GCG AGA CGU

    d. The only codon would be UUU, which encodes for phenylalanine.

    e. Lysine

    f. Serine and leucine in equal quantities. The codons would be alternately UCU and CUC.

11. GENETIC DIVERSITY

1. **a.** substitution, addition, substitution, deletion.

   b. Val, Val, Ser, Thr, Leu

   Val, Ala, Ser, Thr, Leu

   Val, Val, Phe, Tyr, Ser

   Val, Val, Ser, Thr, Leu

   Val, Tyr, Leu, Leu

   c. One amino acid different.

   All amino acids changed except first two.

   No change (CAT and CAG code for same amino acid).

   Only first amino acid the same.
2. **a.** As the quantity of substance Z increases, the activity of enzyme A will decrease. This will lower the quantity of substance Z that is produced. Enzyme A will only become active again once the quantity of substance Z decreases. This is a negative feedback mechanism.

   b. The addition mutation will change all the DNA triplets that follow it. This will completely change the amino acids that are used to build the enzyme. The enzyme will therefore not have the correct three-dimensional shape. Its active site will be the wrong shape to bind with its substrate.

   c. No substance Z will be produced. (However, remember that there are two copies of each gene in a cell. The other gene for the enzyme may be normal, so the cell may be able to produce some normal enzyme B, and therefore be able to produce some substance Z.)

   d. The quantity of substance X will increase, as it is not now being changed to substance Y.

   e. The cell will change substance Y to substance Z, using enzyme C. The quantity of substance Z will therefore increase. Substance X will still build up, however, because it is still not being converted to substance Y.

3. a.

Mitosis	Meiosis
one division	two divisions
two daughter cells produced	four daughter cells produced
daughter cells have the same number of chromosomes as the parent cell	daughter cells have half the number of chromosomes as the parent cell
homologous chromosomes do not pair up	homologous chromosomes pair to form bivalents
no crossing over takes place	crossing over takes place between chromatids of homologous chromosomes
daughter cells are genetically identical with the parent cell and each other	daughter cells are genetically different, containing different combinations of alleles

b. Any three from:

Chromosomes replicate in interphase or before the division starts.

Chromatids are joined by centromeres.

Nuclear membrane disappears at end of prophase.

Centromeres attach to equator or spindle in metaphase.

Chromatids are pulled to opposite poles in anaphase.

4. a. 2^4 (16 combinations)

b. 2^{23} (8388 608 combinations)

5.

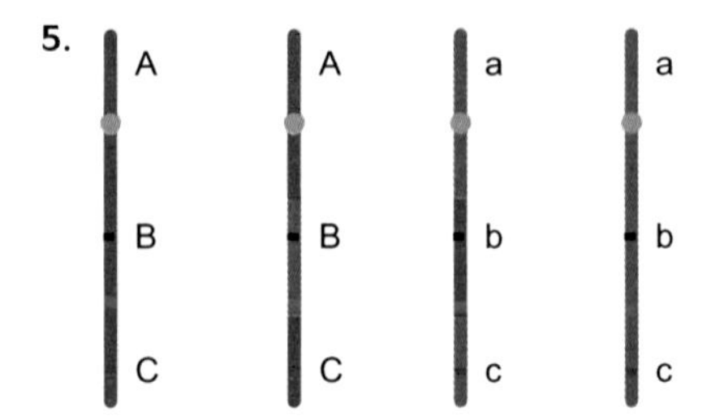

6. a. The amount of food available, and space for example, for breeding sites.

b. The population would increase 10-fold each year. There would soon be severe competition for food.

c. Two

7. a.

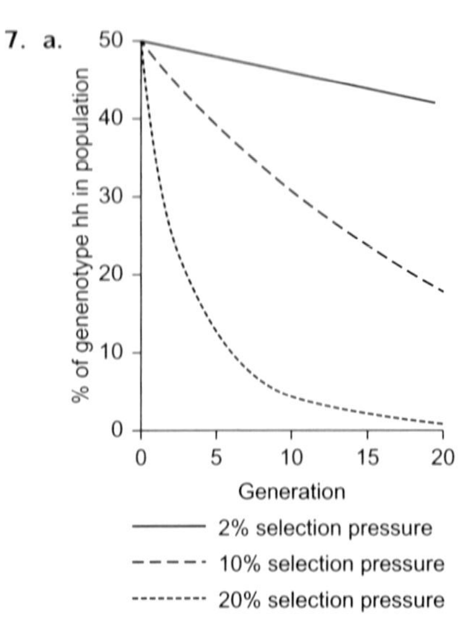

b. At 2%, 4200 individuals. At 10%, 1800. At 50%, 100.

c. It would decrease.

8. Tree trunks in industrial areas became blackened. Melanic moths were better camouflaged. Predators ate fewer of the melanic moths, so a higher proportion survived. Therefore these melanic moths would pass on their melanic allele to their offspring. The process would continue over successive generations until most moths were melanic.

9. A higher proportion of melanic moths were recaptured in the polluted wood, suggesting that black wing colour helped their survival. Conversely, a higher proportion of speckled moths were recaptured in the Dorset wood.

10. The proportion of the speckled form in the population is likely to increase again.

11. The widespread use of antibiotics in farm animals creates an environment in which there is a strong selection pressure favouring the survival of antibiotic-resistant individual bacteria. These reproduce, and perhaps also pass on their resistance alleles to other bacteria of other species. These resistant bacteria could then infect humans.

12. TAXONOMY AND BIODIVERSITY

1. When a gamete from a horse and a gamete from a donkey fuse together, the resulting zygote gets one set of chromosomes from the horse

(33 chromosomes) and one set of chromosomes from the donkey (31 chromosomes). The cell is able to divide by mitosis to produce an embryo and finally an adult mule. However, in order to produce gametes and reproduce, the mule's cells must divide by meiosis. This involves the pairing of homologous chromosomes. Because there are different numbers of chromosomes in the two sets, not every horse chromosome has a donkey chromosome to pair up with. Meiosis cannot take place, so no gametes can be formed.

2. a. phylum, class, order, genus, species
   b. *Erinaceus europaeus* (You should underline your answer to indicate that, when printed, it should be in italics.)

3. *Bombus pratorum* and *Bombus terrestris*, as they both belong to the same genus.

4. a.

Species	n	$n-1$	$n(n-1)$
grass	185	184	34 040
thistles	28	27	756
stinging nettles	35	34	1190
moss	2	1	2
Total $N = 250$		$\Sigma n(n-1) = 35988$	

   b. $d = \frac{(250 \times 249)}{35\,988} = 1.73$

   c. The species diversity in the field is much less than that in the lawn. We can only guess the reasons for this.

   Some possibilities would be:

   - the sheep graze selectively in the field – perhaps they do not eat thistles or nettles, but have grazed so heavily on other plants that they cannot grow;
   - perhaps the field has been sprayed with a selective weedkiller, whereas no spray has been used on the lawn;
   - perhaps the soil in the field and lawn differ in their mineral content or soil type

5. Beta cells specialise in producing insulin. They are therefore constantly producing mRNA by transcription of the insulin gene, and will contain many copies of it. A high proportion of the mRNA in a beta cell will be for the production of insulin, rather than other proteins. There will, however, be only two copies of the DNA for insulin in any of the cells in the organism's body, and this will be only a tiny proportion of all of the DNA in the cell. It can therefore be easier to obtain large quantities of the relevant mRNA than it is to obtain large quantities of the relevant DNA.

6. The greater the differences, the less closely related the animals are to humans.

GLOSSARY

α-glucose Alpha glucose – a form of the simple sugar (monosaccharide) glucose.

β-glucose Beta glucose – a form of the simple sugar (monosaccharide) glucose.

Absorption The process by which nutrients move across the cells of the alimentary canal wall.

Activation energy The minimum amount of energy needed to trigger a reaction.

Active immunity Immunity gained by being exposed to the antigen so the body makes its own antibodies against it.

Active site A 'dent' on an enzyme molecule that binds to a protein during a reaction.

Active transport The movement of molecules through a cell membrane against the concentration gradient; this process requires energy.

Adenosine triphosphate (ATP) A nucleotide derivative – the source of energy for almost all cellular processes.

Adhesion The force that attracts water molecules to the walls of xylem vessels in a plant.

Agglutination A process in which antibodies attach to antigens on the surface of bacteria, causing the bacteria to clump together so they cannot reproduce.

Aleveolar epithelium A single layer of cells on the surface of alveoli.

Alleles Different forms of a gene that control particular characteristics.

Alpha helix A common type of secondary structure in a polypeptide chain.

Alveoli Tiny air-filled sacs in the lungs where the exchange of oxygen and carbon dioxide takes place.

Amino acids Molecules that make up proteins, containing atoms of carbon, hydrogen, oxygen and nitrogen.

Amylase An enzyme that hydrolyses starch into maltose

Anaphase The phase of the cell cycle when the spindle fibres drag the chromatids to opposite ends of the cell.

Anticodon The three unpaired bases in tRNA.

Antigen-presenting cell A cell that displays antigens in a form that T cells can recognise.

Antiparallel The way in which the two strands in a polynucleotide run in opposite directions.

Aorta The artery that carries oxygenated blood from the heart to the rest of the body.

Aquaporins Channel proteins for water molecules in a cell membrane.

Artefacts Unnatural structures that form in a specimen while it is being prepared for electron microscopy.

Atoms Tiny particles which cannot be chemically broken down.

ATP hydrolase The enzyme that catalyses the release of energy in an ATP molecule.

ATP synthase An enzyme that joins the ADP and phosphate to resynthesise ATP.

Atrioventricular valve The valve between the atrium and the ventricle in the heart that keeps blood flowing in the right direction.

Atrium The upper chamber of the heart.

B lymphocytes Lymphocytes involved in the humoral response.

Bacterial chromosome The circular DNA molecules in bacteria.

Bile Juices produced in the liver.

Bile duct The channel that connects the gall bladder to the small intestine.

Bile salts Molecules in bile that have both hydrophilic and hydrophobic regions.

Binary fission The process by which prokaryotic cells divide.

Binomial system A system in which all organisms are identified by a two part name – a generic name and a specific name.

Bivalents The pairing of two homologous chromosomes in an early stage of meiosis.

Bohr effect The shift of the oxyhaemoglobin dissociation curve to the right due to the influence of carbon dioxide.

Bronchi The two branches into which the trachea divides in the thorax.

Bronchioles The smaller branches into which the bronchi divide

Capillary bed The network of capillaries among cells in tissues, forming an exchange surface.

Capsid The protein coat that encloses virus particles.

Capsule Tissue forming a wall around a structure. In bacteria – a slimy protective layer outside the cell wall.

Carbohydrate A chemical containing carbon, hydrogen and oxygen; the hydrogen and oxygen being in the ratio of 2:1.

Carrier proteins Proteins that facilitate the diffusion of different molecules across a cell membrane.

Catalase An enzyme that catalyses the decomposition of hydrogen peroxide to water and oxygen.

Catalysts Substances added to chemical reactions to alter the speed of the reaction.

Cell cycle The process of the division and replication of cells.

Cell-surface antigens Cell-specific proteins and glycoproteins in a cell membrane.

Cell-surface membrane A membrane that separates contents of a cell from its surroundings, while still allowing substances to move in and out of the cell.

Cell Theory The idea that all life exists as cells.

Cellular response An immune response involving T cells.

Cellulose A polysaccharide composed of glucose monomers.

Cell wall A structure that surrounds the cell in plants and some types of microorganism.

Centromere The place on a chromosome where there chromatids are held together.

Channel proteins Proteins that facilitate the movement of ions across a cell membrane.

Chiasmata Breaking and crosswise rejoining of homologous chromatids during meiosis.

Chitin A nitrogen-containing polymer that forms part of the cell walls of fungi.

Chloroplasts The organelles in which photosynthesis takes place, found in some plant and algal cells.

Cholesterol A lipid-like substance in cell membranes that prevents too much movement of other molecules in the membrane.

Chromatids The two strands into which a chromosome splits during cell division.

Chromatin DNA or RNA and proteins that condense to form chromosomes during cell division.

Chromosome mutations Mutations that occur in whole chromosomes or large sections of them.

Chromosomes Thread-like structures in the nucleus of a cell, which carry genetic information.

Chylomicrons Fatty droplets in the blood after absorption from the small intestine.

Cisternae The channels that make up the endoplasmic reticulum in the cytosol of cells.

Clonal selection A process in which T cells divide by mitosis to form a clone of identical T cells.

Codon A sequence of three bases on an mRNA molecule that codes for one amino acid.

Cohesion The way in which molecules 'stick' together.

Cohesion–tension The mechanism that allows water to move up the xylem.

Common ancestor A single species from which all other species in a group arose.

Companion cell A cell closely associated with an individual sieve cell.

Competitive inhibitors Molecules that bind to the active site in an enzyme but no reaction takes place.

Complementary Describing the consistent pairing of bases in a polynucleotide (A with T and C with G).

Condensation The process by which water vapour turns into liquid water.

Condensation reaction A reaction in which two compounds are joined together by removing the elements of a water molecule.

Confocal microscope A microscope that can image live cells in 3D.

Continuous variation The outcome of a number of factors influencing the expression of a characteristic.

Coronary arteries The arteries that supply blood to the heart.

Correlation coefficient test A statistical test for determining correlation between variables.

Covalent bonds Very strong bonds formed when two atoms share electrons with each other.

Critical value A cut-off area outside which a test statistic is unlikely to lie.

Cuticle A waxy outer covering.

Cytokines Cells that stimulate macrophages to carry out phagocytosis of infected cells.

Cytokinesis The process by which the cell splits in half after mitosis.

Cytoplasm The cytosol and all the organelles in a cell (apart from the nucleus).

Cytosol The jelly-like material in a cell, made up of proteins and other substances dissolved in water.

Cytotoxic T cell Cells that destroy cells which have been infected by the pathogen that carried the specific antigen to which it can respond.

Degenerate Describing a triplet that codes for the same amino acid as other triplets.

Denatured Describing an enzyme that has been 'inactivated' when hydrogen bonds are broken through heating.

Deoxyribose The sugar found in the nucleotides of DNA.

Diaphragm The muscles between the thorax and the abdomen.

Diffusion The spread of particles from regions of higher concentration to regions of lower concentration.

Digestion The process in which larger molecules are broken down into smaller molecules to allow nutrients to be absorbed.

Dipeptide A molecule made up of two amino acrid molecules joined by a condensation reaction.

Diploid Cells containing two sets of chromosomes.

Dipole Describing a molecule that has a small positive charge in some areas and a small negative charge in others.

Directional selection A type of natural selection in which the selection pressure is shifted towards one end of a variation range by favouring more extreme phenotypes over others.

Disaccharides Carbohydrates consisting of two single sugars joined together.

Discontinuous gas-exchange cycle (DCG) A respiratory pattern in insects in which the spiracles can be closed for long periods.

Discontinuous variation Where only one or a few genes cause variation in a characteristic, resulting in two or a few classes.

Dispersed Describing a wide spread of data sets.

DNA Deoxyribonucleic acid – a nucleic acid that stores genetic information in every living cell.

DNA helicase An enzyme that 'unwinds' a DNA molecule and breaks the hydrogen bonds that connect the bases.

DNA polymerase An enzyme that catalyses reactions that join nucleotides to form DNA.

Double circulatory system The circulatory system in mammals, in which the blood passes through the heart twice on one complete circuit of the body.

Emulsion A mixture of micelles and water.

Endopeptidase An enzyme that breaks peptide bonds.

Endoplasmic reticulum A network of membrane-bound channels found in the cytosol of all cells.

Endothelium A layer of cells that forms the lining of organs and other parts of the body.

Envelope The two membranes that surround a chloroplast.

Enzymes Catalysts (usually proteins) produced by cells, which control the rate of chemical reactions in cells.

Enzyme–substrate complex The combination of an enzyme and a substrate.

Ester bonds The bonds that join fatty acids to glycerol molecules in triglycerides.

Evaporation The process by which water molecules become a gas.

Exons The sections of a gene that code for amino-acid sequences in a protein.

External intercostal muscles The set of muscles between the ribs.

Facilitated diffusion A passive diffusion process in which a channel protein makes it easy for ions to move through cell membranes.

Fatty acids Acids containing a long hydrocarbon chain attached to a –COOH group.

Fenestrations Pores in endothelial cells that allow a faster rate of diffusion between capillaries and tissues.

Fibrous proteins Proteins that form long, thin molecules.

Fluid-mosaic Describing the unfixed structure of cell membranes in phospholipids.

Fructose A simple sugar (monosaccharide).

Galactose A sugar that joins with glucose to form the disaccharide lactose.

Gall bladder A small, sac-like structure connected to the small intestine, which stores bile to break down fats in partly digested food.

Gastric juice The juice secreted by glands in the walls of the stomach, which destroys potentially harmful bacteria in food.

Gated channels Channels in cells that can be opened or closed to allow ions to move through them.

Gene mutation A change in the base sequence that affects the gene in that part of the DNA sequence.

Genetic code The way in which a sequence of DNA bases represents a sequence of amino acids.

Genetic diversity The different genetic make-up of the organisms in a population.

Genome The complete set of genes in a cell.

Genome sequencing Analysing DNA to find out about genetic diversity within a population.

Genus A taxonomic group coming between family and species.

Gills Structures used for gaseous exchange between an organism and water.

Globular protein A protein that forms a ball-shaped molecule.

Glucose A monosaccharide that exists in two forms: α-glucose and β-glucose.

Glycerol A chemical that combines with three fatty acid molecules to form a lipid.

Glycogen A carbohydrate composed of many α-glucose molecules joined together.

Glycolipids Composite molecules that are part lipid and part carbohydrate.

Glycoproteins Composite molecules that are part protein and part carbohydrate.

Glycosidic bond A bond formed between two sugar molecules.

Golgi apparatus A group of membrane-bound cavities that packages synthesised proteins into membrane-bound vesicles.

Grana Stacks of membranes within the stroma of a chloroplast.

Haemoglobin A red pigment that transports oxygen around the body.

Haploid Cells that only have one set of chromosomes, such as gametes.

Heat capacity The amount of heat energy that has to be transferred to an object to increase its temperature.

Helper T cell A T cell that helps other cells in the immune response by recognising foreign antigens.

Herd immunity Immunity that occurs when a large enough proportion of the population are immune so that a disease is unlikely to spread to those who are not.

Hierarchy A classification system with the largest groups at the top and individual species at the bottom.

Histones Proteins present in chromatin, around which are coiled DNA molecules.

Homologous chromosomes The 'matching' chromosomes in diploid cells.

Humoral response An immune response mediated by an antibody.

Hydrogen bonds Weak chemical bonds involving hydrogen atoms.

Hydrolysis Reactions in which compounds are broken down by reacting with water.

Hydrolysis of ATP A reaction in which chemical energy stored in ATP is released; essential for the process of active transport.

Hydrophilic Water-loving.

Hydrophobic Water-hating.

Immune A state in which the immune system produces so many antibodies so quickly that the antigen is destroyed before it can make us ill.

Immune response The response of the immune system to invasion by foreign cells.

Immunoglobulins Protein molecules produced by B lymphocytes.

Induced-fit model A type of enzyme action in which the substrate binds to the active site of the protein.

Interphase The phase in the cell cycle when DNA replication takes place.

Introns Non-coding sections of DNA.

Iodine Straw coloured iodine/potassium iodide solution that turns blue/black in the presence of starch.

Lactose Disaccharide composed of glucose and galactose.

Latent heat of vaporisation The energy that is lost from liquid water as it evaporates.

Lignin A waterproofing substances found in the walls of xylem cells.

Linear Describing the relationship between two variables shown by a straight line in a graph of results.

Lipase An enzyme in the pancreas that catalyses the breakdown of fats.

Lipids Fats made up of molecules of carbon, hydrogen and oxygen.

Locus The position of a gene on a chromosome.

Lymphocytes Cells that help to defend the body against infection.

Lysosomes Vesicles that contain the digestive enzymes lysozymes.

Lysozyme An enzyme found in tears and saliva that destroys bacterial cell walls.

Maltose A disaccharide composed of two glucose molecules.

Mass flow The movement of a whole mass of water together.

Mass transport The movement of huge numbers of molecules or ions of a substance in the same direction at the same time.

Mature mRNA The state of mRNA after the introns have been removed and exons stuck together (also known as mRNA).

Mean Calculated by taking the sum of all the values then dividing by the number of values.

Median The middle value of all the values in all sets of data.

Memory cells Lymphocytes that respond rapidly to reinfection.

Metaphase The phase in the cell cycle when two spindle fibres attach to the centromeres and pull the chromosomes into a line across the centre of the cell.

Micelles Tiny droplets that are formed when lipids do not fully dissolve in water.

Micrometre One thousandth of a metre.

Microvilli Very thin projections on the surface of the cells of the villi.

Mitochondria The places in a cell where aerobic respiration takes place.

Mitosis A method of eukaryotic cell division in which the new cells have exactly the same number and type of chromosomes as the parent cell.

Modal set of data The most common set in a series of results.

Molecules Small particles made up of atoms that are linked by bonds.

Monoclonal antibodies Antibodies produced by a single clone of white blood cells.

Monoculture The growing of large areas of a single crop.

Monomers Small molecules that can be chemically bonded to form polymers.

Monosaccharides Simple sugars – the monomers from which all larger carbohydrates are made.

mRNA Messenger ribonucleic acid – a molecule of RNA with the code for a protein.

Murein A polymer containing amino acids and sugars, which makes up the cell wall in prokaryotic cells.

Mutagenic agents Factors that increase the risk of gene mutation, such as ionising radiation and X-rays.

Mutations Changes to the base sequence in DNA or to the number or structure of the chromosomes in cells.

Nanometre One thousandth of a micrometre.

Negative correlation As the value of one variable increases, the value of the dependent variable decreases.

Non-competitive inhibitors Molecules that inhibit enzyme reactions by altering the shape of the enzyme molecule.

Non-disjunction A chromosome mutation that occurs during cell division in which one cell gets an extra chromosome, leaving another cell with one less than it should have.

Non-overlapping Describing the triplets in a DNA code that are read sequentially.

Non-polar A state in which the distribution of charge in a molecule is uneven.

Non-reducing sugars Sugars such as sucrose that are not readily reduced in other substances.

Normal distribution A bell-shaped curve that is defined by the mean and the standard distribution of the data set.

Nuclear envelope Two membranes that enclose the nucleus of a cell.

Nucleoli Structures in the nucleus involved in protein synthesis.

Nucleotides Molecules that combine to form strands of DNA and RNA.

Nucleus The largest organelle in a cell, containing most of the cell's DNA.

Null hypothesis A statement that you are expecting to reject.

Optical microscope A microscope that passes rays of light through a specimen to produce an image on the retina of the viewer's eye.

Optimum temperature The temperature at which enzymes work fastest (about 40 °C).

Organ A structure with a specific function made up of different types of tissue.

Organelles Small structures that exist in the cytosol of a cell.

Organ systems Bodily systems that perform particular functions, made up of several different organs.

Osmosis The movement of water from a less concentrated solution to a more concentrated solution through a selectively permeable membrane.

Oxyhaemoglobin The combination of oxygen and haemoglobin in areas where oxygen concentration is high.

Oxyhaemoglobin dissociation curve A graphic representation of how readily haemoglobin gains and releases oxygen molecules.

Passive Describing a biological process that does not require a cell to do anything to make it happen.

Passive immunity Immunity produced by giving an individual an injection containing the appropriate antibodies.

Pectin A carbohydrate found in the cell walls of plants.

Pepsin An enzyme in gastric juices that breaks down proteins in polypeptides.

Peptide bond A bond between two amino acids.

Percentage error A way of expressing uncertainties by dividing the error by the actually measurement then dividing by 100.

Permeable membrane A membrane that contains pores large enough for water molecules to pass through but not large enough for the solute molecules to pass through.

Phloem Living cells in plant stems.

Phloem sieve elements The individual cells that make up phloem tissue; also known as sieve cells.

Phloem tissue The tissue that carries organic substances such as sucrose and amino acids around the bodies of plants.

Phosphodiester bonds The bonds that join sugar and phosphate molecules in a polynucleotide.

Phospholipids Lipids containing two fatty acids, attached to glycerol via ester bonds.

Phylogenetic Based on the natural relationships between organisms.

Plasma cells B cells that secrete antibodies.

Plasmids Circular molecules of DNA found in some prokaryote cells.

Plasmodesmata Gaps in a cell wall that allow the transfer of material between cells in multicellular plants.

Polygenic Describing a characteristic that is affected by many different genes.

Polymer A long chains of chemically bonded smaller molecules called monomers.

Polynucleotides Polymers whose molecules comprise many nucleotides, such as DNA and RNA.

Polypeptides Long chains of amino acids.

Polysaccharides Large molecules made of monosaccharide molecules joined together by condensation reactions.

Positive correlation When one variable increases, another increases as well.

Potassium iodide solution Used to dissolve iodine for the test for starch.

Pre-mRNA The state of mRNA before it leaves the nucleus, when it contains exons and introns.

Primary structure The unique sequence of amino acids in a polypeptide.

Products The molecules produced by a reaction.

Prophase The first stage of mitosis, when the two membranes of the nuclear envelope break down.

Protein A molecule made up of long chains of amino acids.

Proteome The complete set of proteins that are coded for in a cell.

Pulmonary artery The artery that carries blood from the right ventricle to the lungs.

Pulmonary system One part of the double circulatory system, transporting deoxygenated blood to the lungs and then to the heart.

Pulmonary veins The veins that carry oxygenated blood from the lungs to the left atrium of the heart.

Quaternary structure The structure of proteins that contain more than one polypeptide chain.

Random sampling Collecting information from a number of randomly selected sites within an area.

Range A measure of dispersion from the highest value to the lowest value.

Reducing sugars Sugars such as glucose and fructose that readily lose electrons to another substance.

Replication The process of making perfect copies of DNA.

Resolving power A microscope's ability to distinguish between two objects.

Ribosomes The parts of a cell that create proteins from all amino acids and RNA representing the protein.

RNA A nucleic acid involved in protein synthesis by transferring information from DNA to the sites of protein synthesis.

RNA polymerase An enzyme that links RNA nucleotides by forming phosphodiester bonds between their ribose and phosphate groups.

Rough endoplasmic reticulum An endoplasmic reticulum that has ribosomes on its outer surface.

Saturated fatty acid A fatty acid in which all the carbon atoms use two of their bonds to join to other carbon atoms and two to join to hydrogen atoms.

Scanning electron microscope A microscope that produces high resolution images of the surface of an object.

Secondary lamellae Projections inside the gills of fish that provide a large surface area for increased oxygen intake.

Semi-conservative replication The process of copying DNA in which one strand is retained from the original and one new one is made.

Sense strand The strand in DNA that is used as a code for making proteins.

Septum The partition that separates the left and right sides of the heart.

Sieve plates An area of large pores or perforations on the end wall of sieve cells.

Sieve tubes A long column of sieve cells made up of many continuous tubes.

Sink The name given to an area to which sucrose is moving during the transport of organic substances in a plant.

Smooth endoplasmic reticulum An endoplasmic reticulum that has no ribosomes on its outer surface.

Solutes Substances that can be dissolved in a solvent.

Solvent A liquid in which other substances can dissolve.

Source The name given to an area from which sucrose is moving during the transport of organic substances in a plant.

Speciation The production of a new species, when something occurs to prevent two groups of the same species being able to breed together.

Species The taxonomic category comprising individuals with common characteristics that distinguish them from other individuals at the same taxonomic level.

Species diversity The number of species in a habitat, the relative sizes of the populations and their spread throughout the habitat.

Species richness The number of different species that live in a particular habitat.

Spiracles External opening leading to the air tubes (tracheae) inside an insect.

Splicing The process in which introns are removed and exons and stuck together in mRNA.

Stabilising selection A type of natural selection that causes no change due to a lack of change in the environment and an already adapted population.

Standard deviation A measure of the variability in a population.

Stem cells Undifferentiated cells that retain the ability to become specialised.

Stroma The mixture of water, enzymes and other substances contained within a chloroplast.

Substrate The molecule that an enzyme allows to react.

Sucrose A disaccharide composed of a glucose molecule and a fructose molecule.

Surface tension The way in which water behaves as though there is a 'skin' where it meets the air, caused by the net downward attraction of molecules.

Systemic system One part of the double circulatory system, transporting oxygenated blood from the heart to the rest of the body.

Taxonomy A system of classifying organisms.

T lymphocytes Lymphocytes involved in the cellular response.

Telophase The final stage of mitosis, when the chromatids unravel to become strands of DNA again and the nuclear envelope reforms.

Tertiary structure The three-dimensional shape of a polypeptide molecule.

Thylakoids The interconnected spaces formed by a series of membranes within the stroma of a chloroplast.

Tidal ventilation The flow of air through the lungs in both directions (in and out).

Tissue A group of similar cells performing the same specific function.

Tissue fluid A fluid mostly made up of water with ions and small molecules in solution which collects between cells in capillary beds.

Trachea The tube that runs from the mouth and nose down into the thorax in the human gas-exchange system.

Tracheae Branched, gas-filled tubes in the breathing system of an insect.

Tracheoles The finest air tubes in the breathing system of an insect.

Transcription The process of using the coded information in DNA to form mRNA.

Translation The process in which mRNA code is used to construct a polypeptide.

Translocation The process in which organic substances move around a plant by way of phloem tissue.

Transmission electron microscope (TEM) The type of electron microscope where a beam of electrons passes through the specimen to produce a high resolution image.

Transpiration The process in which water vapour is lost from the leaves of plants.

Triglycerides Molecules made up of glycerol molecules each bonded to three fatty acid molecules.

Triplet A sequence of three bases on a DNA molecule coding for one amino acid.

tRNA Transfer ribonucleic acid – a molecule of RNA that transports amino acids to ribosomes.

Trypsin An endopeptidase found in pancreatic juice.

Turgid Describing a cell that has become firm due to water retention.

Universal Describing the way that the genetic code is almost identical in all living organisms.

Unsaturated fatty acid A fatty acid in which some of the carbon atoms have a double bond lining them to a neighbouring carbon atom.

Vacuole A liquid-filled space inside a cell, surrounded by a membrane.

Variation Deviation in the characteristics of an organism.

Venae cavae Two large veins that carry deoxygenated blood into the heart.

Ventilation The active movement of air to the gas-exchange surface.

Ventricle A lower chamber in the heart; pumps blood out of the heart.

Vesicles Small liquid-filled spaces inside a cell, surrounded by a membrane.

Villi Small projections on the wall of the small intestine.

Viruses Pathogens consisting of nucleic acid – surrounded by a protein coat.

Voltage-gated channels Channels in cells that open or close in response to voltage changes.

Water A colourless liquid that is a major component of every living organism.

Water potential The tendency of water to move out of a solution.

Water potential (Ψ) gradient A range of values for water potential.

Xerophytes Plants that have adapted to survive long periods of drought.

Xylem Dead reinforced cells in plant stems.

Xylem vessels Long chains of hollow xylem cells.

INDEX

1 mol dm^{-3} solution 42–43
10% solution 42

A

abnormal body cells 121
absorption
  carbohydrates 160–61
  humans 157–58, 160–62
  small intestine 160–61
  substance exchange 157–58, 160–62
accuracy, data 264
acquired immunodeficiency syndrome (AIDS) 120, 133–35
action mechanisms, enzymes 35–37
activation energy 34–36
active immunity 128, 130, 133
active sites, enzymes 35–37, 39, 44, 46–47
active transport 113–16
activity, enzymes 37–49
addition mutations 216, 219
adenosine diphosphate (ADP) 61, 62
adenosine triphosphate (ATP)
  eukaryotic cells 70, 76, 77
  hydrolysis 113–14
  nucleotides 53, 61–63
  protein synthesis 207, 208, 212
adhesion 188–89, 190
adipose tissue 22–23
ADP *see* adenosine diphosphate
adult haemoglobin 169
aerobic respiration 70, 77
agglutination 126
AIDS *see* acquired immunodeficiency syndrome
air pollution 228–29, 232
air temperatures 7–8
alimentary canal 157–58
alleles 215–38
α-glucose (alpha glucose) 10, 18
alpha helix 27
altitude sickness 153–54
alveolar epithelium 151–52
alveoli 150–52
amino acids 5
  biodiversity 248–50, 256
  chromatography 30–31
  DNA and protein synthesis 199–214
  enzymes 35
  plant transport 191–94
  protein structure 26–32
Amoeba, gas exchange 145
amylase 48–49, 158
analysing data 264–69
anapase cell cycle stage 85–86
ancestors 243, 244
animals
  biodiversity 248–49, 251
  cell structure 68–77
  dissections 183–84
  diversity 240, 251
  genetic diversity 225–27
  mass transport 165
  substance exchange 143–45
antibiotics 138–39, 231–32
antibodies 126–30, 133–38
anticodons 207–12
antigen-presenting cells (APC) 125, 127
antigens
  cell-mediated response 123–25
  genetic diversity 224–25
  immune systems 120, 121–33
  variability 128
antimicrobial substances 233–35
antiparallel strands 55, 58
antiretrovirals (ARV) 120, 134–35
ants 147
aorta 172–73
APC *see* antigen-presenting cells
ape–human relationships 248–49
apparatus/technique practicals
  cell membranes 109–12
  cells 89–91
  dissections 183–84
  enzymes 41–47
  genetic diversity 233–35
  mass transport 183–84
apricot tree diversity 256
aquaporins 105
Archaea 77
artefacts, microscopy 83–84
arteries 176–77, 182
arterioles 176–77, 180, 182
ARV *see* antiretrovirals
aseptic techniques 233–35
atoms 5
ATP *see* adenosine triphosphate
ATP hydrolase 61
ATP synthase 62
atrioventricular valves 172, 173, 176
atrium 172–73, 176
Attacus atlas 147
Avery, Oswald 56

B

bacteria
  antibiotic resistance 231–32
  cell relationships 67, 93
  DNA and protein synthesis 199
  growth calculations 93
  prokaryotic cells 77, 200–201
bar charts 264
base pairing 54–55, 58, 59, 61, 207–8, 212
base sequences 200–202, 212, 216
beetroot 111–12
Benedict's solution 16, 18
β-glucose (beta glucose) 10, 18
beta radiation 193–94

biconcave cells 102–3
bile 159
bile duct 159
bile salts 159, 161
binary fission 85, 92
binomials, taxonomy 242, 244
biochemical tests 16–17
biodiversity
  amino acid sequences 248–50, 256
  animals 248–49, 251
  continuous variation 251–53, 256
  crops 245–48
  cytochromes 249–50
  discontinuous variation 251
  farming 245–48
  haemoglobin 249
  immunology 250
  investigative studies 248–56
  mutations 249–50
  proteins 239, 243, 248–51, 256
  taxonomy 244–59
  variation within species 251–56
biological catalysts 33–35
biological drawings 90
biological molecules 5–6
biological washing powders 38–39
birds 217–18, 240
birth weights 232
Biston betularia 228–29
biuret test 30
bivalent pairs 220–23
blood cells
  genetic diversity 215, 224
  surface area measurements 102
blood plasma 104–5, 166–68, 178, 180–82
blood pressure 252–53
blood transport 166–84
  circulatory system 170–82
  flow 175, 179
  heart 170–82
  mammals 166–82
  oxygen transport 166–70
blood vessels 176–84
  arteries 176–77, 182
  arterioles 176–77, 180, 182
  capillaries 176, 177–78, 180–81, 182
  cardiovascular disease 182
  hydrostatic pressure 181
  lumen 176–77, 182
  structure 176–78
  tissue fluids 178, 180–82
  veins 176, 177–78, 182
  venules 176, 177–78, 182
  water potential 181
blubber insulation properties 24–26
B lymphocytes/B cells 123, 125–28, 136–39
Bohr effect 168–70
boiling points 8
bonding
  disaccharides 12, 18
  DNA 54–55, 58
  enzymes 35–36
  proteins 26–31
  triglycerides 21–23, 24
  water molecules 6–8, 9, 10
  water transport in plants 188
brain cells 87
breast cancer 87–88
breathing 152–54
breath tests 15–16
bronchi 150, 151
bronchioles 150, 151
bulk liquid transport *see* mass transport

C

calibration
  curves 11, 109–10
  eyepiece graticules 90
cancers
  cell cycle control 87–88, 91
  cell division 87–88, 91
  cervical cancer vaccinations 132
  chemotherapy 53
  immune systems 132, 139
  nucleotides 53
capillaries 176, 177–78, 180–81, 182
capillary beds 178, 180–81, 182
capsid 78
capsules, prokaryotic cells 78
carbohydrates 4–6, 10–19
  absorption 160–61
  biochemical tests 16–17
  digestion 158–59, 161
  disaccharides 10, 12, 14–18
  identification 16–17
  lactose intolerance 14–16, 17
  monosaccharides 10–13, 16–18
  polysaccharides 10, 12–13, 18
  solubility 10, 12
  structure 10, 12–13, 18
  types 10–13
  water 4–6, 10–19
carbon-14 193–94
carbon–carbon bonds 22, 24
carbon dioxide molecules 5
carboxyl groups 26
carcinogens 156
cardiac cycles 172–76
cardiac muscles 171–76
cardiac output 174–75
cardiovascular disease (CVD) 182
carrier proteins 103–4, 113–15
catalase 34, 36
catalysts 33–35
causation studies 155–57
cell cycles 85–93
  anapase stage 85–86
  chromatids 85–86
  chromosomes 85–86
  controlling 87–93
  cytokinesis 86
  interphase stage 85, 86, 90
  metaphase stage 85, 86
  prophase stage 85, 86
  telophase stage 86

cell division 68, 85–93
  cancers 87
  eukaryotic cells 85–93
  meiosis 218, 219–23
  nucleotides 53
  prokaryotic cells 85, 92
cell membranes 98–119
  active transport 113–16
  apparatus/technique practicals 109–12
  aquaporins 105
  ATP hydrolysis 113–14
  carrier proteins 113–15
  cell-surface membranes 68–70, 99–100, 111–12, 120, 121–22
  channel proteins 103, 104, 105
  cholesterol 99, 100
  co-transport 113–16
  diffusion 98, 101–16
  facilitated diffusion 101–5, 113–16
  fluid-mosaic structure 99–100
  glucose 98, 104–5, 113–16
  glycolipids 99, 100
  glycoproteins 99, 100
  graphs of transport 115–16
  hydrolysis 113–14
  importance of osmosis 106–7
  lipids 98, 99–100, 103–5, 108, 111
  osmosis 98, 105–12
  phospholipids 98, 99–100, 103–5, 108, 111
  practical studies 109–12
  proteins 99–100, 103–4, 105
  rehydration 98, 108
  structure 99–100
  technique practicals 109–12
  transport 98–119
  water diffusion 98, 101–8, 113–16
cells 67–97
  apparatus/technique practicals 89–91
  bacterial growth 93
  centrifugation 84, 85
  chloroplasts 67, 68, 73–74, 76, 77, 92
  division 68, 85–93, 218, 219–23
  electron microscopes 68, 69, 70, 72, 74, 75, 82–85, 92
  eukaryotic cell structure 68–77
  fractionation 84, 85
  growth calculations 93
  light microscopes 79–82, 89–91
  living organisms 67, 68–77
  mass transport 165–98
  microscopy 68, 69, 70, 72, 74, 75, 79–85, 89–91
  mitosis 85–93
  multicellular organisms 67, 68–77
  optical microscopes 79–82, 89–91
  organelles 67–69, 71–74, 78–85, 87
  practical activities 89–91
  production 68, 85–93
  prokaryotic cells 67, 68, 73, 77, 83, 85, 92
  scanning electron microscopes 83
  structure 67, 68–77
  study methods 79–85
  symbiosis 67, 68
  technique practicals 89–91
  transmission electron microscopes 82–83, 92
  ultracentrifugation 84, 85
  viruses 67, 77, 78
  see also cell...
cell-surface antigens 120, 121–33
cell-surface membranes
  eukaryotic cells 68–70
  immune systems 120, 121–22
  permeability 111–12
  structure 99–100
Cell Theory 78
cell transplants 121
cellular response 123–25, 127
cellulose 13, 18, 74–75, 77, 158
cell vacuoles 75, 77
cell walls 74–75, 77
centrifugation 84, 85
centromere points 85
cervical cancer 91, 132
channel proteins 103, 104, 105
Chargaff, Erwin 56
chemical enzyme reactions 34–49
chemotherapy 53
childhood vaccinations 128–33
chimpanzee–human relationships 248–49
chi-squared tests 229, 230–31, 267–68
chitin 74–75, 77
chlasmata 222–23
chloroplasts
  cells 67, 68, 73–74, 76, 77, 92
  DNA in eukaryotic cells 201, 202
  eukaryotic cells 68, 73–74, 77, 201, 202
  photosynthesis 68, 73–74, 77
  prokaryotic cells 92
  structure 73–74
cholesterol 99, 100
Chrichton's *Jurassic Park* 205
chromatids 85–86, 220
chromatin 69–70
chromatography 30–31
chromosomes
  cell cycles 85–86
  cross over 222–23
  DNA and protein synthesis 200–202, 204
  eukaryotic cells 69–70, 200, 202
  gene positions 202, 204
  independent assortment 221–22, 223

chromosomes (*continued*)
meiosis 218, 219–23
micrographs 202
mutations 216, 218–19, 236
prokaryotic cells 77, 200–201, 202
random fertilisation 223
chylomicrons 160–61
circulatory system 170–82
cirl buntings 246
cisternae 70–72
classification 239–59
clonal selection 125, 127, 136
codons 199, 207–12
cohesion 9, 10, 188–89, 190
cohesion–tension 188–89, 190
collagen 29–30
colorimeters 112
colour changes 39–43
common ancestors 243, 244
companion cells 191
competitive inhibitors 46, 47
complementary base pairing 55, 59, 61, 207–8, 212
compound optical microscopes 79
concentration
enzyme–activity relationship 43–45, 47
gradients 101
condensation reactions
disaccharides 12, 18
nucleotides 54, 57
triglycerides 21–23, 24
water 9–10
confocal microscopes 80
continuous variation 251–53, 256
controlling cell cycles 87–93
cooling 8, 10
coronary arteries 172
correlation 155–57, 267
coefficient tests 25–26
co-transport 113–16
countercurrent flow 148–49
covalent bonds 6–7
Crick, Francis 54, 56, 58
cristae 77
critical values 268
Crohn's disease 136
crop biodiversity 245–48
cross over, meiosis 222–23
culture
incubation 235
transfers 234–35
curcumin 88
cuticle 146
CVD *see* cardiovascular disease
cyclin D1 87–88
cytochrome biodiversity 249–50
cytokines 125
cytokinesis 86
cytoplasm 68–69, 106–7
cytosol 68–69, 105
cytotoxic T cells 125, 127

D

dandelion stem lengths 253–54
Darwin, Charles 228
data analysis/interpretation 264–69
data display/recording 262–64
daughter cells
cell division 86
genetic diversity 216, 217, 218–19, 221, 223
decimal places 261
deep sea divers 181–82
defining a species 241
degenerate feature of triplet codes 203
degrees of freedom 268–69
dehydration 108
deletion mutations 216, 219
denatured enzymes 37
deoxyribonucleic acid (DNA) 5
base sequences 200–202, 212, 216
cell division 85–93
chromosomes 200–202, 204
eukaryotic cells 69–70, 77, 200, 202
gene mutation 216–18
genes 200–202, 216–18
genetic code 199, 200, 203–14
genetic diversity 248, 256
immune systems 121, 133–35
information 55
meiosis 220
mitochondria 67, 70, 71
nucleotides 53, 54–61
polypeptides 201–2, 204, 207, 209–12
prokaryotic cells 77–78, 92, 200–201, 202
protein synthesis 199–214
proteomes 205–12
replication 53, 59–61, 85
strands, triplet code 203
structure 54–61
transcription to mRNA 206–7, 212
triplet code 203–4, 209–10
viruses 78
deoxyribose 54
deserts, water transport 189–90
DGC *see* discontinuous gas-exchange cycles
diagnostics, monoclonal antibodies 136
dialysis 115
diaphragm, lungs 152
diarrhoea 98
diffusion
cell membranes 98, 101–16
gas exchange in humans 151–52
glucose 104–5
mass flow 166
digestion
carbohydrates 158–59, 161
humans 157–62
lactose 14–16
lipids 159, 161
proteins 159–60, 161
substance exchange 157–62
digestive enzymes 48–49, 73

dilution solutions 43, 109–10, 235
dinoflagellates 221
dinosaur recreation 205
dipeptides 26–27
diploid cells 219, 221, 223
dipoles, water molecules 7, 10
directional selection 231–32
disaccharides 10, 12, 14–18, 158–59
discontinuous gas-exchange cycles (DGC) 147
discontinuous variation 251
disease *see* immune systems
dispersed data sets 266
displaying data 262–64
dissections 183–84
dissociation curves 167–70
dissolving substances in water 8–9
distribution curves 251–53
divers 181–82
diversity
  immunology 250
  proteins 239, 243, 248–51, 256
  see also biodiversity; genetic diversity
division, cell division 68, 85–93, 218, 219–23
DNA helicase 59–61
DNA polymerase 59, 61
DNA *see also* deoxyribonucleic acid
Dodds, Diane 229–31
dogs 240, 251
donkeys 241
double circulatory systems 170–82
double helices 55, 58, 61
Down's syndrome 218–19
drawings, usage tips 90
Drosophila pseudoobscura 229–31

E

Ecstasy 108
effectiveness of vaccinations 130
electron microscopy/micrographs
  artefacts 83–84
  cells 68, 69, 70, 72, 74, 75, 82–85, 92
  chromosomes 202
  DNA in prokaryotic cells 201
  eukaryotic cells 68, 69, 70, 72, 74, 75
  gas exchange in plants 149
  immune systems 122
  scanning electron 83, 149
  transmission electron 82–83, 92, 122, 201
  triglycerides lipids 23
ELISA tests *see* enzyme-linked immunosorbant assays
Emberiza cirlus 246
embryos 87
emulsions, lipids 24, 159
endopeptidase 159–60
endoplasmic reticulum (ER) 70–73, 77
endosymbiotic theory 67, 92, 201
endothellum 176
energy
  ATP 61–62
  triglycerides 22
envelopes
  chloroplast structure 73–74
  eukaryotic cells 73–74, 77
environment
  gas exchange 142–57
  nutrient exchange 157–62
enzyme-linked immunosorbant assays (ELISA tests) 136–37
enzymes 33–52
  action mechanisms 35–37
  activation energy 34–36
  active sites 35–37, 39, 44, 46–47
  activity 37–49
  amylase activity 48–49
  apparatus/technique practicals 41–47
  ATP 61–62
  biological catalysts 33–35
  bonds 35–36
  carbohydrate digestion 158–59
  catalysts 33–35
  chemical reactions 34–49
  color changes 39–43
  competitive inhibitor–activity relationship 46, 47
  concentration–activity relationship 43–45, 47
  DNA replications 59–61
  Golgi apparatus 73
  induced fit model 36
  inhibition 46–47
  lipid digestion 159
  mechanisms 35–37
  non-competitive inhibitor–activity relationship 46, 47
  optimum pH 39–43, 47
  optimum temperature 37, 47
  pH–activity relationship 39–43, 47
  practical activities 41–47
  products 35, 36
  protein digestion 159–60
  protein structure/function 29
  rate of reaction 41–47
  structure 35–36
  substrates 35–36, 43–45, 47
  temperature-activity relationship 37–39, 41, 47
  time–activity relationship 41
  washing powders 37–39
  see also proteins
enzyme–substrate complex 35–36
epidemiology 155–57
epithelial cells 160–61
equation of a straight line 15
ER *see* endoplasmic reticulum
ester bonds 21–23, 24

eukaryotic cells 68–77
aerobic respiration 70, 77
ATP 70, 76, 77
cell division 85–93
cell-surface membranes 68–70
cell walls 74–75, 77
chloroplasts 68, 73–74, 77, 201, 202
chromosomes 200, 202
DNA and protein synthesis 200, 202
DNA 69–70, 77
endoplasmic reticulum 70–73, 77
envelopes 73–74, 77
genetic code 204
Golgi apparatus 72–73, 77
lysosomes 72–73, 77
membranes 68–70, 73–74, 77
mitochondria 70, 71, 76, 77, 201, 202
nucleoli /nucleus 69–70, 77
organs 76, 77
ribosomes 70–72, 77
specialised cells 75–77
structure 68–77
tissue 76, 77
vacuoles 75, 77
walls 74–75, 77
evaporation 8, 10
exchange with the environment 142–64
exercise, blood flow 175, 179
exons 204, 210–11, 212
expected numbers 267–68
external intercostal muscles 152
eyepiece graticules 89–90

F

facilitated diffusion 101–5, 113–16
false-color transmission electron micrographs 201
farming biodiversity 245–48
fats/lipids 6
fatty acids 21–23, 24, 160–61
fenestrations 177
fermentation 17
fertilisation 219–23
fetal haemoglobin 169
fibrous proteins 29, 30
fish, gas exchange 148–49
flagella 78
flaming inoculating loops 234
fluid-mosaic structure 99–100
fluids, mass transport 165–98
fractionation, cells 84, 85
Franklin, Rosalind 54, 56
frequency diagrams 263–64
frequency distributions 252–54
frogs 142
fructose 10, 16
fruit flies 229–31
functions
cells 67
proteins 29–31

G

galactose 10, 14
gall bladder 159
gametes 217–18, 219–23
gases, mass flow 166
gas exchange
dissections 183–84
environment 142–57
fish 148–49
humans 150–57
insects 146–47
lungs 150–57
photosynthesis 142, 149–50
plants 143, 149–50
respiration 142–57
unicellular organisms 145
uptake/losses 142–57
gastric juices 159–60
gated channels 103
gene alleles 215–38
genes
DNA 200–202, 216–18
mutations 216–18, 219, 236
position on chromosomes 202, 204
protein synthesis 200–202
RNA and protein synthesis 201–2
genetically modified organisms 199
genetic code
DNA and protein synthesis 199, 200, 203–14
eukaryotic cells 204
exons 204, 210–11, 212
introns 204, 210–11, 212
multiple repeats 204
genetic diversity 215–38, 248, 256
alleles 215–38
antigens 224–25
apparatus/technique practicals 233–35
apricot trees 256
chromosome mutation 216, 218–19, 236
daughter cells 216, 217, 218–19, 221, 223
directional selection 231–32
DNA 248, 256
gene mutations 216–18, 219
genome sequencing 248, 250
human migration 224–25
meiosis 219–23
migration patterns 224–25
mitosis 215, 216, 219, 220
mRNA 248
mutations 216–19, 236
natural selection 223–36
phenotypes 215, 226–27, 231, 236
practical investigations 233–35
RNA 248
stabilising selection 232
statistical tests 229, 230–31
technique practicals 233–35
genetic information 55
genomes 204, 212, 248, 250

genus names 242, 244
giant frogs 142
gills 148–49
globular proteins 29, 30, 215
glucose
addition to saline solutions 98
calibration curves 11
carbohydrates 6, 10, 11, 14, 16, 18
cell membranes 98, 104–5, 113–16
concentration 11
diffusion 104–5
monomers 6
reducing sugars 16
uptake 104–5, 113–16
glycerols 21–23
glycogens 13, 18
glycolipids 99, 100
glycoproteins 99, 100, 120, 121–22, 125, 133
glycosidic bonds 12, 18
Golgi apparatus 72–73, 77
graded pipettes 234
gradient, rate of change 15
grana 73
graphs
maths techniques 15, 263, 264–66
transport across membranes 115–16
grasses 187
graticules 89–90
gravity 188–89, 190
greening the desert 189–90
groups, amino acids 26

H

haemoglobin
biodiversity 249
genetic diversity 215
oxygen transport 167–70
protein structure 28–29
types 169
handling numbers 260–62
haploids 219–23
HCG *see* human chorionic gonadotrophin
HD *see* Huntingdon's disease
health and safety precautions 233
heart
beat rate 174–76
blood transport 170–82
cardiac cycles 172–76
circulatory system 170–82
dissections 183–84
structure 171–76
heat
insulation 24–26
loss 143–44
molecular motion 7–8
heat capacity 7–8
hedgehogs 243–44
hedges, crop biodiversity 246, 247
helper T cells 125, 127, 133–34
herd immunity 131, 133
hierarchies, taxonomy 242
histograms 252–54, 263–64
histones 77, 200, 202
HIV *see* human immunodeficiency virus
HLA *see* human leucocyte antigens
homogenesis, cells 84, 85
homologous chromosomes 218, 219–23
horses 241
HPV (Human Papilloma Virus) vaccinations 132
human chorionic gonadotrophin (HCG) 136
human genome 204
human immunodeficiency virus (HIV) 120, 133–35, 137
human leucocyte antigens (HLA) 224
Human Papilloma Virus (HPV) vaccinations 132
humans
absorption 157–58, 160–62
birth weights 232
chimpanzee relationships 248–49
digestion 157–62
gas exchange 150–57
lung ventilation 152–54
migration 224–25
nutrient exchange 157–62
humoral response 123, 125–27
Huntingdon's disease (HD) 217
hydrogen bonds 6–8, 10, 54–55, 58, 188
hydrogen breath tests 15–16
hydrogen gas 47
hydrogen peroxide 34, 36
hydrolysis
ATP 61–62
cell membranes 113–14
disaccharides 12
water 9–10
hydrophilic molecules 23, 24
hydrophobic molecules 23, 24
hydrostatic pressure 181
hyperbaric oxygen 181–82
hypothesis, statistics 266–69

I

immune response 121, 123–33
immune systems 120–41
abnormal body cells 121
AIDS 120, 133–35
antibiotic usage reduction 138–39
antibodies 126–30, 133–38
antigen-presenting cells 125, 127
antigens 120, 121–33
antigen variability 128
B lymphocytes/B cells 123, 125–28, 136–39
cancers 132, 139
cell-surface antigens 120, 121–33
cell-surface membranes 120, 121–22
cell transplants 121
cellular response 123–25
clonal selection 125, 127, 136

immune systems (*continued*)
glycoproteins 120, 121–22, 125, 133
HIV 120, 133–35, 137
humoral response 123, 125–27
immune response 121, 123–33
lymphocytes 123–28, 136–39
macrophages 122–23, 127
monoclonal antibodies 126, 136–39
pathogens 121, 122, 123–25, 127–28
phagocytosis 122–23
plasma cells 126, 127, 136–39
primary responses 127–28, 133
probability 139
proteins 120, 121–22, 125, 133
retroviruses 120, 133–35
RNA 133–35
secondary responses 127–28, 133
T lymphocytes/T cells 123–25, 127, 133–34
toxins 121–22
transplants 121
vaccination 128–33
viruses 120, 127–28, 133–35
white blood cells 122–23
immunoglobulins 126
immunology, diversity 250
incubating cultures 235
independent assortment 221–22, 223
induced fit model 36
information, DNA 55
inhibiting enzymes 46–47
inoculating loops 234
insects 146–47, 165, 228–29, 231
insulation properties of blubber 24–26
insulin 29, 203, 209
interphase cell cycle stage 85, 86, 90
interpreting data 264–69
introns 204, 210–11, 212
investigative studies, biodiversity 248–56
iodine addition to potassium iodide solution 16, 18
isomers 10
isotonic drinks 108

J

Jenner, Edward 129–30
Jurassic Park 205

K

karotypes 218–19
keratin 29
Kettlewell, Henry 229
kidney disease 115
kilopascals 107, 113
kinetic energy 7

L

lactase 158–59
lactose 12, 14–16, 17, 18
lactose intolerance 14–16, 17
Lake Titicaca 142
latent heat of vaporisation 8, 10
leaves 165, 184, 185–87, 190–94
leg muscles 178
lentiviruses 134–35
leucocytes 224
Levene 56
life cycles, meiosis 221
life on distant planets 4
light microscopes 23, 79–82, 89–91
lignins 186
linear relationships 15, 265
line graphs 15, 263, 264–66
Linnaeus, Carl 242
lipase enzymes 159
lipids 6
blubber insulation 24–26
cell membranes 98, 99–100, 103–5, 108, 111
digestion 159, 161
phospholipids 23–24
proteins 26–32
triglycerides 21–23, 24–25
liquid transport *see* mass transport
locus, gene position on chromosomes 202, 204
logarithms, cell growth calculations 93
lumen 176–77, 182
Lundy Island 255
lungs 150–57, 184
lymphatic system 180
lymphocytes 123–28, 136–39
lysosomes 72–73, 77, 122–23
lysozymes 73, 77, 122–23

M

macrophages 122–23, 127
magnification 79–82
malonate 46–47
maltase 158–59
maltose 12, 16, 18, 158–59
mammals
blood transport 166–82
oxygen transport 166–70
mammary cancer 87–88
margins, farming biodiversity 246
Margulis, Lynn 67, 92, 201
marine life 255
Mars, water 4
marula trees 189–90
mass flow 9, 166, 193
mass transport 165–98
apparatus/technique practicals 183–84
blood 166–84
blood vessels 176–84
circulatory system 170–82
dissection apparatus/ techniques 183–84
heart and circulatory system 170–82
organic substances in plants 191–94
oxygen in mammals 166–70
plants 165, 184, 185–94

practical investigations 183–84
technique practicals 183–84
water in plants 185–90
matching chromosomes 218, 219–23
maths techniques 260–69
correlation 267
data analysis/interpretation 264–69
data display/recording 262–64
graphs 15, 263, 264–66
handling numbers 260–62
probability 262, 268
statistics 266–69
matrices, mitochondria 70, 71
mature mRNA 210, 212
meadows 247
mean 251–53, 254, 268
measles, mumps and rubella (MMR) vaccine 129, 131
mechanisms
enzymes 35–37
natural selection 227–31
median 254, 263–64
meiosis
cell division 218, 219–23
chromosomes 218, 219–23
cross over 222–23
genetic diversity 219–23
homologous chromosomes 218, 219–23
independent assortment 221–22, 223
random fertilisation 223
variation sources 221–23
melanic moths 228–29, 231
melanin 217
membranes
cell membranes 98–119
cell-surface membranes 68–70, 99–100, 111–12, 120, 121–22
chloroplasts 73–74
eukaryotic cells 68–70, 73–74, 77
mitochondria 70, 71
partially permeable 105–6
memory cells 126, 127, 133
Meselson, Matthew 60–61
messenger ribonucleic acid (mRNA) 205–12, 248
metabolites 9–10
metaphase cell cycle stage 85, 86
methotrexate 53
micelles 159, 160–61
microbial growth 233–35
micrometres 79
microscopy
optical microscopes 23, 79–82, 89–91
photomicrographs 122
see also electron microscopy/micrographs
microvilli 158–59
Miescher, Friederich 56
migration patterns 224–25
mitochondria
eukaryotic cells 70, 71, 76, 77, 201, 202
prokaryotic cells 67, 68, 92
mitochondrial DNA (mtDNA) 225
mitosis
cell division 85–93
chromosome mutations 216
genetic diversity 215, 216, 219, 220
meiosis 221
monoclonal antibodies 136
mitotic index 89, 90–91
MMR (measles, mumps and rubella) vaccine 129, 131
modal set of data 263
mode, continuous variation 254
molecular motion 7–8
molecular size factors 101
monoclonal antibodies 126, 136–39
monocultures 245
monoglycerides 160–61
monomers 6
monosaccharides 10–13, 16–18
mosses 187
moths 147, 228–29, 231
mRNA *see* messenger ribonucleic acid
mtDNA *see* mitochondrial DNA
mules 241
multicellular organisms 67, 68–77
multiple repeats 204
murein 78
muscles 178
cardiac muscles 171–76
external intercostal muscles 152
mutagenic agents 219
mutations
biodiversity 249–50
chromosome mutation 216, 218–19, 236
gene mutations 216–18, 219, 236
genetic diversity 216–19, 236

N

naming species 242
nanometres 79
natural selection
evidence for 228–29
genetic diversity 223–36
mechanisms 227–31
process of 227–31
rodents 225–27
taxonomy 239, 240–42, 244
negative correlation 25–26
neutrophils 122–23
non-competitive inhibitors 46, 47
non-disjunction mutations 218–19
non-overlapping DNA strands 203
non-reducing sugars 16, 17
non-self antigens 120
normal distribution curves 251–53
normal lactose digestion 14
nuclear divisions 219–23
nuclear envelopes 69–70
nucleic acids 54–61

nucleoli 69–70, 77
nucleotides 5, 53–66
 ATP 53, 61–63
 cancers 53
 derivatives (ATP) 53, 61–63
 DNA 53, 54–61
 polymers 6
 protein synthesis 199
 RNA 53, 54, 57–58
 structure 54–61
nucleus, eukaryotic cells 69–70, 77
null hypothesis 266–69
nutrient exchange 142, 157–62

O

optical microscopes 23, 79–82, 89–91
optimum pH 39–43, 47
optimum temperature 37, 47
oral rehydration therapy (ORT) 98
ordinary form 260
organ dissections 183–84
organelles 67–69, 71–74, 78–85, 87
organic bases 54, 57, 61
organic substance transport 191–94
organisms
 cells 67, 68–77
 classification 239–59
 cooling 8, 10
 gas exchange 142–64
 genetic diversity 215–38
organs 76, 77
origins of AIDS/HIV 134–35
ORT *see* oral rehydration therapy
osmosis
 cell membranes 98, 105–12
 importance of to cells 106–7
 lactose intolerance 14
 mass flow 166
oxygen, hyperbaric oxygen 181–82
oxygen transport 166–70
oxyhaemoglobin dissociation curves 167–70

P

pairs
 base pairing 54–55, 58, 59, 61, 207–8, 212
 bivalent pairs 220–23
 homologous chromosomes 220–23
partially permeable membranes 105–6
partial pressures 167–70
passive immunity 130, 133
passive processes 101
pastuer pipettes 234
pathogens 121, 122, 123–25, 127–28
pathway lengths 101
pectin 74–75
peppered moths 228–29, 231
pepsin 159–60
peptide bonds 26–31
percentages 261–62
 change 261–62
 difference 261–62
 errors 16, 266
 glucose uptake 104–5
 probability values 139
permeability 105–6, 111–12
pH 39–43, 47
phagocytes 122–23
phagocytosis 122–23
phagosomes 122–23
phenotypes 215, 226–27, 231, 236
phloem sieve elements 191
phloem tissue 191–94
phloem transport 193–94
phosphodiester bonds 54, 57
phospholipids 23–24, 98, 99–100, 103–5, 108, 111
phosphorylated nucleotides 61–63
photomicrographs 122
photosynthesis
 chloroplasts 68, 73–74, 77
 gas exchange 142, 149–50
 organic substance transport 191
phylogenetic classification 242–44
pipettes 234
planets 4
plants 68
 cell-surface membrane permeability 111–12
 gas exchange 143, 149–50
 mass transport 165, 184, 185–94
 organic substance transport 191–94
 phloem tissue 191–94
 stem dissections 184
 translocation 191–93
 transpiration rates 185–86
 water potential 109–10
 water transport 165, 184, 185–90
 xerophytes 186–90
 xylem tissue 185, 186–90, 192
 xylem vessels 186–90
plasma cells 126, 127, 136–39
plasmids 78, 200–201, 202
plasmodesmata 75, 191
Pogonomyrmex barbatus 147
pollution 228–29, 232
polyclonal antibodies 137
polygenic characteristics 232
polymerase 59, 61
polymers 6
 DNA 53, 54–61
 nucleotides 53–66
 polysaccharides 10, 12–13, 18
 proteins 26–32
 RNA 53, 54, 57–58
polynucleotides 54–61, 199
polypeptides
 protein structure 26–30
 protein synthesis 199, 201–2, 204, 207–12
polysaccharides 10, 12–13, 18
polyunsaturated fatty acids 22
positive correlation 25–26
potassium iodide solution 16, 18
potassium permanganate solution 11

practical investigations
  cell membranes 109–12
  cells 89–91
  dissections 183–84
  enzymes 41–47
  genetic diversity 233–35
  mass transport 183–84
precision, data 264
pre-mRNA 210
pressure changes 172–74, 192
pressure units 107, 113
primary responses 127–28, 133
primary structure 26–27, 30
probability 139, 262, 268
process of natural selection 227–31
production, cells 68, 85–93
products, enzymes 35, 36
prokaryotic cells 68, 73, 77, 83, 85, 92
  binary fission 85, 92
  cell division 85, 92
  chloroplasts 92
  chromosomes 200–201, 202
  DNA and protein synthesis 200–201, 202
  DNA 77–78, 92
  mitochondria 67, 68, 92
prophase cell cycle stage 85, 86
proteases 37–39, 159–60
protein coat (capsid) 78
proteins 5, 26–32
  biodiversity 239, 243, 248–51, 256
  biuret test 30
  bonds 26–31
  cell membranes 99–100, 103–4, 105, 113–15
  diffusion 103–4, 113–15
  digestion 159–60, 161
  diversity 239, 243, 248–51, 256
  function 29–31
  haemoglobin 215
  immune systems 120, 121–22, 125, 133
  lipids 26–32
  peptide bonds 26–31
  polymers 6
  structure 26–31
  see also enzymes; protein...
protein synthesis
  anticodons 207–12
  ATP 207, 208, 212
  chromosomes 200–202, 204
  codons 199, 207–12
  DNA 199–214
  eukaryotic cells 200, 202
  genes 200–202
  genetic code 199, 200, 203–14
  messenger ribonucleic acid 205–12
  polypeptides 199, 201–2, 204, 207–12
  prokaryotic cells 200–201, 202
  proteomes 205–12
  ribosomal RNA 211–12
  ribosomes 205, 207–12
  RNA 199, 201–2, 205–12
  single-stranded RNA 205–7, 212
  transcription 206–7, 212
  transfer RNA 207–10, 212
  triplet code 203–4, 209–10
proteomes 205–12
Prunus armeniaca diversity 256
pulmonary artery 172, 176
pulmonary circulation system 170, 172, 176
pulmonary veins 172
Punnett squares 139

Q

quaternary structure 28–29, 30

R

random fertilisation 223
random sampling 253–54
range 254, 266
rank correlation 267
rate calculations 264–66
rate of change, gradient 15
rate of diffusion 101–3
rate of reaction 41–47
ratios 262
recording data 262–64
recreating dinosaurs 205
red blood cells 104–5, 106, 215
redox reactions 16
reducing sugars 16, 17
rehydration 98, 108
replication
  DNA 53, 59–61, 85
  HIV 133–34
RER *see* rough endoplasmic reticulum
reshuffling genes 219–23
resolving power 79
respiration 142–57
results tables 262
retroviruses 120, 133–35
R_f values 30–31
R groups 26, 35
rheumatoid arthritis 136
ribonucleic acid (RNA) 5
  codons 199, 207–12
  eukaryotic cells 69–70
  genetic diversity 248
  immune systems 133–35
  nucleotides 53, 54, 57–58
  protein synthesis 199, 207–12
  splicing 210–11, 212
  structure 54, 57–58
  transcription 206–7, 212
  translation 207–10
  viruses 78
ribosomal RNA (rRNA) 211–12
ribosomes
  eukaryotic cells 70–72, 77
  mitochondria 67
  prokaryotic cells 77–78
  protein synthesis 205, 207–12
  RNA 54, 57
RNA polymerase 207
RNA *see also* ribonucleic acid
rodents 225–27

root rip squashes 89–91
roots
organic substance transport 191–92
water transport 165, 184, 185–87, 190
rough endoplasmic reticulum (RER) 71, 72, 77
rRNA *see* ribosomal RNA

S

safety precautions 233
saline solutions 98
sampling 253–54
sand dunes 187
saturated fatty acids 22, 24
saturation, haemoglobin with oxygen 167–70
scanning electron microscopes (SEM) 83, 149
scattergraphs 25–26, 264
SCT *see* sickle cell trait
SD *see* standard deviation
sea life diversity 255
secondary lamellae 148–49
secondary responses 127–28, 133
secondary structure 27, 30
SEM *see* scanning electron microscopes
semi-conservative DNA replication 59–61
sense strands 203
septum 172
sequences, DNA 55
SER *see* smooth endoplasmic reticulum
serial dilutions 43, 109–10, 235
SFV *see* Simian Foamy Virus
shipworms 39–40
sickle cell alleles 215
sickle cell trait (SCT) 215
sieve cells 191
sieve plates 191, 193
sieve tubes 191–92
significant figures 261
silkworm spiracles 147
Simian Foamy Virus (SFV) 135
Simian Immunodeficiency Virus (SIV) 134–35
single cell organisms 67, 68, 73, 77, 83, 85, 92, 145
single-stranded RNA 205–7, 212
sink areas 191, 192
SIV *see* Simian Immunodeficiency Virus
size
microscopy 80–82
surface area relationships 143–45
small intestine 160–61
smallpox 128, 129
smooth endoplasmic reticulum (SER) 71, 72, 77
sodium chloride 8–9
sodium ions 113–16
solubility 10, 12, 24
solutes 8–9, 106–7, 109–10
solutions, mass flow 166
solvents 8–9, 10, 112
source areas 191, 192
Spearman's rank correlation 267
specialised cells 75–77
speciation 242–44
species concept 240–44
species diversity 244
species names 242, 244
species richness 244
speckled moths 228–29
spherical cell diffusion 102–3
spindle fibres 85–86
spiracles 146
splicing RNA 210–11, 212
squashes 89–91
S-shaped dissociation curves 168–69
stabilising selection 232
stage micrometers 90
Stahl, Franklin 60–61
stained squashes 89–91
standard deviation (SD) 251–53, 254, 266, 268
standard form 62–63, 260–61
starch
biochemical tests 16, 17
digestion 158–59
polysaccharides 12–13, 18
starch-digesting enzymes 48–49
statistics
maths techniques 266–69
null hypothesis 266–69
tests 25–26, 229, 230–31, 267–69
stem cells 76
stems
dissections 184
length of dandelions 253–54
organic substance transport 191–94
water transport 186, 187
sterilising equipment 234
strands 55, 58
streak plates 234–35
stroke volume 174–75
stroma 73
structure
animal cells 68–77
antibodies 126–27
ATP 61
blood vessels 176–78
carbohydrates 10, 12–13, 18
cell membranes 99–100
cells 67, 68–77
cellulose 13, 18
chloroplasts 73–74
disaccharides 12
DNA 54–61
enzymes 35–36
eukaryotic cells 68–77
gas exchange in humans 150, 151
glucose 10
heart 171–76
HIV 133
monosaccharides 10
nucleic acids 54–61
nucleotides 54–61
phospholipids 23–24
polysaccharides 12–13

proteins 26–31
RNA 54, 57–58
starch 12–13
triglycerides 21–23, 24
viruses 78
water molecules 6–10
Student's t-test 267, 268–69
study methods, cells 79–85
substance exchange
absorption 157–58, 160–62
digestion 157–62
environment 142–64
gases 142–57
human gas exchange 150–57
nutrients 142, 157–62
respiration 142–57
substitution mutations 216
substrates 35–36, 43–45, 47
succinate 46–47
sucrase 158–59
sucrose 12, 17, 191–94
sugars 9, 10–11, 12, 16
surface areas 102, 143–45, 146, 150
surface tension 9
symbiosis 67, 68
systemic circulation system 170

T

taxonomy
binomials 242, 244
biodiversity 244–59
natural selection 239, 240–42, 244
species concept 240–44
TB *see* tuberculosis
technique practicals
cell membranes 109–12
cells 89–91
dissections 183–84
enzymes 41–47
genetic diversity 233–35
mass transport 183–84
Telmatobius culeus 142
telophase cell cycle stage 86
TEM *see* transmission electron microscopes
temperature
cell-surface membrane permeability 112
diffusion rates 101
enzyme–activity relationship 37–39, 41, 47
molecular motion 7–8
stability 7–8
tension 188–89, 190
tertiary structure 27–28, 30
tests, statistics 25–26, 229, 230–31, 267–69
thermoplasmacidophilum 47
thylakoids 73
tidal ventilation 152
time, enzyme-activity relationship 41
tissue
cell production 87
eukaryotic cells 76, 77
tissue fluids 178, 180–82
T lymphocytes/T cells 123–25, 127, 133–34
TNF-α, monoclonal antibodies 136
toxins 121–22
trachea 150, 151
tracheae 146
tracheoles 146
transcription 206–7, 212
transferring cultures 234–35
transfer RNA (tRNA) 207–10, 212
translation, RNA 207–10
translocation 191–93
transmission electron microscopes (TEM) 82–83, 92, 122, 201
transpiration 185
transpiration rates 185–86
transplants, immune systems 121
transport
blood transport 166–84
cell membranes 98–119
diffusion 98, 101–16, 151–52, 166
gas exchange 142–57
mass transport 165–98
nutrient exchange 142, 157–62
organic substances 191–94
oxygen transport 166–70
treating HIV 134–35
tree diversity 256
triglycerides 21–23, 24–25
triplet codes 203–4, 209–10, 248
tRNA *see* transfer RNA
trypsin 159–60
t-tests 267, 268–69
tuberculosis (TB) 232
Tubifex dissociation curves 169
turgidity 74–75

U

UAG codons 199
ultracentrifugation 84, 85
uncertainties in measurements 266
unicellular organisms 67, 68, 73, 77, 83, 85, 92, 145
units, water potential 107, 113
universal genetic code 200
unsaturated fatty acids 22, 24
uptake
gas exchange 142–57
nutrients 142, 157–62

V

vaccinations 128–33
vacuoles 75, 77
variation
natural selection 224, 225–27
sources 221–23
within species 251–56
veins 176, 177–78, 182
venae cavae 172
ventilation 146, 152–54
ventricles 172–73, 176
venules 176, 177–78, 182
villi 158–59
viruses
cells 67, 77, 78
DNA and protein synthesis 199
immune systems 120, 127–28, 133–35
structure 78

Visking tubing 105
voltage-gated channels 103
volume ratios 143–45, 146, 150

W

Wallace, Alfred Russel 228
walls *see* cell walls
washing powders 37–39
water 4–10
  bonding 6–8, 9, 10
  carbohydrates 4–6, 10–19
  cooling 8, 10
  covalent bonds 6–7
  diffusion 98, 101–8, 113–16
  evaporation 8, 10
  hydrogen bonding 6–8, 10
  lactose intolerance 14–16
  latent heat of vaporisation 8, 10
  metabolites 9–10
  solubility 10, 12
  solvents 8–9, 10
  temperature stability 7–8
  tissue fluids 180–82
  transport in plants 165, 184, 185–90
water molecules
  bonding 6–8, 9, 10
  dipoles 7, 10
  structure 6–10
water potential
  blood vessels 181
  gradients 106–7, 113
  osmosis in cell membranes 106–14
  plant tissue 109–10
  units 107
Watson, James 54, 56, 58
wheat field biodiversity 246
white cells 120, 122–23, 224
whole animal dissections 184
working with large numbers 62–63

X

xerophytes 186–87
xylem tissue 185, 186–90, 192
xylem vessels 186–90

Y

yields 246, 247

Z

zygotes 76, 221–22

ACKNOWLEDGEMENTS

The Publishers gratefully acknowledge the permissions granted to reproduce copyright material in this book. Every effort has been made to contact the holders of copyright material, but if any have been inadvertently overlooked, the Publisher will be pleased to make the necessary arrangements at the first opportunity.

Practical work in biology

p1: Miles Studio/Shutterstock; p2 top: Imfoto/Shutterstock, middle: sfam_photo/Shutterstock, bottom: Miles Studio/Shutterstock; p3 top: Photographee.eu/Shutterstock, bottom: Ozgur Coskun/Shutterstock

Chapter 1

p4 background: DLR/FU Berlin (G. Neukum)/European Space Agency/ Science Photo Library; p7: jordache/ Shutterstock; p9: John Griffiths/ Shutterstock

Chapter 2

p20 background: photo_journey/ Shutterstock; p23: Jose Luis Calvo/ Shutterstock; p28: Scott Camazine/ Alamy

Chapter 3

p33 background: Phototribe/ Shutterstock; p34: Martyn F. Chillmaid/Science Photo Library; p37: Malota/Shutterstock; p38: Paul John Fearn/Alamy; p47: format4/Alamy

Chapter 4

p53 background: Apples Eyes Studio/Shutterstock; p54: Will & Deni Mcintyre/Science Photo Library; p56: A. Barrington Brown/ Science Photo Library

Chapter 5

p67 background: Wim Van Egmond/ Visuals Unlimited, inc./Science Photo Library; p68 top: John Durham/Science Photo Library, bottom: blickwinkel/Alamy; p69: Biophoto Associates/Science Photo Library; p70 left: Biophoto Associates/Science Photo Library, right: CNRI/Science Photo Library; p72 top: Microscape/Science Photo Library, bottom: Biophoto Associates/Science Photo Library; p74 left: Dr Jeremy Burgess/ Science Photo Library, right: Biophoto Associates/Science Photo Library; p75: Dr David Furness, Keele University/Science Photo Library; p69 top: Gerd Guenther/ Science Photo Library, bottom: Steve Gschmeissner/Science Photo Library; p79: Science Photo Library /Alamy; p80: David Scharf/ Science Photo Library; p81 top: Dr Jeremy Burgess/Science Photo Library, bottom left: A.B. Dowsett/ Science Photo Library, bottom right: Eye Of Science/Science Photo Library; p82: K.R. Porter/Science Photo Library; p84 left: Marilyn Schaller/Science Photo Library, right: Eye Of Science/Science Photo Library; p87: royaltystockphoto. com/Shutterstock; p91: Steve Gschmeissner/Science Photo Library; p92: CNRI/Science Photo Library

Chapter 6

p98 background: Godong Alamy; p99: Don W. Fawcett/Science Photo Library; p101: Andrew Lambert Photography/Science Photo Library; p108: Anthony Mooney/ Shutterstock; p111: omphoto/ Shutterstock; p115: gopixa/ Shutterstock

Chapter 7

p120 background: Africa Media Online Alamy; p122 left: Science Photo Library Alamy, right: National Institute of Allergy and Infectious Diseases (NIAID)/ National Institutes Of Health/ Science Photo Library; p126 top: Steve Gschmeissner/Science Photo Library, bottom: Steve Gschmeissner/Science Photo Library; p129 top: Monkey Business Images/Shutterstock, bottom left: Georgios Kollidas/ Shutterstock, bottom right: NYPL/ Science Source/Science Photo Library; p131: Lowell Georgia/ Science Photo Library; p132: Image Point Fr/Shutterstock; p134: Patrick Rolands/Shutterstock; p135: Sebastian Kaulitzki/ Shutterstock; p136 top: Dr P. Marazzi/Science Photo Library, bottom: mast3r/Shutterstock; p137: Hank Morgan/Science Photo Library; p138 left Gladskikh Tatiana/Shutterstock, right: bagwold/Shutterstock

Chapter 8

p142 background: Nature Picture Library / Alamy; p143 left: Stephaniellen/Shutterstock, right: Mike Dexter/Shutterstock; p144 left: Wayne Johnson/Shutterstock, right: BMJ/Shutterstock; p145: Melba Photo Agency Alamy; p147: Jubal Harshaw/Shutterstock; p149: Power and Syred/Science Photo Library; p154: Michael McLaughlin/ Flickr Commons; p158: Eye of Science/Science Photo Library

Chapter 9

p165 background: Paul Broadbent/ Alamy; p166: Roxana Bashyrova/ Shutterstock; p169: Steve Gschmeissner/Science Photo Library; p171: John Radcliffe Hospital/Science Photo Library; p175: wavebreakmedia/ Shutterstock; p179: Lightspring/ Shutterstock; p182: imageshunter/ Shutterstock; p184 top middle: Gridin Alex shutterstock, top right: Adamantios CC By-Sa 3.0; top left: Evan-Amos public domain; bottom right: domnitsky Shutterstock; p184 top left: kzww / Shutterstock, bottom left: Pöllö CC BY-SA 3.0, Right; Nigel Cattlin, Visuals Unlimited/Science Photo Library; p186 left: Hannes Thirion/ Shutterstock, right: Buddy Mays/ Alamy; p188: Dr David Furness, Keele University/Science Photo Library; p189: Andre Coetzer/ Shutterstock

Chapter 10

p199 background: Eye of Science/ Science Photo Library; p200: Power And Syred/Science Photo Library; P201 top left: Alfred Pasieka/Science Photo Library, bottom left: Dr Gopal Murti/Science Photo Library, right: Alain Pol, ISM/Science Photo Library; p202 top: CNRI/Science Photo Library, bottom: L Willatt, East Anglian Regional Genetics Service/Science Photo Library; p205: hjschneider/

Shutterstock; p209: Dr Elena Kiseleva/Science Photo Library

Chapter 11

p215 background: Szabo/ Shutterstock; p217 top: Pete Jenkins/Alamy, bottom: Bruce Coleman Inc. Alamy; p218: CNRI/Science Photo Library; p224: Estudi M6/Shutterstock; p228 left: Bildagentur Zoonar GmbH/Shutterstock, top right: o.leillinger@web.de CC By-Sa 3.0, bottom right: o.leillinger@web.de CC By-Sa 3.0; p233 top: wacpan Shutterstock, bottom: Nadina Wiórkiewicz CC BY-SA 3.0; p234 top left: 33333 Shutterstock, top right: Paket Shutterstock, bottom left: Martyn F. Chillmaid/ Science Photo Library, bottom right: angellodeco Shutterstock; p235: Zaharia Bogdan Rares Shutterstock

Chapter 12

P239 background: Lukas Gojda/ Shutterstock; p240 left: Animal Photography Alamy, far right: Nigel Pye Alamy, central right: Nature Photographers Ltd Alamy, bottom: Andrew Parkinson Alamy; p241 left: Juniors Bildarchiv GmbH Alamy, centre: Age Fotostock Alamy, right: MGrebler Alamy; p246 left: Science Photo Library Alamy, top right: picturesbyrob Alamy, bottom right: Chris Gomersall Alamy; p247: Geoff Jones/www.ghwstudios.com, reproduced with kind permission; p249: Martin Harvey Alamy; p251: Life on white Alamy; p253: Nigel Cattlin Alamy; p255 left: Diana Mower Shutterstock; right: DavidYoung Shutterstock; p256: Fotokostic Shutterstock

Chapter 13

p260: Alexeysun/Shutterstock

NOTES